International Review of **Cytology**

A Survey of **Cell Biology**

SEXUAL REPRODUCTION IN FLOWERING PLANTS

VOLUME 140

International Review of **A Survey of**
Cytology **Cell Biology**

Guest Edited by

Scott D. Russell
Department of Botany and Microbiology
University of Oklahoma
Norman, Oklahoma

Christian Dumas
Ecole Normale Supérieure de Lyon
Lyon, France

SEXUAL REPRODUCTION IN FLOWERING PLANTS

VOLUME 140

Academic Press, Inc.
Harcourt Brace Jovanovich, Publishers
San Diego   New York   Boston   London   Sydney   Tokyo   Toronto

Academic Press, Inc.
1250 Sixth Avenue, San Diego, California 92101-4311

United Kingdom Edition published by
Academic Press Limited
24–28 Oval Road, London NW1 7DX

Library of Congress Catalog Number: 52-5203

International Standard Book Number: 0-12-364543-3

PRINTED IN THE UNITED STATES OF AMERICA
92 93 94 95 96 97   EB   9 8 7 6 5 4 3 2 1

CONTENTS

PART I. POLLEN GRAIN AND TUBE

Pollen Gene Expression: Molecular Evidence

Joseph P. Mascarenhas

Identification and *in Situ* Localization of Pollen-Specific Genes

S. P. Davies, M. B. Singh, and R. B. Knox

Pollen Wall and Sporopollenin

R. Wiermann and S. Gubatz

Cytoskeleton and Cytoplasmic Organization of Pollen and Pollen Tubes

Elisabeth S. Pierson and Mauro Cresti

PART II. GAMETES

The Male Germ Unit: Concept, Composition, and Significance

H. Lloyd Mogensen

Organization, Composition, and Function of the Generative Cell and Sperm Cytoskeleton

Barry A. Palevitz and Antonio Tiezzi

Freeze Fracture of Male Reproductive Cells

Darlene Southworth

Generative Cells and Male Gametes: Isolation, Physiology, and Biochemistry

Annie Chaboud and Réjane Perez

Female Germ Unit: Organization, Isolation, and Function

Bing-Quan Huang and Scott D. Russell

PART III. PROGAMIC PHASE AND FERTILIZATION

A Dynamic Role for the Stylar Matrix in Pollen Tube Extension

Luraynne C. Sanders and Elizabeth M. Lord

Double Fertilization in Nonflowering Seed Plants and Its Relevance to the Origin of Flowering Plants

William E. Friedman

Double Fertilization

Scott D. Russell

PART IV. MANIPULATION IN POLLINATION AND FERTILIZATION MECHANISMS

In Vitro Pollination: A New Tool for Analyzing Environmental Stress

Isabelle Dupuis

In Vitro Fusion of Gametes and Production of Zygotes

Erhard Kranz, Horst Lörz, Catherine Digonnet, and Jean-Emmanuel Faure

Plant Transformation Using the Sexual Route

Patricia Roeckel, Maurice M. Moloney, and Joël R. Drevet

PART V. SELF-INCOMPATIBILITY BETWEEN POLLEN AND STIGMA

Gametophytic Self-Incompatibility: Biochemical, Molecular Genetic, and Evolutionary Aspects

Anuradha Singh and Teh-Hui Kao

Sporophytic Self-Incompatibility Systems: *Brassica S* Gene Family

Martin Trick and Philippe Heizmann

Sporophytic Self-Incompatibility Systems: *S* Gene Products

H. G. Dickinson, M. J. C. Crabbe, and T. Gaude

PART VI. POSTSCRIPT

Plant Reproductive Biology: Trends

Christian Dumas and Scott D. Russell

CONTRIBUTORS

Numbers in parentheses indicate the pages on which the authors' contributions begin.

Annie Chaboud (205), *Reconnaissance Cellulaire et Amélioration des Plantes, Université Claude Bernard-Lyon 1, Villeurbanne, France*

M. J. C. Crabbe (525), *Department of Microbiology, School of Animal and Microbial Sciences, University of Reading, Whiteknights, Reading RG6 2AS, England*

Mauro Cresti (73), *Dipartimento di Biologia Ambientale, Università di Siena, Siena, Italy*

S. P. Davies (19), *School of Botany, University of Melbourne, Parkville, Victoria 3052, Australia*

H. G. Dickinson (525), *Department of Plant Sciences, University of Oxford, Oxford OX1 3RB, England*

Catherine Digonnet (407), *Reconnaissance Cellulaire et Amélioration des Plantes, Université Claude Bernard-Lyon 1, Villeurbanne, France*

Joël R. Drevet (425), *Faculty of Medicine, Calgary, T2N 4N1, Canada*

Christian Dumas (565), *Ecole Normale Supérieure de Lyon, Lyon, France and Reconnaissance Cellulaire et Amélioration des Plantes, Université Claude Bernard-Lyon 1, Villeurbanne, France*

Isabelle Dupuis (391), *CIBA-GEIGY Biotechnology Research, Research Triangle Park, North Carolina 27709*

Jean-Emmanuel Faure (407), *Reconnaissance Cellulaire et Amélioration des Plantes, Université Claude Bernard-Lyon 1, Villeurbanne, France*

William E. Friedman (319), *Department of Botany, University of Georgia, Athens, Georgia 30602*

T. Gaude (525), *Reconnaissance Cellulaire et Amélioration des Plantes, Université Claude Bernard-Lyon 1, Villeurbanne, France*

S. Gubatz (35), *Institut für Botanik der Westfälischen Wilhelms-Universität, D-4400 Münster/Westfall, Germany*

Philippe Heizmann (485), *Reconnaissance Cellulaire et Amélioration des Plantes, Université Claude Bernard-Lyon 1, Villeurbanne, France*

Bing-Quan Huang (233), *Department of Botany and Microbiology, University of Oklahoma, Norman, Oklahoma 73019*

Teh-Hui Kao (449), *Department of Molecular and Cell Biology, Pennsylvania State University, University Park, Pennsylvania 16802*

R. B. Knox (19), *School of Botany, University of Melbourne, Parkville, Victoria 3052, Australia*

Erhard Kranz (407), *Universität Hamburg, Institut für Allgemeine Botanik, Angewandte Molekularbiologie der Pflanzen II, Hamburg, Germany*

Elizabeth M. Lord (297), *Department of Botany and Plant Sciences, University of California, Riverside, Riverside, California 92521*

Horst Lörz (407), *Universität Hamburg, Institut für Allgemeine Botanik, Angewandte Molekularbiologie der Pflanzen II, Hamburg, Germany*

Joseph P. Mascarenhas (3), *Department of Biological Sciences, State University of New York at Albany, Albany, New York 12222*

H. Lloyd Mogensen (129), *Department of Biological Sciences, Northern Arizona University, Flagstaff, Arizona 86011*

Maurice M. Moloney (425), *Department of Biological Sciences, University of Calgary, Calgary T2N 1N4, Canada*

Barry A. Palevitz (149), *Department of Botany, University of Georgia, Athens, Georgia 30602*

Réjane Perez (205), *Reconnaissance Cellulaire et Amélioration des Plantes, Université Claude Bernard-Lyon 1, Villeurbanne, France*

Elisabeth S. Pierson (73), *Dipartimento di Biologia Ambientale, Università di Siena, Siena, Italy*

Patricia Roeckel (425), *Department of Biological Sciences, University of Calgary, Calgary, T2N 1N4, Canada*

Scott D. Russell (233, 357, 565), *Department of Botany and Microbiology, University of Oklahoma, Norman, Oklahoma 73019*

Luraynne C. Sanders (297), *Department of Botany and Plant Sciences, University of California, Riverside, Riverside, California 92521*

Anuradha Singh (449), *Department of Molecular and Cell Biology, Pennsylvania State University, University Park, Pennsylvania 16802*

M. B. Singh (19), *School of Botany, University of Melbourne, Parkville, Victoria 3052, Australia*

Darlene Southworth (187), *Department of Biology, Southern Oregon State College, Ashland, Oregon 97520*

Antonio Tiezzi (149), *Dipartimento di Biologia Ambientale, Università di Siena, Siena, Italy*

Martin Trick (485), *Cambridge Laboratory, Institute of Plant Science Research, John Innes Centre, Norwich NR4 7UJ, England*

R. Wiermann (35), *Institut für Botanik der Westfälischen Wilhelms-Universität, D-4400 Münster/Westfall, Germany*

PREFACE

In 1987, K. L. Giles and J. Prakesh edited a special volume of the *International Review of Cytology* entitled "Pollen: Cytology and Development" in which the state of pollen research at that time was described. Since then, sexual reproduction in flowering plants has undergone a surprising series of revolutions. Tremendous progress has been made in understanding self-incompatibility in angiosperms through cloning, sequencing, and localizing both genes and gene products related to this phenomenon (see contributions by Dickinson *et al.*, Singh and Kao, and Trick and Heizmann). This work has also provided evidence of a membrane-bound protein kinase containing an extracellular domain closely homologous to the *S*-gene that is involved in pollen–pistil incompatibility that also seems to be associated with a more general role of signal transduction in plants (at least in sporophytic self-incompatibility systems). Similar progress has been made in understanding gametophytic gene expression, including both "early" genes, expressed from meiosis to the microspore stage, and a class of "late" genes, typically linked with gametophytic function (see Mascarenhas, and Davies *et al.*).

Other areas of revolution include new insights into the involvement of transmitting tissue as a dynamic matrix during pollen tube extension (see Sanders and Lord), the organization of sporopollenin (see Gubatz and Wiermann), and the importance of the cytoskeleton in pollen tube growth and conveyance of gametes (see Pierson and Cresti). Information about the gametes themselves, i.e., the egg and sperm cells, and their isolation and physiology have also been new topics of research (see Chaboud and Perez, and Huang and Russell). Insights on fertilization in angiosperms, both *in vivo* (see Russell) and *in vitro* (see Dupuis, Kranz *et al.*), are emerging that will lead to the establishment of models for gamete recognition that may allow experimental manipulation. Functional units of reproduction, i.e., the male and female germ units, are now well established in the embryological literature, and their importance in gametic dimor-

phism, cellular associations, cytoplasmic transmission, and preferential fertilization is becoming apparent. This work has extended to painstaking three-dimensional reconstructions with transmission electron microscopy (see Mogensen), freeze fracture (see Southworth), gamete isolation (see Chaboud and Perez), and the involvement of cytoskeletal elements in the unusual sequence of mitotic events during the formation of some male gametes (see Palevitz and Tiezzi).

Research published since the previous review has changed and challenged some long-held ideas of angiosperm reproduction. One of the most central of these is that double fertilization is restricted to angiosperms. Although some historical data have contested this view, the most recent data provide evidence that the ''fertilization'' of the ventral canal nucleus with the second sperm nucleus in *Ephedra* results in true embryos (see Friedman). This may also influence our interpretations of the origin of endosperm. The development of sexual reproductive cells as model systems is indicated by recent experiments involving both *in* and *ex ovulo* fertilization *in vitro* (see Dupuis, Kranz *et al.*) and the use of sexual vectors to transform embryonic lineages (see Roeckel *et al.*). The breadth of these recent developments impelled us to broaden our coverage from pollen to *Sexual Reproduction in Flowering Plants*, which is the basis of our current review.

We have attempted to review these and other recent areas of development that were outside of the scope of the previous review, and therefore chose the 17 topics included in this volume and organized a summary chapter at the conclusion.

We sincerely thank each of the contributors for their diligent efforts in assembling this volume and for suggesting improvements of the volume as it emerged, and to each of the researchers who were consulted for advice and provided material from original publications. We thank Norman Lin and Wendy S. Brooks for valued assistance in proofreading and organizing the submitted manuscripts, and the series editors and Academic Press for allowing us to publish this special volume.

Scott D. Russell
Christian Dumas

Part I
Pollen Grain and Tube

Pollen Gene Expression: Molecular Evidence

Joseph P. Mascarenhas
Department of Biological Sciences, State University of New York at Albany
Albany, New York 12222

I. Introduction

There has been a dramatic increase in the number of laboratories studying genes that are expressed in the male gametophyte and in the isolation and analysis of pollen-specific promoters. Potential perceived applications of this information in plant biotechnology have been largely responsible for this heightened interest in pollen.

This review covers current advances with respect to genes expressed in pollen that have been isolated, the identification of these genes with respect to function, and our current knowledge of the makeup of promoters that direct pollen-specific expression of the genes under their regulation. Tapetum and other anther tissue-expressed genes are not covered in this review. Various other aspects of the study of pollen development and earlier work on gene expression have been reviewed recently,and these should be consulted for information complementing that presented here (Mascarenhas, 1988, 1989, 1990a,b; Giles and Prakash, 1987; Pacini, 1990; Cresti and Tiezzi, 1990; Ottaviano and Mulcahy, 1989; Ottaviano et al., 1990; Roeckel et al., 1990; Evans et al., 1990).

II. Evidence for Haploid Transcription

Transcription of genes occurs from the haploid genome following meiosis of microsporocytes and during the subsequent differentiation of the microspores into mature male gametophytes. The first conclusive evidence that transcription and translation from the haploid genome occur during male

gametophyte development was obtained from studies of the inheritance of dimeric enzymes. In maize the *Adh1* gene specifies a dimeric enzyme that is responsible for the alcohol dehydrogenase (ADH) activity seen in the pollen grain (Felder *et al.,* 1973; Freeling and Schwartz, 1973). Pollen extracts made from plants heterozygous for two *Adh1* electrophoretic variants, fast (F) and slow (S), showed only the two homodimeric enzymes (FF and SS) and no heterodimers (Schwartz, 1971). Because two alleles were present at each locus in somatic cells, but only one or the other allele was present in individual pollen grains, such results would only be expected if haploid transcription and translation occurred and if in extracts the dimer did not dissociate and reassociate to form hybrids. All three enzymes, FF, FS, and SS, would, however, be expected in each haploid pollen grain if the enzyme had been synthesized in the pollen mother cells prior to meiosis, as is found in sporophytic tissues. In plants in which individual pollen grains carried two *Adh1* alleles, as was the case with the *Adh-FCM* duplication, heterodimers were found in the pollen. These results indicate in a conclusive manner that the synthesis of the ADH enzyme in maize pollen depends entirely on the genotype of the pollen nuclei and is not influenced by the genotype of the diploid sporophyte (Schwartz, 1971). Similar kinds of genetic evidence for haploid transcription have been obtained for several other enzymes in several different plants (Brink and MacGillivray, 1924; Demerec, 1924; Frova, 1990; Miller and Mulcahy, 1983; Nelson, 1958; Sari-Gorla *et al.,* 1986; Singh *et al.,* 1985; Tanksley *et al.,* 1981; Weeden and Gottlieb, 1979).

In addition to this genetic evidence, more recent information obtained by cloning pollen-expressed genes and studying the patterns of transcription of these genes confirms the occurrence of haploid gene transcription. These studies have utilized pollen-expressed and pollen-specific clones to show the activation of specific genes after meiosis and at specific periods during microspore and pollen development (Acevedo and Scandalios, 1990; Albani *et al.,* 1990; Brown and Crouch, 1990; Griffith *et al.,* 1991; Hanson *et al.,* 1989; Roberts *et al.,* 1991; Scott *et al.,* 1991; Singh *et al.,* 1991; Stinson *et al.,* 1987; Twell *et al.,* 1989; Ursin *et al.,* 1989).

A large number of different genes are transcribed during postmeiosis microspore and pollen development. We do not yet, however, have an accurate estimate of the total number of genes involved. From an analysis of the kinetics of hybridization of ^{3}H-cDNA with poly(A)RNA in excess, the messenger RNAs in mature pollen grains of *Tradescantia* and maize have been found to be the products of about 20,000–24,000 different genes. In comparison, maize and *Tradescantia* shoots contain about 30,000 different mRNAs. The different mRNA abundance classes in pollen are much more abundant than the corresponding classes in shoots (Willing and Mascarenhas, 1984; Willing *et al.,* 1988). These 20,000–24,000 different mRNAs that are found in mature pollen represent genes primarily acti-

vated late in pollen development (Mascarenhas, 1990a). Only a small fraction of these genes are expressed specifically in pollen but not in sporophytic tissues of the plant. The majority of the pollen-expressed genes, however, appear also to be active in vegetative tissues (Mascarenhas, 1990a). There is evidence that additional genes are activated after meiosis and during the early stages of microspore development. We do not currently have any estimates of the total number of these early genes and what fraction of them might be identical to the genes expressed later in development. Male gametophyte development, at least in terms of the number of genes required, is thus a complex process.

III. Transcription of Specific Genes during Pollen Development

Several laboratories have constructed cDNA libraries made to poly (A)RNA from pollen at different stages of development and from different plant species. Clones from these libraries have been used to determine the time during male gametophyte development when the genes are first activated and the pattern of accumulation and decay of the respective mRNAs. Such studies are important because they indicate critical stages in pollen development when new genes get activated.

Several pollen-specific cDNA clones made to mRNAs from mature pollen of *Tradescantia paludosa* (*pTpc44, pTpc70*) and maize (*pZmc30, pZmc13)* have been used as probes in RNA blot hybridizations to determine the pattern of activity of these genes (Guerrero *et al.*, 1990; Mascarenhas *et al.*, 1985; Stinson *et al.*, 1987). The mRNAs complementary to all the probes both in maize and *Tradescantia* were first detectable in the young pollen grain after microspore mitosis. The mRNAs continued to accumulate thereafter and reached their maximum concentrations in the pollen grain just before anthesis. In contrast, using an actin clone as a probe, actin mRNA was detectable in *Tradescantia* microspores soon after release from the tetrads. The actin mRNA accumulated thereafter, reached a maximum concentration at late pollen interphase, and decreased substantially in the mature pollen grain (Mascarenhas *et al.*, 1985; Stinson *et al.*, 1987).

Based on these results and other information in the literature, it was suggested that there were at least two sets of genes that were activated at different times during male gametophyte development. Genes in the first set termed the early genes, such as actin, become active soon after meiosis is completed. Genes of the second group, the late genes, are turned on after microspore mitosis (Stinson *et al.*, 1987; Mascarenhas, 1990a). More

recent data indicate that the actual situation is more complex, as will be apparent from the discussion that follows.

Several clones from a cDNA library of mature tomato pollen have been isolated and characterized. Five of these clones, *LAT51*, *LAT52*, *LAT56*, *LAT58*, and *LAT59*, correspond to the late class of genes. Their mRNAs are first detectable following microspore mitosis and increase progressively in concentration until anthesis (Twell *et al.*, 1989; Ursin *et al.*, 1989).

In *Oenothera* the mRNAs for three pollen-specific clones *(P1, P2,* and *P3)* follow a pattern of accumulation similar to the tomato clones, that is, they are activated late. The mRNA for one other pollen-specific clone, *P6*, does not accumulate to detectable levels until just before anthesis. The mRNAs for two other clones, *P4* and *P5*, which are not pollen-specific but are expressed also in leaves and ovaries, are present at all stages of pollen development (Brown, 1988; Brown and Crouch, 1990).

Antibodies prepared to the polypeptide encoded by the *Oenothera P2* clone were used in immunoblots to determine the timing of the appearance and accumulation of the *P2* family of proteins during pollen development. The *P2* proteins were present in pollen at later stages of development and in germinating pollen tubes. The mRNAs for the *P2* proteins are thus translated in developing pollen before anthesis and not just stored in the grain for translation later during germination and pollen tube growth (Brown and Crouch, 1990).

Antibodies made to a synthetic peptide of the putative sequence of the maize *Zm13* gene product were used in Western blots to determine when the *Zm13* protein was synthesized. As with the *Oenothera P2* protein, the *Zm13* protein was also translated late in pollen maturation. ^{35}S-Methionine labeling experiments moreover indicated that the *Zm13* mRNA was, in addition, translated during pollen germination (A. Scheewe, T. Reynolds, M. Gutensohn, and J. P. Mascarenhas, unpublished data).

A pollen-specific clone from *Brassica napus*, *Bp4*, representing a multigene family of 10–15 closely related pollen-expressed genes, is expressed early in microspore development, although exactly when it is activated was not reported. The mRNA is present at high concentrations until after generative cell division is completed and then decreases drastically just before pollen maturity (Albani *et al.*, 1990).

Another clone, *Bp19*, isolated from *Brassica napus*, is highly expressed in pollen but not in leaves, stems, or seeds. It seems also to be expressed in much lower levels in pistils and petals (Albani *et al.*, 1991). The accumulation of *Bp19* mRNA begins in the early uninucleate microspore, reaches a peak at later stages, and decreases substantially in mature pollen.

A different pollen-specific *B. napus* cDNA clone, *I3*, detects an mRNA that accumulates to high levels around microspore mitosis and finally

undergoes a dramatic decrease in the maturing pollen grain after generative cell division, although low levels were still present in the mature pollen grain (Roberts *et al.*, 1991). The expression patterns of three other *B. napus* microspore-expressed clones, *#17, E2,* and *F2S,* were studied by dot blot analyses of RNA isolated from anthers of different sizes (Scott *et al.*, 1991; from the data presented in this article it is unclear whether any of these clones is pollen-specific). The *E2* transcripts were detected after microspore release from the tetrads, and they maintained their high levels until just before generative cell division, when they were no longer present. The *F2S* transcripts were first detectable during premitotic interphase in the microspore. The levels of *F2S* mRNA remained high until after generative cell division and then decreased rapidly so that no transcripts were found in the mature pollen grain. The mRNA complementary to clone *#17* appears to be first detectable after meiosis, reaches a maximum during and following microspore mitosis, decreases thereafter, and is not detectable in the maturing pollen grain some time before generative cell division (Scott *et al.*, 1991). The temporal pattern of expression of these clones overlaps; although each clone appears to have a unique pattern of expression, together they span the entire period of microspore and pollen development.

The expression pattern of the different genes active in pollen is thus fairly complex. We need additional studies of the temporal activation and expression of many more pollen-expressed genes before any accurate picture is obtained concerning whether small or large groups of genes are turned on at specific stages during male gametophyte development or whether there is, in addition, a cascade of temporally overlapping genes. Such studies might be useful in the identification of critical regulatory points in the pollen development pathway.

IV. Characterization and Identification of Pollen-Expressed Genes

A number of cDNA clones corresponding to genes expressed in the developing male gametophyte have been isolated, several sequenced, and a few tentatively identified as to function. In addition, a few of the corresponding genomic clones have been isolated and some sequenced.

The pollen-specific cDNA clone from maize, *Zm13,* is a full-length copy of the mRNA, which is 929 nucleotides long. *Zm13* codes for a predicted polypeptide of 170 amino acid residues with a molecular mass of 18.3 kDa (Hanson *et al.*, 1989). The hydropathy profile of the polypeptide suggests a possible signal sequence at the amino terminus. The mRNA contains a 5'-untranslated region of 127 nucleotides and a 3'-untranslated region of

292 nucleotides to the polyadenylation site. The putative polyadenylation signal is spaced unusually distant from the actual site of poly(A) addition. The consensus AATAAA polyadenylation signal motif is located 180 nucleotides upstream from the site of poly(A) addition and 110 nucleotides downstream from the presumptive stop codon (Hanson *et al.*, 1989). The *Zm13* gene has been isolated and sequenced. It does not contain any introns (Hamilton *et al.*, 1989). Southern blot analysis indicates that the *Zm13* gene is present in one or two copies in the maize genome (Stinson *et al.*, 1987).

The tomato *LAT52* gene codes for an 800-nucleotide mRNA that is detectable in pollen, anthers, and at 20- to 50-fold lower levels in petals (Twell *et al.*, 1989). It encodes a putative protein of 17.8 kDa that has an amino-terminal hydrophobic region with characteristics of a signal sequence. *LAT52* and the *Zm13* sequence from maize exhibit substantial amino acid sequence homology. *LAT52* shows 32% amino acid identity to the predicted polypeptide sequence of *Zm13*, including the presence of six conserved cysteine residues. The *LAT52* gene, unlike that of *Zm13*, contains a single intron with an unusual structure. It is composed of direct repeats of a 46-nucleotide sequence that occurs nine times as a tandem array that is flanked by a small number of unique sequence nucleotides at the 5' and 3' ends of the intron (Twell *et al.*, 1989). The *LAT52* gene appears to be a single copy gene in the tomato genome (Twell *et al.*, 1989).

Two other pollen- and anther-expressed tomato cDNA clones, *LAT56* and *LAT59*, have been sequenced and their corresponding genomic clones isolated and characterized (Wing *et al.*, 1990). DNA blot analyses indicate that these genes are present in single copies in the genome. The two genes are moreover linked on chromosome 3, approximately 5 centimorgans (cM) apart. Two small introns are found in each of the genes but not in homologous positions. The deduced amino acid sequences of *LAT56* and *LAT59* show 54% amino acid identity. *LAT56* codes for a 356-amino acid polypeptide of 40.6 kDa, whereas *LAT59* encodes a putative polypeptide of 449 amino acids with a molecular mass of 50.9 kDa. Both polypeptides appear to have signal sequences that could target the proteins for secretion. The proteins for both *LAT56* and *LAT59* show significant homology to pectate lyases from the bacterial plant pathogen *Erwinia chrysanthemi* and homology to a 20-amino acid amino-terminal sequence (Tanai *et al.*, 1988) of the major pollen allergen of Japanese cedar (Wing *et al.*, 1990). A pollen-specific cDNA clone from maize, *Zm58*, also shows sequence homology to *Erwinia* pectate lyases and to *LAT56* and *LAT59* (D. Hamilton, D. Bashe, and J. P. Mascarenhas, unpublished results).

Two cDNA clones from the *P2* gene family expressed in *Oenothera* pollen have been sequenced. This gene family contains six to eight members, which are expressed at high levels in pollen and not in other tissues.

The two cDNAs (*P2* and *P22*) share 87% sequence identity. The putative amino acid sequence of the *P22* protein shows significant homology to the amino acid sequence of a polygalacturonase from tomato fruit (Brown and Crouch, 1990).

The complete nucleotide sequence of a genomic clone from *Brassica napus, Bp4*, and of three homologous cDNA clones *(cBp401, cBp405,* and *cBp408)* has been reported (Albani *et al.*, 1990). The cloned genomic fragment contains three genes, *Bp4A, Bp4B,* and *Bp4C*. The *Bp4* transcripts are about 450 nucleotides in length and code for small polypeptides, as determined by the open reading frames. *Bp4C* and *cBp405* encode putative proteins containing 73 amino acids (about 8 kDa). *Bp4A* and *cBp401* code for proteins of 63 (6.9 kDa) and 52 (5.6 kDa) amino acids, respectively. The proteins are unique and contain a high number of lysine and cysteine residues. The polypeptides coded by *Bp4C* and *cBp405* contain 11 lysine residues and eight cysteine residues. The functions of these extremely small but potentially interesting polypeptides are not known. Gene *Bp4B* in the genomic clone is thought to be nonfunctional because of critical sequence rearrangements (Albani *et al.*, 1990). These genes that code for relatively small mRNAs surprisingly contain large introns. The two genes *Bp4A* and *Bp4C* contain a single intron each, at lengths of 765 and 766 nucleotides, respectively (Albani *et al.*, 1990).

The nucleotide sequences of the genomic clone and several homologous cDNA clones for the *Bp19* gene from *B. napus* have been determined (Albani *et al.*, 1991). The *Bp19* gene contains a single intron of 137 bp and codes for a messenger RNA of about 1900 nucleotides in length. The *Bp19* protein, as deduced from its nucleotide sequence, codes for a polypeptide of 584 amino acids with a molecular mass of 63 kDa. The protein has a signal peptide at the amino terminus. The carboxyl portion of the *Bp19* protein, starting with amino acid 269, has strong sequence homology to pectin esterases of tomato and of the bacterial plant pathogen *Erwinia chrysanthemi* (Albani *et al.*, 1991).

Pollen allergens are major causes of Type I allergic reactions in humans. Recombinant DNA technology has become the method of choice for the characterization of allergens and the large-scale synthesis of the allergens or their peptides (Silvanovich *et al.*, 1991; Davies *et al.*, this volume). The amino acid sequence of several pollen allergens has now been determined.

Using an expression cDNA library to mRNA from Kentucky bluegrass *(Poa pratensis)* pollen and screening the library with human allergic serum, several cDNA clones encoding human IgE-binding proteins have been obtained. It is interesting that the cDNA library was made from commercially available dry pollen from which translatable messenger RNA could be obtained. One of the isolated clones *(KBG7.2)* was characterized by sequence analysis (Mohapatra *et al.*, 1990). It hybridizes to a messenger

RNA 1500 nucleotides in length from Kentucky bluegrass pollen, but no hybridization was seen to RNA from leaves. Transcripts hybridizing to the *(KBG7.2)* probe were found in pollens of eight other grass species (Mohapatra *et al.*, 1990).

The complete nucleotide and deduced amino acid sequences have been determined for three other cDNA clones from Kentucky bluegrass, *KBG41, KBG60,* and *KBG31,* which encode a major group of Kentucky bluegrass pollen allergens. These three clones are similar to one another, having 95% identity within a consensus region of 771 bp but have only minor homology to other known allergens and have been designated as the *Poap*IX group of isoallergenic proteins (Silvanovich *et al.*, 1991). *KBG41* contains a characteristic 32-nucleotide sequence in phase internal repeat at the 3' end. Probes to all three clones hybridize to messenger RNAs of 1320–1420 nucleotides. From Northern blot analyses the clones appear to be pollen-specific in their expression. Clones *KBG31, KBG41,* and *KBG60* have 373, 333, and 303 codons with deduced molecular masses of 37.8, 32.7, and 30.5 kDa, respectively. All three peptides have signal sequences at their amino termini that have features typical of eukaryotic signal peptides. A search of protein data bank sequences revealed no meaningful homologies. Southern blot analyses suggest that the Kentucky bluegrass allergenic proteins are encoded by a multigene family (Silvano-vich *et al.*, 1991).

Ragweed is the major cause of late summer hayfever in the Eastern United States and Canada (Rafner *et al.*, 1991). Antigen E or *Amba*I is considered to be the most important allergen in ragweed pollen because 95% of ragweed-sensitive persons react to it. It is an abundant 38-kDa protein in pollen, making up about 6% of the total protein in neutral aqueous extracts of pollen. Messenger RNA was isolated and cDNA libraries prepared from short ragweed *(Ambrosia artemisiifolia)* pollen and flowers. Screening the libraries with antibodies to *Amba*I resulted in the isolation of several clones. Based on sequence analysis of these clones, they could be divided into three groups according to their sequence similar-ity: *Amba*I.1, *Amba*I.2, and *Amba*I.3. Greater than 99% identity at the nucleotide level was found within a group and 85–90% among groups. Southern blot analyses indicate that *Amba*I is a family of closely related proteins (Rafner *et al.*, 1991). The clones hybridize to messenger RNAs of about 1500 nucleotides; deduced amino acid sequence analyses indicate polypeptides of 398 amino acids. A hydrophobic stretch of amino acids at the amino terminus has the properties of a signal sequence, indicating the possible secretion of the *Amba*I proteins (Rafner *et al.*, 1991). No consen-sus AATAAA polyadenylation signal was found at the 3' ends of the clones. A variant sequence AATGAA was, however, located 45 nucleo-tides from the site of poly(A) addition. A variant sequence, AATAAT,

was located 101 nucleotides upstream of the polyadenylation site in clone *Amba*I.2 (Rafner *et al.*, 1991).

A full-length cDNA clone *(Lolp*Ia) has been isolated from ryegrass *(Lolium perenne)* pollen, which codes for the major glycoprotein allergen, *Lolp*I, with a molecular mass of 35 kDa (Griffith *et al.*, 1991). The *Lolp*I gene is expressed in pollen but not in seeds, leaves, or roots. The nucleotide sequence of *Lolp*Ia indicates an open reading frame of 263 amino acids (29.1 kDa). A putative amino-terminal signal sequence of 23 amino acids would indicate a mature processed protein of 240 amino acids with a molecular mass of 26.6 kDa. A single potential *N*-glycosylation site is present in the molecule (Griffith *et al.*, 1991).

A different cDNA clone designated *Lolp*Ib, which codes for an allergen distinct from that of *Lolp*I *(Lolp*Ia), has also been isolated. Based on its deduced amino acid sequence, *Lolp*Ib is a protein of 308 amino acids (34.1 kDa). The predicted protein is rich in alanine (23%) and proline (13%) and appears to contain a signal peptide at the amino terminus. The estimated molecular mass of the mature processed protein is about 31 kDa (Singh *et al.*, 1991). The *Lolp*Ib clone hybridizes to a messenger RNA of 1200 nucleotides from pollen but does not hybridize to RNAs from leaf or root tissue. Unlike *Lolp*Ia (Staff *et al.*, 1990), which occurs in the pollen cytoplasm, *Lolp*Ib is located mainly in the starch granules (Singh *et al.*, 1991). This was shown by double labeling with immunogold probes of two different sizes (Singh *et al.*, 1991). It has been hypothesized that the *Lolp*Ib allergen is synthesized in the pollen cytoplasm as a precursor with a transit peptide that targets the protein to amyloplasts. This fact suggests a possible role for the *Lolp*Ib protein in starch mobilization during pollen germination and tube growth. The *Lolp*Ib sequence, however, shares no identity with other sequences in sequence data banks (Singh *et al.*, 1991).

Pollen of the white birch *(Betula verrucosa)* is a major cause of Type I allergic reactions in temperate regions of the world. The complete amino acid sequence of *Betv*I, the major allergen of white birch pollen, has been deduced from the nucleotide sequence of a cDNA clone obtained to poly(A)RNA from mature birch pollen (Breiteneder *et al.*, 1989). Sequence comparisons showed a strong similarity between the *Betv*I protein and a disease resistance response gene from pea (Fristensky *et al.*, 1988). The two proteins, *Betv*I and the pea disease resistance gene, are of similar size (160 and 158 amino acids, respectively) and show 55% sequence identity and 70% sequence similarity, including conservative exchanges (Breiteneder *et al.*, 1989).

Two cDNAs encoding a pollen allergen from white birch that is distinct from the major birch pollen allergen, *Betv*I, have also been cloned and sequenced (Valenta *et al.*, 1991). The cDNA library was screened using a serum from an individual allergic to birch pollen. The clones, when used

as probes, identified a messenger RNA of about 800 nucleotides in birch pollen; they also cross-react with a similar size messenger RNA from alder and hazel pollen. The deduced amino acid sequence of the cDNA clones showed homology to profilins from slime mold, ameba, yeast, mouse, calf, and human. In addition, other functional tests also support the identification as profilin. Profilins are molecules that regulate the polymerization of actin (Pollard and Cooper, 1986). They participate in signal transduction via the phosphoinositide-signaling pathway (Goldschmidt-Clermont *et al.*, 1990). In view of the large amount of actin present in pollen (Heslop-Harrison and Heslop-Harrison, 1989; Piersen and Cresti, this volume) and the function of the actin in cytoplasmic streaming in the pollen tube (Mascarenhas and Lafountain, 1972; Heslop-Harrison and Heslop-Harrison, 1989), the identification of profilin in pollen could potentially be a significant finding in our understanding of the regulation of growth of the pollen tube. Large amounts of profilin were found in pollens of trees (birch, *Alder glutinosa;* hazel, *Corylus avellana*), in grasses (timothy grass, *Phleum pratense;* rye, *Secale cereale*), and other plants, such as *Artemisia vulgaris* (Valenta *et al.*, 1991). It is interesting that a molecule from pollen that is a potent human allergen appears to have a critical role in pollen function and development.

V. Structure of Promoters Responsible for Regulating Pollen Specificity and Temporal Specificity of Transcription of Pollen-Expressed Genes

Analyses in various degrees of detail have been made with the promoters of several pollen-expressed genes. This has been done by constructing promoter fusions with reporter genes and introducing these constructs into plants by *Agrobacterium*-mediated transformation or by transient assays after microprojectile bombardment of pollen. In general, the *cis*-active sequence elements responsible for tissue specificity and temporal specificity of transcription of pollen-expressed genes appear to reside in the 5′ flanking region relatively near the start site of transcription.

Genomic and cDNA clones have been isolated from *Petunia* that code for the flavonoid biosynthetic enzyme chalcone flavonone isomerase (CHI). There are two CHI genes, *chiA* and *chiB*. The *chiA* gene does not contain any introns, whereas the *chiB* gene is interrupted by three introns (Van Tunen *et al.*, 1988, 1989). The *chiA* gene appears to be regulated by two different promoters arranged in tandem. The promoters are differentially used. The more proximal 5′ promoter, P_{A1}, is active in petals, whereas the more distal promoter, P_{A2}, is active in mature pollen. The P_{A1}

promoter gives rise to a 1000-nucleotide messenger RNA in petals and the P_{A2} promoter to a 1500-nucleotide messenger RNA in pollen grains at late stages of development (Van Tunen *et al.*, 1989, 1990). The *chiB* gene is regulated by a single promoter, P_B, which produces a 1000-nucleotide transcript in immature anther tissue (Van Tunen *et al.*, 1989, 1990).

A detailed analysis of various *chi* promoter segments has been made by fusing them to a β-glucuronidase (GUS) reporter gene (Jefferson *et al.*, 1986, 1987) and assaying them in transgenic petunia and tobacco plants (Van Tunen *et al.*, 1990). The P_{A1} and the P_{A2} start sites of transcription were separated, and the resulting *chiA* promoter fragments were then fused to the GUS-coding region. Each promoter fragment was able to drive the GUS gene, providing direct evidence for the existence of two distinct promoters for the *chiA* gene. The 0.44-kb *chiA* P_{A2} promoter is active in pollen late in anther development, but no GUS activity was detected in the corolla of the flower, ovary, leaf, or stem. This pattern of activity of the P_{A2} promoter in driving GUS activity parallels the accumulation of the 1500-nucleotide *chiA* P_{A2} transcript in normal plants. The P_{A2} DNA segment is thus a pollen-specific promoter that is activated late in anther development (Van Tunen *et al.*, 1990).

The 0.6-kb *chiA* P_{A1} promoter GUS-transformed plants, however, expressed GUS activity in the corolla, sepals, seeds, stems, and leaves but not in anthers. Hence, sequences upstream of the P_{A2} start of transcription do not seem to be necessary for the proper tissue-specific and temporal activities of the P_{A1} promoter.

The gene *Po* regulates CHI expression. *PoPo* petunia plants show CHI enzyme activity in both corolla and anthers, whereas *PoPo* plants have CHI enzyme activity in corolla but not in anthers. The *Po* gene is apparently identical to the *chiA* gene and the *Po* mutation appears to be a mutation in the promoter region of *chiA*, probably in the P_{A2} region, which abolishes the activity of the gene in anthers but not in the corolla. Deletions and additions have been found in the *chiA* promoter region that seem to have inactivated the pollen component of the promoter (Van Tunen *et al.*, 1991).

It would be interesting to determine whether this sort of tandem location of promoters is found in other genes that are expressed both in pollen and in vegetative tissues. Most of the genes expressed in pollen development are of this type (i.e., they are not pollen-specific), but they are transcribed in addition in sporophytic tissues (Mascarenhas, 1990a).

A 1.75-kb *chiB* promoter fragment fused to the GUS gene was introduced by transformation into both petunia and tobacco plants. GUS activity driven by the P_B promoter was detected in the tapetal cells and pollen grains of anthers in early stages of development, in similarity with the pattern of accumulation of the normal *chiB* transcript in wild-type plants.

The *chiB* P_B promoter is anther-specific, and GUS activity was not detected in any of the other floral or vegetative tissues tested (Van Tunen *et al.*, 1990).

The 8.5-kb genomic clone, *Bp4* from *Brassica*, has three tandemly arranged genes that represent members of a small gene family. In the promoters of all three genes (*Bp4A, Bp4B,* and *Bp4C*) two 11-bp direct repeats (TAAATTAGATT) are found about 20 and 60 nucleotides upstream of the TATA box (Albani *et al.*, 1990). Transgenic tobacco plants containing a 5.8-kb fragment of the genomic clone produce a transcript in the immature anthers of the transformed plants, corresponding most likely to that for the *Bp4A* gene. This is the pattern of expression of the *Bp4* gene in *Brassica;* hence, the 235 nucleotides upstream of the transcription start site present in the 5.8 kb of genomic DNA thus appear to be sufficient for the correct temporal and pollen-specific expression of the gene in tobacco. There were, however, 3' flanking sequences and introns in the construct that could possibly also contain regulatory elements that could be responsible for the results obtained (Albani *et al.*, 1990).

Sequence comparisons in the 5' flanking region of the *Bp19* gene with the promoter sequences of several other pollen-expressed genes show several short elements that appear to be conserved. These could be candidates for the pollen specificity component of the promoter, but no direct experimental evidence is yet available (Albani *et al.*, 1991). Several 5' promoter deletions from a genomic clone of the pollen-specific gene of maize, *Zm13*, were transcriptionally fused to a GUS reporter gene joined to the nopaline synthase (NOS) polyadenylation region and introduced into tobacco by *Agrobacterium*-mediated transformation (Guerrero *et al.*, 1990). Analysis of transgenic plants containing promoter constructs of 3100, 1063, 646, and 375 bp of *Zm13* promoter DNA (all the promoter fragments ended at +61 relative to the start of transcription at + 1) showed GUS mRNA in mature pollen when assayed with a probe of GUS-coding sequences. In leaf tissue, however, no GUS RNA was detectable. A promoter region from -314 to +61 is thus sufficient to drive the pollen-specific expression of the GUS gene. A promoter fragment from -184 to -61 lacking the putative *Zm13* TATA box (Hamilton *et al.*, 1989) did not give GUS-positive pollen grains in transgenic plants containing this construct (Guerrero *et al.*, 1990). The cauliflower mosaic virus 35S promoter is not expressed or expressed only poorly in transgenic tobacco plants (Guerrero *et al.*, 1990) or in *Tradescantia* pollen in transient assays after particle gun bombardment (Hamilton *et al.*, 1992).

The *Zm13* gene is expressed late in maize pollen development, only after microspore mitosis. Transgenic plants containing the 1063-bp and 375-bp promoter constructs fused to the GUS gene showed GUS activity in the developing pollen grains only after the first mitotic division, and this

activity increased thereafter in the developing grain (Guerrero *et al.*, 1990). The *cis*-acting elements necessary for the correct temporal pattern of expression of the *Zm13* gene in maize are thus retained in the -314 to +61 DNA fragment.

Moreover, the promoter sequences from a gene from a monocot plant, maize, are able to correctly direct genetically stable, pollen-specific gene expression in a transgenic dicot plant, tobacco (Guerrero *et al.*, 1990).

A more detailed analysis of the *Zm13* 5' regions responsible for expression in pollen has been made by using a transient expression system. Constructs containing the GUS gene under the control of various sized fragments of the *Zm13* 5' flanking region were introduced into *Tradescantia* pollen via high velocity microprojectile bombardment and monitored both visually and with a fluorescence quantitative assay (Hamilton *et al.*, 1992). These results suggest that the nucleotide sequences necessary for expression in pollen are present in a region from -100 to -54, whereas other sequences that enhance that expression reside between -260 and -100. These transient assays have provided evidence for a negative regulatory element in the -1001 to -260 promoter region. The normal NOS 3' terminator in the GUS gene construct was replaced with a portion of the *Zm13* 3' region containing the putative polyadenylation signal and site of this maize gene. Substitution of the NOS terminator with the *Zm13* 3' terminator resulted in a substantial increase in transient GUS activity in pollen. The reasons for this enhanced activity are not clear but are under further study (Hamilton *et al.*, 1992).

The most detailed analysis of the fine structure of pollen promoters to date has been done by Twell *et al.* (1989, 1990, 1991). The *cis* elements involved in promoter strength and specificity were identified and characterized for three tomato pollen-expressed genes by constructing 5' and internal deletion mutants of the promoters and linking them to the GUS gene. These constructs were assayed for activity by transferring them into tomato plants by transformation and also by a transient assay system using tobacco pollen (Twell *et al.*, 1991). When the *LAT52* promoter was deleted from -3000 to -492 bp, no significant change in promoter activity in pollen was detected in transgenic plants. Deletion to -145 bp reduced promoter activity about 10-fold. Further deletion to -86 bp or -71 bp reduced activity 100-fold, although the activity was still about 20-fold above background levels. The *LAT52* promoter contains at least three distinct regions that regulate gene expression in pollen. The -492 bp to -145 bp region and the -124 bp to -86 bp regions contain sequences that enhance expression in pollen. The region between -71 and +110 bp is, however, sufficient to drive gene expression in pollen, although at a low level (Twell *et al.*, 1991).

Transient assays defined further subregions of importance in the *LAT52* promoter. The -225 to -145 bp region was able to enhance the activity of

a truncated -89 bp *CaM* V 35S promoter in pollen. A single copy of this region in reverse orientation enhanced the activity of the minimal *CaM* V 35S promoter 14-fold, and two tandem copies gave an approximately 25-fold enhancement.

Deletion analysis with the *LAT59* promoter in transgenic plants did not identify strong upstream activating sequences, as was found for the *LAT52* promoter. Regulatory elements located between -115 and -45 bp appear to be mainly responsible for the activity of the *LAT59* promoter. Evidence for a negative regulatory element between -804 and -418 bp and a positive regulatory element present between -1305 and -804 bp was also obtained (Twell *et al.*, 1991).

The analyses of the *LAT52*, *LAT56*, and *LAT59* promoters have provided evidence for the presence of two regulatory elements, the 52/56 box (TGTGGTTATATA) and the 56/59 box (GAATTTGTGA), which are shared between two of the three *LAT* promoters (Twell *et al.*, 1991). Twell *et al.* (1991) have proposed a quantitative role for the 52/56 box. Presumably, transcription factors that recognize this sequence enhance the basal levels of *LAT52* and *LAT56* promoter activity in pollen. The 56/59 box may also have similar functions in enhancing the basal activity of the *LAT56* and *LAT59* promoters and/or it may regulate the turning on of the expression of these two genes (Twell *et al.*, 1991).

In summary, substantial progress has recently been made in attempts to identity *cis* elements and *trans*-acting factors that are involved in pollen specificity of expression of genes. To understand the full complexities of the regulatory mechanisms will require further efforts, and these are ongoing in several laboratories.

Acknowledgment

Work in the author's laboratory has been supported by grants from the National Science Foundation.

References

Acevedo, A., and Scandalios, J. G. (1990). *Theor. Appl. Genet.* **80**, 705–711.
Albani, D., Robert, L. S., Donaldson P. A., Altosaar, I., Arnison, P. G., and Fabijanski, S. F. (1990). *Plant Mol. Biol.* **15**, 605–622.
Albani, D., Altosaar, I., Arnson, P. G., and Fabijanski, S. F. (1991). *Plant Mol. Biol.* **16**, 501–513.
Breiteneder, H., Pettenburger, K., Bito, A., Valenta, R., Kraft, D., Rumpold, H., Scheiner, O., and Breitenbach, M. (1989). *EMBO J.* **8**, 1935–1938.
Brink, R. A., and MacGillivray, J. H. (1924). *Am. J. Bot.* **11**, 465–469.

Brown, S. M. G. (1988). Ph.D. thesis, Indiana University, Bloomington.

Brown, S. M., and Crouch, M. L. (1990). *Plant Cell* **2**, 263–274.

Cresti, M., and Tiezzi, A. (1990). *In* "Microspores, Evolution and Ontogeny" (S. Blackmore and R. B. Knox, eds.), pp. 239–263. Academic Press, London.

Demerec, M. (1924). *Am. J. Bot.* **11**, 461–464.

Evans, D. E., Singh, M. B., and Knox, R. B. (1990). *In* "Microspores, Evolution and Ontogeny" (S. Blackmore and R. B. Knox, eds.), pp. 309–338. Academic Press, London.

Felder, M. R., Scandalios, J. R., and Liu, E. H. (1973). *Biochim. Biophys. Acta* **317**, 149–159.

Freeling, M., and Schwartz, D. (1973). *Biochem. Genet.* **8**, 27–36.

Fristensky, B., Horovitz, D., and Hadwiger, L. A. (1988). *Plant Mol. Biol.* **11**, 713–715.

Frova, C. (1990). *Sex. Plant Reprod.* **3**, 200–206.

Giles, K. L., and Prakash, J. eds. (1987). "Pollen: Cytology and Development. International Review of Cytology," Vol. 107. Academic Press, San Diego.

Goldschmidt-Clermont, P. J., Machesky, L. M., Baldassare, J. J., and Pollard, T. D. (1990). *Science* **247**, 1575–1578.

Griffith, I. J., Smith, P. M., Pollock, J., Theerakulpisut, P., Avjioglu, A., Davies, S., Hough, T., Singh, M. B., Simpson, R. J., Ward, L. D., and Knox, R. B. (1991). *FEBS Lett.* **279**, 210–215.

Guerrero, F. D., Crossland, L., Smutzer, G. S., Hamilton, D. A., and Mascarenhas, J. P. (1990). *Mol. Gen. Genet.* **224**, 161–168.

Hamilton, D. A., Bashe, D. M., Stinson, J. R., and Mascarenhas, J. P. (1989). *Sex. Plant Reprod.* **2**, 208–212.

Hamilton, D. A., Roy, M., Rueda, J., Sindhu, R. K., Sanford, J., and Mascarenhas, J. P. (1992). *Plant Mol. Biol.* **18**, 211–218.

Hanson, D. D., Hamilton, D. A., Travis, J. L., Bashe, D. M., and Mascarenhas, J. P. (1989). *Plant Cell* **1**, 173–179.

Heslop-Harrison, J., and Heslop-Harrison, Y. (1989). *Sex. Plant Reprod.* **2**, 199–207.

Jefferson, R. A., Burgess, S. M., and Hirsch, D. (1986). *Proc. Natl. Acad. Sci. U.S.A.* **83**, 8447–8451.

Jefferson, R. A., Kavanagh, T. A., and Bevan, M. V. (1987). *EMBO J.* **6**, 3901–3907.

Mascarenhas, J. P. (1988). *In* "Temporal and Spatial Regulation of Plant Genes" (D. P. S. Verma and R. B. Goldberg, eds.), pp. 97–115. Springer-Verlag, New York.

Mascarenhas, J. P. (1989). *Plant Cell* **1**, 657–664.

Mascarenhas, J. P. (1990a). *Annu. Rev. Plant Physiol. Plant Mol. Biol.* **41**, 317–338.

Mascarenhas, J. P. (1990b). *In* "Microspores, Evolution and Ontogeny" (S. Blackmore and R. B. Knox, eds.), pp. 265–280. Academic Press, London.

Mascarenhas, J. P., and Lafountain, J. (1972). *Tissue Cell* **4**, 11–14.

Mascarenhas, J. P., Eisenberg, A., Stinson, J. R., Willing, R. P., and Pe, M. E. (1985). *In* "Plant Cell/Cell Interactions" (I. Sussex, A. Ellingboe, M. Crouch, and E. Malmberg, eds.), pp. 19–23. Cold Spring Harbor Press, New York.

Miller, J. C., and Mulcahy, D. L. (1983). *In* "Pollen: Biology and Applications in Plant Breeding" (D. L. Mulcahy and E. Ottaviano, eds.), pp. 317–321. Elsevier/North Holland, New York.

Mohapatra, S., Hill, R., Astwood, J., Ekramoddoulah, A. K. M., Olson, E., Silvanovich, A., Hatton, T., Kisil, F. T., and Sehon, A. (1990). *Int. Arch. Allergy Appl. Immunol.* **91**, 362–368.

Nelson, O. E. (1958). *Science* **130**, 794–795.

Ottaviano, E., and Mulcahy, D. L. (1989). *Adv. Genet.* **26**, 1–64. Academic Press, San Diego.

Ottaviano, E., Gorla, M. S., and Mulcahy, D. L. (1990). *In* "Isozymes: Structure, Function and Use in Biology and Medicine" (Z. I. Ogita and C. L. Markert, eds.), pp. 575–588. Wiley–Liss, New York.

Pacini, E. (1990). *In* "Microspores, Evolution and Ontogeny" (S. Blackmore and R. B. Knox, eds.), pp. 213–237. Academic Press, London.

Pollard, T. D., and Cooper, T. A. (1986). *Annu. Rev. Biochem.* **55,** 987–1035.

Rafner, T., Griffith, I. J., Kuo, M. C., Bond, J. F., Rogers, B. L., and Klapper, D. G. (1991). *J. Biol. Chem.* **266,** 1229–1236.

Roberts, M. R., Robson, F., Foster, G. D., Draper, J., and Scott, R. J. (1991). *Plant Mol. Biol.* **17,** 295–299.

Roeckel, P., Chaboud, A., Matthys-Rochon, E., Russell, S., and Dumas, C. (1990). *In* "Microspores, Evolution and Ontogeny" (S. Blackmore and R. B. Knox, eds.), pp. 281–307. Academic Press, London.

Sari-Gorla, M., Frova, C., Binelli, G., and Ottaviano, E. (1986). *Theor. Appl. Genet.* **72,** 42–47.

Schwartz, D. (1971). *Genetics* **67,** 411–425.

Scott, R., Dagless, E., Hodge, R., Paul, W., Soufleri, I., and Draper, J. (1991). *Plant Mol. Biol.* **17,** 195–207.

Silvanovich, A., Astwood, J., Zhang, L., Olsen, E., Kisil, F., Sehon, A., Mohapatra, S., and Hill, R. (1991). *J. Biol. Chem.* **266,** 1204–1210.

Singh, M. B., O'Neill, P., and Knox, R. B. (1985). *Plant Physiol.* **77,** 225–228.

Singh, M. B., Hough, T., Theerakulpisut, P., Avjioglu, A., Davies, S., Smith, P. M., Taylor, P., Simpson, R. J., Ward, L. D., McCluskey, J., Puy, R., and Knox, R. B. (1991). *Proc. Natl. Acad. Sci. U.S.A.* **88,** 1384–1388.

Staff, I. A., Taylor, P. E., Smith, P. M., Singh, M. B., and Knox, R. B. (1990). *Histochem. J.* **22,** 276–290.

Stinson, J. R., Eisenberg, A. J., Willing, R. P., Pe, M. E., Hanson, D. D., and Mascarenhas, J. P. (1987). *Plant Physiol.* **83,** 442–447.

Tanai, M., Ando, S., Usui, M., Kurimoto, M., Sakagushi, M., Inouye, S., and Matuhasi, T. (1988). *FEBS Lett.* **239,** 329–332.

Tanksley, S. D., Zamir, D., and Rick, C. M. (1981). *Science* **213,** 454–455.

Twell, D., Wing, R., Yamaguchi, J., and McCormick, S. (1989). *Mol. Gen. Genet.* **247,** 240–245.

Twell, D., Yamaguchi, J., and McCormick, S. (1990). *Development* **109,** 705–713.

Twell, D., Yamaguchi, J., Wing, R. A., Ushiba, J., and McCormick, S. (1991). *Genes Dev.* **5,** 496–507.

Ursin, V. M., Yamaguchi, J., and McCormick, S. (1989). *Plant Cell* **1,** 727–736.

Valenta, R., Duchene, M., Pettenburger, K., Sillaber, C., Valent, P., Bettelheim, P., Breitenbach, M., Rumpold, H., Kraft, D., and Scheiner, O. (1991). *Science* **253,** 557–560.

Van Tunen, A. J., Koes, R. E., Spelt, C. E., Van der Krol, A. R., Stuitze, A. R., and Mol, J. N. M. (1988). *EMBO J.* **7,** 1257–1263.

Van Tunen, A. J., Hartman, S. A., Mur, L. A., and Mol, J. N. M. (1989). *Plant Mol. Biol.* **12,** 539–551.

Van Tunen, A. J., Mur, L. A., Brouns, G. S., Rienstra, J. D., Koes, R. E., and Mol, J. N. M. (1990). *Plant Cell* **2,** 393–401.

Van Tunen, A. J., Mur, L. A., Recourt, K., Gerats, A. G. M., and Mol, J. N. M. (1991). *Plant Cell* **3,** 39–48.

Weeden, F., and Gottlieb, L. D. (1979). *Biochem. Genet.* **17,** 287–296.

Willing, R. P., and Mascarenhas, J. P. (1984). *Plant Physiol.* **75,** 865–868.

Willing, R. P., Bashe, D., and Mascarenhas, J. P. (1988). *Theor. Appl. Genet.* **75,** 751–753.

Wing, R. A., Yamaguchi, J., Larabell, S. K., Ursin, V. M., and McCormick, S. (1990). *Plant Mol. Biol.* **14,** 17–28.

Identification and *in Situ* Localization of Pollen-Specific Genes

S. P. Davies, M. B. Singh, and R. B. Knox
School of Botany, University of Melbourne
Parkville, Victoria 3052, Australia

I. Introduction

By isolating genes that are expressed exclusively during pollen development, a detailed analysis of cell-specific gene regulation and cellular differentiation has become a reality. The previous section (Mascarenhas, this volume) has provided an overview of pollen development. Also, the molecular evidence available suggests that haploid gene expression not only occurs in the pollen grain but some of the genes expressed are specific to this developmental process. In this chapter we will examine ways in which the expression of such genes and their products can be localized.

One way of identifying genes that are transcriptionally active in a particular cell is to examine the mRNA population present in that cell. By isolating poly(A)RNA from pollen or anthers, cDNA libraries have been successfully constructed in which each unique cDNA clone represents the transcript of a gene that is active in these tissues. The identification of cDNA clones corresponding to mRNAs that are uniquely transcribed in pollen or anthers can then be attempted by a process of differential hybridization. Total RNA is isolated from pollen or anthers along with RNA from a number of sporophytic tissues, such as stems, roots, and leaves. In most cases a cDNA clone is said to encode a pollen- or anther-specific transcript if hybridization occurs with pollen or anther RNA but is not detectable by standard RNA blot techniques with RNA from stems, roots, etc.

How these genes are regulated with respect to their temporal and spatial expression patterns can be analyzed. In other words, we can answer the questions: when are these genes expressed and in which tissue or cell does expression occur? In addition, the expression patterns of such genes can also elucidate their function in development of the male gametophyte.

II. Two Groups of Genes Are Expressed during Pollen Development

Many of the genes expressed during pollen development are not unique to this process (Mascarenhas, this volume). The expression patterns of some "housekeeping genes" has been studied. For example, genes such as actin, ADH, and β-galactosidase are expressed at microspore release, soon after meiosis is complete. Such genes have been described as the early set of pollen-expressed genes. Their expression increases during microspore development but diminishes substantially before anthesis.

Studies that have determined the timing of expression of pollen- or anther-specific genes (Section III) indicate that most belong to the group described as the late expressed pollen genes. These genes become active after microspore mitosis and their mRNAs increase in abundance up to maturity. The mRNAs for these genes continue to accumulate during the maturation phases. This accumulation of late mRNAs suggests an important role for these genes during one or more of the later stages of pollen maturation, pollen germination or pollen tube growth.

Recently, two early expressed pollen-specific genes, *Bp4* and *Bp19*, have been identified from *Brassica napus* (Albani *et al.*, 1990, 1991). Activation of the *Bp4* and *Bp19* genes occurs particularly early during microspore development prior to the first mitotic division and might start during some of the meiotic stages. The levels of *Bp4* and *Bp19* transcripts decrease significantly when most of the microspores are close to maturity, indicating a specific role for this gene early during pollen development.

The majority of pollen- and anther-specific genes have been identified from cDNA libraries constructed from RNA isolated from anther tissue late during pollen development, for example, from mature pollen. It could be argued that it is for this reason that most of the genes isolated fall into the late expressed group of genes. Isolation of early expressed genes would therefore require the construction of cDNA libraries from early stages, (e.g., from microspores.)

Alternatively, it has been suggested that microspore mitosis is an important deterministic switch in pollen development with the unequal partitioning of nuclear and/or cytoplasmic factors acting to regulate transcription factors that may control these genes (Twell *et al.*, 1991). The asymmetric mitotic division of the haploid microspore (microspore mitosis) results in the formation of two dimorphic cells (the vegetative cell and the generative cell) with different developmental fates (Sunderland and Huang, 1987). The vegetative cell constitutes the bulk of the young pollen grain and supports pollen maturation and pollen tube growth. The genera-

tive cell, which inherits a small amount of the microspore cytoplasm, lies wholly within the vegetative cell—a cell within a cell.

cDNA clones encoding transcripts expressed exclusively during pollen development and in some cases their corresponding genomic clone have been isolated from an increasingly large number of plant species; Table I provides a summary of these genes.

TABLE I

Selected List of Anther- and Pollen-Specific cDNA Clones Isolated from Flowering Plants, with Their Putative Function and Specificity Indicated

Clone name	Species	Putative function	Specificity	Reference
AmbI.1.2.3	Ragweed	Pectate lyase	Pollen	Rafner *et al.*, 1991
Bcp1	*Brassica*	—	Pollen, tapetum	Theerakulpisut *et al.*, 1991
Bp4	*Brassica*	—	Microspores	Albani *et al.*, 1990
Bp19	*Brassica*	—	Microspores	Albani *et al.*, 1991
*Cyn d*I	Bermuda grass	—	Pollen	Smith (unpublished)
KBG 41,60,30	Kentucky bluegrass	—	Pollen	Silvanovich *et al.*, 1991
LAT52	Tomato	Kunitz trypsin inhibitor	Pollen	McCormick, 1991
LAT56	Tomato	Pectate lyase	Pollen, anther wall	Twell *et al.*, 1991
LAT59	Tomato	Pectate lyase	Pollen, anther wall	Twell, *et al.*, 1991
*Lolp*I	Ryegrass	—	Cytosol–pollen grain	Griffith *et al.*, 1991
*Lolp*Ib	Ryegrass	Amylase inhibitor	Starch grains, pollen	Singh *et al.*, 1991
P2	*Oenothera*	Poly-galacturonase	Pollen, pollen tube	Brown and Crouch, 1990
TA29	Tobacco	Glycine-rich	Tapetum	Koltunow *et al.*, 1990
TA56	Tobacco	Thiol endopeptidase	Anther wall	Koltunow *et al.*, 1990
TA36	Tobacco	Lipid transfer protein	Tapetum	Koltunow *et al.*, 1990
Zm13	Corn	Kunitz trypsin inhibitor	Pollen	Hamilton *et al.*, 1989

III. Temporal Expression Patterns of Pollen-Expressed Genes

Pollen development is correlated with a number of morphological markers, including floral bud length, hypanthium length, or anther length, and many authors have used such markers to describe the various stages of microsporogenesis (Evans *et al.*, 1990). The temporal expression patterns of pollen- or anther-specific genes can therefore be determined by isolating RNA from flower buds of different lengths to identify when the gene transcript first appears. This can then be related back to a particular stage of pollen development.

Bud length proved a convenient marker in the case of *B. campestris* (Theerakulpisut *et al.*, 1991). The timing of gene expression encoding the anther-specific transcript represented by the *B. campestris* cDNA clone *Bcp1* is shown in Figure 1A. Total RNA isolated from *B. campestris* flower buds of lengths from 2 to 6 mm and from mature pollen was electrophoresed on an RNA gel, blotted, and then probed with labeled *Bcp1* cDNA. The developmental Northern blot shown in Figure 1A shows that RNA transcripts complementary to *Bcp1* do not appear in flower buds until the 5-mm bud length stage of development, indicating that the gene is not transcriptionally active until this stage. At the 5-mm bud length stage of

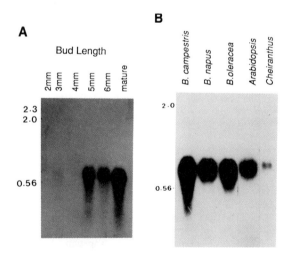

FIG. 1   RNA gel blot analysis showing hybridization of *Bcp1*. (A) Transcripts present at different stages of anther development in *Brassica campestris*, represented by the various bud lengths of the flowers, (B) Homologous transcripts in pollen of five members of the family *Brassicaceae*. Reproduced with permission from Theerakulpisut *et al.*, (1991).

development microspore mitosis is complete (Evans *et al.*, 1990). Furthermore, Figure 1A shows that transcripts for this gene accumulate and are maximal in mature pollen grains. The size of the *Bcp1* transcript (750 nucleotides) is also indicated. Transcripts homologous to *Bcp1* (Fig. 1B) are only detectable in pollen from other species of *Brassica* and from other genera of the family Brassicaceae, including *Arabidopsis*.

RNA blot experiments have identified a similar expression pattern for the tomato pollen-specific genes *LAT52*, *LAT56*, and *LAT59* (Table I). These genes also encode transcripts that accumulate late during anther development and are maximal in mature pollen (Twell *et al.*, 1989).

By using cDNA clones to probe RNA isolated from flower buds of 11 different lengths, Koltunow *et al.* (1990) were able to identify three sets of anther-specific genes in *Nicotiana tabacum*, which are temporally regulated. The largest group encoded transcripts that appear early in development, just after microspore meiosis, and accumulate up to the start of tapetum degeneration, then decline significantly after the microspore nucleus divides (this group includes the *TA29* and *TA36* genes listed in Table I). The transcripts from a second group were present at similar levels in all stages tested and the third group of transcripts accumulated up to microspore nucleus division, remaining constant up to the separation of the pollen sac and then declining to low levels before dehiscence (this group includes the *TA56* gene included in Table I).

The *Bp4* and *Bp19* pollen-specific genes identified in *B. napus* by Albani *et al.* (1990) have expression patterns that differ from most pollen- or anther-specific genes. *Bp4* transcripts first appear in RNA isolated from 1-mm flower buds, at which stage the formation of tetrads is occurring (Albani *et al.*, 1990). The amount of *Bp4* transcripts decreases by the time the tricellular pollen is close to maturity. The highest levels of *Bp4* mRNA are present in developing microspores and not in mature pollen grains. A similar pattern has been observed for *Bp19* (Albani *et al.*, 1991).

IV. Allergenic Proteins Are Encoded by a Unique Class of Pollen-Specific Genes

The vast majority of pollen- or anther-specific genes have been isolated from cDNA and genomic libraries by differential hybridization. A different approach has been used to isolate genes, the protein products of which elicit an allergenic response (hay fever and asthma) in susceptible humans.

Allergen genes expressed specifically in pollen can initially be isolated by virtue of the allergenicity of their encoded proteins. Allergenic pollen proteins elicit the formation of a specific class of antibodies in sensitized

humans, immunoglobulin E (IgE). In addition, monoclonal antibodies (mAbs) have been prepared specifically against allergenic proteins from ryegrass *(Lolium perenne)* (Singh *et al.,* 1991). Thus, both antibodies IgE and mAbs have been used to detect recombinant allergen proteins in *L. perenne* cDNA expression libraries, and two pollen-specific genes encoding the major 35-kDa allergenic protein, *Lolp*I (Griffith *et al.,* 1991), and a new 31-kDa allergenic protein, *Lolp*Ib (Singh *et al.,* 1991), have been isolated (Table I). Both IgE antibodies and mAbs have been used to test the tissue specificity of these allergen proteins by immunoblotting, a process in which total protein from different tissues are isolated and probed with specific antibodies. Figure 2 presents such an immunoblot for *Lolp*Ib in which total protein was isolated from the indicated tissues and probed with *Lolp*Ib-specific mAbs or IgE as shown. Figure 2 also presents a conventional RNA blot in which RNA from the tissues indicated was

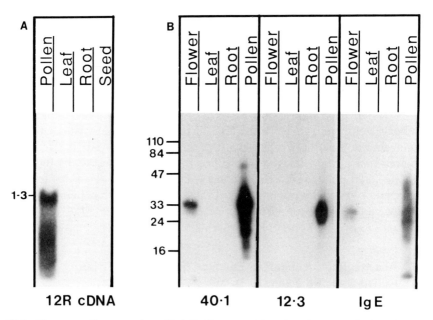

FIG. 2   Tissue-specific expression of *Lolp*Ib allergen. (A) Northern blot analysis of total RNA from ryegrass pollen, leaves, roots, and seeds with 12R cDNA as a probe. (B) Immunoblot analysis of proteins from the same tissues, showing tissue-specific distribution of *Lolp*I antigens. Twenty micrograms of soluble protein was loaded per lane. Probes were mAbs 40.1 (exhibiting preferential binding for *Lolp*Ia) and 12.3 (specific for *Lolp*Ib), detected with ^{125}I-labeled antimouse immunoglobulin (Amersham, Arlington Heights, Illionis). IgE from pooled grass pollen-allergic patients' sera was detected with ^{125}I-labeled antihuman IgE (Kallestad, Austin, Texas). Reproduced with permission from Singh *et al.* (1991).

probed with *Lolp*Ib cDNA. This allergen shows pollen specificity of expression because neither the mRNA transcript of the gene nor its protein product is present in tissue other than pollen.

Similar experiments have successfully isolated allergenic pollen-specific genes from both Bermuda grass and Johnsongrass (unpublished results). In addition, pollen-specific allergen genes from Kentucky bluegrass (Silvanovich *et al.*, 1991) and ragweed (Rafner *et al.*, 1991) have been isolated. Although the functions of these allergenic proteins remain unknown, they are usually abundant proteins conserved across taxa (Singh and Knox, 1985).

V. Localization of Pollen-Specific Transcripts by *in Situ* Hybridization

The specificity and timing of expression of many of the pollen-specific clones isolated by differential hybridization of cDNA libraries have initially been tested by Northern blot analysis. By definition, pollen- or anther-specific cDNA clones encode RNA transcripts that are present only in pollen or anthers, respectively. Therefore, in experimental terms this means that pollen- or anther-specific cDNA clones will not hybridize to RNA isolated from tissue other than pollen or anther. Similarly, the timing of expression of such genes is considered to coincide with the presence of transcripts complementary to the cDNA clone in RNA extracted from developing pollen or anthers.

These methods of isolating total RNA from pollen or anther tissue do not, however, allow the testing of cross-hybridization with specific components of the anther. For example, Northern analysis will not differentiate between clones expressed exclusively in pollen and those expressed in both pollen and tapetum.

A more accurate way of determining exactly where expression of these genes is occurring is by performing *in situ* hybridizations using labeled cDNA to probe sections taken from pollen grains or developing anthers. Hybridization of the probe to the endogenous transcript in the tissue section gives a direct indication of the spatial expression pattern of that gene. This is a far more sensitive method of determining the specificity and spatial expression patterns of genes expressed during pollen development and an important way of correlating the temporal patterns of gene expression with the presence of specific anther tissues and cell types. By hybridizing single-stranded radiolabeled cDNA probes with anther and/ or pollen sections *in situ,* the spatial expression patterns of the mRNA transcripts of a number of the genes listed in Table I have been determined.

In situ hybridization studies using the *B. campestris* cDNA clone *Bcp1* are able to more accurately localize expression of this gene (Theerakulpisut *et al.*, 1991). Developmental Northern blots indicated that expression of the gene was specific to the pollen grain at the later stages of its development. However, by probing anther sections from flower buds of different lengths, expression of *Bcp1*-specific mRNAs was revealed not only in the cytoplasm of pollen grains, in particular the vegetative cell, but also in the tapetum. Figure 3 shows the localization pattern of *Bcp1*

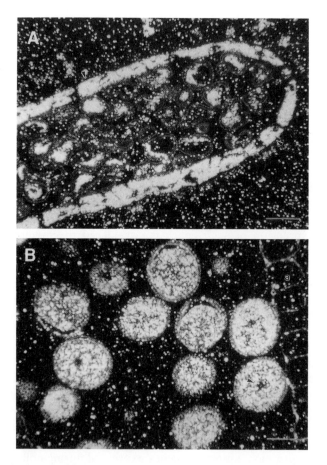

FIG. 3   Localization of mRNAs in *Brassica campestris* with clone *Bcp*I, as detected by *in situ* hybridization. (A) Gene expression in the early stages of development is observed in both the tapetum and the microspore. (B) After tapetum degeneration the hybridization signal is detected only in the mature pollen grains. Reproduced with permission from Theerakulpisut *et al.* (1991).

transcripts. As indicated in Figure 3A, gene expression is observed in both the tapetum and the microspores; however, as the tapetum degenerates (Fig. 3B), the hybridization signal is detected only in the mature pollen grains. Although these *in situ* hybridizations did reveal that *Bcp1* was specific to the tapetum and pollen rather than just the pollen, the results extend the temporal expression pattern of *Bcp1* as determined by Northern blots.

Probing anther sections from three different stages of development allowed Koltunow *et al.* (1990) to localize RNA transcripts from the anther-specific genes *TA29, TA56,* and *TA36*. These *in situ* hybridizations identified the *TA29* and *TA36* transcripts as being tapetum-specific. Strong hybridization signals are present over the tapetum of anther sections at a stage just after microspore meiosis (confirming the temporal expression patterns determined by RNA blots) but not over any other anther region. However, *TA56* transcripts demonstrated a more complex temporal and spatial pattern of expression; both the location and the intensity of the hybridization signals changed with the developmental state of the anther. Early during anther development, *TA56* transcripts are localized over a circular cluster of cells between the stomium and the connective; this signal intensified as development continued. Late in anther development, *TA56* mRNA becomes localized over both stomium regions and the connective tissue separating the pollen sacs.

Localization of mRNA transcripts corresponding to the maize cDNA clone *Zmc13* by *in situ* hybridization to anther sections revealed a similar temporal expression pattern to *Bcp1,* with transcripts present in the vegetative cell cytoplasm of the pollen grain and tube (Hanson *et al.*, 1989).

Expression of *Bcp1* in both the pollen grain and the tapetum indicates that the gene is expressed both in gametophytic (pollen) and sporophytic (tapetum) tissue. The same can also be said for the *LAT* series of cDNA clones, *LAT52, LAT56,* and *LAT59,* which have been identified in tomato (McCormick *et al.*, 1987). *In situ* hybridization using these clones to probe anther sections from three developmental stages shows that RNA corresponding to these clones is present in microspores, pollen, and in the endothecium and epidermis of the anther wall (Ursin *et al.*, 1989). It is possible that the transcripts detected in sporophytic tissue are, in both cases, a closely related cross-hybridizing gene. However, both the *Brassica* and tomato genes may utilize different promoters or regulatory sequences to control expression in the two different tissues. This is certainly the case in *Petunia hybrida,* in which expression of the chalcone flavanone isomerase gene in corolla, anthers, and pollen is regulated by the use of alternative promoters (Van Tunen *et al.*, 1989; Mascarenhas, this volume).

VI. Localization of Pollen- and Anther-Specific Transcripts in Transgenic Plants

Hybridization studies can also assess the expression of pollen- and anther-specific genes in genetically transformed plants. Expression of the *B. napus* pollen-specific gene *Bp4* has been analyzed in transformed tobacco plants. Total RNA isolated from small flower buds and small, medium, and large anthers was probed with *Bp4* cDNA. Complementary transcripts were evident only in the small flower buds and in the smallest anthers of the transformed plants. Further studies have indicated that this expression is restricted to the microspores (Albani *et al.*, 1990).

Whereas the hybridization results obtained with *Bp4* transgenic plants were consistent with those obtained with blots of endogenous *B. napus* RNA, the tomato clones *LAT52* and *LAT59* appear to show a different pattern of expression following transformation (Twell *et al.*, 1990). When the regulatory regions of the *LAT52* and *LAT59* are fused to the reporter gene β-glucuronidase (GUS) and transformed into tomato, GUS activity is restricted to the pollen grain. This expression pattern for the introduced construct is inconsistent with expression of the endogenous gene, which is also expressed in the anther wall (Twell *et al.*, 1989; Ursin *et al.*, 1989). One explanation for this discrepancy may relate to the fact that the translated product (GUS protein) is being detected in the transformed plant rather than the *LAT* mRNA. Antibodies to the native *LAT* proteins could resolve this question.

VII. Localization of Pollen- and Anther-Specific Proteins: Intracellular Targeting

Many pollen proteins have been located in the walls of the pollen grain. The origin of many of these proteins is unclear. Some are suspected to have a sporophytic origin, being present in the pollen coat, which is derived from the tapetum (Evans *et al.*, 1990).

The production of polyclonal antibodies against the polypeptide encoded by the *P2* pollen-specific cDNA from *Oenothera organensis* allowed the expression pattern of this protein to be determined. *P2* proteins were first detectable in 3–7 cm hypanthiums but not in 1–3 cm hypanthiums even though *P2* RNA transcripts could be detected at the earlier stage, suggesting posttranscriptional control of this gene (Brown and Crouch, 1990). *P2* antiserum was also used to localize *P2* proteins to the pollen grain and pollen tube by immunoblotting (Brown and Crouch, 1990).

A method with far greater resolution has been developed to provide information regarding the intracellular locations of the pollen-specific gene products of the ryegrass allergen genes *Lolp*I and *Lolp*Ib. By using an anhydrous fixation method it has been possible to attach specific antibodies to the water-soluble ryegrass allergens and then probe these antibodies with immunogold labels to localize the genes (Staff *et al.*, 1990). A series of mAbs probed with specific immunogold labels was used to localize the *Lolp*I and *Lolp*Ib proteins *in situ*. As shown in Figure 4A, the *Lolp*I protein has been located in the cytosol. In contrast, mAbs specific for *Lolp*Ib bound mostly to the starch granules as shown in Figure 4B. Mature grass pollen is filled with starch granules that originate in the amyloplasts.

The localization of *Lolp*Ib in the starch granules implies that this protein should be transported from the cytosol to the lumen of the amyloplast during development. Consistent with this, the *Lolp*Ib protein has a hydrophobic leader sequence of 25 amino acids with motifs similar to chloroplast transit peptides, indicating that the protein is synthesized as a preallergen in the cytosol and is then transported to the amyloplast for posttranslational modification (Singh *et al.*, 1991). The *Lolp*I protein, however, has a leader sequence that is similar to the signal peptide sequences of barley amylases, which are secreted enzymes (Griffith *et al.*, 1991). Confirmation of these intracellular locations will require the production of radioactively labeled precursor proteins by *in vitro* transcription and translation so that transport of the protein can be followed after *in vitro* uptake.

VIII. What Is the Function of Pollen- and Anther-Specific Genes?

Little is known concerning the nature and role of the proteins expressed by pollen- and anther-specific genes. Direct efforts have been made to assign a function to the pollen-specific allergen genes from *L. perenne*. A novel immunodepletion technique has been used to determine whether these allergen genes have enzymic activity. This work suggested a function for this allergenic protein as an amylase inhibitor (Lavithis, 1990). The function of many other pollen-specific allergen genes has, however, only been inferred from their expression patterns or by amino acid similarities with known proteins of known function.

The time of expression of a pollen-specific gene can give a general indication of that gene's role in pollen development. For example, the *B. campestris* cDNA clone *Bcp1* is, like a number of pollen- and anther-specific genes, expressed late in the development process and accumulates

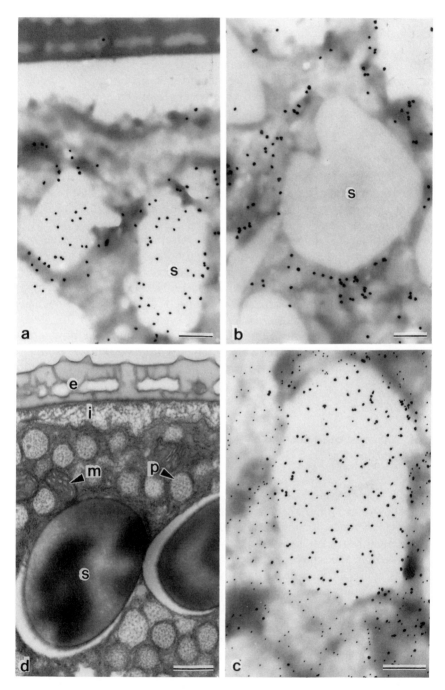

in mature pollen (Theerakulpisut *et al.*, 1991). It is likely that such a gene plays a role during one or more of the later stages of pollen maturation, pollen germination, or pollen tube growth. However, genes expressed early in the development process, such as the *B. napus* pollen-specific clones *Bp4* and *Bp19* (Albani *et al.*, 1990, 1991), are likely to play a much different role in microsporogenesis.

Most of the information concerning the function of pollen- or anther-specific gene products has come from data base searches for sequence similarities. The tomato cDNA clones *LAT56* and *LAT59* show striking similarities to a limited region of the pectate lyases from the bacterial plant pathogen *Erwinia* (Wing *et al.*, 1989) and even more extensive sequence similarity to the major ragweed pollen allergen (*Amba*I) (McCormick, 1991). If indeed *LAT56* and *LAT59* encode functional pectate lyases, then their presence in the pollen tube would be consistent with suggestions that pectin degradation may be involved in pollen germination and/or pollen tube growth through the pistil (Wing *et al.*, 1989). Similarly, pollen-specific genes from two other species encode pectin-degrading enzymes. The *P2* gene from *O. organensis* shows homology with polygalacturonase from tomato, and this gene is also expressed in the pollen tube (Brown and Crouch, 1990). The pollen-specific gene *Bp19* from *B. napus* shows striking similarity to the pectin esterases of tomato; however, its expression is limited to developing microspores where its possible involvement in pectin degradation is less clear (Albani *et al.*, 1991).

The *TA* series of anther-specific genes from tobacco correlate well with the known function of cells in which they are localized (Koltunow *et al.*, 1990). The *TA29* and *TA36* tapetal-specific mRNAs encode putative glycine-rich cell wall proteins and lipid transfer proteins, respectively (Koltunow *et al.*, 1990). This is suggestive of a critical role in the tapetum where proteins and lipids are produced and secreted into the pollen sac. Similarly, the *TA56* mRNA encodes a protein similar to the thiol endopeptidases present in germinating cotyledons and senescing leaves (Koltunow

---

FIG. 4 Detection of *Lolp*Ia and *Lolp*Ib in mature pollen of ryegrass using anhydrous fixation, specific mAbs, and immunogold probes. (a–c) Localization of allergens using immunogold probes on unstained thin sections. (a) Typical binding of mAb 12.3, specific for *Lolp*Ib, predominantly to starch granules (s). (b) typical binding of mixed mAbs 3.2, 21.3, and 40.1, specific for *Lolp*Ia, predominantly to the cytosol (the electron-opaque regions of the cytoplasm). (c) Double labeling to detect cellular sites of *Lolp*Ia and *Lolp*Ib on the same section. The large gold particles show binding of mAb 12.3 to *Lolp*Ib, whereas smaller particles indicate binding to *Lolp*Ia. (d) Transmission electron micrograph of thin section of ryegrass pollen grain poststained with uranyl acetate and lead citrate showing structural features: starch granule (s), polysaccharide particles (p), mitochondrion (m), exine (e), and intine (i). Bar, 0.5 μm. Reproduced with permission from Singh *et al.* (1991).

et al., 1990). The presence of *TA56* mRNAs in the connective, stomium, and circular cell cluster just before their degeneration suggests that the putative *TA56* thiol endopeptidase may aid in the degeneration process (Koltunow *et al.*, 1990). Little is known about the function of the remaining genes in Table I except for the two similar genes from tomato and maize. The maize pollen-specific gene *Zmg13* shows extensive amino acid identity with the tomato anther-specific gene *LAT52* (Hamilton *et al.*, 1989). Recent studies have indicated that *LAT52* is a Kunitz trypsin inhibitor (McCormick, 1991).

IX. Future Directions

A number of pollen- and anther-specific genes from a variety of plant species have been characterized. The temporal and spatial patterns of expression of these genes has, for the most part, been determined. However, to fully understand the molecular processes involved in pollen development, a number of important questions remain unanswered. What, for example, is the biological role of these developmentally regulated genes? Are the protein products expressed in the same temporal and spatial patterns as the mRNA transcripts or does some posttranscriptional regulation occur? How is the tissue-specific expression of these genes regulated?

One direct way of determining the biological role of pollen- and anther-specific genes is by creating antisense constructs of these genes. An antisense copy of the gene consists of an RNA sequence that is complementary to the mRNA transcript and, if provided (by transformation) in large enough numbers, will bind to the transcript and prevent its translation. The antisense RNA approach can be applied to pollen genes to create mutant lines that will specifically define the biochemical significance of the gene concerned. The function of the gene can be determined by analyzing what happens when that gene is "switched off."

Determining the time of expression and site of action of pollen- and anther-specific gene products has been completed for some genes, in particular, the allergen genes of ryegrass (Singh *et al.*, 1991; Griffith *et al.*, 1991). In all cases this has involved isolating either polyclonal or monoclonal antibodies raised to the protein gene product. Such antibodies can then be used to determine the following: (1) the specificity of the protein product, (2) when during development it is expressed, and (3) the site where protein synthesis and/or function occurs (by detecting the protein *in situ*).

The regulation of developmentally expressed pollen- and anther-specific genes has been studied by gene transformation experiments using both

stable and transient expression systems (i.e., microprojectile bombardment; McCormick, 1991; Mascarenhas, this volume). This analysis has revealed that there are, at best, only limited regions of sequence similarity between promoters that direct similar patterns of gene expression, even within the same species of plant. It is likely that, as in other tissues, a combination of general and tissue-specific factors and *cis*-acting elements will be shown to regulate pollen and anther gene expression.

Apart from a more detailed understanding of tissue-specific gene regulation during pollen development, the knowledge gained from the study of pollen- and anther-specific genes has many potential applications. In particular, such knowledge can be used in the production of value-added agricultural products through the use of recombinant DNA technology. Identification of the DNA components (promoters) that control the specificity, timing, and cellular location of gene expression will allow the correct regulation of desirable genes transferred into pollen (e.g., genes conferring an increased tolerance to environmental stress) and the pollen-specific switching on/off of genes to produce desirable traits (e.g., nuclear male sterility, low-allergen lines, or enhanced pollen viability following storage).

Two groups have illustrated how genes, the temporal and spatial expression patterns of which are tissue-specific can be manipulated. By fusing a 5' fragment from the tapetum-specific tobacco gene *TA29* to an RNAse gene, male sterile lines of *N. tabacum* and *B. campestris* were produced in which the chimeric gene directed selective ablation of the tapetum (Mariani *et al.,* 1990). This is the first example of nuclear male sterility as the result of recombinant DNA manipulation of a plant. In a similar experiment, expression of the diphtheria toxin gene controlled by the *Brassica* S-glycoprotein gene promoter resulted in ablation of both pollen and pistil cells in transgenic tobacco (Thorsness *et al.,* 1991). The ability to so precisely manipulate transgenic gene expression has the potential to revolutionize agricultural practices that are centuries old.

References

Albani, D., Robert, L. S., Donaldson, P. E., Altosaar, I., Arnison, P. G., and Fabijanski, S. F. (1990). *Plant Mol. Biol.* **15,** 605–622.

Albani, D., Altosaar, I., Arnison, P. G., and Fabijanski, S. F. (1991). *Plant Mol. Biol.* **16,** 501–513.

Brown, S. M., and Crouch, M. L. (1990). *Plant Cell* **2,** 263–274.

Evans, D. E., Singh, M. B., and Knox, R. B. (1990). *In* "Microspores, Evolution and Ontogeny" (S. Blackmore and R. B. Knox, eds.) pp. 309–329. Academic Press, London.

Griffith, I. J., Smith, P. M., Pollock, J., Theerakulpisut, P., Avjioglu, A., Davies, S. P., Hough, T., Singh, M. B., Simpson, R. J., Ward, L. D., and Knox, R. B. (1991). *FEBS Lett.* **279,** 210–215.

Hamilton, D. A., Bashe, D. M., Stinson, J. R., and Mascarenhas, J. P. (1989). *Sex. Plant Reprod.* **2**, 208–212.

Hanson, D. D., Hamilton, D. A., Travis, J. L., Bashe, D. M., and Mascarenhas, J. P. (1989). *Plant Cell* **1**, 173–179.

Koltunow, A. M., Truettner, J., Cox, K. H., Wallroth, M., and Goldberg, R. B. (1990). *Plant Cell* **2**, 1201–1224.

Lavithis, M. (1990). Honours Thesis. University of Melbourne. Parkville, Australia.

Mariani, C., Beuckeleer, M. De., Truettner, J., Leemans, J., and Goldberg, R. B. (1990). *Nature* **347**, 737–741.

McCormick, S. (1991). *Trends Genet.* **7**, 298–303.

McCormick, S., Smith, A., Gasser, C., Sachs, K., Hinchee, M., Horsch, R., and Fraley, R. (1987). *In* "Tomato Biotechnology" (D. Nevins and R. Jones, eds.), pp. 255–265. Liss, New York.

Rafner, T., Griffith, I. J., Kuo, M. C., Bond, J. F., Rogers, B. L., and Klapper, D. G. (1991). *J. Biol. Chem.* **226**, 1229–1236.

Silvanovich, A., Astwood, J., Zhang, L., Olsen, E., Kisil, K., Sehon, A., Mohapatra, S., and Hill, R. (1991). *J. Biol. Chem.* **266**, 1204–1210.

Singh, M. B., and Knox, R. B. (1985). *Int. Arch. Allergy Appl. Immunol.* **78**, 300–304.

Singh, M. B., Hough, T., Theerakulpisut, P., Avjioglu, A., Davies, S. P., Smith, P. M., Taylor, P., Simpson, R. J., Ward, L. D., McCluskey, J., Puy, R., and Knox, R. B. (1991). *Proc. Natl. Acad. Sci. U.S.A.* **88**, 1384–1388.

Staff, I. A., Taylor, P. E., Smith, P. M., Singh, M. B., and Knox, R. B. (1990). *Histochem. J.* **22**, 276–290.

Sunderland, N., and Huang, B. (1987). *Int. Rev. Cytol.* **107**, 175–220.

Theerakulpisut, P., Singh, M. B., Xu, H. L., Pettitt, J. M., and Knox, R. B. (1991). *Plant Cell* **3**, 1073–1084.

Thorsness, M. K., Kandasamy, M. K., Nasrallah, M. E., and Nasrallah, J. B. (1991). *Dev. Biol.* **143**, 173–184.

Twell, D., Wing, R. A., Yamaguchi, J., and McCormick, S. (1989). *Mol. Gen. Genet.* **217**, 240–245.

Twell, D., Yamaguchi, J., and McCormick, S. (1990). *Development* **109**, 705–713.

Twell, D., Yamaguchi, J., Wing, R. A., Ushiba, J., and McCormick, S. (1991). *Genes Dev.* **5**, 496–507.

Ursin, V. M., Yamaguchi, J., and McCormick, S. (1989). *Plant Cell* **1**, 727–736.

Van Tunen, A. J., Hartman, S. A., Mur, L. A., and Mol, J. N. M. (1989). *Plant Mol. Biol.* **12**, 539–551.

Wing, R. A., Yamaguchi, J., Larabell, S. K., Ursin, V. M., and McCormick, S. (1989). *Plant Mol. Biol.* **14**, 17–28.

Pollen Wall and Sporopollenin

R. Wiermann and S. Gubatz
Institut für Botanik der Westfälischen Wilhelms-Universität
D-4400 Münster/Westfall, Germany

I. Introduction

The pollen grain is the carrier of the male gametes or their progenitor cell (Knox, 1984a). During dispersal the pollen constitutes a system that is exposed to various environmental factors, such as wind and water or animal vectors, and finally interacts with the pistil, the site of the female gametes. Form, structure, and the biochemical machinery of pollen, especially of the pollen wall, have therefore undergone a remarkable adaptation. The pollen grain is provided with a heterogeneously developed wall, and pollen grains are distinguished by a multitude of sculptures and structures. The pollen wall, the sporoderm, is the most complex wall system in the life cycle of higher plants.

The pollen wall includes three main domains that are highly different in their chemical composition, morphological structure, and their physiological and biological significance (Knox, 1984a). These domains are intine, exine, and pollen coat.

The exine (the outer, resistant and rigid layer) is richly sculptured and ornamented. It is composed of sporopollenin, an organic biopolymer of extremely high stability. Ultrastructural and cytochemical studies indicate that the exine itself commonly comprises two layers: ectexine and endexine, which exhibit differences in chemical resistance (e.g., ethanolamine), autofluorescence, and staining capacity. It must be assumed that these differences are caused by a different composition of the two layers, and likely by variations in the sporopollenin composition itself (Jonker, 1971; Southworth, 1974; 1985a,b; Knox, 1984a; Kress, 1986; Sreedevi et al., 1990).

Underlying the exine is a layer rich in polysaccharides, the intine, which is present in all pollen types and envelopes the protoplast of the male gametophyte. Concerning its chemical composition, the intine is similar to primary walls of common plant cells. The intine has been shown to

contain primarily substances such as cellulose, hemicellulose, and pectic polymers (Heslop-Harrison, 1975; Kress and Stone, 1983; Knox, 1984a). In addition to this, substances of proteinaceous nature have fairly often been detected in the intine.

Another domain of the pollen wall is formed by the *surface substances* that are accumulated in and/or on the exine structures. The complex exine structures are storage sites for carbohydrates, glycoproteins, proteins, lipids, terpenoids, and phenolics, all compounds of different chemical structure.

On the basis of ultrastructural, cytochemical, immunological, and chemical studies, detailed knowledge about organization and chemistry of the three domains was obtained. The ultrastructural and cytochemical relevance of the pollen wall has been the subject of excellent reviews (Buchen and Sievers, 1981; Knox, 1984a,b; Shivanna and Johri, 1985; Blackmore and Ferguson, 1986; Iwanami *et al.,* 1988; Hesse and Ehrendorfer, 1990; Hesse, 1991). This chapter, however, will emphasize the chemical composition of the substances comprising the surface of pollen grains as well as on the present state of knowledge concerning the biosynthesis and structure of sporopollenin.

II. Outer Pollen Wall as an Accumulation Site of Chemicals of High Diversity: The Pollen Coat Substances

The outer pollen wall with its structures and surface substances takes part in the process of adaptation between pollen and environment. The exine is a storage layer for materials synthesized mainly by the anther tapetum. As studies in recent years have shown, the pollen coat substances differ widely in chemical structure and function. They comprise volatiles, surface lipids, pigments, and various secondary plant products, such as phenolic conjugates.

The aforementioned coat substances are often accompanied by carbohydrates, proteins, and glycoproteins. The occurrence and the biological and medical significance of the wall proteins have been the subject of extensive and excellent discussions, inter alia, by Heslop-Harrison (1975), Knox *et al.* (1975), Mascarenhas (1975), Knox (1984a), Shivanna and Johri (1985), Sarker *et al.* (1988), Dulberger (1990), and Elleman and Dickinson (1990), and thus, will not be further discussed in this review.

A. Pollen Volatiles and Surface Lipids

Pollen grains, especially those of entomophilous plants (but not exclusively, Pacini *et al., 1985*) are typically coated with an oily, sticky, and

frequently colored material, commonly termed *pollenkitt* (Knoll, 1930; Heslop-Harrison, 1975; Knox, 1984a; Hesse, 1978a,b,c, 1979a,b, 1980a,b, 1991; Keijzer and Cresti, 1987; Dobson, 1989). Recent studies on pollen of two *Rosa* species showed that pollenkitt is a main source of volatiles. Many kinds of pollen have characteristic odors, which possibly play an important role in guiding pollen-foraging insects to flowers (Dobson, 1987, 1988). Volatiles isolated from the pollen of *Rosa rugosa* were compared with those of pollenkitt extracts. In volatiles from the pollen diverse compounds were identified, including terpenoids, aliphatics, and aromatics (Dobson *et al.*, 1987, 1990; Dobson 1991; Table I). The pollenkitt extract of this pollen contained half of the identified substances found among the extracted volatiles of the whole pollen. Furthermore, it could

TABLE I

Chemicals Identified in Volatiles from Pollen and Pollenkitt of *Rosa rugosa* and *R. canina*[a]

Compound	*R. rugosa* pollen	Pollenkitt[b]	*R. canina* pollen[b]
Terpenoids			
6-Methyl-5-hepten-2-one	xx		x
Geranyl acetone	xxx	x	xx
Neral	xx		
Geranial	xxx	x	
Nerol	t		
Geraniol	xx		
Citronellyl acetate	xx	x	
Neryl acetate	xx		
Geranyl acetate	xxx	x	
Aliphatics			
Pentadecane	xx	x	x
2-Undecanone	xxx		
2-Tridecanone	xxx	xx	
2-Pentadecanone	xx		
Tetradecanal	xxx	xx	xxx
Hexadecanal	x		xx
Acetic acid	t		
Tetradecyl acetate	xxx	xxx	
Hexadecyl acetate	t	xxx	
Aromatics			
β-Phenylethanol	xxx		
Methyleugenol	xxx	x	
Eugenol	xx		
β-Phenylethyl acetate	xx	x	

[a] Data from Dobson *et al.* (1987).

[b] Includes only compounds detected also in *R. rugosa* pollen volatiles. t, trace quantities, identification tentative; x, ≤4%; xx, ≥4% and ≤20%; xxx, ≥20% of largest peak.

be shown that the volatiles of *R. rugosa* and *R. canina* differ in their composition.

Intensive studies on the presence and chemical composition of pollen lipids have been carried out. The outer pollen wall was clearly recognized as an important accumulation site of diverse lipids. The topic of pollen lipids in general has been adequately covered by Stanley and Linskens (1974). This review will therefore focus on the lipids that are demonstrated as or are suggested to be soluble compounds of the outer pollen wall.

Exine-bound lipids have been studied in several systems. For pollen of *Brassica napus,* linolenic acid together with palmitic, stearic, and myristic acids, as well as paraffins, were the predominant constituents of the lipid fraction obtained from the pollen coat domain (Evans *et al.,* 1987, 1990). External lipids and paraffins were determined to be in the range of 9.8% of pollen dry weight (Caffrey *et al.,* 1987).

The wings of *Pinus* exclusively consist of exine material (Martens and Waterkeyn, 1962; Shaw and Yeadon, 1964; Rowley and Walles, 1985); therefore, substances extracted from this material are unequivocally accumulated on and/or in the structures of the exine. The exine-bound location of lipids was shown by experiments in which lipids from the whole pollen and from the wings of the pollen from *Pinus mugo* were analyzed separately (for isolation and fractionation of the wing material, Section III,B,5). The composition of lipids from *Pinus* pollen and their wings is depicted in Figure 1 (Niester *et al.,* 1987; for comparison see Scott and Strohl, 1962). Surface waxes constitute between 1.3% and 1.6% of the dry weight of the pollen, and the wings contained 3.5% of these substances. The chloroform extracts contained hydrocarbons, aldehydes, wax esters, fatty acids, and

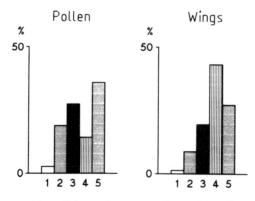

FIG. 1   Percent composition of the surface waxes from pine pollen and wing material: 1, hydrocarbons; 2, aldehydes; 3, wax esters; 4, free fatty acids; 5, primary alcohols. Modified with permission from Niester *et al.* (1987).

primary alcohols in varying quantities and, which is not depicted, possibly also triterpenols, such as amyrins. The substances were present as a homologous series.

A comparison concerning the composition of the surface lipids from the whole pollen and from the pollen wings shows that although differences in the quantitative composition may occur, the distribution patterns of chain lengths are similar in general. The composition of the individual lipid classes is shown in Table II (Niester *et al.*, 1987).

Hydrocarbons in these extracts are constituted of homologous n-alkanes with chain lengths ranging from C_{20} to C_{33}. This is demonstrated in Table III for whole *Pinus* pollen in comparison with results obtained from pollen of various gymnosperms and angiosperms (Hagenberg *et al.*, 1990).

TABLE II

Distribution Patterns of Surface Waxes (in Percantage of the Total Weight of Each Lipid Class) Isolated from *Pinus* Pollen and Wing Material[a]

	No. of carbon atoms	*Pinus* pollen (1984)	Wings (1984)
Aldehydes	20	1.66	—
	22	1.60	1.23
	24	15.90	15.14
	26	35.14	32.94
	28	43.10	41.11
	30	1.60	1.72
Wax esters	38	0.44	—
	40	4.52	4.84
	42	15.32	13.58
	44	31.83	27.92
	46	25.57	30.36
	48	15.02	18.17
	50	3.21	4.09
	52	1.31	1.73
Fatty acids	16	13.32	22.44
	18	43.40	40.69
	20	21.93	8.97
	22	2.97	2.00
	24	3.35	2.83
	26	2.92	2.06
	28	1.78	1.21
Primary alcohols	22	—	1.83
	24	20.16	13.62
	26	39.64	41.37
	28	42.67	33.21
	30	11.28	9.14

[a] Data from Niester *et al.* (1987).

TABLE III

Composition (Percentage of Peak Area) of N-Alkanes from Different Pollen Species[a]

Gymnosperms

No. of carbon atoms	Pinus mugo	Taxus baccata	Picea abies	Torreya nucifera
20			2.2	
21			3.4	
22	3.9	8.8	5.7	9.0
23	7.9	9.5	9.9	9.0
24	9.7	14.7	8.1	8.3
25	11.3	16.4	12.3	12.1
26	10.5	14.4	9.1	7.9
27	16.2	12.8	16.8	15.6
28	10.3	9.7	7.7	6.5
29	12.6	9.3	8.3	10.9
30	7.4	4.4	6.1	6.9
31	10.2		5.5	9.3
32			2.9	4.5
33			2.0	

Angiosperms

No. of carbon atoms	Cucurbita maxima	Corylus avellana	Narcissus pseudonarcissus	Zea mays B37[b]
20			0.3	
21			13.9	
22		2.0	9.7	
23		18.5	69.4	2
24		12.9	1.4	1
25	14.4	17.9	2.8	30
26	2.3	9.1	1.8	2
27	37.9	9.4	0.4	46
28	4.2	7.6	0.1	2
29	34.8	8.0	0.2	10
30	0.7	5.5		–
31	5.7	4.6		7
32		2.6		
33		1.9		

[a] Data from Hagenberg et al. (1990).
[b] Extended according to Bianchi et al. (1990).

Unusual alkanes like iso-heptocosane and iso-pentacosane are prominent compounds of extracts from pollen of tulip and lily (Tsuda et al., 1981; Stránsky et al., 1991).

The composition of lipids from four maize genotypes was analyzed by Bianchi et al. (1990). These substances were readily extracted by

TABLE IV

Composition (Percentage) of Classes of Compounds Comprising the
Lipids from Maize Pollen[a,b]

Components	B37	K55	Mo506	NY821
Alkanes	6	18	13	16
Alkenes	9	22	18	19
Triterpene esters	34	5	27	25
Linear esters	1	2	10	1
Triglycerides	31	18	17	19
Triterpenols	8	11	tr	tr
Alcohols	tr	tr	tr	tr
Free fatty acids	8	22	10	18
Unidentified	3	2	5	2

[a] Data from Bianchi *et al.* (1990).
[b] Genotypes *B37, K55, Mo506,* and *NY821:* tr,
traces.

chloroform, suggesting that the compounds were accumulated on the surface of the pollen grains. Alkanes, alkenes, fatty acids, triterpene esters,
and triglycerides were found as the main compounds of the lipid extracts
(Table IV). The unusually high percentage of alkenes making up the hydro-
carbon fraction is remarkable. In addition to triglycerides, long-chain fatty
esters of triterpenols constituted the dominant class of pollen lipids. The
gas chromatography–mass spectroscopy (GC–MS) analysis of these esters
permitted the identification of compounds, such as 24-methylenecholes-
terol, cyclolaudenol, campesterol, and stigmasterol esterified with C_{16} and
C_{18} fatty acids. From two genotypes free triterpenols could be isolated and
identified as stigmasterol, campesterol, sitosterol, α- and β-amyrin, 24-
methylenelophenol, cyclolaudenol, cycloartenol, lupeol, and taraxasterol
(Fig. 2).

B. Carotenoids, Flavonoids, and Phenolic Conjugates

The outer pollen wall represents an important accumulation site for pigments. This has for a long time been accepted for carotenoids and was
recently confirmed for various phenolic compounds (Section II,B,3).

1. Carotenoids

Carotenoids are commonly involved in pollen pigmentation, primarily in
plants pollinated by insects. Extensive studies on the carotenoid patterns

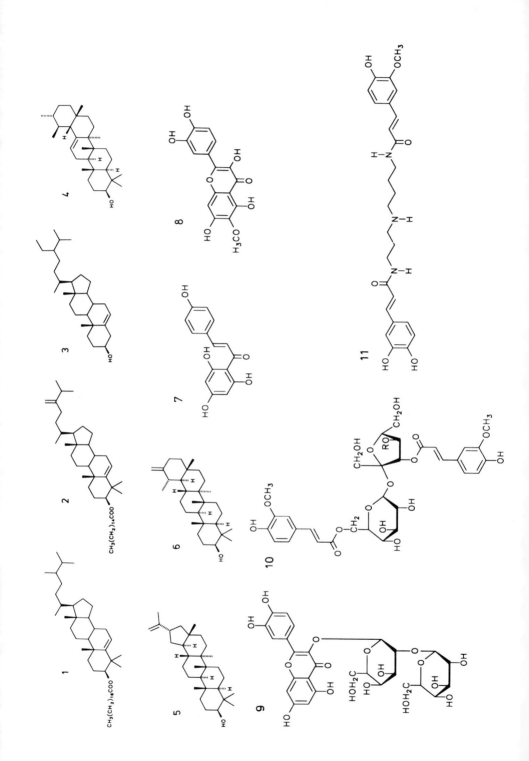

of pollen do not exist. Moreover, some accounts that had been cited in the literature focused on extracts of anthers. Accounts about the occurrence of carotenoids that were exclusively obtained from pollen grains are scarce. Carotenoids were discussed in detail by Stanley and Linskens (1974). Prahl *et al.* (1985) confirmed their discourse; they showed that carotenes and xanthophylls, such as β-carotene, lutein, violaxanthin, and antheraxanthin, are widespread and major components of the pollen coat. The synthesis of these compounds takes place in the tapetum before its final dissolution. Beginning with the release of the microspores from the tetrad, the pigments accumulate in osmiophilic globuli until the dissolution of the tapetum as a tissue. This has been shown in detailed studies previously (Heslop-Harrison, 1968; Heslop-Harrison and Dickinson, 1969). The carotenoids are then transferred into the loculus and trapped on and/or in the cavities within the exine (Keijzer, 1987).

2. Flavonoids

Phenolics are constituents of pollen of many angiosperm and gymnosperm species (Stanley and Linskens, 1974). Many new pollen compounds, especially flavonoids, have been described. For some of these their biosynthesis and localization have been intensively studied.

The results of investigations on this group of phenols have shown that the principal flavonoids in pollen were frequently 3-diglycosides and 3-triglycosides of kaempferol, quercetin, and isorhamnetin (Strack *et al.,* 1981a, 1984; Styles and Ceska, 1981; Zerback *et al.,* 1989a; Table V, Fig. 2). The flavonoid pattern is often of high diversity, as has been shown for pollen of *Petunia hybrida, Tulipa* cultivar Apeldoorn and *Zea mays,* for example.

In contrast to the aforementioned flavonols (kaempferol, quercetin, and isorhamnetin), the yellow flavonols (Harborne, 1969) are seldom encountered in the flavonoid pattern of pollen. These flavonols were found in hydrolyzed extracts from pollen of only a few plants (Wiermann *et al.* 1981). The aglycons isolated from these pollen species are summarized in Table VI (Wollenweber and Wiermann, 1979; Wiermann *et al.,* 1981; Ferreres *et al.,* 1989).

FIG. 2   Selected examples of chemicals isolated as pollen surface substances. Sterols: 1, campestosterol stearic acid; 2, 24-methylenecholesterol palmatic acid; 3, β-sitosterol; 4, α-myrin; 5, lupeol; 6, taraxasterol. Phenols: 7, 2',4',6',4-tetrahydroxychalcone; 8, patuletin; 9, quercitin-3-glucosylgalactoside; 10, diferuloylsucrose [R = H], triferuloylsucrose [R = feruloyl]; 11, caffeoylferuloylspermidine.

TABLE V

Flavonolglycosides Isolated from Pollen

---

[a]*Fagus sylvatica*
  Kaempferol-3-*p*-coumaroylglucoside

[b,c]*Corylus avellana*
  Quercetin-3-glucosylgalactoside
  [a]Quercetin-3-sophoroside
    [a]As a major constituent in pollen from *Juglans cordiformis, Juglans siboldiana, Alnus cordata, Alnus incana, Betula medwediewii, Carpinus carpinizza, Ostrya carpinifolia,* and *Fraxinus sogdiana*

[d]*Populus yunnanesis*
  Kaempferol-3-rhamnodiglucoside

[e]*Prunus amygdalus*
  8-Methoxykaempferol-3-diglucoside
    in trace amount:
      Kaempferol-3-diglucoside
      Quercetin-3-diglucoside
      Kaempferol-3-*p*-coumaroylglycoside

[f]*Cucurbita maxima*
  Isorhamnetin-3-rutinoide

Cucurbita moschata
  Kaempferol-3-rutinoside
  Kaempferol-3-robinobioside
  Other flavonoids in minor quantities

Cucurbita ficifolia
  Isorhamnetin-3-rutinoside
  Quercetin-3-rutinoside
  Other flavonoids in minor quantities

Cucurbita pepo
  Isorhamnetin-3-rutinoside
  Other flavonoids in minor quantities

[g]*Petunia hybrida*
  Kaempferol-3-glucosylgalactoside
  Quercetin-3-glucosylgalactoside
  Dihydroquercetin
  Other flavonoids in minor quantities

[h]*Tulipa* cultivar Apeldoorn
  Quercetin-3-rhamnosylglucoside
  Isorhamnetin-3-rhamnosylglucoside
  Quercetin-3-glucosylrhamnosylglucoside,
  as major compounds:
    Kaempferol-3-rhamnosylglucoside
    Kaempferol-3-xylosylrhamnosylglucoside
    Isorhamnetin-3-xylosylrhamnosylglucoside,
  as minor compounds:
    Delphinidin-3-rutinoside

TABLE V (Continued)

---

[i]Zea Mays
  Kaempferol-3-glucoside
  Quercetin-3-glucoside
  Quercetin-3,3'-diglucoside
  Quercetin-3,7-diglucoside
  Quercetin-3-neohesperidoside
  Quercetin-3-glucoside 3'-diglucoside
  Isorhamnetin-3-glucoside
  Isorhamnetin-3,4'-diglucoside
  Isorhamnetin-3-neohesperidoside
  Isorhamentin-3-glucoside-4'-diglucoside

---

[a] Data from Pratviel-Sosa and Percheron (1972).
[b] Data from Strack et al. (1984).
[c] Data from Meurer et al. (1988).
[d] Data from Sosa and Percheron (1970).
[e] Data from Ferreres et al. (1989).
[f] Data from Imperato (1979).
[g] Data from Zerback et al. (1989a).
[h] Data from Strack et al. (1981a).
[i] Data from Ceska and Styles (1984).

The occurrence of chalcone aglyca is remarkable (Table VII). Usually these compounds are found only in low concentrations. Thus, it has to be assumed that their occurrence has often been overlooked. Chalcone aglyca are compounds of an intense yellow color, which must, therefore, like the already mentioned "yellow flavonols", often be of great importance for the pigmentation of pollen.

The biosynthesis of pollen flavonoids has been intensively studied in *Tulipa* cultivar Apeldoorn. During pollen development in this system, a highly differentiated, phase-specific accumulation of various phenylpropanoids occurs. The diverse compounds involved in flavonoid biosynthesis accumulate transiently and sequentially during morphological differentiation. After the disintegration of the tetrads, the accumulation of simple phenylpropanes (phase A), chalcones (phase B), and flavonols (phase C) takes place consecutively at various stages of pollen maturation (Fig. 3). This phase-dependent accumulation of various intermediates correlates to a considerable extent with the development of the relevant enzymes required in the different steps of the biosynthetic pathway (Wiermann, 1981).

3. Phenolic Conjugates

Hydroxycinnamic acid and benzoic acid conjugates and their derivatives are found in the pollen of many species of higher plants (Strohl and

TABLE VI

Occurrence of "Yellow Flavonols" in Various Pollen Types[a,b]

	1	2	3	4	5	6	7	8
Paeonia daurica				+	+		+	+
Paeonia delavay				+	+		+	+
Paeonia lutea var. ludlowii				+	+		+	+
Paeonia mlokosewitschii				+	+		(+)	+
Paeonia tenuifolia				+	+			+
Rumex acetosa				+			+	+
Rumex acetosella				+			+	+
Nothofagus antarctica				+		+	+	+
Spinacia oleracea	+	+	+					

1, 6-Methoxykaempferol; 2, Patuletin; 3, Spinacetin; 4, Sexangularetin; 5, Limocitrin; 6, Isosalipurpol; 7, Kaempferol; 8, Quercitrin.

[a] Data from Wollenweber and Wiermann (1979) and Wiermann *et al.* (1981).

[b] Kaempferol, quercetin, and isosalipurpol (2', 4', 6', 4-tetrahydroxychalcone) are included to give the complete flavonoid pattern identified in the pollen material investigated.

Seikel, 1965; Stanley and Linskens, 1974). Aromatic amides, such as di-p-coumaroyl-, di-feruloyl-, and feruloyl-caffeoylspermidines, represent widespread compounds of pollen from the *Fagales* and some other taxa of the *Hamamelididae* (Meurer *et al.*, 1986; 1988a,b; Fig. 2). The distribution of these compounds contributes useful characteristics to chemosystematic studies (Meurer *et al.*, 1988b).

Two very rare hydroxycinnamic conjugates were isolated from pollen and loculus material of immature anthers of *Tulipa* cultivar Apeldoorn and identified as 6,3'-di- and 6,3',4'-triferuloylsucrose by chromatography, electron mass spectroscopy, and nuclear magnetic resonance (NMR) spectroscopy (Fig. 2; Strack *et al.*, 1981b; Bäumker *et al.*, 1988). A chemical survey of anthers from 175 dicotyledons and monocotyledons showed that the conjugates exclusively occur in anthers of *Liliaceae (Tulipeae, Lilieae, Lloydieae;* Meurer *et al.*, 1984).

A series of novel p-coumaric esters was obtained from a dichloromethane-soluble neutral fraction of *Pinus densiflora* pollen and identified as *cis*- and *trans*-isomers of 1,16-dioxo-1-hydroxy-16-oxo- and 1,16-dihydroxy-hexadecan-7-yl p-coumaric acid (Shibuya *et al.*, 1978; Caldicott and Eglinton, 1975). Because pollen is regarded as a system without a cuticle, these compounds are of special interest in that some are comparable to structures that participate in the formation of cutin.

TABLE VII

Occurrence of Free 2′, 4′, 6′, 4-Tetrahydroxychalcone in Pollen
Extracts of Various Plants[a]

Betula verrucosa	tr
Corylus avellana	tr
Nothofagus antarctica	+ +
Fraxinus ornus	+ +
Lilium regale	+
Tulipa cultivar Apeldoorn	tr[b]
Tulipa cultivar Andes	tr[b]
Tulipa cultivar Preludium	tr[b]
Typha angustifolia	+

[a] Relative amounts: + +, medium; +, small; tr, trace.

[b] Additional traces of 2′, 4′, 6′, 3,4-penta-hydroxychalcone.

Many studies have concentrated on elucidating the structure of the aforementioned soluble compounds. The localization of some of these compounds on and/or in the pollen wall could also be convincingly demonstrated.

Wiermann and Vieth (1983) showed that flavonoids of pollen, in contrast to vacuole-localized flavonoids from cells of suspension cultures, can easily be removed from the grains by dipping the pollen into hydrous solutions. These results correspond to data obtained from fluorescence analysis of the pollen of *Petunia hybrida* (Zerback *et al.*, 1989b).

The aforementioned hydroxycinnamic acid spermidine amides isolated from *Corylus avellana* pollen are unequivocally located either in or both in and on the exine. This was shown by analyzing an extract of an exine fraction, which was obtained by homogenization of the pollen grains. The homogenate was fractionated with a glycerol gradient (Section III,B,5) and subsequently treated with a series of different enzymes removing contaminating materials, such as remains of the cytosol and intine (Gubatz *et al.*, 1986).

In extracts of separated wings from pollen of *Pinus mugo,* different phenolics could be detected, as previously described for diverse pollen of pine species by Strohl and Seikel (1965). The occurrence of these phenolics in extracts of wing material, a structure lacking cytosol and intine, indicates that they are abundant in the exine.

Thus, a number of the aforementioned substances have been demonstrated to be located in the outer pollen wall. Whether this is true for all components has to be examined in detail.

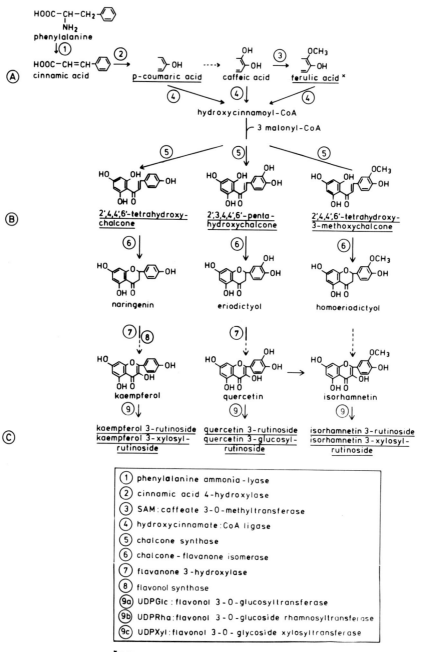

① phenylalanine ammonia-lyase

② cinnamic acid 4-hydroxylase

③ SAM:caffeate 3-O-methyltransferase

④ hydroxycinnamate:CoA ligase

⑤ chalcone synthase

⑥ chalcone-flavanone isomerase

⑦ flavanone 3-hydroxylase

⑧ flavonol synthase

⑨a UDPGlc:flavonol 3-O-glucosyltransferase

⑨b UDPRha:flavonol 3-O-glucoside rhamnosyltransferase

⑨c UDPXyl:flavonol 3-O-glycoside xylosyltransferase

*6,3'-diferuloylsucrose, 6,3',4'-triferuloylsucrose

4. Significance of the Anther Tapetum for Biosynthesis of Pollen Wall Phenolics

The biosynthetic pathway of pollen phenolics, namely, the flavonoids, has been clarified to a large extent. It has been well documented that flavonoids are formed from cinnamic acids and acetate units via a chalcone as the first C-15 intermediate (Heller and Forkmann, 1988; Stafford, 1990).

L-Phenylalanine ammonia-lyase (PAL) is the first enzyme in flavonoid biosynthesis; it catalyzes the formation of *trans*-cinnamic acid from phenylalanine. Aromatic hydroxylation of cinnamic acid leads to *p*-coumaric acid, which is then transformed to *p*-coumaroyl-CoA by catalysis of *p*-coumroyl:CoA-ligase.

Chalcone synthase (CHS) is the key enzyme in flavonoid biosynthesis. The enzyme catalyzes the formation of 2',4',6',4-tetrahydroxychalcone by condensation of one molecule of *p*-coumaroyl-CoA and three acetate units from malonyl-CoA. Chalcones are central intermediates in the biosynthesis pathway of the various flavonoid classes (Heller and Forkmann, 1988; Stafford, 1990).

The anther tapetum plays an important role in the formation of pollen wall flavonoids. This assumption is based upon results of enzymic, immunological, and genetic experiments.

In enzymic studies the pollen and tapetum fractions of the anthers were compared. The latter was obtained after squeezing the contents from the anther and separating the contents using filters. The tapetum fraction showed the highest specific activities for the key enzyme of the phenylpropanoid metabolism (PAL) and for flavonoid metabolism (CHS), respectively (Table VIII; Sütfeld and Wiermann, 1974; Herdt *et al.*, 1978; Kleinehollenhorst *et al.*, 1982; Beerhues *et al.*, 1989). These results could be confirmed by enzymatic studies on isolated tapetum tissues (Rittscher and Wiermann, 1983), and by immunofluorescence studies using antibodies directed to PAL and CHS (Kehrel and Wiermann, 1985; Beerhues and Wiermann, unpublished). By means of these methods, the predominant distribution of these two enzymes in the anther tapetum could unequivocally be demonstrated (for the function of the tapetum in general see Chapman, 1987; Bhandari, 1984; Pacini *et al.*, 1985).

To test whether the *Tulipa* anther is a representative model system, the distribution of PAL and CHS was determined in anthers of various plants

---

FIG. 3 Scheme of the biosynthetic pathway postulated for flavonolglycosides (C) isolated from *Tulipa* pollen including the intermediates appearing during pollen development in phase A and B and the relevant enzymes. Compounds actually isolated are underlined; delphinidin-3-rutinoside is not shown in this figure. Modified with permission from Barz and Wiermann (1981).

TABLE VIII

Distribution of Phenylalanine Ammonia-Lyase (PAL) and Chalcone Synthase (CHS) Activities after Pollen–Tapetum Fractionation[a]

	Total activity (cpm $\times$ 10^4)			Distribution (%)	
Species	Anther content s	Tapetum	Pollen	Tapetum	Pollen
PAL					
Tulipa cultivar Lustige Witwe	2374.4	2371.7	36.4	98.5	1.5
Iris pseudacorus	1031.9	907.9	31.7	96.7	3.3
Lilium croceum cultivar umbellatum	1110.7	1274.5	15.7	98.8	1.2
Hemerocallis fulva	690.4	481.1	50.2	90.5	9.5
Narcissus pseudonarcissus	74.2	69.8	3.45	95.3	4.7
CHS					
Tulipa cultivar Lustige Witwe	342.0	240.0	2.0	99.2	0.8
Iris pseudacorus	89.5	73.1	0.7	99.0	1.0
Lilium croceum cultivar umbellatum	21.6	21.4	0.2	99.1	0.9
Hemerocallis fulva	10.1	7.0	0.0	100.0	0.0
Narcissus pseudonarcissus	8.9	7.2	0.7	90.6	9.4

[a] Data from Beerhues *et al.* (1989).

(Table VIII). In all species that have been examined, both enzymes were predominantly present in the tapetum fraction. This tight association of PAL and CHS, two key enzymes in the biosynthesis of pollen wall flavonoids, thus appears to represent a widespread phenomenon.

Genetic data achieved from the pollen of different maize genotypes confirm the control exerted by the tapetum in the biosynthesis of pollen flavonoids. The formation of the various flavonoid patterns can best be explained by assuming that the genotype of the sporophyte bearing the pollen determines the kind of accumulation (Styles and Ceska 1977, 1981; Coe *et al.*, 1981; Koes *et al.*, 1990).

The biochemical mechanism of flavonoid accumulation at the pollen wall itself is unknown at present. It is hypothesized that flavonoid biosynthetic enzymes may be transported from tapetum cells into the free space of the loculus to the pollen grain where biosynthesis takes place (Herdt *et al.*, 1978; Koes *et al.*, 1990).

5. The Biological Functions of the Pollen Wall Substances

The surface substances are of unusual diversity; they are synthesized through the shikimate/amino acid-, acetate/malonate-, and acetate/

mevalonate metabolism. Their physiological, biochemical, and/or ecological functions have been the subject of speculation.

With exceptions, such as the wall proteins and their significance in the recognition system, only few clear, experimentally assured functions have yet been proved for the coat substances.

With respect to the general role of pollen in the life cycle of higher plants, it is reasonable to assume that the chemicals in and/or on the structures of the outer pollen wall may contribute to the ability to withstand environmental stress and to affect pollen–pistil interactions. Secondary plant products play important roles in the selection of host plants by phytophagous insects (Visser, 1983). Therefore, some of the pollen compounds may likewise serve as recognition cues to bees foraging pollen (Dobson, 1987). For example, the importance of certain chemicals in attracting honeybees has been demonstrated for several kinds of pollen (Dobson, 1988).

About the biological significance of the surface lipids research one can merely speculate at present. Polar lipids accumulated as surface substances may possibly be involved in pollen–pistil interactions as assumed by Mattsson (1983). As a consequence of the accumulation of lipids as surface substances, the pollen grain has acquired an additional and effective mechanism against environmental influences. According to Schönherr (1976) and Riederer and Schneider (1990), surface waxes may possibly substantially increase the barrier against water diffusion.

Both flavonoids and carotenoids have a key function in the pigmentation of pollen. The accumulation of various flavonoids and carotenoids might be involved in optical attraction of pollinators. Furthermore, it may be assumed that these pigments help to protect against ultraviolet light and photo-oxidative damage (Stanley and Linskens, 1974; Flint and Caldwell, 1983; Schmelzer et al., 1988).

Hydroxycinnamic acid amides with the aliphatic amines, putrescine, spermidine, and spermine, have been reported from a large number of families throughout the plant kingdom (Martin-Tanguy et al., 1978). Their accumulation was shown to be linked to cell multiplication (Martin et al., 1985) or fertility of reproductive organs (Martin-Tanguy et al., 1979; Cabanne et al., 1981; Martin-Tanguy et al., 1982). Moreover, they may also protect the plant from viral, bacterial, or fungal infections (Martin et al., 1978; Pandey et al., 1983; Tripathi et al., 1985). As has been pointed out, the hydroxycinnamic acid amides are accumulated in or at the exine. Their actual functions are speculative, however.

A comprehensive review of the complex surface chemicals should provide an impetus for further intensive studies of their true significance and biological roles. The results imparted by Sarker et al. (1988) are remarkable in this context and offer an interesting aspect for further experiments.

III. Biosynthesis and Structure of Sporopollenin

Sporopollenin represents an unusual biopolymer of extremely high chemical, physical, and biological stability. It forms the basic structure of the exine of pollen from higher plants as well as of spores from mosses and ferns.

For several reasons sporopollenin lacks a chemical definition. Mainly due to difficulties in purification, sporopollenin is often either accompanied by contaminating substances or altered in its structure by harsh treatments. In addition, sporopollenin itself is insoluble in most solvents and therefore not readily amenable to conventional methods of analysis.

A. The Occurrence of Sporopollenin in Higher Plants

In the case of higher plants the accumulation of sporopollenin is not restricted to the outer pollen wall alone. Other sporopollenin-containing elements include the following:

1. Ubisch Bodies (Orbicules)

These spherical bodies originate from the tapetum cells and are frequently attached to both the inner (tangential and/or radial) tapetum wall and the pollen exine. Their development and structure have been intensively studied (Reznickowa and Willemse, 1980; Audran, 1981; Pacini et al., 1985; El-Ghazaly and Jensen, 1986, 1987; Bhandari, 1984; Shivanna and Johri, 1985; Pacini et al., 1985; Hesse, 1986a, 1991). Ubisch bodies have been considered as sporopollenin concretions homologous with the sporopollenin of the exine. The orbicules originate in the cytoplasm within the tapetal cells as lipoidal pro-Ubisch bodies. They accumulate below the plasma membrane and are extruded to the cell surface in the loculus where they are coated with sporopollenin. The formation of the Ubisch bodies closely follows the pattern described for the pollen wall (Hesse, 1986b). They act neither as a transport form of sporopollenin nor as intermediates utilized in exine development. The sporopollenin polymerized on the bodies is unlikely to be remobilized.

2. Tapetal and Peritapetal Cell Walls

During the meiotic and postmeiotic stage of development, tapetal and peritapetal cell walls (Reznickowa and Willemse, 1980) are organized. They proved to be resistant to all hydrolytic enzymes used and to aceto-

lysis, respectively. Therefore, it is widely accepted that sporopollenin is involved in the formation of these walls (Banerjee, 1967; Dickinson, 1970; Banerjee and Barghoorn, 1971; Gupta and Nanda, 1972; Bhandari and Kishori, 1973; Cousin, 1979; Bhandari, 1984; Colhoun *et al.*, 1984; Shivanna and Johri, 1985; Vishnjakova and Lebsky, 1986; Ogorodnikova, 1986; Chen *et al.*, 1987; Keijzer, 1987; Chapman, 1987; Parkinson, 1988; Mu *et al.*, 1988).

3. Viscin Threads

In a few cases the adhesive properties of pollen grains are strengthened by viscin strands, which especially occur on pollen grains from the *Onagraceae* and *Ericaceae*. Their cord-shaped structures consist of a material cytochemically similar to sporopollenin (Hesse, 1981a,b, 1986a,b; Knox, 1984a; Waha, 1984; Pacini *et al.*, 1985; Takahashi and Skvarla, 1990). Comparable thread-like structures have been observed on grains of some *Leguminosae*. The ends of these structures are attached to the surface of different pollen grains ("exine connections"; Arora, 1985; Iwanami *et al.*, 1988).

B. Isolation and Purification Procedures of Exines and Sporopollenin

The isolation and enrichment of a highly purified sporopollenin fraction uncontaminated by cytoplasm, intine material, and/or soluble exinous substances are fundamental prerequisites for biochemical and chemical approaches to studying the biosynthesis and structure of sporopollenin. To obtain large quantities of sporopollenin material, various methods have been applied and are described herein in greater detail. Merely a precise knowledge of the respective procedures allows a reasonable assessment of chemically and biochemically determined results. The exhaustive purification of the sporopollenin fraction under as gentle conditions as possible ought to receive particular attention in the near future.

1. Isolation of Exines by the Acetolysis Method

One of the most common methods to isolate exine material, especially for palynological purposes, is the acetolysis method introduced by Erdtman (1960), which is still frequently used (Hesse and Waha, 1989). Acetolysis destroys all pollen material except for sporopollenin. It is a harsh treatment, which may result in changes in the structure of the sporopollenin.

2. Isolation of Sporopollenin by the Extraction–
Hydrolysis Method

Chemists often use treatments involving hot potassium hydroxide and phosphoric acid solutions to extract sporopollenin. This procedure was originally developed and applied by Zetzsche and his group (Zetzsche and Huggler, 1928; Zetzsche and Vicari, 1931) and has been widely modified by Green (1973) and Shaw and Yeadon (1966).

The procedure comprises complete extraction with boiling ethanol and ether, saponification in hot 6% aqueous potassium hydroxide solution, extraction with hot solvents, and, finally, a several day treatment with 85% phosphoric acid to remove intine material.

Recently, this conventional, aggressive isolation and purification procedure was applied in combination with high resolution solid-state ^{13}C-NMR spectroscopy to elucidate the chemical composition of sporopollenin from diverse plant sources by Guilford et al. (1988).

3. Isolation of Exines Using 4-Methylmorpholine N-Oxide

The application of this chemical to pollen grains of different sources produced a purified fraction of exine, and subsequently of sporopollenin, by a gentle treatment (Loewus et al., 1985; Baldi et al., 1986; Baldi et al.,1987; Espelie et al., 1989).The pollen material was completely extracted with acetone (Loewus et al., 1985). Afterwards the pollen was suspended in aqueous 4-methylmorpholine N-oxide and 1 M sucrose solution that had been adjusted to an alkaline pH. The suspension was heated to 70°C or 20°C, respectively, and stirred for 1 hour. This procedure promoted the release of both exines and sporoplasts. The exine fraction was recovered and further purified by a series of differential centrifugations with a discontinuous sucrose gradient and finally washed. The method was applied in several modified ways (Baldi et al., 1987).

4. Isolation of Exines from Hydrated and Autoclaved Pollen
of Gymnosperms

A gentle procedure to isolate exines from various gymnosperm pollen was described by Duhoux (1980), Southworth (1988), and Southworth et al. (1988). Hydration in distilled water caused the exine to rupture and to be shed spontaneously. The yielded suspension was autoclaved and exines were collected by fractionation on a discontinuous sucrose gradient. They were further purified by extraction with distilled water, acetone, and diethyl ether. The purification procedure does not require the use of harsh acids to remove the intine and therefore minimizes possible chemical

alteration of the sporopollenin. Exine material purified by this method was used to prepare antibodies to pollen exines (Southworth *et al.*, 1988).

The dehiscence of the exine by force of pollen hydration was successfully applied to pollen of *Iris tectorum, Zephyranthes grandiflora,* and *Hemerocallis fulva* (Zhou, 1988).

5. Isolation of Sporopollenin from Separated Wings of *Pinus* Pollen

The isolation and purification of sporopollenin from the wings of pine pollen was carried out with a gentle method, as described by Schulze Osthoff and Wiermann (1987). The individual steps of this method are summarized in Figure 4. In the first step exine material was isolated. Initially, the pollen was ultrasonicated. Following the disruption of the wings, the suspension was fractionated with a three-step gradient of glycerol and water. In this way the wings could be separated from pollen and other pollen fragments. A final purification of the wing fraction was performed by filtration through a filter cascade of nylon meshes.

With respect to chemical analysis of sporopollenin, the further steps in the purification procedure of the wing material by solvent extractions and enzymatic hydrolysis are of fundamental importance. Therefore, the wings were sequentially extracted with solvents in an increasing polarity sequence and incubated with a series of hydrolytic enzymes (Fig. 4) to eliminate possible contaminants. Material enriched and purified by this method was applied to degradation experiments (Section III,D,1).

A modified method was developed to receive highly purified sporopollenin from *Corylus avellana* pollen (Gubatz *et al.*, 1986). The isolation and purification procedure of the pollen walls was carried out by a potter treatment and ultrasonication, followed by a fractionation of the yielded homogenate on a glycerol–water gradient, subsequent incubation with a series of hydrolytic enzymes, and a complete extraction of noncovalently bound substances (Herminghaus *et al.*, 1988). After modification this procedure was successfully adapted to other systems in connection with tracer experiments (Rittscher and Wiermann, 1988a,b; Gubatz and Wiermann, unpublished data).

C. Biochemical Studies on Sporopollenin Biosynthesis

Since the 1960s sporopollenin has generally been considered to be a biopolymer derived from carotenoids and carotenoid esters. This assumption has been substantiated by the results of tracer experiments, the similarity of IR spectroscopic and elemental analyses obtained from naturally oc-

ISOLATION AND PURIFICATION OF WINGS FROM PINUS MUGO POLLEN

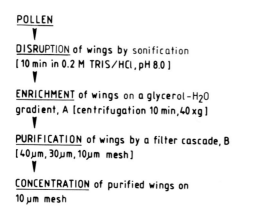
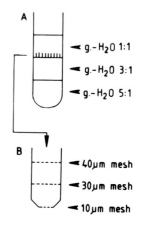

POLLEN
▼
DISRUPTION of wings by sonication
[10 min in 0.2 M TRIS/HCl, pH 8.0]
▼
ENRICHMENT of wings on a glycerol-H₂0
gradient, A [centrifugation 10 min, 40 x g]
▼
PURIFICATION of wings by a filter cascade, B
[40μm, 30μm, 10μm mesh]
▼
CONCENTRATION of purified wings on
10 μm mesh

A ◄ g.-H₂0 1:1
◄ g.-H₂0 3:1
◄ g.-H₂0 5:1

B ◄ 40μm mesh
◄ 30μm mesh
◄ 10μm mesh

EXTRACTIONS WITH SOLVENTS

Wings were sequentially extracted with solvents in an increasing polarity
sequence: petrolether, diethylether, acetone, methanol, ethylene-glycol-
monomethylether, distilled water

ENZYMATIC HYDROLYSIS OF NON-EXINE MATERIAL

Incubation with: pronase, lipase, Onozuka R-10 cellulase[3x], amylase,
cellulysin, cellulase TC, pectinase, hemicellulase

▼

SPOROPOLLENIN FRACTION

FIG. 4 Scheme of the isolation and purification procedure of wings from *Pinus* pollen:
g, glycerol. Reproduced with permission from Schulze Osthoff and Wiermann (1987).

curring sporopollenin and a polymer synthesized from carotenoids, and
from the resistance of both polymers to chemical reagents including acetolysis (Brooks and Shaw, 1968, 1971, 1977; Shaw, 1970, 1971).

Tracer experiments are particularly suited for clarifying biosynthetic
pathways. This technique has therefore been applied several times in
studies on sporopollenin biosynthesis in higher plants (Southworth, 1971;
Shaw, 1971; Green, 1973; Brooks and Shaw, 1977).

It must be stressed, however, that the results of the tracer experiments, which were carried out on higher plants and which were instrumental in establishing the carotenoid hypothesis, comprise only a few unequivocal pieces of data.

In 1971 Brooks and Shaw reported that [U-^{14}C] and [2-^{14}C] acetate as well as [1-^{14}C]palmitate applied to *Cucurbita* were incorporated into sporopollenin. [2-^{14}C]Mevalonic acid was incorporated into antheraxanthin, which is accumulated in the anthers of *Lilium henryi,* but it was not incorporated into sporopollenin. The labeled and subsequently isolated antheraxanthin applied to the same system was likewise not incorporated into sporopollenin. In summary, the results presented do not convincingly prove the hypothesis that terpenoid/carotenoid metabolism is involved in sporopollenin biosynthesis. Furthermore, the incubation time in the experiments, which was up to 3 weeks (Shaw, 1971), was much too long for the results to be interpreted meaningfully. Therefore, it is not surprising that there still exists great uncertainty concerning the metabolic pathway(s) that lead(s) to sporopollenin formation.

1. Application of Inhibitors

Inhibitors of carotenoid biosynthesis have been used to test the working hypothesis that carotenoids and/or carotenoid esters are involved in sporopollenin biosynthesis and to determine whether an intact carotenoid biosynthesis system is a prerequisite for undisturbed sporopollenin accumulation (Prahl *et al.,* 1985).

Application of Sandoz 9789 (Norflurazon), an inhibitor of the terminal steps of the carotenoid biosynthesis pathway in *Cucurbita,* did not significantly affect sporopollenin accumulation (Fig. 5). However, a control showed that Sandoz drastically interferes with carotenoid metabolism in the loculus of the anthers. As shown by an analysis of the composition of the carotenoids from pollen grains, the application of the inhibitor hindered the desaturation step in carotenoid biosynthesis. Saturated carotenoids, such as phytoene and phytofluene, were found instead of desaturated pigments (Prahl *et al.,* 1985, 1986). These results offer a convincing marker for the uptake and transport of the inhibitor into the anther loculus, the site of sporopollenin biosynthesis.

The results of these inhibitor experiments indicate that severe disruptions in carotenoid biosynthesis do not result in a subsequent, drastic inhibition of sporopollenin accumulation.

These observations correspond well to results of previously described studies by Heslop-Harrison and Dickinson (1969). The authors were unable to detect carotenoids during the stages of pollen ripening when sporopollenin undergoes intensive accumulation.

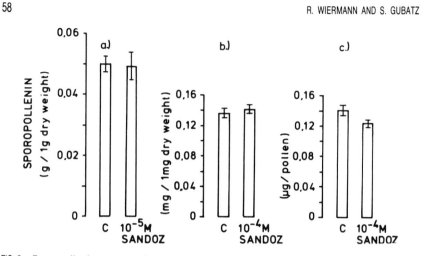

FIG. 5   Sporopollenin content of anthers (a) and pollen (b,c) from *Cucurbita maxima* after application of Sandoz. Modified with permission from Prahl *et al.* (1985).

Summarizing the aforementioned results, apparently carotenoids must play a minor part, if any, in sporopollenin biosynthesis. This assumption corresponds well to results of Guilford *et al.* (1988). ^{13}C-NMR Analysis of sporopollenin from several sources has shown that carotenoids do not play a fundamental role in sporopollenin biosynthesis and that the widely held carotenoid hypothesis is incorrect.

2. Tracer Experiments

Tracer experiments were carried out with anthers of *Tulipa* cultivar Apeldoorn and *Cucurbita maxima*. Initially, an appropriate application technique was developed to obtain the highest degree of labeling of sporopollenin within the shortest possible incubation time (Rittscher and Wiermann, 1988a). In addition, the range of precursors that have so far been applied, as mentioned, was broadened. Results from a number of experiments showed that the radioactive precursors must be applied as closely as possible to the site of sporopollenin biosynthesis. Thus, the anthers were incubated in a medium directly with the precursor. The best incorporation of radioactivity in the sporopollenin fraction was observed when the application was carried out with anthers without filaments and with a longitudinal section in the middle of the connective tissue (Rittscher and Wiermann, 1988a).

In these studies the sporopollenin fraction was isolated and purified by using either a gentle method (method A: hydrolyzing enzymes and alkaline hydrolysis) or a conventional, aggressive procedure (method B) introduced

by Zetzsche and Huggler (1928) and Zetzsche and Vicari (1931), and modified by Green (1973) (Section III,B,2). The results are summarized in Figure 6. The precursors were [^{14}C]-labeled mevalonate, glucose, acetate, malonic acid, phenylalanine, tyrosine, and p-coumaric acid. Regardless of the method of enrichment employed, the level of incorporation into the sporopollenin fractions was always higher with [U-^{14}C]phenylalanine than with the other substances. Only low levels of incorporation into sporopollenin were achieved after the application of mevalonic acid. The level of radioactivity found in sporopollenin after the application of [^{14}C]phenylalanine and [^{14}C]malonate was sufficiently high for the labeled polymer to lead to labeled products after degradation by potash fusion. In the case of phenylalanine-labeled sporopollenin, it could be shown for the first time that p-hydroxybenzoic acid, the main degradation product of phenolic nature (Fig. 7), was the most heavily labeled substance (Table IX). By comparison, little radioactivity was detected in p-hydroxybenzoic acid when sporopollenin labeled with malonic acid was broken down.

Because [U-^{14}C]phenylalanine was used in these experiments, the results do not unambiguously prove the integral incorporation of the aromatic ring system; however, this has subsequently been proven to be the case by application of [ring-^{14}C]phenylalanine in subsequent studies. This precursor is also efficiently incorporated into sporopollenin (Fig. 8; Gubatz and Wiermann, unpublished data). In the course of breaking down the sporopollenin fraction labeled with [ring-^{14}C]phenylalanine by potash fusion, radioactively labeled p-hydroxybenzoic acid was released as the major compound again.

The two findings—(1) the higher level of incorporation into the sporopollenin fraction with [^{14}C]phenylalanine than with other substances and

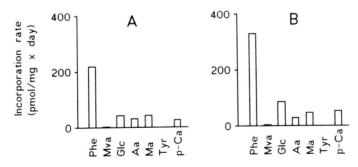

FIG. 6 The extent of incorporation of different precursors into sporopollenin isolated and purified by a gentle method (A) or the conventional aggressive procedure (B). Phe, uniformly labeled phenylalanine; Mva, mevalonic acid; Glc, glucose; Aa, acetate; Ma, malonic acid; Tyr, tyrosine; p-Ca, p-coumaric acid. Reproduced with permission from Rittscher and Wiermann (1988b).

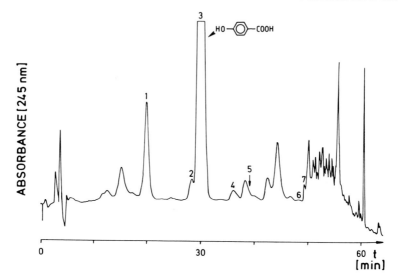

FIG. 7   High-pressure liquid chromatography analysis of ether-soluble acids released during degradation of a sporopollenin fraction using potash fusion. 1, protocatechuic acid; 2, unknown compound; 3, *p*-hydroxybenzoic acid; 4, *m*-hydroxybenzoic acid; 5, vanillic acid; 6, syringic acid; 7, solvent. The value of radioactivity eluted from peak 3 is shown in Table IX, in comparison with the radioactivity of the unsoluble residue. Modified with permission from Rittscher and Wiermann (1988b).

(2) the high degree of radioactivity in *p*-hydroxybenzoic acid as degradation product of sporopollenin labeled with [U-^{14}C]phenylalanine as well as with [ring-^{14}C]phenylalanine—are interpreted as indications that phenylpropanoid metabolism via PAL is involved in sporopollenin biosynthesis.

In further tracer experiments it has yet to be clarified whether the incorporation of phenylalanine into the sporopollenin fraction of *Tulipa* cultivar Apeldoorn is a unique phenomenon or if it is also valid for other systems. Therefore, [^{14}C]phenylalanine was applied to anthers of *Cucur-*

TABLE IX

The Radioactivity of Sporopollenin and Insoluble Residues and *p*-Hydroxybenzoic Acid Obtained after the Degradation of the Sporopollenin Fraction Labeled with [U-^{14}C]Phenylalanine[a]

Sporopollenin		*p*-Hydroxybenzoic acid		Insoluble residues	
9.63 mg (100%)		0.36 mg (3.8%)		1.02 mg (16.6%)	
10,102	dpm (100%)	4,356	dpm (43.1%)	275	dpm (2.7%)
1,049	dpm/mg	12,030	dpm/mg	269	dpm/mg

[a] Data from Rittscher and Wiermann (1988b).

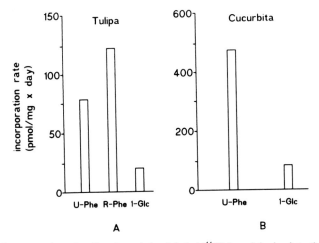

FIG. 8   The incorporation of uniformly and ring-labeled [^{14}C]phenylalanine into the sporopollenin fractions of *Tulipa* and *Cucurbita* in comparison with the incorporation of [^{14}C]glucose. U-Phe, uniformly labeled phenylalanine; R-Phe, ring-labeled phenylalanine; 1-Glc, [1-^{14}C]glucose.

bita maxima and compared with [^{14}C]glucose as a marker for the metabolic activity of the system. The results shown in Figure 8 clearly demonstrate that it also holds true for this system that phenylalanine is efficiently incorporated into the sporopollenin. The results achieved with *Tulipa* are therefore verified.

D. Chemical Studies: Degradation and Spectroscopic Analysis of Sporopollenin

1. Degradation Experiments

From the beginning of research into the chemistry of sporopollenin, attempts have been made to degrade this resistant biopolymer and analyze the released structures. These studies were supposed to yield insight into the composition of sporopollenin, but the methods employed were usually procedures with rather vigorous conditions, namely, potash fusion, ozonization, and other oxidative techniques, including sodium hypochloride, chromic acid, and strong nitric acid.

After degradation of sporopollenin by ozonization, diverse simple monocarboxylic and dicarboxylic acids and longer-chained dicarboxylic acids appear, whereas after potash fusion these acids along with a mixture of phenolic acids, namely, *p*-hydroxybenzoic acid, *m*-hydroxybenzoic acid,

and protocatechuic acid, are obtained. According to the interpretation of Shaw (1970), the phenols would originate from the postulated carotenoid skeletal units when forcing conditions, such as potash fusion, were applied. The results obtained by these methods are not discussed here in detail; the reader is instead referred to the articles by Zetzsche *et al.* (1937), Shaw (1971), and Brooks and Shaw (1977).

More recently, degradation experiments were continued to address the question of whether aromatic compounds appear as degradation products. These products were isolated after saponification, potash fusion, or nitrobenzene oxidation and are summarized in Table X. Under the chosen experimental conditions, *p*-coumaric acid appeared as the main compound after saponification and nitrobenzene oxidation (Schulze Osthoff and Wiermann, 1987). Following potash fusion, *p*-hydroxybenzoic acid was the main degradation product. Similar phenolic compounds were obtained after degradation of a highly purified sporopollenin fraction (Section III,B,5) from *Corylus avellana* pollen (Herminghaus *et al.,* 1988). In this case the degradation by potash fusion resulted in the release of *p*-hydroxybenzoic acid as a main component, whereas the degradation by nitrobenzene oxidation yielded *p*-hydroxybenzaldehyde; in addition, phenolics such as *p*-coumaric acid, ferulic acid, vanillin, and vanillic acid, were formed to varying degrees.

TABLE X

Degradation Products of *Pinus* Pollen Wing Material Isolated after Saponification, Potash Fusion, and Nitrobenzene Oxidation[a]

Treatment	Products
Saponification	*p*-Hydroxybenzoic acid
	p-Hydroxybenzaldehyde
	Vanillic acid
	Vanillin
	p-Coumaric acid (main compound)
	Ferulic acid
Potash fusion	Unidentified products
	p-Hydroxybenzoic acid (main compound)
Nitrobenzene oxidation	Unidentified products
	p-Hydroxybenzoic acid
	p-Hydroxybenzaldehyde (main compound)
	Vanillic acid
	Vanillin
	p-Coumaric acid (main compound)
	Ferulic acid

[a] Data from Schulze Osthoff and Wiermann (1987).

According to the interpretation by Brooks and Shaw (1968, 1971), degradation products, such as *p*- and *m*-hydroxybenzoic acid, are generated from a carotenoid skeleton as a result of the harsh experimental conditions following potash fusion. To verify this assumption, "synthetic sporopollenin" synthesized from β-carotene was degraded in control experiments in the same manner as naturally occurring sporopollenin (Herminghaus *et al.*, 1988). The results clearly demonstrated that phenolic compounds, if they really resulted from the degradation of the synthetic carotene polymer, are generated only in extremely small quantities in comparison with those released by degradation of *Corylus* sporopollenin. For this reason, it was concluded that phenolic compounds are integral constituents of sporopollenin (Schulze Osthoff and Wiermann, 1987; Herminghaus *et al.*, 1988). These findings are supported by results obtained by means of high resolution solid-state ^{13}C-NMR spectroscopy or pyrolysis GC–MS on sporopollenin from different sources (Guilford *et al.*, 1988; Espelie *et al.*, 1989; Wehling *et al.*, 1989).

It cannot be assumed that all compounds that have been released from sporopollenin under these harsh experimental conditions occur in sporopollenin in its natural form. The procedures used undoubtedly produced artifacts. This was shown in control experiments in which model substances of different natures were treated the same way as sporopollenin. The results demonstrated that changes in the structure of phenolic compounds may occur to a varying degree.

However, in the case of sporopollenin from wings of *Pinus* pollen, it is postulated that *p*-coumaric acid is a genuine structural unit of sporopollenin. After pyrolysis and mass spectroscopy, the main mass peak, m/z 164, as well as the peaks m/z 120 and 147, are characteristic of *p*-coumaric acid (Fig. 9). The main peak, m/z 164, is a molecular ion of *p*-coumaric acid, which decarboxylates during ionization to m/z 120. The main degradation compound when the wing material was treated by a gentle method using AlJ$_3$ was *p*-coumaric acid (Wehling *et al.*, 1989). In addition, the effects of AlJ$_3$ treatment indicate that *p*-coumaric acid might possibly be partially bound by ether linkages (Bhatt and Kulkarni, 1983). Thus, it was shown by two independent analyses that *p*-coumaric acid is a genuine structural unit of sporopollenin.

2. Spectroscopic Analyses

Shaw *et al.* introduced new analytical techniques of chromatography and spectroscopy that revealed further information about the structure of sporopollenin. A detailed description of these results is given by Shaw (1971) and Brooks and Shaw (1971, 1977). IR spectroscopy, which was first applied by the authors, was often used for characterization and identifica-

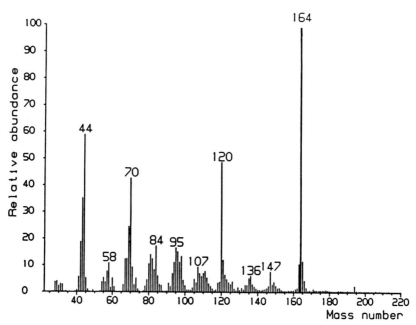

FIG. 9   Curie-point pyrolysis low-voltage electron-impact spectra of a highly purified sporopollenin fraction obtained from pollen wings. Reproduced with permission from Wehling *et al.* (1989).

tion of sporopollenin (König and Peveling, 1980; Rittscher and Wiermann, 1983; Brunner and Honegger, 1985; Schraudolf and Haag, 1985). This method is suitable for comparative studies and has contributed interesting findings about the structure of sporopollenin.

High resolution solid-state ^{13}C-NMR spectroscopy, applied to elucidate sporopollenin structure, has been a fundamentally new approach (Guilford *et al.*, 1988; Espelie *et al.*, 1989). Guilford *et al.* (1988) analyzed sporopollenin material gathered from a broad phylogenetic range of plant sources. The carbon resonances in the spectra occur primarily in four distinct regions but vary in peak intensity and shape. According to the authors, sporopollenin is primarily an aliphatic polymer, in case of *Lycopodium* perhaps of terpenoid nature, and contains oxygen in the form of ether, hydroxyl, carboxylic acid, ester, and ketone groups in varying amounts depending on the origin of the sporopollenin used (Fig. 10). Aromatic rings occur to a minor degree in some samples.

As previously shown, hexadecanedioic acid, 6,11-dioxohexadecanedioic acid, 7-hydroxyhexadecanedioic acid, and a mixture of alkanoic acids were detected after the degradation of sporopollenin by a gentle ozoniza-

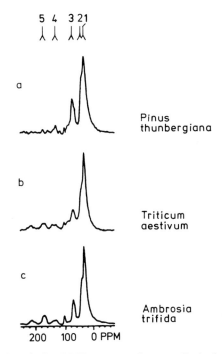

FIG. 10 ^{13}C-Magic angle spinning NMR spectra of sporopollenin fractions from different sources. 1, aliphatic carbons; 2, shoulder attributed to carbon adjacent to an oxygen-bearing carbon; 3, carbon-bearing oxygen; 4, olefinic resonance; 5, carbonyl resonance indicative of an ester or a carboxylic acid group. Reproduced with permission from Guilford *et al.* (1988).

tion and subsequent treatment of the solid and liquid fractions produced (Shaw and Yeadon, 1966). It must be assumed that these compounds originate from the aliphatics shown as the major structures of sporopollenin by ^{13}C-NMR spectroscopy.

According to Guilford *et al.* (1988), the results obtained by ^{13}C-NMR spectroscopy indicate that sporopollenin formed by plants of different systematic categories has a distinct and individual composition. This view is also supported by combined ultrastructural and cytochemical studies, which have shown that there is a wide diversity in form and composition of this structure. This was demonstrated in studies by Southworth (1973, 1974) and other authors studying the cytochemical reactivity and solubility of pollen walls. From the two distinct strata of the outer pollen wall, exine 1 and exine 2, only exine 1 was affected by 2-aminoethanol (Southworth, 1974, 1985a,b; Kress, 1986; Gabaraeva, 1987; Kedves and Rojik, 1989; Sreedevi *et al.*, 1990).

Fundamental progress was made in studies on the substructural organization of the exine. In this context, the reader is referred to reports by the

following authors: Southworth (1985a,b, 1986a,b), Hesse (1985), Rowley *et al.* (1987), Claugher and Rowley (1987, 1990), Kedves (1989), Kedves and Rojik (1989), Rowley and Dunbar (1990), and Rowley (1990).

It is well accepted that sporopollenin-containing structures arise from both the tapetum cell and the microspores. Rowley and co-workers hypothesized that the sporopollenin-receptive areas correspond to a glycocalyx (Rowley and Skvarla, 1975, 1986; Abadie *et al.*, 1987). However, the primary biochemical mechanism of sporopollenin synthesis and accumulation, namely, the pathway of the polymerization process, is not understood at present.

What is sporopollenin? Neither a proposition about its structure that is assured to some extent nor a plausible scheme of the biosynthetic pathway can be presented at present. Summarizing the results obtained from tracer and degradation experiments and from spectroscopic analysis, respectively, it has to be assumed that sporopollenin is a mixed polymer with a large amount of long-chain aliphatics containing additional compounds, for example, phenols to varying degrees. If and to what extent further structural elements, such as polysaccharide compounds (Traverse 1968, 1988; Espelie *et al.*, 1989), are involved in the structure of sporopollenin, or whether they should be considered as low levels of contaminating material from the intine, will have to be clarified in further studies.

Wall systems of high chemical resistance have also been described for algae and fungi. Based on the results obtained from *Botryococcus braunii* and other systems (Geisert *et al.*, 1987; Burczyk and Dworzanski, 1988), those walls are presumed to contain polymers of primarily aliphatic nature (Berkaloff *et al.*, 1983; Derenne *et al.*, 1989; Kadouri *et al.*, 1988). At present it cannot be resolved whether these biopolymers partially correspond to the sporopollenin of higher plants in their chemical composition or if they are very different and show only a poor relation. The same applies to the nonsaponifiable, highly aliphatic, and resistant biopolymer that was recently detected in plant cuticles (Schmidt and Schönherr, 1982; Nip *et al.*, 1986). To resolve these questions, further research will be necessary.

IV. Concluding Remarks

The pollen wall is a fascinating system, which differs in its sculpture, substructure, and chemical composition from most cell walls of higher plants. The knowledge of the chemical composition of the three domains of a pollen wall (intine, exine, and surface substances) is rather different. Important findings about the chemical composition of the surface sub-

stances have been achieved, and analyses showed that they have a very inhomogenous chemical structure and are formed in completely different metabolic pathways. Some pollen coat substances have been clearly shown to be involved in the recognition system, but the massive accumulation of different substances at the surface of pollen grains is not yet understood in terms of its physiological and biological role.

The substructure of the exine in general has attracted wide interest; however, the chemical structure of sporopollenin, which is the basic chemical substance of the exine, is still unknown in many important respects. Fortunately, the biosynthesis and structure of sporopollenin have generated greater interest. Advanced methods and new analytical techniques have contributed to some important progress in elucidating the biosynthesis and structure of this extraordinary biopolymer.

Despite this progress, future studies are required to gain new insight into sporopollenin structure and to allow a general proposal concerning the biosynthesis of sporopollenin. To resolve different problems of structure and biosynthesis of sporopollenin, the methods for isolation and purification of exine material and sporopollenin should receive particular attention; only then will it be possible to obtain detailed and comparable results.

Moreover, intensive investigations are required to elucidate the biochemical mechanism of sporopollenin synthesis and accumulation. Many other areas of scant knowledge regarding sporopollenin formation could be cited, and for that reason studies on this unusual biopolymer of plants will be worthwhile for many years.

Acknowledgments

The authors thank Sybille Arendt and Eva Hansen for their help in preparing the manuscript and Dr. Gregory Armstrong for reading the manuscript. Research of the senior author has generously been supported by the Deutsche Forschungsgemeinschaft and the Fonds der Chemischen Industrie.

References

This review covers literature until Spring 1991. Unfortunately, it was impossible to include important articles and reviews (such as D. Southworth (1990): Exine Biochemistry. In "Microspores: Evolution and ontogeny" (S. Blackmore and R. B. Knox, eds.) pp. 193–212. Academic Press) published after completing the manuscript.

Abadie, M., Hideux, M., and Rowley, J. R. (1987). *Ann. Sci. Nat., Bot. Biol. Veg.* **8**, 1–16.
Arora, S. (1985). *Phytomorphology* **35**, 161–166.

Audran, J. C. (1981). *Rev. Palae. P.* **33**, 315–346.

Baldi, B. G., Franceschi, V. R., and Loewus, F. A. (1986). *In* "Biotechnology and Ecology of Pollen" (D. L. Mulcahy, G. B. Mulcahy, and E. Ottaviano, eds.), pp. 77–82. Springer-Verlag, New York.

Baldi, B. G., Franceschi, V. R., and Loewus, F. A. (1987). *Protoplasma* **141**, 47–55.

Banerjee, U. C. (1967). *Grana Palynologica* **7**, 365–371.

Banerjee, U. C., and Barghoorn, E. S. (1971). *In* "Pollen: Development and Physiology" (J. Heslop-Harrison, ed.), pp. 126–127. Butterworth Press, London.

Barz, W., and Wiermann, R. (1981). *Proceedings of the International Bioflavonoid Symposium Munich, FRG, 1981,* pp. 185–211.

Bäumker, P. A., Arendt, S., and Wiermann, R. (1988). *Z. Naturforsch. [C]* **43**, 641–646.

Beerhues, L., Forkmann, G., Schöpker, H., Stotz, G., and Wiermann, R. (1989). *J. Plant Physiol.* **133**, 743–746.

Berkaloff, C., Casadevall, E., Largeau, C., Metzger, P., Peracca, S., and Virlet, J. (1983). *Phytochemistry* **22**, 389–397.

Bhandari, N. N. (1984). *In* "Embryology of Angiosperms" (B. M. Johri, ed.), pp. 53–121. Springer-Verlag, Berlin.

Bhandari, N. N., and Kishori, R. (1973). *Beitr. Biol. Pflanz.* **49**, 59–72.

Bhatt, M. V., and Kulkarni, S. U. (1983). *Synthesis* **1983**, 249–282.

Bianchi, G., Murelli, C., and Ottaviano, E. (1990). *Phytochemistry* **29**, 739–744.

Blackmore, S., and Ferguson, I. K., eds. (1986). "Pollen and Spores Papers from International Symposium 1985," *Linn. Soc. Symp. Ser, 12.* Academic Press, London.

Brooks, J., and Shaw, G. (1968). *Nature (London)* **219**, 532–533.

Brooks, J., and Shaw, G. (1971). *In* "Pollen: Development and Physiology" (J. Heslop-Harrison, ed.), pp. 99–114. Butterworth Press, London.

Brooks, J., and Shaw, G. (1977). *Trans. Bose Res. Inst. (Calcutta)* **40**, 19–38.

Brunner, U., and Honegger, R. (1985). *Can. J. Bot.* **63**, 2221–2230.

Buchen, B., and Sievers, A. (1981). *In* "Cytomorphogenesis in Plants" (O. Kiermayer, ed.), Cell Biology Monographs **8**, pp. 349–376. Springer-Verlag, Wien.

Burczyk, J., and Dworzanski, J. (1988). *Phytochemistry* **27**, 2151–2153.

Cabanne, F., Dalebroux, M. A., Martin-Tanguy, J., and Martin, C. (1981). *Physiol. Plant.* **53**, 399–404.

Caffrey, M., Werner, B. G., and Priestley, D. A. (1987). *Biochim. Biophys. Acta* **921**, 124–134.

Caldicott, A. B., and Eglinton, G. (1975). *Phytochemistry* **14**, 1799–1801.

Ceska, O., and Styles, E. D. (1984). *Phytochemistry* **23**, 1822–1823.

Chapman, G. P. (1987). *In* "Pollen: Cytology and Development" (K. L. Giles and J. Prakash, eds.), *Int. Rev. Cytol.* **107**, pp. 111–125. Academic Press, San Diego.

Chen, Z. K., Wang, F. X., and Zhou, F. (1987). *Acta Bot. Sin.* **29**, 486–491.

Claugher, D., and Rowley, J. R. (1987). *Pollen Spores* **29**, 5–20.

Claugher, D., and Rowley, J. R. (1990). *Can. J. Bot.* **68**, 2195–2200.

Coe, E. H., McCormick, S. M., and Modena, St. A. (1981). *J. Hered.* **72**, 318–320.

Colhoun, C. W., Herd, Y. R., and Steer, M. W. (1984). *Caryologia* **37**, 309–322.

Cousin, M. (1979). *Grana* **18**, 115–128.

Derenne, S., Largeau, C., Casadevall, E., and Berkaloff, C. (1989). *Phytochemistry* **28**, 1137–1142.

Dickinson, H. G. (1970). *New Phytol.* **69**, 1065–1068.

Dobson, H. E. M. (1987). *Oecologia* **72**, 618–623.

Dobson, H. E. M. (1988). *Am. J. Bot.* **75**, 170–182.

Dobson, H. E. M. (1989). *In* "The Evolutionary Ecology of Plants" (J. H. Bock and Y. B. Linhart, eds.), pp. 227–246. Westview Press, Boulder, Colorado.

Dobson, H. E. M. (1991). *In* "Sixth Pollination Symposium" (C. van Heemert and A. de Ruijter, eds.), *Acta Hortic.* **288**, 313–320.

Dobson, H. E. M., Bergström, J., Bergström, G., and Groth, I. (1987). *Phytochemistry* **26**, 3171–3173.

Dobson, H. E. M., Bergström, G., and Groth, I. (1990). *Isr. J. Bot.* **39**, 143–156.

Duhoux, E. (1980). *Z. Pflanzenphysiol.* **99**, 207–214.

Dulberger, R. (1990). *Sex. Plant Reprod.* **3**, 18–22.

El-Ghazaly, G., and Jensen, W. A. (1986). *Grana* **25**, 1–29.

El-Ghazaly, G., and Jensen, W. A. (1987). *Am. J. Bot.* **74**, 1396–1418.

Elleman, C. J., and Dickinson, H. G. (1990). *New Phytol.* **114**, 511–518.

Erdtman, G. (1960). *Svensk. Bot. Tidskr.* **54**, 561–564.

Espelie, K. E., Loewus, F. A., Pugmire, R. J., Woolfenden, W. R., Baldi, B. G., and Given, P. H. (1989). *Phytochemistry* **28**, 751–753.

Evans, D. E., Rothnie, N. E., Palmer, M. V., Burke, D. G., Sang, J. P., Knox, R. B., Williams, E. G., Hilliard, E. P., and Salisbury, P. A. (1987). *Phytochemistry* **26**, 1895–1897.

Evans, D. E., Sang, J. P., Cominos, X., Rothnie, N. E., and Knox, R. B. (1990). *Plant Physiol.* **92**, 418–424.

Ferreres, F., Tomás-Barberán, F. A., Tomás-Lorente, F., Nieto, J. L., Rumbero, A., and Oliás, J. M. (1989). *Phytochemistry* **28**, 1901–1903.

Flint, St. D., and Caldwell, M. M. (1983). *Am. J. Bot.* **70**, 1416–1419.

Gabaraeva, N. J. (1987). *Bot. Zh. (Leningrad)* **72**, 1310–1317.

Geisert, M., Rose, T., Bauer, W., and Zahn, R. K. (1987). *Biosystems* **20**, 133–142.

Green, D. (1973). Ph.D. thesis, University of Bradford. Bradford, England.

Gubatz, S., Herminghaus, S., Meurer, B., Strack, D., and Wiermann, R. (1986). *Pollen Spores* **28**, 347–354.

Guilford, W. J., Schneider, D. M., Labovitz, J., and Opella, S. J. (1988). *Plant Physiol.* **86**, 134–136.

Gupta, S. C., and Nanda, K. (1972). *Grana* **12**, 99–104.

Hagenberg, S., Wehling, K., and Wiermann, R. (1990). *Z. Naturforsch. [C]* **45**, 1090–1092.

Harborne, J. B. (1969). *Phytochemistry* **8**, 177–183.

Heller, W., and Forkmann, G. (1988). *In* "The Flavonoids" (J. B. Harborne, ed.), pp. 399–425. Chapman and Hall, London.

Herdt, E., Sütfeld, R., and Wiermann, R. (1978). *Eur. J. Cell. Biol.* **17**, 433–441.

Herminghaus, S., Gubatz, S., Arendt, S., and Wiermann, R. (1988). *Z. Naturforsch. [C]* **43**, 491–500.

Heslop-Harrison, J. (1968). *New Phytol.* **67**, 779–786.

Heslop-Harrison, J. (1975). *Proc. R. Soc. Lond. [Biol.]* **190**, 275–299.

Heslop-Harrison, J., and Dickinson, H. G. (1969). *Planta* **84**, 199–214.

Hesse, M. (1978a). *Linzer Biologische Beiträge* **9(2)**, 181–201.

Hesse, M. (1978b). *Plant Syst. Evol.* **130**, 13–42.

Hesse, M. (1978c). *Linzer Biologische Beiträge* **9(2)**, 237–258.

Hesse, M. (1979a). *Flora (Jena)* **168**, 540–557.

Hesse, M. (1979b). *Flora (Jena)* **168**, 558–577.

Hesse, M. (1980a). *Plant Syst. Evol.* **134**, 229–267.

Hesse, M. (1980b). *Plant Syst. Evol.* **135**, 253–263.

Hesse, M. (1981a). *Grana* **20**, 145–152.

Hesse, M. (1981b). *Mikroskopie* **38**, 85–89.

Hesse, M. (1985). *Grana* **24**, 93–98.

Hesse, M. (1986a). *Plant Syst. Evol.* **153**, 37–48.

Hesse, M. (1986b). Linn. Soc. Symp. Ser. **12**, 109–118.

Hesse, M. (1991). *Prog. Bot.* **52**, 19–34.

Hesse, M., and Ehrendorfer, F. (eds.) (1990). *In* "Morphology, Development, and Systematic Relevance of Pollen and Spores," *Plant Syst. Evol. (Suppl. 5).* Springer-Verlag, Vienna.

Hesse, M., and Waha, M. (1989). *Plant Syst. Evol.* **163,** 147–152.

Imperato, F. (1979). *Experientia* **35,** 13–14.

Iwanami, Y., Sasakuma, T., and Yamada, Y. (1988). *In* "Pollen: Illustrations and Scanning Electronmicrographs." Springer-Verlag, Berlin.

Jonker, F. P. (1971). *In* "Sporopollenin, Proc. Symp. 1970" (J. Brooks, P. R. Grant, M. Muir, P. van Gijzel, and G. Shaw, eds.), pp. 686–707. Academic Press, London.

Kadouri, A., Derenne, S., Largeau, C., Casadevall, E., and Berkaloff, C. (1988). *Phytochemistry* **27,** 551–557.

Kedves, M. (1989). *Acta Biol. Szeged.* **35,** 59–70.

Kedves, M., and Rojik, I. (1989). *Acta Biol. Szeged.* **35,** 71–80.

Kehrel, B., and Wiermann, R. (1985). *Planta* **163,** 183–90.

Keijzer, C. J. (1987). *New Phytol.* **105,** 499–507.

Keijzer, C. J., and Cresti, M. (1987). *Ann. Bot.* **59,** 533–542.

Kleinehollenhorst, G., Behrens, H., Pegels, G., Strunk, N., and Wiermann, R. (1982). *Z. Naturforsch. [C]* **37,** 587–599.

Knoll, F. (1930). *Z. Bot.* **23,** 609–675.

Knox, R. B. (1984a). *In* "Cellular Interactions" (H. F. Linskens, and J. Heslop-Harrison, eds.), Encycl. Plant Physiol., New Ser. Vol. **17,** pp. 508–608. Springer-Verlag, Berlin.

Knox, R. B. (1984b). *In* "Embryology of Angiosperms" (B. M. Johri, ed.), pp. 197–271. Springer-Verlag, Berlin.

Knox, R. B., Heslop-Harrison, J., and Heslop-Harrison, Y. (1975). *Biol. J. Linn. Soc. (Suppl. 1)* **7,** 177–187.

Koes, R. E., van Blokland, R., Quattrocchio, F., van Tunen, A. J., and Mol, J. N. M. (1990). *Plant Cell* **2,** 379–392.

König, J., and Peveling, E. (1980). *Z. Pflanzenphysiol.* **98,** 459–464.

Kress, W. J. (1986). *Grana* **25,** 31–40.

Kress, W. J., and Stone, D. E. (1983). *In* "Pollen: Biology and Implications for Plant Breeding" (D. L. Mulcahy and E. Ottaviano, eds.), pp. 159–163. Elsevier Biomedical, New York.

Loewus, F. A., Baldi, B. G., Franceschi, V. R., Meinert, L. D., and McCollum, J. J. (1985). *Plant Physiol.* **78,** 652–654.

Martens, P., and Waterkeyn, L. (1962). *Cellule* **62,** 172–222.

Martin, C., Kunesch, G., Martin-Tanguy, J., Negrel, J., Paynot, M., and Carre, M. (1985). *Plant Cell Reprod.* **4,** 158–160.

Martin, J. H., Kunstmann, M. P., Barbatschi, F., Hertz, M., Ellestad, G. A., Dann, M., Redin, G. S., Dornbush, A. C., and Kuck, N. A. (1978). *J. Antibiot.(Tokyo)* **31,** 398–404.

Martin-Tanguy, J., Cabanne, F., Perdrizet, E., and Martin, C. (1978). *Phytochemistry* **17,** 1927–1928.

Martin-Tanguy, J., Deshayes, A., Perdrizet, E., and Martin, C. (1979). *FEBS Lett.* **108,** 176–178.

Martin-Tanguy, J., Perdrizet, E., Prevost, J., and Martin, C. (1982). *Phytochemistry* **21,** 1939–1945.

Mascarenhas, J. P. (1975). *Bot. Rev.* **41,** 259–314.

Mattsson, O. (1983). *In* "Pollen: Biology and Implications for Plant Breeding" (D. L. Mulcahy and E. Ottaviano, eds.), pp. 257–264. Elsevier Biomedical, New York.

Meurer, B., Strack, D., and Wiermann, R. (1984). *Planta Med.* **1984,** 376–380.

Meurer, B., Wray, V., Grotjahn, L., Wiermann, R., and Strack, D. (1986). *Phytochemistry* **25,** 433–435.

Meurer, B., Wray, V., Wiermann, R., and Strack, D. (1988a). *Phytochemistry* **27**, 839–843.
Meurer, B., Wiermann, R., and Strack, D. (1988b). *Phytochemistry* **27**, 823–828.
Mu, X. J., Wang, F. H., and Wang, W. L. (1988). *Acta Bot. Sin.* **30**, 6–13.
Niester, C., Gülz, P. G., and Wiermann, R. (1987). *Z. Naturforsch.* [*C*] **42**, 858–862.
Nip, M., Tegelaar, E. W., de Leeuw, J. W., and Schenk, P. A. (1986). *Naturwissenschaften* **73**, 579–585.
Ogorodnikova, V. F. (1986). *Bot. Zh.* (*Leningrad*) **71**, 1366–1371.
Pacini, E., Franchi, G. G., and Hesse, M. (1985). *Plant Syst. Evol.* **149**, 155–185.
Pandey, D. K., Tripathi, R. N., Tripathi, R. D., and Dixit, S. N. (1983). *Grana* **22**, 31–33.
Parkinson, B. M. (1988). *Ann. Bot.* **61**, 695–703.
Prahl, A. K., Springstubbe, H., Grumbach, K., and Wiermann, R. (1985). *Z. Naturforsch.* [*C*] **40**, 621–626.
Prahl, A. K., Rittscher, M., and Wiermann, R. (1986). *In* "Biotechnology and Ecology of Pollen" (D. L. Mulcahy, G. B. Mulcahy, and E. Ottaviano, eds.), pp. 313–318. Springer-Verlag, New York.
Pratviel-Sosa, F., and Percheron, F. (1972). *Phytochemistry* **11**, 1809–1813.
Reznickowa, S. A., and Willemse, M. T. M. (1980). *Acta Bot. Neerl.* **29**, 141–156.
Riederer, M., and Schneider, G. (1990). *Planta* **180**, 154–165.
Rittscher, M., and Wiermann, R. (1983). *Protoplasma* **118**, 219–224.
Rittscher, M., and Wiermann, R. (1988a). *Sex. Plant Reprod.* **1**, 125–131.
Rittscher, M., and Wiermann, R. (1988b). *Sex. Plant Reprod.* **1**, 132–139.
Rowley, J. R. (1990). *In* " Morphology, Development, and Systematic Relevance of Pollen and Spores" (M. Hesse and F. Ehrendorfer, eds.), *Plant Syst. Evol.* (*Suppl. 5*), pp. 13–29. Springer-Verlag, Vienna.
Rowley, J. R., and Dunbar, A. (1990). *Bot. Acta* **103**, 355–359.
Rowley, J. R., and Skvarla, J. J. (1975). *Am. J. Bot.* **62**, 479–485.
Rowley, J. R., and Skvarla, J. J. (1986). *Nord. J. Bot.* **6**, 39–65.
Rowley, J. R., and Walles, B. (1985). *In* "Proceedings of the 8th International Symposium on Sexual Reproduction in Seed Plants, Ferns and Mosses" (M. T. M. Willemse and J. C. van Went, eds.), p. 56. Pudoc, Wageningen.
Rowley, J. R., El-Ghazaly, G., and Rowley, J. S. (1987). *Palynology* **11**, 1–21.
Sarker, R. H., Elleman, C. J., and Dickinson, H. G. (1988). *Proc. Natl. Acad. Sci. U.S.A.* **85**, 4340–4344.
Schmelzer, E., Jahnen, W., and Hahlbrock, K. (1988). *Proc. Natl. Acad. Sci. U.S.A.* **85**, 2989–2993.
Schmidt, H. W., and Schönherr, J. (1982). *Planta* **156**, 380–384.
Schönherr, J. (1976). *Planta* **131**, 159–164.
Schraudolf, H., and Haag, R. (1985). *Naturwissenschaften* **72**, 433–434.
Schulze Osthoff, K., and Wiermann, R. (1987). *J. Plant Physiol.* **131**, 5–15.
Scott, R. W., and Strohl, M. J. (1962). *Phytochemistry* **1**, 189–193.
Shaw, G. (1970). *In* "Phytochemical Phylogeny" (J. Harborne, ed.), pp. 31–57. Academic Press, London.
Shaw, G. (1971). *In* "Sporopollenin, Proceedings of Symposium 1970" (J. Brooks, P. R. Grant, M. Muir, P. van Gijzel, and G. Shaw, eds.), pp. 305–348. Academic Press, London.
Shaw, G., and Yeadon, A. (1964). *Grana Palynologica* **5**, 247–252.
Shaw, G., and Yeadon, A. (1966). *J. Chem. Soc.* 16–22.
Shibuya, T., Funamizu, M., and Kitahara, Y. (1978). *Phytochemistry* **17**, 979–981.
Shivanna, K. R., and Johri, B. M. (1985). "The Angiosperm Pollen. Structure and Function" Wiley Eastern Limited, New Delhi.
Sosa, F., and Percheron, F. (1970). *Phytochemistry* **9**, 441–446.
Southworth, D. (1971). *In* "Pollen: Development and Physiology" (J. Heslop-Harrison, ed.), pp. 115–120. Butterworth Press, London.

Southworth, D. (1973). *J. Histochem. Cytochem.* **21,** 73–80.
Southworth, D. (1974). *Am. J. Bot.* **61,** 36–44.
Southworth, D. (1985a). *Am. J. Bot.* **72,** 1274–1283.
Southworth, D. (1985b). *Grana* **24,** 161–166.
Southworth, D. (1986a). *Linn. Soc. Symp. Ser.* **12,** 61–69.
Southworth, D. (1986b). *Can. J. Bot.* **64,** 983–987.
Southworth, D. (1988). *Am. J. Bot.* **75,** 15–21.
Southworth, D., Singh, M. B., Hough, T., Smart, I. J., Taylor, P., and Knox, R. B. (1988). *Planta* **176,** 482–487.
Sreedevi, P., Pillai, G. S., and Namboodiri, A. N. (1990). *Curr. Sci.* **59,** 324–325.
Stafford, H. A. (1990). "Flavonoid Metabolism." CRC Press, Boca Raton, Florida.
Stanley, R. G., and Linskens, H. F., eds. (1974). "Pollen Biology Biochemistry Management." Springer-Verlag, Berlin.
Strack, D., Sachs, G., and Wiermann, R. (1981a). *Z. Pflanzenphysiol.* **103,** 291–296.
Strack, D., Sachs, G., Römer, A., and Wiermann, R. (1981b). *Z. Naturforsch. [C]* **36,** 721–723.
Strack, D., Meurer, B., Wray, V., Grotjahn, L., Austenfeld, F.A., and Wiermann, R. (1984). *Phytochemistry* **23,** 2970–2971.
Stránsky, K., Streibl, M., and Ubik, K. (1991). *Collect. Czech. Chem. Commun.* **56,** 1123–1129.
Strohl, M. J., and Seikel, M. K. (1965). *Phytochemistry* **4,** 383–399.
Styles, E. D., and Ceska, O. (1977). *Maize Genetic Cooperation Newsletters* **51,** 87.
Styles, E. D., and Ceska, O. (1981). *Maydica* **26,** 141–152.
Sütfeld, R., and Wiermann, R. (1974). *Ber. Dtsch. Bot. Ges.* **87,** 167–174.
Takahashi, M., and Skvarla, J. J. (1990). *Am. J. Bot.* **77,** 1142–1148.
Traverse, A. (1968). *Am. J. Bot.* **55,** 722.
Traverse, A. (1988). "Paleopalynology." Unwin Hyman, Boston.
Tripathi, R. N., Dubey, N. K., and Dixit, S. N. (1985). *Grana* **24,** 61–63.
Tsuda, Y., Kaneda, M., Sato, K., and Kokaji, M. (1981). *Phytochemistry* **20,** 505–506.
Vishnjakova, M. A., and Lebsky, V. K. (1986). *Bot. Zh. (Leningrad)* **71,** 754–759.
Visser, J. H. (1983). *In* "Plant Resistance to Insects" (P. A. Hedin, ed.), pp. 215–230. American Chemical Society, Washington, D.C.
Waha, M. (1984). *Plant Syst. Evol.* **147,** 189–203.
Wehling, K., Niester, C., Boon, J. J., Willemse, M. T. M., and Wiermann, R. (1989). *Planta* **179,** 376–380.
Wiermann, R. (1981). *In* "The Biochemistry of Plants" (P. K. Stumpf and E. E. Conn, eds.), Vol. 7, pp. 85–116. Academic Press, New York.
Wiermann, R., and Vieth, K. (1983). *Protoplasma* **118,** 230–233.
Wiermann, R., Wollenweber, E., and Rhese, C. (1981). *Z. Naturforsch. [C]* **36,** 204-206.
Wollenweber, E., and Wiermann, R. (1979). *Z. Naturforsch. [C]* **34,** 1289–1291.
Zerback, R., Bokel, M., Geiger, H., and Hess, D. (1989a) *Phytochemistry* **28,** 897–899.
Zerback, R., Dressler, K., and Hess, D. (1989b). *Plant Sci.* **62,** 83–91.
Zetzsche, F., and Huggler, K. (1928). *Justus Liebigs Ann. Chem.* **461,** 89–108.
Zetzsche, F., and Vicari, H. (1931). *Helv. Chim. Acta* **14,** 58–62.
Zetzsche, F., Kalt, P., Liechti, J., and Ziegler, E. (1937). *J. Prakt. Chem.* **148,** 266–286
Zhou, C. (1988). *Acta Bot. Sin.* **30,** 362–367.

Cytoskeleton and Cytoplasmic Organization of Pollen and Pollen Tubes

Elisabeth S. Pierson and Mauro Cresti

Dipartimento di Biologia Ambientale, Università di Siena, Siena, Italy

I. Introduction

A. Objectives

The pollen grain and the pollen tube fulfill important functions in the sexual reproduction of higher plants. The pollen grain is concerned with conveying the two sperm cells or their progenitor, the generative cell, to the female gametophyte. The pollen tube serves as a guide and a pathway for the sperm cells on their course to the embryo sac, so that they can complete double fertilization. In this chapter we will not go into the details of the physiology of pollen germination, pollen tube growth, and pollen–pistil interaction, which have been the subjects of many detailed reviews (Stanley and Linskens, 1974; Mascarenhas, 1975; De Nettancourt, 1977; Knox, 1984; Heslop-Harrison, 1983, 1987; Knox and Singh, 1987; Shivanna and Johri, 1985), but we will mainly concentrate on the cytoskeleton of the vegetative cytoplasm of pollen grains and pollen tubes.

The first part of this review is a summary of the current views on the morphology of dry and activated pollen and growing pollen tubes. The distribution of calcium in pollen tubes is discussed in relationship to pollen tube growth, intracellular movement, and the activity of the cytoskeleton. The movements of the organelles and nuclei inside the pollen grain and the pollen tube are described in Section II,F. These morphological and cytological data should clarify the context in which the cytoskeleton of the vegetative cell is operating.

The second part of the review provides a synopsis of the characteristics of the major elements of the pollen cytoskeleton, for example, microtubules, actin filaments, a few cytoskeleton-associated proteins, myosin, and a kinesin-like protein. Intermediate filaments (Nagle, 1988), the third class of cytoskeletal elements in animal and human cells, are not included

73

in this survey, because they have only rarely been detected in plant cells (Staiger and Lloyd, 1991), and, as far as we know, never in pollen or related material. Background information on the structure of the elements of the pollen cytoskeleton can be found in Sections III,A,1, III,B,1, and III,C,1. For more details on the cytoskeleton we refer to the excellent reviews and handbooks that exist on specific topics: cytoskeletal components (Dustin, 1984; Bershadsky and Vasiliev, 1988), the biochemistry, genetics, and cytology of plant microtubules and tubulins (Hepler and Palevitz, 1974; Gunning and Hardham, 1982; Lloyd, 1987, 1988, 1989; Kristen, 1986; Morejohn and Fosket, 1986; Fosket, 1989; Traas, 1990; Derksen and Emons, 1990; Derksen *et al.*, 1990; Staiger and Lloyd, 1991), actin filaments in plants (Hepler and Palevitz, 1974; Kristen, 1987; Traas, 1990; Steer, 1990; Staiger and Schliwa, 1987; Derksen *et al.*, 1990; Staiger and Lloyd, 1991) and fungi (Heath, 1990), and the locomotory systems involved in cell motility in flagellated organisms and animals (Lackie, 1986; Warner and McIntosh, 1989; Warner *et al.*, 1989; Cross and Kendrick-Jones, 1991).

This review also gives a critical evaluation of the various microscopic methods used for the visualization of the cytoskeleton of pollen. The literature referring to the localization of actin filaments and microtubules in pollen grains and pollen tubes is summarized in two tables. We focus on the microscopic localization of microtubules, actin filaments, and their associated proteins in the vegetative cell in Section IV. Brief reference is made to the cytoskeleton of other components of the male gametophytic apparatus, for example, meiocytes, microspores, generative cells, and sperm cells, the biology of which is closely related to that of mature pollen. The cytoskeleton of generative and sperm cells and its significance for the division of the generative cell are reviewed in the chapter by Palevitz and Tiezzi. Finally, the effects of treatment of pollen grains and pollen tubes with specific inhibitors of the cytoskeleton are discussed in detail (Section V) because these results are basic for understanding the functions of the cytoskeleton. Concluding comments on the state of the art in pollen cytoskeleton research and perspectives for the near future are expounded in Section VI and VII.

B. Hypotheses on the Functions of the Cytoskeleton in the Male Gametophyte

A great deal of our present knowledge on the organization and function of the cytoskeleton in higher plants has been acquired from experiments on pollen grains and tubes (Pierson, 1989; Cresti and Tiezzi, 1990; Tiezzi and

Cresti, 1990; Tiezzi, 1991). The cytoskeleton is believed to be involved in numerous intracellular processes: organelle movement, organization of the cytoplasm, endocytosis and exocytosis, cytomorphogenesis, meiotic and mitotic division, and cell wall deposition (Dustin, 1984; Bershadsky and Vasiliev, 1988). Male gametophytic cells exhibit all these features: vigorous movement of the organelles and translocation of the nuclei in pollen tubes (Heslop-Harrison, 1988; Kuroda, 1990; Emons *et al.*, 1991; Section II,F), polar organization of the vegetative cytoplasm in germinating pollen and pollen tubes (Schnepf, 1986; Sections II,B, II,C, II,D, and II,E) accompanied by rapid growth at the tip and the formation of the tube-like protrusion (Steer, 1990; Steer and Steer, 1989; Derksen and Emons, 1990), male meiosis and microsporogenesis, mitotic division of the generative cell (reviewed by Palevitz and Tiezzi, this volume), and the existence of a complex cell wall (Wiermann and Gubatz, this volume).

Numerous original articles (Sanger and Jackson, 1971b; Clapham and Östergren, 1984; Dickinson and Sheldon, 1984; Van Lammeren *et al.*, 1985, 1988, 1989; Murgia *et al.*, 1986; Sheldon and Dickinson, 1986; Hogan, 1987; Brown and Lemmon, 1988, 1990, 1991a, 1991b, 1991c; Sheldon and Hawes, 1988; Pierson, 1989; Tiwari, 1989; Traas *et al.*, 1989; Perez-Munoz and Webster 1990; Staiger and Cande, 1990; Simmonds, 1990, in press; Tanaka, 1991; Tiezzi *et al.*, in press) demonstrate the presence of microfilaments, actin filaments, or microtubules during male meiosis and microsporogenesis. Some of these studies (e.g., Traas *et al.*, 1989) indicate that actin filaments and microtubules are jointly involved in the spatial coordination of the meiotic division process. Microtubules radiating from around the nuclei seem to exclude the plastids during microsporogenesis in *Lilium longiflorum* (Tanaka, 1991). From experiments with anticytoskeleton drugs, neither Sheldon and Dickinson (1986) nor Tiwari (1989) found firm evidence that the microtubular cytoskeleton determines the generation of cell wall patterns in *Lilium* and *Tradescantia*. A contradictory conclusion was drawn by Perez-Munoz and Webster (1990) in *Vigna vexillata,* on account of the spatial and temporal development of the cortical cytoskeleton. Sheldon and Dickinson (1986) proposed a model based on the self-assembly of patterning imprints within the plasma membrane, whereas Tiwari (1989) suggested that the pattern of exine formation is induced by microfilament–plasma membrane interactions. These concepts need to be further verified and evaluated in the context of current opinions on the significance of the cytoskeleton for cell wall deposition in plant cells (Seagull, 1989, 1991).

The generative cell or the sperm cells of all species investigated contain a prominent system of bundles of microtubules. Except for a few articles (Zhu *et al.*, 1980; Russell and Cass, 1981; Taylor *et al.*, 1989), the over-

whelming majority of the studies show the absence of actin filaments–microfilaments in the generative and in the sperm cell (Palevitz and Tiezzi, this volume).

II. Vegetative Cytoplasm of Pollen Grains and Pollen Tubes

A. Dry Pollen

The scarcity of ultrastructural studies dealing with dry pollen probably derives from the technical difficulties frequently encountered at the moment of fixation of this coarse, desiccated material. Elleman *et al.* (1987) found that the most satisfactory procedure for preserving the features of dry or partially hydrated pollen of *Brassica oleracea* was vapor fixation with osmium tetroxide, followed by postfixation in 2.5% glutaraldehyde and embedding in Epon 812. Lowering the temperature of the first fixative to less than 4°C had a positive effect on the preservation of the cytoplasm. Tiwari *et al.* (1990) and Cresti *et al.* (unpublished data) have obtained well-preserved ultrastructural features in dry pollen grains by applying the rapid freezing and the freeze substitution technique (RF–FS) described by Lancelle *et al.* (1986). Because this technique does not involve aqueous, chemical fixatives, rapid water uptake of the desiccated pollen grains is virtually excluded. The cytoplasm of dry *Pyrus communis* pollen (Tiwari *et al.*, 1990) is characterized by tightly packed multilamellate membranous proliferations and densely osmophilic bodies, whereas the cytoplasm of dry *Arabidopsis thaliana* pollen (Cresti *et al.*, unpublished data) appears to be rich in distinct endoplasmic reticulum (ER) lamellae that surround lipid globuli.

The cytoplasm of fresh pollen is compact, due to the low cell water content. As currently described, the vegetative cytoplasm of unhydrated pollen grains is characterized by an extensive system of rough ER, often organized in stacks, and the presence of storage elements, such as lipid droplets, protein-rich bodies, proplastids containing phytoferritin aggregates, and amyloplasts (Jensen *et al.*, 1974; Cresti *et al.*, 1975, 1977, 1984, 1985; Charzynska *et al.*, 1989; Shivanna and Johri, 1985; Cresti and Tiezzi, 1990). Furthermore, it contains a wide variety of organelles, such as dictyosomes, ribosomes, mitochondria, and plastids. The generative cell (in bicellular pollen grains) or the two sperm cells (in tricellular pollen grains) are usually located in the central part of the pollen grain. They are sometimes connected to the vegetative nucleus, forming the so-called male germ unit (Mogensen, this volume). The inner wall of the pollen grain, the intine, is constituted by a stratified layer, composed of a callosic matrix,

a cellulosic middle layer, and an outer coating of pectin (illustrated in Nakamura, 1979, for *Lilium longiflorum*; Heslop-Harrison, 1987). The outer wall of the pollen grain, the exine, consists of sculptured and resistant structures and contains sporopollenin (Heslop-Harrison, 1987; Wiermann and Gubatz, this volume).

B. Pollen Hydration, Activation, and Germination

Pollen germination and pollen tube growth were observed as early as 1824 by Amici. The process that leads to the formation of the pollen tube begins after pollination, when the pollen grain has reached the stigma; the pollen grain begins to rehydrate, leading to a considerable increase in the volume of the vegetative cell (Heslop-Harrison, 1979b, 1987; Ciampolini *et al.*, 1988; Pacini, 1990), as well as to a more modest enlargement of the nuclei (Wagner *et al.*, 1990). Complete hydration can take from a few minutes to hours, depending on the species. During the lag period between the beginning of pollen hydration and germination, the synthetic machinery of the cell is activated (Mascarenhas, 1975; Shivanna and Johri, 1985). Proteins and other substances held in the intine are released (Heslop-Harrison *et al.*, 1986b; Que and Tang, 1988). Release of partly *de novo* synthesized cytoplasmic proteins also continues during the first phase of germination in some species as, for example, in *Curcubita moschata* (Que and Tang, 1988). Rapid starch synthesis coincides with the first period of high respiration at the beginning of germination in *Lilium* (Dickinson, 1968).

Major morphological changes occur in the vegetative cytoplasm during pollen hydration, activation, and germination. These changes include the restoration of the bilayer organization of membrane lipids, the aggregation of ribosomes to form polysomes, the formation of lamellae and starch grains in the plastids, and the dilatation and scattering of the rough ER. Other developmental changes include the decline of the number of fibrillar bodies, a decrease in the number of nuclear pores, and an increase in secretory activity of the Golgi apparatus, resulting in the accumulation of vesicles near the germination pore (Southworth and Dickinson, 1981; Clarke and Steer, 1983; Cresti *et al.*, 1977, 1980, 1983a, 1983b, 1985, 1986b, 1988, 1990a; Miki-Hirosige and Nakamura, 1982; Lancelle *et al.*, 1986; Ciampolini *et al.*, 1988; Weber, 1989; Tiwari *et al.*, 1990; Wagner *et al.*, 1990). These changes are accompanied by the beginning of a rotational movement of the organelles in the grain (Iwanami, 1956; Venema and Koopmans, 1962). The modifications in the external morphology that occur at the moment of germination begin with the opening of the pore at one of the apertures of the grain. The intine protrudes through the exine, a kind of hinge is formed like a porthole door, and the pollen tube emerges. The

aperture is not enlarged during germination. In some species the emerging tube is therefore constricted where it emerges through the exine (Cresti *et al.*, 1977; Southworth and Dickinson, 1981; Clarke and Steer, 1983; Cresti and Keijzer, 1985; Heslop-Harrison *et al.*, 1986c; Tiwari *et al.*, 1990).

In many species it is possible to induce pollen activation and germination *in vitro* by allowing the pollen grains to imbibe water in a moist chamber (e.g., Gilissen, 1977) and by incubating them in a medium, usually containing boric acid, calcium nitrate, and an osmoticum, such as sucrose or polyethylene glycol (Shivanna and Rangaswamy, in press). In tobacco S. M. Read (personal communication) obtained improved pollen tube growth by enriching the medium with a buffer, a balanced concentration of Ca^{2+}, and a mixture of amino acids. Unless explicitly stated, the cytoskeletal data reported next will deal with pollen grains and tubes activated and grown *in vitro*.

C. Growing Pollen Tubes

Since the early articles by Rosen *et al.* (1964) and Sassen (1964) on the fine structure of pollen tubes of *Lilium* and *Petunia*, it has been known that these cells contain a wide variety of organelles (Fig. 1) but that their distribution is not uniform over the length of the tube. Four zones can be distinguished starting from the pollen tube tip toward the pollen grain (Rosen *et al.*, 1964; Sassen, 1964; Cresti *et al.*, 1976, 1977): (1) an apical zone populated by numerous vesicles, (2) a subapical zone rich in organelles, especially dictyosomes and mitochondria, (3) a nuclear zone, with the vegetative nucleus and the generative cell, and (4) a zone with large vacuoles and a thin layer of cortical cytoplasm, separated from the more apical part of the tube by callose plugs. This zonated structure is a common feature of the pollen tubes of many species, but it is not a universal property, because, for example, in grasses no clear zonation at the tip exists (Heslop-Harrison, 1979a; Heslop-Harrison and Heslop-Harrison, 1982).

Whereas much work has been performed on the structure and distribution of organelles in pollen tubes, rather little is known about the characteristics of the cytosol. Steer *et al.* (1984) made attempts to estimate the viscosity of the cytosol of living pollen tubes by using a laser Doppler microscope; they found highly variable values for the viscosity, ranging between 2 and 70 mPa, with a median value of about 50 mPa.

The polar organization of the cytoplasm reflects the unipolar growth of pollen tubes (early works: *Veronica* by Schoch-Bodmer, 1932; and *Lilium*, Rosen, 1961), which is exclusively restricted to the tip region, also called the cap-block or hyaline zone (Steer, 1990; Steer and Steer, 1989; Derksen

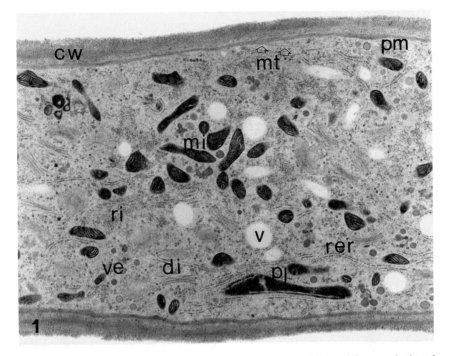

FIG. 1 Longitudinal section through the central zone of a cryofixed and freeze-substituted pollen tube of *Nicotiana tabacum,* in which the cell wall (cw), plasma membrane (pm), and various components of the vegetative cytoplasm can be seen: mitochondria (mi), dictyosomes (di), rough endoplasmic reticulum (rer), ribosomes (ri), a plastid (pl), microtubules (mt, open arrows), small vacuoles (v), and vesicular structures (ve). Micrograph courtesy of C. Milanesi and M. Cresti. ×11,400.

and Emons, 1990). Briefly, the dictyosomes produce secretory vesicles that display a net flow in the direction of the plasma membrane at the tip of the tube where growth takes place. The vesicles provide the new membrane and cell wall components necessary for extension by fusing with the plasma membrane. The growth of pollen tubes can be regarded as an example of high secretory activity (Steer, 1990; Steer and Steer, 1989; Derksen and Emons, 1990; Steer and O'Driscoll, 1991) because pollen tubes may grow with rates exceeding 2 mm/hr. Based on the calculation that much more membrane is delivered at the tip than is needed to support the formation of the new plasma membrane, some scientists posit that internalization and recycling of plasma membrane must occur in pollen tubes (Steer and O'Driscoll, 1991). The uptake of membrane may be realized through clathrin-coated vesicles and coated pits (Robinson and Depta, 1988; Robinson and Hillmer, 1990), which are conspicuous compo-

nents of the plasma membrane of pollen tubes (Derksen *et al.*, 1985; Pierson *et al.*, 1986a). The ER is considered to act as a reservoir of membrane material within the cell.

D. Calcium Gradients and Calcium—Cytoskeleton Interactions in Pollen Tubes

Calcium is widely recognized as an important cation for the regulation of various intracellular processes in plant cells, either directly or indirectly via calcium-dependent regulatory proteins, for example, calmodulin or protein kinases (Hepler and Wayne, 1985; Gilroy *et al.*, 1987; Hepler, 1988; Callaham and Hepler, 1991). Also, the organization of the zonated cytoplasm, cytoplasmic streaming, cell wall biogenesis, and the growth of pollen tubes are considered to be, at least in part, regulated by spatial and temporal variations in the level of free, cytosolic calcium (Tirlapur, 1988; Steer and Steer, 1989; Herth *et al.*, 1990).

The importance of calcium ions for pollen germination and pollen tube growth was demonstrated by Brewbaker and Kwack (1963). Jaffe *et al.* (1975) showed by ^{45}Ca analysis that pollen tube growth can proceed optimally only when a balance is maintained between uptake and utilization of calcium. Picton and Steer (1983a) demonstrated that pollen tube growth in *Tradescantia virginiana* is inhibited by both suboptimal ($<10^{-4} M$) and supraoptimal($>10^{-3} M$) Ca^{2+} concentrations in culture test solutions. The use of Ca^{2+} ionophores on growing pollen tubes of *Lilium longiflorum* induces drastic inhibition of pollen tube growth and a disturbance of the ultrastructural zonation at concentrations as low as $1.0 \times 10^{-7} M$ for A23187 (Herth, 1978; Reiss and Herth, 1979) or $5.0 \times 10^{-5} M$ for X-537A (Reiss and Herth, 1980). Moreover, the presence of ionophore in the medium (Herth, 1978; Kohno and Shimmen, 1987; Kohno and Shimmen, 1988a) and the elevation of the free cellular calcium level to values of $5–10 \times 10^{-6} M$ (Kohno and Shimmen, 1987; 1988a) causes cyclosis in pollen tubes to cease. The movement of organelles isolated from pollen tubes along actin filament bundles of tonoplast-free characean cells is inhibited by calcium concentrations above $4.5 \times 10^{-6} M$ (Kohno and Shimmen, 1988b). Because such an elevated calcium concentration also induces irreversible fragmentation of the actin cytoskeleton of the pollen tube (Kohno and Shimmen, 1987, 1988a), it seems reasonable to explain the cessation of movement as being due to actin disruption. However, because the movement of myosin-coated beads remains unaffected by pollen tube homogenates containing such a concentration of calcium, it is also plausible that myosin inactivation is a major reason for the calcium effect on organelle movement *in vitro* (Kohno and Shimmen, 1988a). This explana-

tion is compatible with the presence of a myosin coating on the surface of pollen tube organelles (Section IV,C,1) and with the report that myosin is inactivated at a critical calcium concentration of $1.0 \times 10^{-5} M$ (Kohno and Shimmen, 1988a, 1988b). The organization of the cytoplasm is sensitive to ruthenium red, a compound that may elevate endogenous calcium levels, but it is not completely clear whether the effects are related to inhibition of free Ca^{2+} uptake by mitochondria, as presumed by some authors (Picton and Steer, 1985; Bednarska, 1989).

Interestingly, the existence of a gradient of free calcium was demonstrated along the pollen tube of *Lilium longiflorum* and *Vinca rosea* by addition of 50 μM acetomethyl (AM) esters of Quin-2 to the culture medium (Reiss and Nobiling, 1986; Tirlapur, 1987). Confocal laser scanning microscopy (CLSM) of 2 μM Fluo-3 (Knebel and Reiss, 1990) revealed greater fluorescence in the subapical and cortical regions than in the central part of the tube. Nobiling and Reiss (1987) also estimated the free calcium levels in the cell by measuring the fluorescence intensity and globally correcting these values for cell thickness. These adjusted values ranged from about 100 nM Ca^{2+} in the area just behind the tip and 90 nM Ca^{2+} near the apical dome to about 20 nM Ca^{2+} in the area behind the apex (Nobiling and Reiss, 1987). Similar results were obtained using the dual-wavelength method after addition of Fura-2 AM to the medium (Herth *et al.*, 1990). This approach is preferable to single-wavelength methods because it circumvents the possible artifacts caused by variability in cell shape and cytosol volume. However, plant cells seem to be particularly resistant to loading with Fura-2 AM. Long incubation times, such as those in the works reported by Herth *et al.* (1990), indicate that the probe tends to be sequestered in noncytoplasmic compartments, especially in vacuoles and in the ER, leading to an underestimation of the cytosolic calcium concentration (Haughland, 1989; Callaham and Hepler, 1991). Miller *et al.* (see note added page 116) more recently reinvestigated the problem using Fura-2 coupled to a 10-kDa dextran molecule that they then microinjected into growing pollen tubes of *Lilium*. The dextran conjugate prevents compartmentalization of the Fura-2 complex. Miller *et al.* measured a steeper and more apically focused gradient than Nobiling and Reiss (1987); they found a maximum of 420 nM Ca^{2+} along the edge of the plasma membrane in the extreme tip of the pollen tube and a basic value of about 170 nM Ca^{2+} at a distance of 20 μm away from the tip.

A tip-to-base gradient of membrane-associated calcium was shown in growing pollen tubes of *Lilium* (Reiss and Herth, 1978; Reiss *et al.*, 1985a; Nobiling and Reiss, 1987), *Najas marina* (Reiss and McConchie, 1988), *Vinca rosea* (Tirlapur, 1987), and *Nicotiana* (Tirlapur, unpublished) by chlorotetracycline (CTC) fluorescence (final concentration about $1 \times 10^{-4} M$), an indicator for membrane-bound calcium, including organelle-

stored calcium. However, in pollen tubes of *Zea mays* and *Pennisetum americanum*, no tapering of calcium distribution was found along the tube, either by CTC staining or by energy-dispersive x-ray analysis (Heslop-Harrison *et al.*, 1985). It may be that these differences between the two *Gramineae*, *Zea* and *Pennisetum*, and *Lilium*, *Vinca* and *Nicotiana* are due to dissimilarities in apical organization and growth physiology of the pollen tubes (Heslop-Harrison and Heslop-Harrison, 1982). Prior to germination a bright CTC fluorescence was revealed near the apertures of the pollen grain in *Nicotiana tabacum* by video microscopy and digital analysis of video images (U. K. Tirlapur, unpublished data).

The total amount of cellular calcium in pollen tubes mainly depends on the amount of membrane-bound calcium. A maximum was demonstrated in the inner tip of young pollen tubes of *Vinca rosea* grown in the style using the antimonate precipitation technique (Tirlapur and Shiggaon, 1987) and in the subapical zone of pollen tubes of *Lilium longiflorum* using the proton-induced x-ray emission (PIXE) technique (Reiss *et al.*, 1983, 1985b). The resolution obtained by this technique is poor, however, and the primary data need to be corrected for cell shape and cell density.

The influx of Ca^{2+} from the extracellular milieu into the pollen tube has been demonstrated by Jaffe *et al.* (1975). Bednarska (1991) showed that the germinating pollen grains of *Primula officinalis* and *Ruscus aculatus* take up calcium from the stigma. However, so far only indirect evidence exists on the nature of the transients. In this regard, the possibility of calcium channels was tested by studying the effects of calcium channel blockers, such as nifedipine (Reiss and Herth, 1985) and verapamil (Bednarska, 1989), in cultures of *Lilium*, *Haemanthus*, and *Oenothera* pollen tubes. Both pharmacological agents affected pollen germination and pollen tube growth; nifedipine (optimal at $1.0 \times 10^{-5} M$) affected pollen tubes more drastically than verapamil ($1.0 \times 10^{-4} M$ for 2.0–2.5 hours), suggesting that calcium channels are indeed functional in pollen tubes.

The Ca^{2+}-binding protein calmodulin (CaM) has been indirectly demonstrated in pollen tubes based on fluorescence of the CaM-binding compound phenothiazine (Hausser *et al.*, 1984) and by observing effects of the CaM inhibitors trifluoperazine and chlorpromazine (Polito, 1983; Picton and Steer, 1985) on pollen tube growth. The findings of Picton and Steer (1985) on the pollen of *Tradescantia* indicate that calmodulin is not directly involved in the process of vesicle fusion and tip extension. Steer (1990) is inclined to assume that the actin cytoskeleton acts as a mediator for the calcium message. Further studies using more specific CaM antagonists that bind to calmodulin and inhibit Ca^{2+} calmodulin-regulated enzyme activities, such as W-7 and W-5 (Seikagaku Kogyo Co., Tokyo), need to be carried out to elucidate the possible role of this protein during pollen germination and tube growth.

Recently, a calcium-dependent but calmodulin- and phospholipid-independent protein kinase (CDPK) with a molecular weight around 51–55 kDa has been shown to co-localize with rhodamine–phalloidin-stained actin filaments in pollen tubes of *Tradescantia* (Putnam-Evans *et al.*, 1989). Antibodies against the CDPK from soybean do not recognize actin from rabbit muscle or *Dictyostelium*, but gizzard myosin light chain and a synthetic myosin peptide can be phosphorylated in an Ca^{2+}-dependent manner by this protein kinase (Putnam-Evans *et al.*, 1989, 1990).

E. Ionic Compounds and Intracellular pH in Pollen Tubes

Using vibrating microelectrodes, Weisenseel *et al.* (1975) were able to measure an electric current entering living pollen tubes and leaving from the side of the grains, which they considered to be related to a flow of positive ions. Little is known about the distribution and significance of ionic compounds other than calcium in growing pollen tubes, although low concentrations of some elements, such as boron (Vasil, 1964; Fähnrich, 1964; Peter and Stanley, 1974) and copper (S. M. Read, unpublished data), may be of essential importance for normal growth. Semiquantitative techniques have enabled rough estimations of the distribution of phosphorus, sulphur, carbon, chlorine, magnesium, oxygen, manganese, potassium, iron, copper, and zinc in pollen tubes (Ender *et al.*, 1983; Reiss *et al.*, 1985b; Heslop-Harrison *et al.*, 1985), germinated pollen grains (Baldi *et al.*, 1987), and isolated sperm cells (Russell, 1990).

The importance of intracellular pH for the regulation of many enzyme activities, including ATPase activity of molecules such as myosin for the stability of the cytoskeleton and for many other metabolic aspects is beyond doubt. The first evaluations of this parameter in pollen tubes were acquired from staining with the pH indicators bromocresol purple, bromocresol green, bromocresol blue, Congo red, and 4-methylesculetin on germinated pollen of *Narcissus pseudonarcissus* and *Hippeastrum vittatum* (Turian, 1981). The results indicated that the most acidic region is located in the tip zone occupied by Golgi vesicles (pH 4.5–5.0). The subapical zone, rich in mitochondria, has a pH of about 6.0–6.5. Preliminary observations obtained by CLSM of *Nicotiana* pollen tubes loaded with 2′,7′-bis-(2-carboxyethyl)-5(6)-carboxyfluorescein (principles of pH measurements with BCECF in Dixon *et al.*, 1989) confirm the presence of a pH gradient along the tube with a minimum in the tip region (U. K. Tirlapur, personal communication). The low pH in the tip of the pollen tube could be a reason for the nonvectorial behavior of the vesicles present here, compared with other organelles showing distinct cyclosis.

F. Movements of Organelles and Nuclei

Intracellular movement of organelles and particles, often referred to as cytoplasmic streaming or flow (Shimmen, 1988; Kuroda, 1990), is one of the most fascinating features of pollen grains and pollen tubes. Iwanami (1956) described this phenomenon in detail in *Lilium auratum*. He described the first sign of cytoplasmic movement after hydration of the pollen grain as agitation and the following movement as circulation and rotation. In early stages of pollen tube growth the protoplasm was seen to stream toward the tip along the wall of the tube and to turn back from near the so-called cap-block to the center of the tube (reverse fountain streaming pattern). In another pioneering work on *Tradescantia virginiana* (Venema and Koopmans, 1962), the formation of a small vacuole was observed in the side of the grain opposite the germination pore, just before the pollen tube emerged. In the first quarter of an hour after germination, the growing pollen tube was filled with relatively slowly rotating cytoplasm of a fine texture, which contained many mitochondria. Later, large cytoplasmic inclusions, such as starch grains, oil droplets, and carotenoid crystals, also participated in the streaming in the pollen tube. Similar observations were made in *Tradescantia bracteata* by Lewandowska and Charzynska (1977).

The introduction of video-enhanced contrast (VEC) light microscopy techniques and digital image improvement has made it possible to visualize and track single organelles. Two major observations arise from these video microscopy investigations made on pollen tubes of *Secale cereale* (Heslop-Harrison and Heslop-Harrison, 1987), *Epilobium angustifolium* (Heslop-Harrison and Heslop-Harrison, 1990), *Lilium longiflorum* (Herth, 1989; Pierson, 1989; Pierson *et al.*, 1990), *Ornithogalum virens,* and *Nicotiana tabacum* (Strömgren-Allen, Cresti, Derksen, Lichtscheidl, Pierson, *et al.*, unpublished results). First, the dense population of small particles in the apex of the tube, evidently corresponding to the Golgi vesicles or polysaccharide particles, shows intense activity, resembling random vibrations. The net flux of vesicles to the plasma membrane is thus a turbulent process. Second, the organelles in the pollen tube move independently. The representation of a passive bulk flow or streaming of cytoplasm does not fit well with these observations.

Whereas most large organelles are excluded from the tip, the small vesicles tend to remain in the apex, where they are often distributed in the form of a cone (example, *Ornithogalum virens*). In some species, such as *Amaryllis* and *Lilium longiflorum,* some vesicles seem to recoil basipetally and join the track of cytoplasmic components in the central part of the tube. In the median part of the pollen tube, the organelles move in cytoplasmic lanes. The vacuoles are also mobile and show continuous reshaping, suggesting high plasticity of the tonoplast (Pierson *et al.*, 1990). Mitochondria appear as flexible rods with rounded ends. In pollen tubes

of *Amaryllis* (Strömgren-Allen, 1983), *Iris pseudacorus, Iris xiphioides* (Heslop-Harrison and Heslop-Harrison, 1988a, 1989a), and *Lilium longiflorum* (Pierson *et al.*, 1990) vectorial movement of organelles, especially in the older stretches, was shown to be associated with straight fibrils, which possibly represent bundles of cytoskeletal elements. The movement of organelles in the pollen tube can be reversibly inhibited by cytochalasins (Franke *et al.*, 1972; Heslop-Harrison and Heslop-Harrison, 1989a; Section V). In the thin layer of cortical cytoplasm in the oldest part of the pollen tube of *Lilium longiflorum*, networks of undulating tubular structures, probably tubular cisternae of ER, have been detected by VEC and ultraviolet microscopy, and using low light video imaging after staining with a 5 μg/ml concentration of 3,3'-dihexyloxacarbocyanine iodide (DiOC$_6$; Pierson *et al.*, 1990). These networks are highly polymorphic. Their configuration varies between isodiametric polygons and parallel arrays. The tubules occasionally fuse with each other or separate, suggesting membrane fluency. Organelles preferably move along the putative ER tubules in a saltatory manner.

Inside the pollen grain, the vegetative nucleus is usually elongated and the generative cell or sperm cells are spindle-shaped. Displacement of the vegetative nucleus and the generative cell is accompanied by a rotational movement in germinating pollen of *Clivia nobilis* (Tang, 1973, 1988). Discordant results have been obtained on the question of which nucleus enters the pollen tube first. As a rule, in *Lilium* (Maheshwari, 1949) and *Tradescantia paludosa* (Hesemann, 1972), the generative cell enters the pollen grain first, followed by the vegetative nucleus, but variations have also been reported (Poddubnaja-Arnoldi, 1936). The opposite was observed by Tang (1973) in *Clivia nobilis*. Tang (1973, 1988) also observed that the vegetative nucleus and the generative cell became closely associated with each other for a period of 2–3 hours in the pollen tube but that they separate before generative cell division is completed. From a statistical study of 4',6-diamidino-2-phenylindole (DAPI) stained nuclei in pollen tubes of *Alopecurus pratensis,* Heslop-Harrison and Heslop-Harrison (1984) concluded that the type of nucleus leading through the tube was a matter of chance. Real-time analysis demonstrated that the male gametes in actively growing pollen tubes of *Secale cereale* move independently of each other (Heslop-Harrison and Heslop-Harrison, 1987), but results from tubulin staining suggest that the gametes nevertheless remain physically linked to each other during their passage through the tube (Heslop-Harrison and Heslop-Harrison, 1988c). The streamlined shape of the generative cell–sperm cells and vegetative nucleus is even more evident in the pollen tube than in the pollen grain. This change in form certainly facilitates the passage through the narrow pollen tube. In *Galanthus nivalis* (Steffen, 1953), *Amaryllis* sp. (Polunia and Sveshnikov,

1963), *Clivia nobilis* (Tang, 1973), and *Secale cereale* (Heslop-Harrison and Heslop-Harrison, 1987), it was demonstrated by cinematographic or video records that the vegetative nucleus and the sperm cells progress with discontinuous velocity through the pollen tube, sometimes displaying short basipetal migration. The generative cell is remarkably static compared with the surrounding vegetative cytoplasm (Tang, 1973; Pierson *et al.*, 1990). Strong distortion of the shape of the nuclei (Venema and Koopmans, 1962; Tang, 1973, 1988; Heslop-Harrison and Heslop-Harrison, 1984, 1989c; Heslop-Harrison *et al.*, 1986a, 1988) and re-arrangement of the microtubular cytoskeleton of the generative cell (Derksen *et al.*, 1985; Heslop-Harrison *et al.*, 1988) occurs when one nucleus passes the other and during generative cell mitosis and cytokinesis (reviewed by Palevitz and Tiezzi, this volume). Steffen (1953) described how, in a vibration-free setup, a generative cell continued to show ameboid movements after isolation from pollen tubes of *Galanthus* in an isotonic medium. The validity of this single observation may be questioned because none of the many subsequent works published on isolated generative cells (GC) and sperm cells (SC) confirms this finding (Theunis *et al.*, 1991; Palevitz and Tiezzi, this volume; Chaboud and Perez, this volume). Soon after isolation from the pollen grain or tube, GCs or SCs in general become spherical. This does not automatically imply that viability (Theunis *et al.*, 1991) or the potency to divide is lost (Zee and Aziz-Un-Nisa, 1991).

III. Cytoskeletal Proteins and Methods of Visualization

A. Actin, Actin Filaments, and Microfilaments

1. Structure of Actin Molecules and Actin Filaments

The globular form of actin (G-actin) is composed of a single polypeptide chain of about 375 amino acids. The molecular weight of animal actin is about 42–45 kDa and its isoelectric point is about 5.4. Plant proteins with molecular weights of 41, 45, and 58 kDa have been reported to react with antiactin on Western blots (Staiger and Schliwa, 1987; Tang *et al.*, 1988, 1989a). By x-ray diffraction of actin–bovine DNase I complexes and by electron microscopy of crystalline actin sheets obtained by the addition of gadolinium, it has been shown that the actin molecule has a wedge-shaped structure with two asymmetric lobes. The dimensions of the molecule are approximately $5.6 \times 3.3 \times 4.0$ nm (Kristen, 1987).

Actin filaments (F-actin) consist of identical actin monomers (G-actin) with polar alignment. At high microscopic magnification the actin filaments

appear as helical structures, which can be described either as a single strand helix or as a double-stranded left-handed helix. *In vitro* polymerization of G-actin to F-actin can be induced by the addition of ionic buffers (containing, e.g., 1 mM Mg^{2+}, 1 mM Ca^{2+}, or 100 mM K$^+$) to a solution of monomeric actin. Analysis of the time course of actin polymerization suggests that the assembly process occurs in two stages: a slow nucleation step and a more rapid elongation phase requiring the hydrolysis of ATP to ADP. At present, three models exist for the assembly and disassembly equilibria between polymerized and unpolymerized actin: true equilibrium, dynamic instability, and treadmilling (Pollard and Cooper, 1986; Bershadsky and Vasiliev, 1988). Actin filaments, often referred to as microfilaments in the literature, can be visualized by x-ray diffraction (apparent width of microfilament 9–10 nm) or transmission electron microscopy of sections and replicas (apparent diameter about 7 nm).

2. Biochemical Characteristics of Pollen Actins

In 1974 John Condeelis identified actin filaments in pollen tubes of *Amaryllis* based on heavy meromyosin binding and arrowhead formation. In 1986 a group from Beijing Agricultural University published a study, in which the presence of actin in pollen of *Brassica pekinensis* and *Nicotiana tabacum* was postulated on account of a similar migration of pollen proteins and rabbit muscle actin in SDS-PAGE (Yen *et al.*, 1986). More persuasive biochemical support for the presence of actin in extracts of pollen was provided by Tang *et al.* (1989b). They separated soluble protein extracts of *Nicotiana* pollen by electrophoresis, according to the method of Laemmli (1970), and blotted the products with a monoclonal antibody to chicken gizzard actin (IgM, Tang *et al.*, 1989b). The antiactin antibody recognized a polypeptide of 45 kDa, which co-migrated with a muscle actin standard, a weakly labeled 51 kDa unit, and three diffuse bands of less than 45 kDa, possibly representing proteolytic fragments of actin (Tang *et al.*, 1989b). The same year, in pollen of *Zea mays*, Ma and Yen (1989) described a protein with the same molecular weight as rabbit skeletal muscle actin, 42 kDa, which cross-reacted with the serum of rabbit against chicken gizzard actin. This corn pollen protein had the same C-terminal amino sequence (Lys-Cys-Phe) as the actin from rabbit muscle (Ma and Yen, 1989), and the profile of its secondary structure showed many common points with that of muscle actin. The authors concluded that the protein from corn pollen was an actin that perhaps shares common ancestors with the actin from rabbit muscle. Yen *et al.* (in press) applied a sequence of protein isolation and purification techniques, such as acetone precipitation, ammonium sulfate fractionation, DEAE-cellulose chromatography, rounds of actin polymerization and depolymerization, and S-200 gel filtra-

tion, and succeeded to obtain 6.5-mg portions of purified putative actin from 30 g of *Zea mays* pollen. They analyzed the biochemical characteristics of this final product and found a molecular weight of 43 kDa, a strong enhancement of muscle myosin ATPase activity in the presence of the protein, and the capacity to polymerize into filaments of 7 nm diameter in the presence of KCl and $MgCl_2$, resulting in increased ultraviolet absorbance and increased viscosity. Recently, Åström *et al.* (1991) localized one strong band in the 43–45 molecular weight region, using antiactin in Western blots of extracts from *Nicotiana tabacum* pollen tubes grown *in vitro*. Åström *et al.* (1991) also grew pollen tubes under various conditions of culture (3–5 hours or 12–14 hours at constant temperature of 22°C or 2 hours at 4°C after 3 or 12 hours of preculture at 22°C); in all treatments they found comparable levels of bulk actin polypeptide. Staining with fluorescein-conjugated phalloidin and NBD (4-chloro-7-nitrobenz-2-oxa-1,3-diazole)–phallacidin confirmed that the actin filaments, unlike microtubules (Section III,B,2), were stable to cold treatment.

From a survey of the genes encoding for actin of fungal, plant, insect, and mammalian actins (Hightower and Meagher, 1986), it appears that globally two-thirds of the actin molecule has been conserved across these lines of evolution. Northern blot analysis of hybridization of total RNA from developing *Tradescantia paludosa* pollen with soybean actin clone *psAC3* revealed that the actin mRNA is first seen during microspore interphase prior to microspore mitosis. The actin mRNA gradually accumulates until the formation of the generative cell is concluded and suddenly declines when the pollen reaches maturation (Stinson *et al.*, 1987). Similar observations have been made in *Brassica napus* (D. Albani, personal communication).

3. Methods to Visualize Actin Filaments and Microfilaments in Pollen

Myosin decoration allows the location, polarity, and actin composition of microfilaments to be established. In the absence of ATP, so-called rigor complexes in the shape of arrowheads may be formed by the attachment of myosin molecules, or their S1 or HMM fragments (Section III,C,1), to actin filaments. The periodicity of these arrowheads ranges between 34 and 40 nm. Their orientation makes it possible to distinguish the two ends of the actin filament, known as the pointed and the barbed end. A serious disadvantage of the myosin decoration method is that it requires severe pretreatment of the material, leading to partial degradation of the original cellular structure and possibly rearrangement of the actin filaments.

Glutaraldehyde fixation, osmium tetroxide, lead citrate, uranyl acetate, and hafnium tetrachloride provide deep fixation of the material and sharp

contrast. However, most of these chemicals significantly disrupt the fine cytoskeleton system, especially actin filaments (Maupin-Szamier and Pollard, 1978; Lehrer, 1981). The damage can be only partly reduced by pretreatment with 0.2 mg/ml tropomyosin (Kakimoto and Shibaoka, 1987) or 100 μM m-maleidobenzoyl N-hydroxysuccinimide ester (MBS; Sonobe and Shibaoka, 1989), and the addition of tannic acid to the buffers (Seagull and Heath, 1979), which may in turn produce other artifacts in the form of precipitation.

A real advance in this field was made with the introduction of the rapid freeze immobilization and freeze substitution technique (RF–FS, for plant cells: Tiwari et al., 1984; Lancelle et al., 1986; Robards and Sleytr, 1985), which allows excellent preservation of fine structure, especially membranes and cytoskeletal elements. Although this technique provides the best preservation of microfilaments so far, it, too, is not without problems. Ice crystals that continue to form during substitution at -80°C may disrupt fine structure, including single actin microfilaments (Steponkus, unpublished). Another cold technique, rapid-freezing and deep-etching, has permitted the examination of the fine structure of actin bundles in *Chara* (McLean and Juniper, 1988), but it has not yet been applied to pollen.

The choice of resins with low viscosity, for example, Epon-Araldite (Picton and Steer, 1981), vinyl cyclohexene dioxide (ERL-4206, introduced by Spurr, 1969), and Transmit resin (Heslop-Harrison et al., 1991) offers satisfactory penetration in pollen grains and tubes, but these epoxy resins are not well suited for antibody labeling (Tiezzi, 1991; Shivanna and Rangaswamy, in press). Immunogold antiactin labeling of pollen tubes has been successfully combined with RF–FS after inclusion in London white resin polymerized by heat or longwave ultraviolet light (Lancelle and Hepler, 1989, Tang et al., 1989b).

Actin filaments have also been shown by light microscopy in wholemount preparations. The use of phallotoxins labeled with the fluorophores rhodamine or 4-chloro-7-nitrobenz-2-oxa-1,3-diazole (NBD) has been most popular (molecular weight fluorescent phallotoxins less than 1200 daltons; Haughland, 1989). Phallotoxins are cyclic oligopeptides from the poisonous mushroom *Amanita phalloides* that tightly bind to actin filaments. The major representatives of this group are phallacidin and phalloidin (Wieland and Faulstich, 1978; Cooper, 1987; Tewinkel et al., 1989). Another approach is indirect immunofluorescence microscopy (Dewey et al., 1991). It has been applied to pollen using a monoclonal antibody from Amersham raised against chicken gizzard actin (Taylor et al., 1989; Tang et al., 1989b; Åström et al., 1991). This type of general labeling of F-actin has the great advantage of allowing the actin cytoskeleton to be observed in three dimensions, both alone and in combination with other specific labels, for example, for microtubules and/or nuclei.

Pollen is rather impermeable to large molecules (Heslop-Harrison and Heslop-Harrison, 1988b). To let antibodies penetrate into pollen grains and pollen tubes, drastic preliminary measures are required, for example, strong enzyme degradation of the cell wall or mechanical removal of the wall by scratching air-dried cells (e.g. dry-cleaving, Derksen *et al.*, 1985). The standard procedures further include either fixation with paraformaldehyde (protocol for pollen, Parthasarathy, 1987) or glutaraldehyde (Kohno and Shimmen, 1988a), or permeabilization of the cell with detergents (0.01% saponin, Kohno and Shimmen, 1988a; 1% Triton X-100, Tiwari and Polito, 1988b) or dimethylsulfoxide (up to 5% DMSO; original method described by Traas *et al.*, 1987, used for pollen by Pierson, 1988; Heslop-Harrison and Heslop-Harrison, 1989c, 1991a, 1991b; Zhu *et al.*, 1991). Like glutaraldehyde (Lehrer, 1981), formaldehyde has a harmful effect on actin filaments. Detergents and DMSO may induce preferential stabilization of certain cellular components, including cytoskeletal elements (Mesland and Spiele, 1984; Schroeder *et al.*, 1985; Raudaskoski *et al.*, 1986; Tewinkel *et al.*, 1989). Microwave-accelerated DMSO permeabilization has been judged to be favorable in some cases, but the technique needs to be standardized (Heslop-Harrison and Heslop-Harrison, 1991a). Another limitation of these procedures is that they involve repeated washing steps, during which important components of the cytoplasm may be extracted.

Alternative methods have therefore been sought for the insertion of marked phallotoxins, actin analogs, or antibodies inside the cell. Microinjection (Jokusch *et al.*, 1986) of diluted rhodamine–phalloidin is an elegant solution for labeling the actin cytoskeleton in living plant cells (Schmit and Lambert, 1990; Hepler and Zhang, unpublished), but is a not easily applicable to turgescent, cell wall-bearing cells. For optimal imaging of the weak labeling signal, it is advisable to use powerful light microscope detection techniques, for example, CLSM (Shotton, 1989; Pawley, 1989) and low light video microscopy (Inoué, 1986; Weiss, 1986), for the detection of fluorescent labeling, and video-enhanced contrast microscopy (Inoué, 1986; Weiss, 1986) for silver-enhanced gold labeling. Under favorable conditions, VEC phase-contrast differential interference contrast, and ultraviolet microscopy can provide enough contrast and resolution for direct visualization of bundles of actin filaments in living plant cells (Lichtscheidl and Url, 1987; Lichtscheidl and Weiss, 1988).

B. Tubulin and Microtubules

1. Structure of Tubulin Molecules and Microtubules

Microtubules consist of 13 polar protofilaments arranged in circular array. The basic components of protofilaments are dimers of α- and β-tubulin,

two evolutionarily related proteins. Both types of tubulin have an apparent molecular weight of around 55 kDa in SDS-PAGE and around 50 kDa when determined from amino acid sequences (about 450 amino acids; Gunning and Hardham, 1982; Kristen, 1986; Bershadsky and Vasiliev, 1988; Fosket, 1989; Warner and McIntosh, 1989; Warner et al., 1989). Multiple tubulin isoforms have also been reported in higher plants (Morejohn and Fosket, 1986; Fosket, 1989). Cross reactions between antibodies raised against animal tubulins and plant polypeptides of similar molecular weight have been frequently observed, indicating homology of antigenic sites between animal and plant tubulins. Tubulins and microtubules from plants are often more resistant to cold than their analogs from animal sources (Seagull, 1989; Fosket, 1989). They also show different sensitivity to specific inhibitors, particularly to colchicine (Section V).

X-ray diffraction indicates values of 14 nm for the internal diameter of microtubules and 30 nm for the external diameter, whereas estimations from electron-microscopy studies give measurements of 19 and 27 nm, respectively. Their length, as established by electron microscopy of dry-cleaved preparations of the cortical cytoplasm from plant cells, appears to exceed 10 μm (e.g., Kengen and Derksen, 1991). Microtubules are dynamic elements that undergo cyclic rounds of assembly and disassembly. Mg^{2+} and GTP are necessary for assembly in vitro. Two scenarios have been proposed for the reorganization of the microtubular cytoskeleton in the cell: disassembly followed by de novo assembly and/or reorientation of stable elements. The best studied proteins associated with cytoplasmic microtubules (MAPs) are those from the brain. They have been classified in the main groups: MAP 1 (300–350 kDa), MAP 2 (270–285kDa), and proteins from the tau group (about 60 kDa). Advances in plants have been made by the biochemical identification of MAPs in carrot suspension cells (Cyr and Palevitz, 1989), and investigations on the binding capacity of pollen tubulin to MAPs are in progress (Moscatelli et al., unpublished data).

2. Biochemical Characteristics of Pollen Tubulins

Raudaskoski et al. (1987) separated polypeptides from extracts of 17-hour-old pollen tubes of Nicotiana tabacum by SDS-PAGE, according to the method of Laemmli (1970) and blotted the products on nitrocellulose filters with monoclonal antibodies directed against chick brain tubulins. They obtained a distinct single band that cross-reacted in the region of 52–54 kDa with the anti-α-tubulin antibody, and another strong single band around 54–56 kDa in reaction with the anti-β-tubulin antibody. The α-subunit of pollen tube extracts migrated slightly faster than the β-subunit, which is in reverse order with respect to that found for polypeptides of

mouse fibroblasts in the same gels. Similar levels of α- and β-tubulin were found on Western blots in pollen tubes continuously grown at a temperature of 22°C or transferred to a cold (4°C) chamber, but cold treatment has a clear depolymerizing effect on the microtubules (Åström et al., 1991).

Complete maps of the sequences for tubulin genes have been published for only a few species of flowering plants: *Arabidopsis thaliana, Glycine max,* and *Zea mays* (Fosket, 1989). Ludwig *et al.* (1987, 1988) could determine the primary structure of the α-tubulin genes of *Arabidopsis thaliana* and show that it has strong homology to the α-tubulin of animals and protists. The α3-tubulin gene transcript is present in flowers of all stages of development, whereas the α1-tubulin gene is preferentially expressed in unopened flowers with pollen, open flowers, and flowers with elongating carpels (Ludwig *et al.*, 1988; R. H. Goddard and S. M. Wick, personal communication). The predicted amino acid sequence of both genes differs only in two amino acids (Ludwig *et al.*, 1988). Goddard *et al.* (personal communication) isolated and sequenced many tubulin clones from cDNA libraries of *Zea mays* and *Arabidopsis*. They raised antibodies in chicken to several α- and β-tubulin isotypes from these plant species and found that all of their antibodies reacted with endogenous plant tubulin on protein blots and were capable of labeling microtubule arrays in fixed cells. In *Zea mays* 2D Western blots indicate that different proportions of tubulin isoforms are found in pollen than in some of the other tissues and organs analyzed.

3. Methods to Visualize Microtubules in Pollen

Because microtubules have a characteristic structure and size, they can be relatively easily recognized by transmission electron microscopy (TEM) of thin sections or dry-cleaved preparations (Traas, 1984). Dry-cleaving has been successfully combined with immunogold labeling (Beesley, 1989) for incontestable identification of microtubules in plant cells (Traas and Kengen, 1986). Antibodies against brain or yeast tubulin have been most frequently employed for immunolabeling of plant microtubules (Wick *et al.*, 1981; Kilmartin *et al.*, 1982), including those of male gametophytic cells. The secondary antibody is usually a rhodamine or fluorescein conjugate, but occasionally protein-A gold has been used (Clapham and Östergren, 1984). Standard immunofluorescence labeling permits visualization of the general microtubular organization in the cell, as well as fine bundles of microtubules as, for example, those in the cortical cytoplasm of pollen tubes. The critical remarks on the preparation procedures used for the visualization of actin filaments are also valid for microtubules, except that microtubules seem to be a little more resistant than microfilaments to

conventional chemical fixation. Exposure of growing pollen tubes of *Nicotiana tabacum* to 1% DMSO or 1 μg/ml taxol, an alkaloid extracted from *Taxus brevifolia* (Bershadsky and Vasiliev, 1988), has been judged to improve the visualization of microtubules in the pollen tube and the generative cell (Raudaskoski *et al.*, 1986). This observation contrasts with the conclusions drawn by De Brabander *et al.* (1986) in their review on taxol. These authors mentioned that taxol certainly induces the assembly of aberrant microtubules *in vitro*, but that paradoxically it often causes depolymerization of existing, organized microtubules *in vivo* (De Brabander *et al.*, 1986).

Although we recognize that the conventional methods for the visualization of the microtubular skeleton are harsh, it is true that the main results obtained by different approaches agree with each other and that these methods have largely contributed to the first descriptions of the organization of the cytoskeleton. We are convinced that in the near future more direct, *in vivo* approaches enabling the study of the dynamics of microtubules will be widely applied, for example, microinjection of fluorescent tubulin and tracing its incorporation in the living cell (Zhang *et al.*, 1990), or microinjection of an immunologically distinct, exogenous tubulin into the cell, which, after incorporation in microtubules, can be discriminated from the native pool of tubulin by specific antibodies (Vantard *et al.*, 1990).

C. Myosin, Kinesin, and Profilin

1. Structure of Myosin and Kinesin Molecules

a. Myosin Myosins belong to a rather heterogenous family, but they all bind to actin and have ATPase activity stimulated by actin binding. In the presence of ATP, force may be generated by cyclic associations and dissociations between actin filaments and myosin (Bershadsky and Vasiliev, 1988).

Myosin II is typical for striated muscles. Simplified, the molecule consists of one long tail and two pear-shaped heads in which ATPase activity is located. The tail can occur in an extended or looped configuration. There are two points of flexion, also called hinges: one in the tail and the other between the tail and the heads. The structural units of the myosin II molecule are as follows: two coiled light meromyosin chains (LMM) in the distal part of the tail, two other coiled chains (S2) in the proximal part of the tail, two pear-shaped heads each formed by a S1 unit, and one pair of light chains. The total molecular weight of a myosin II molecule is about 480 kDa (Bershadsky and Vasiliev, 1988; Cooke, 1989).

A myosin I and myosin-like 110-kDa complex were first identified biochemically from *Acanthamoeba* and *Dictyostelium* and chicken intestine brush border. These myosins are rather globular proteins without much tail. An interesting property attributed to some of the myosin I proteins is the capacity to bind to calmodulin. Even more exciting is the observation that myosin I proteins directly associate to the membranes of organelles or to the plasma membrane in a manner that would allow them to promote movement (Adams and Pollard, 1989).

In vitro reconstituted movement of myosin-coated beads along actin filaments shows a migration of the beads from the pointed end to the barbed end. Actin-myosin–generated movement can be specifically inhibited when head fragments of the molecule are modified by *N*-ethylmaleimide (NEM), which induces the formation of rigor complexes. Modification of -SH groups of myosin with *p*-chloromercuribenzoic acid decreases the dissociation constant of myosin from actin filaments in the presence of ATP, again resulting in a rigid actomyosin complex (Shibata-Sekiya and Tonomura, 1975).

b. Kinesin   According to Sheetz (1989), kinesins are microtubule-dependent motors that move anionic beads toward the plus end of microtubules and bind strongly to microtubules. Kinesins have a native molecular weight of 300–400 kDa. The consensus is that kinesin is a dimer consisting of heavy chains (90–135 kDa) and more variable light chains (55–80 kDa). In platinum replicas kinesin molecules isolated from chick brain or embryonic fibroblasts appear as an elongated central rod (about 70–80 nm long and 3–4 nm in diameter), with two small (5–6 nm diameter) globules on one side and two V-shaped protrusions at the other end (about 6 nm wide and 10–12 nm or more long). The sedimentation coefficient of kinesins is around 9.5S.

2. Biochemical Characteristics of Myosin, a Kinesin-like Protein, and Profilin from Pollen

a. Myosin   In his article on the identification of F-actin in pollen tubes of *Amaryllis belladonna,* Condeelis (1974) already mentioned the likelihood of the presence of a myosin-like protein responsible for the transduction of chemical energy into movement in pollen tubes. In 1986 a molecule with a molecular weight of 220 kDa, which exhibited ATPase activity according to the methods of Horak and Hill (1972) and Kirkeby and Moe (1983), was separated from raw extracts of pollen of *Luffa cylindrica* (Yen *et al.,* 1986). In 1988 using similar materials, ATPase activity of the total protein extract was demonstrated and a 165-kDa band was found in gradient gels (Ma and Yen, 1988). It was suggested that the first molecule was

myosin and the second its heavy chain subunit. Stronger support for the existence of pollen myosin was given by Tang *et al.* (1989a), who applied Western blotting using a monoclonal antibody directed against fast skeletal muscle myosin (anti-S1) and another against pancytoplasmic myosin (anti-LMM), on pollen tube extracts of *Nicotiana alata*. In both cases a polypeptide of approximately 175 kDa was labeled from the total pool of soluble proteins. The work was completed by immunolocalization *in situ*.

There is accumulating evidence that organelle movement in pollen tubes is mediated by myosin (Shimmen, 1988). Kohno and Shimmen (1988b) succeeded in reconstituting the movement of organelles isolated from pollen tubes of *Lilium longiflorum* along characean actin bundles. Movement was remarkably sensitive to calcium (75% of the organelles moved in buffer at <0.18 μM Ca^{2+}, versus 10% at 2.1 μM Ca^{2+}). Like skeletal muscle myosin, the translocator was labile to heat treatment (50°C for 2 minutes) or treatment with 2 mM NEM for 5 minutes (Kohno and Shimmen, 1988b). The direction of movement of pollen tube organelles along the characean actin filaments coincided with that of the native cytoplasmic streaming in this alga. The reconstituted movement required a supply of both Mg^{2+} and ATP. Addition of 100 μM vanadate in the presence of 6 mM $MgCl_2$ and 1 mM ATP had almost no effect, which indicates that neither kinesin nor dynein is involved because at this concentration vanadate would be inhibitory for these two locomotor molecules (Vale, 1987; Sheetz, 1989). However, in *Nicotiana tabacum* preliminary results indicate that the presence of as low a concentration as 25 μM vanadate in the culture medium of pollen already slightly inhibits pollen germination and pollen tube growth and that a 100-μM concentration clearly alters both processes (K. R. Shivanna and A. Tiezzi, personal communication). KCl at a concentration of 0.44 M, which is below the levels inducing degradation of characean actin, stopped organelle movement completely on the system employed by Kohno and Shimmen, suggesting that the translocator was detached from the organelles by KCl. Furthermore, when 3 mg/ml (final concentration) of heavy meromyosin prepared from rabbit skeletal muscle myosin was added to the organelle preparation, partial inhibition of movement was observed. Inhibition was complete when a p-CMB–modified heavy meromyosin (also 3 mg/ml) was mixed with the organelles (Kohno *et al.*, 1990). More recently, the presence of a myosin-like translocator in pollen tubes of *Lilium longiflorum* has been demonstrated by showing that FITC-phalloidin–labeled filaments of breast muscle actin visualized by low light video microscopy can move on the surface of coverslips coated with a crude extract of pollen tubes (Kohno *et al.*, 1991).

b. Kinesin   The first biochemical data indicating the presence of a kinesin-like protein in plants also came from a study on pollen tubes. Moscatelli

et al. (1988b) first elicited antibodies to the heavy chain of calf brain kinesin, purified according to the procedure of Vale *et al.* (1985). Then, they selected a hybridoma cell line (K71S23) that produced a supernatant that stained a single band of 116 kDa on extracts of calf brain. The supernatant of K71S23 also labeled a single 116 kDa and a second, stronger 105 kDa band, when extracts of pollen tubes of *Nicotiana tabacum* were used. The two molecular weights correspond to those of two polypeptides from pollen tube extracts, which bind to calf brain microtubules in the presence of AMP-PNP, a characteristic feature of kinesin (Vale *et al.*, 1985). The sedimentation coefficient of the protein fraction from pollen recognized by the K71S23 antibody ranges between 8.5S and 9.5S and the ATPase activity of the protein is stimulated in the presence of microtubules (Cai *et al.*, in press; Tiezzi *et al.*, 1992). K71S23 further stains specific domains in the vegetative cytoplasm of pollen tubes (Section IV,C,1).

c. Profilin   For the first time in plants, Valenta *et al.* (1991, 1992) reported the presence of profilin in the pollen of several distantly related species. Proteins of this family are typically sequestering molecules. They have the common property of forming 1:1 complexes with G-actin, which seriously diminishes the ability of the actin monomers to polymerize, but they do not seem to interact with F-actin (Bershadsky and Vasiliev, 1988). Profilins play a role during the acrosomal reaction of echinoderm sperms, but their function in plants is still unclear. We can only guess that in pollen profilin is involved in the control of the polymerization of actin by regulating the amount of freely available actin. It is tempting to speculate that profilin could participate in the transformation of actin from storage to filamentous forms during the processes of pollen germination, pollen tube tip growth, and at the moment of fertilization (Valenta, *et al.*, 1992).

IV. Localization of Cytoskeletal Elements in Pollen Grains and Pollen Tubes

A. Actin Filaments in Pollen Grains

1. Overview

Table I is an overview of the literature reporting the organization of actin filaments and microfilaments in pollen grains and pollen tubes at the light and electron-microscopic level.

2. Pollen Grains

Tiwari and Polito (1988b, 1988d, 1990a) have reported on a quantitative analysis of the dynamics of the organization of the actin skeleton in pollen

grains of *Pyrus communis* during the process of pollen hydration, activation, and germination, using rhodamine–phalloidin. They observed a sequence of F-actin patterns, beginning with circular profiles in the peripheral vegetative cytoplasm and coarse granules around the vegetative nucleus in unactivated pollen. These patterns were followed by the occurrence of granular and fusiform, intermediate patterns. The temporal presence of conspicuous masses of F-actin or fibrillar bodies in pollen grains is reported by other authors, too. In incompletely hydrated pollen grains of *Endymion nonscriptus* fixed in 4% paraformaldehyde (Heslop-Harrison *et al.*, 1986b) and ungerminated but hydrated pollen grains of *Lilium longiflorum* (Pierson, 1988) and *Hosta caerulea* (Zhu *et al.*, 1991) permeabilized by the addition of 5% DMSO, distinct fusiform phalloidin-binding inclusions were shown throughout the vegetative cytoplasm. These spicules are probably related to the dense crystalline masses of fine fibrils that have been found at the ultrastructural level in the vegetative cytoplasm of inactivated pollen of *Aloe ciliaris* (Ciampolini *et al.*, 1988), and *Nicotiana alata* (Cresti *et al.*, 1985) and activated pollen of *Linaria vulgaris* and *Nicotiana tabacum* (Cresti *et al.*, 1988). Granules or short rods of actin have been shown in hydrated pollen grains of *Zea mays* (Zhou *et al.*, 1990a). Such aggregations of F-actin microfilaments may constitute a storage form of actin that becomes progressively functional during activation. In hydrated pollen grains of *Lilium longiflorum* (Pierson, 1988, 1989), fine web-like structures have been shown by rhodamine–phalloidin staining. The gradual appearance of filamentous arrays in the stages prior to germination in *Pyrus communis* culminates with the convergence of these arrays toward the germinal aperture at the moment of germination (Miki-Hirosige and Nakamura, 1982; Tiwari and Polito, 1988b, 1990a), one of the earliest signs of polarization in the vegetative cytoplasm. It is noteworthy to consider that the succession of actin patterns manifested during pollen hydration and activation shows great similarity to the reorganization process in subprotoplasts of pollen tubes of *Nicotiana tabacum* (Rutten and Derksen, 1990). Because the patterns of actin distribution were identical in both karyoplasts and cytoplasts, the authors concluded that the pollen cytoplasm may possibly have an intrinsic capacity to reorganize actin filaments in well-defined polar patterns.

All the relevant reports indicate that after germination long arrays of actin filaments remain persistent through the vegetative cytoplasm of pollen grains, constituting a continuous system with the actin skeleton of the pollen tube (Miki-Hirosige and Nakamura, 1982; Perdue and Parthasarathy, 1985; Heslop-Harrison *et al.*, 1986b; Pierson *et al.*, 1986a; Pierson, 1988, 1989; Kohno and Shimmen, 1988a; Tiwari and Polito, 1988b; Heslop-Harrison and Heslop-Harrison, 1989c; Zhou *et al.*, 1990a; Zhu *et al.*, 1991), (Fig. 2a, 2b). Ultrastructural studies on germinated pollen grains of

TABLE I

Observations on Microfilaments–F-Actin in Male Gametophytic Cells[a,b]

Species	Light microscopy				Electron microscopy			
	ME/MI	PG	PT	GC/SC	ME/MI	PG	PT	GC/SC
Aloe ciliaris						(37)		
Alstroemeria sp.		(8)	(8,20)					
Amaryllis belladonna						(38)	(38)	
Beta vulgaris						(39)		
Clivia miniata						(35)	(35)	
Endymion nonscriptus		(9)	(21,22)					
Galanthus nivalis		(10)	(10)					
Gasteria verrucosa	(1,2)							
Gladiolus gandavensis		(11)	(11)					
Helleborus foetidus		(9)						
Hordeum bulbosum		(9)						
Hosta caerulaea		(12)	(12)					
Hyacinthus orientalis		(13)						
Impatiens walleriana			(8,20)					
Iris pseudacorus			(13)					
Linaria vulgaris						(40)		
Lilium longiflorum		(8,14,15)	(8,14,15,20,23,24,25,26)			(35)	(35)	
Luffa cylindra							(44)	
Narcissus poeticus		(9)						
Narcissus pseudonarcissus		(9,10)	(10,22,27)					

Species	ME–MI	PG	PT	CG–SC
Nicotiana alata		(28,29)	(41)	(28,29,45,46)
Nicotiana sp.		(30)		
Nicotiana tabacum		(15,26,31)	(42)	(44)
Petunia hybrida	(8)	(8,20,32)		(47)
Phalaenopsis sp.	(3–6)	(8)		
Plumbago zeylanica				(48)
Pyrus communis	(16–19)	(18,19)	(17)	(16,17)
Rhododendron laetum			(34)	
Solanum melongana	(7)			
Tillandsia caput-medusae				(49)
Tradescantia virginiana		(33)		
Triticum aestivum	(11)		(36)	(43)
Zea mays	(11)			(50)

[a] Distinction has been made between location in male meiocytes and microspores (ME–MI), pollen grains (PG), pollen tubes (PT), and generative cells and sperm cells (CG–SC).

[b] Numbers in parentheses refer to the following references: (1) Van Lammeren et al. (1988); (2) Van Lammeren et al. (1989); (3) Brown and Lemmon (1990); (4–6) Brown and Lemmon (1991a,b,c); (7) Traas et al. (1987); (8) Perdue and Parthasarathy (1985); (9) Heslop-Harrison et al. (1986b); (10) Heslop-Harrison and Heslop-Harrison (1989c); (11) Zhou et al. (1990a); (12) Zhou et al. (1991); (13) Heslop-Harrison and Heslop-Harrison (1989b); (14) Kohno and Shimmen (1988a); (15) Pierson (1988); (16) Tiwari and Polito (1988a); (17) Tiwari and Polito (1990a); (18) Tiwari and Polito (1988b); (19) Tiwari and Polito (1990b); (20) Parthasarathy et al. (1985); (21) Heslop-Harrison and Heslop-Harrison (1989b); (22) Heslop-Harrison and Heslop-Harrison (1989b); (23) Kohno and Shimmen (1987); (24) Pierson et al. (1988); (25) Pierson et al. (1986a); (26) Pierson et al. (1986b); (27) Heslop-Harrison and Heslop-Harrison (1991a); (28) Lancelle and Hepler (1988); (29) Tang et al. (1989a); (30) Moscatelli et al. (1988a); (31) Åström et al. (1991); (32) Rutten and Derksen (1990); (33) Putnam-Evans et al. (1989); (34) Taylor et al. (1989); (35) Franke et al. (1972); (36) Tiwari (1989); (37) Ciampolini et al. (1988); (38) Condeelis (1974); (39) Hoefert (1969); (40) Cresti et al. (1988); (41) Cresti et al. (1985); (42) Cresti et al. (1986b); (43) Lancelle et al. (1986); (44) Liu and Yen (1987); (45) Lancelle and Hepler (1989); (46) Lancelle et al. (1987); (47) Cresti et al. (1976); (48) Russell and Cass (1981); (49) Brighigna et al. (1980); (50) Zhu et al. (1980).

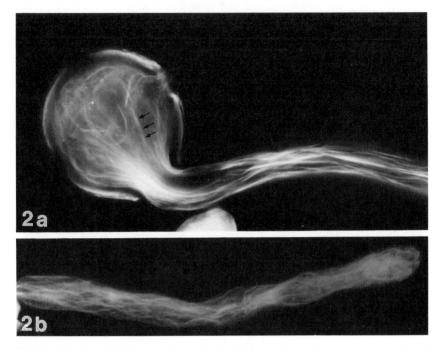

FIG. 2   Rhodamine–phalloidin staining of actin filaments in whole-mount preparations of pollen tubes grown *in vitro:* (a) germinated pollen grain and tube of *Nicotiana tabacum.* Micrograph by E. S. Pierson. ×900, (b) Whole pollen tube of *Vinca rosea.* Micrograph courtesy of U. K. Tirlapur. ×900. The bundles of F-actin (e.g., small arrows) of the pollen tube are continuous with those of the pollen grain and actin filaments are also present in the tip in preparation.

Lilium longiflorum and *Clivia miniata* (Franke *et al.*, 1972; Miki-Hirosige and Nakamura, 1982) show dense arrays of electron-opaque microfilaments in association with Golgi vesicles, periodic acid–Schiff (PAS)-positive particles, ER elements, and other organelles. The morphological data on the actin microfilament system of germinated pollen grains are consistent with the assumption that organelles move along actin filaments in pollen grains and tubes and that they are involved in the transport of vesicles and polysaccharide (p) particles to the tip.

3. Pollen Tubes

The demonstration by Franke *et al.* (1972) that the vegetative cytoplasm of pollen tubes of *Clivia* and *Lilium* contains ~5-nm wide microfilaments is one of the most exciting discoveries concerning the plant cytoskeleton.

The assumption that microfilaments, consisting of actin, are a common component of pollen tubes was confirmed by light microscopic immunolocalization (Taylor *et al.*, 1989; Tang *et al.*, 1989b; Åström *et al.*, 1991), fluorochrome-labeled phalloidin staining (Parthasarathy *et al.*, 1985; Perdue and Parthasarathy, 1985; Pierson *et al.*, 1985, 1986a, 1986b, 1989; Parthasarathy, 1987; Pierson, 1988, 1989; Kohno and Shimmen, 1987, 1988a; Moscatelli *et al.*, 1988a; Tiwari and Polito, 1988a, 1988b, 1988c, 1988d, 1990a; Tang *et al.*, 1989b; Heslop-Harrison and Heslop-Harrison, 1989b, 1989c, 1991a, 1991b; Heslop-Harrison *et al.*, 1986b; Zhou *et al.*, 1990a; Åström *et al.*, 1991; Zhu *et al.*, 1991), and electron microscopy after RF–FS (Lancelle and Hepler, 1988, 1989; Lancelle *et al.*, 1987; Tiwari and Polito, 1988c; Raudaskoski *et al.*, 1987), including immunogold antiactin labeling (Lancelle and Hepler, 1989; Tang *et al.*, 1989b). Heavy meromyosin decoration was applied to pollen tubes or their protoplasts, for the first time in material originating from *Amaryllis belladonna* pollen (Condeelis, 1974), and later in *Luffa cylindrica* (Liu and Yen, 1987; Liu *et al.*, 1990) and *Nicotiana tabacum* (Liu and Yen, 1987). The arrowhead complexes shown by Condeelis (1974) were MgATP-dissociable. Recently, Liu and Yen (1991) reported that beads coated with a fluorescent conjugate to rabbit myosin specifically bind to cytoplasmic fibrils isolated from *Lilium longiflorum* pollen, whereas bovine serum albumin (BSA)-coated beads did not. This method opens interesting perspectives for *in vitro* assays of actomyosin-based motility system in portions of cytoplasm isolated from pollen tubes.

Essentially, actin filaments are distributed throughout the entire pollen tube in longitudinal orientation. When the material is well preserved, the actin filaments appear as fine, sometimes anastomosing arrays, which often pass close to organelles, vacuoles, and the plasma membrane. The results obtained after RF–FS further indicate that the microfilaments seem to form a network in the extreme tip of the pollen tube in *Nicotiana alata* (Lancelle *et al.*, 1987) and *Pyrus communis* (Tiwari and Polito, 1988c). Without contesting the superiority of the RF–FS method for high resolution two-dimensional images using TEM, we would like to emphasize that optical microscopy methods have also revealed the presence of a fine network of actin filaments in the apical region in samples of pollen tubes from various genera (Pierson, 1988, 1989; Tiwari and Polito, 1988c; Heslop-Harrison and Heslop-Harrison, 1991a). In the apical zone of tubes of *Lilium longiflorum*, numerous flat cisternae of ER or undefined membranous material have been visualized between Golgi vesicles by conventional electron microscopy (Franke *et al.*, 1972) and RF–FS (Lancelle and Hepler; Derksen *et al.*, unpublished data), but in *Nicotiana* ER is not as abundant in the apical region (Lancelle and Hepler). In RF–FS pollen tubes of *Nicotiana tabacum*, some of the Golgi vesicles at the tip appear

to be coated (Tiezzi, 1991), possibly representing protein material. In pollen tubes of *Lilium longiflorum* grown in the style after self-incompatible or cross-compatible pollination, brightly stained foci of F-actin have been visualized by rhodamine–phalloidin labeling (Pierson *et al.*, 1986a). It was theorized that these foci might be organizing centers for actin filaments.

In addition, distinct masses of fibrillar structures have been observed in intimate association with the vegetative nucleus, the generative cell, or the sperm cell using TEM in *Petunia hybrida* (Cresti *et al.*, 1976), *Beta vulgaris* (Hoefert, 1969), *Galanthus nivalis* and *Narcissus pseudonarcissus* (Heslop-Harrison and Heslop-Harrison, 1989c), *Pyrus communis* (Tiwari and Polito, 1988c), and after rhodamine–phalloidin staining in *Pyrus communis* (Tiwari and Polito, 1988c) and *Nicotiana tabacum* after 24 hours growth on agar medium (Pierson, 1989). This remarkable organization of the actin cytoskeleton suggests that this system is involved in the transport of the vegetative nucleus and the generative cell–sperm cells.

B. Microtubules

1. Overview

Table II is an overview of the literature reporting the organization of microtubules in pollen grains and pollen tubes at the light and electron-microscopic level.

2. Pollen Grains

In protoplasts of dry *Pyrus communis* pollen obtained by freeze-fracture removal of the cell wall, indirect antitubulin fluorescence microscopy has revealed the presence of distinct axial bundles of microtubules in the generative cell; however, no microtubules could be detected in the vegetative cytoplasm in the same preparations (Tiwari and Polito, 1990b). The absence of microtubules in dry pollen grains coincides with the situation encountered at the ultrastructural level in the same species (Tiwari *et al.*, 1990) and others (Cresti *et al.*, 1977, 1980, 1985). This observation implies that microtubules present at stages prior to pollen maturity (Tiwari, 1989; Table II) have depolymerized. In *Pyrus communis*, microtubules reappear in the vegetative cell soon after incubation in normal culture medium (Tiwari and Polito, 1990b). Initially, microtubules form short arrays located in the cortex near domains with diffuse fluorescence. This could indicate a high concentration of unpolymerized tubulin in these areas. Later stages of pollen activation are characterized by the appearance of

more numerous, longer, and increasingly branched patterns of microtubules (Tiwari and Polito, 1990b). None of the micrographs presented by Tiwari and Polito (1990b) shows any foci-like structures. We agree with the interpretation given by these authors that microtubule nucleation in the pollen of this species is most likely triggered by a high local concentration of unpolymerized tubulin, rather than emerging from typical microtubule-organizing centers (MTOCs, Dustin, 1984).

After germination microtubules are arranged in network-like arrays through most of the pollen grain in *Pyrus communis*, but their general direction is toward the germination aperture (Tiwari and Polito, 1988c, 1990b). In germinating pollen grains of *Amaryllis vitatta* the direction of the microtubule bundles is parallel to that of the emergent tube (Zhu and Liu, 1990).

3. Pollen Tubes

a. Location of Microtubules The studies of Jensen and Fisher (1970) and Franke *et al.* (1972) demonstrated the presence of microtubules for the first time in pollen tubes of the species *Gossypium hirsutum, Clivia miniata,* and *Lilium longiflorum.* Further observations obtained by electron microscopy of thin sections after chemical fixation (Franke *et al.,* 1972; Zhu and Liu, 1990; Cresti *et al.,* 1984, 1986a; Raudaskoski *et al.,* 1987; Tiwari and Polito, 1988c) and RF–FS (Lancelle *et al.,* 1987; Raudaskoski *et al.,* 1987; Lancelle and Hepler, 1988; Tiwari and Polito, 1988c), by electron microscopy of dry-cleaving preparations (Derksen *et al.,* 1985; Pierson *et al.,* 1986a; Tiezzi *et al.,* 1986), and by immunofluorescence microscopy (Derksen *et al.,* 1985; Tiezzi *et al.,* 1986; Raudaskoski *et al.,* 1987; Heslop-Harrison and Heslop-Harrison, 1988b, 1988c; Heslop-Harrison *et al.,* 1988; Tiwari and Polito, 1988c; Pierson *et al.,* 1989; Zhu and Liu, 1990; Åström *et al.,* 1991) together show that microtubules are mainly located in the cortical cytoplasm in pollen tubes grown *in vitro* or in the style (Pierson *et al.,* 1986b) and that the microtubules persist in the oldest part of the pollen tube. Heslop-Harrison and Heslop-Harrison (1988c) noted that in *Lilium auratum* the arrays of microtubules could even be found in the oldest segments of cytoplasm isolated from the central zone by callose plugs. The presence of numerous microtubules in this zone of the tube was also confirmed by our own electron microscopy work on *Iris* spp. (Ciampolini and Pierson, unpublished). Therefore, it is suggested that tubulin is not recycled in pollen tubes and that the assembly of microtubules may depend on a supply of monomers from existing reserves or *de novo* synthesis.

Only a few ultrastructural investigations have successfully revealed the presence of microtubules in the vesicular region, either organized in

TABLE II

Observations on Microtubules in Male Gametophytic Cells[a,b]

Species	Light microscopy				Electron microscopy			
	ME–MI	PG	PT	GC–SC	ME–MI	PG	PT	GC–SC
Allemanda neriifolia				(33)				
Aloe ciliaris								(39,62,63)
Alopecurus pratensis			(20)	(20)				
Amaryllis vitatta		(15)	(15)	(15)			(15)	(15)
Beta vulgaris								(64)
Brassica oleracea								(65,66)
Calanthe discolor				(34)				
Calanthe sieboldii				(34)				
Canna indica							(54)	
Clivia miniata								(67)
Crinum moorei				(35,36)				
Cyphomandra betacea								(68)
Endymion nonscriptus		(16)	(16)	(16)				(69)
Euphorbia dulcis					(50)			(50)
Gasteria verrucosa	(1)							
Gossypium hirsutum							(55)	
Haemanthus katherinae					(51)			(51,70)
Hordeum vulgare								(71–73)
Hyacinthus orientalis				(37,38)			(37)	(37)

Species							
Impatiens balsamina			(34)				
Impatiens sultani	(2)						
Leucojum verum			(39)				
Lilium auratum		(21)					
Lilium henryi	(3,4)			(3,4)			
Lilium longiflorum	(5)	(17)	(17,22–24) (40,86)			(17,22,25,54)	(74)
Lilium sp.			(41)				
Linaria vulgaris							(41,75)
Lonicera japonica	(2)						
Luffa cylindrica						(56)	
Lycopersicum esculentum	(6)						(57)
Lycopersicum peruvianum						(57)	(57)
Malus domestica			(41)			(57)	
Nicotiana alata	(16)	(16)				(58,59)	(16,57,59,75,76)
Nicotiana tabacum		(17,22,25–30)	(22,25–30,42,43)		(52)	(17,22,24,28,29,60)	(28–30,39,60,77,78)
Olea europaea	(6)						(57)
Ornithogalum virens	(6)						
Petunia hybrida				(7–10)		(61)	(57,61,79)
Phalaenopsis sp.	(7–10)						(7)
Plumbago zeylanica							(80)
Prunus avium						(57)	(57)
Pyrus communis	(18,19)	(18)	(19)		(18,53)	(18,53)	
Rhododendron laetum			(44)				(44)
Rhododendron macgregoriae							(44)
Rhododendron sp.							(81)
Secale cereale							
Solanum melongana	(11)						(82)
Spinacia oleracea			(45)				

(*continued*)

TABLE II (*continued*)

Species	Light microscopy				Electron microscopy			
	ME–MI	PG	PT	GC–SC	ME–MI	PG	PT	GC–SC
Tradescantia paludosa				(46)				
Tradescantia virginiana	(12)		(31)	(31,47–49)		(12)		
Tillandsia caput-medusae								(83)
Trillium kamtschaticum								(74)
Triticum aestivum								(84)
Vigna vexillata	(13)							
Zea mays	(14)			(32)				(85)
Zephyranthes grandiflora			(32)					

[a] Distinction has been made between location in male meiocytes and microspores (ME–MI), pollen grains (PG), pollen tubes (PT), and generative cells and sperm cells (GC–SC).

[b] Numbers in parentheses refer to the following references: (1) Van Lammeren *et al.* (1985); (2) Brown and Lemmon (1988); (3) Dickinson and Sheldon (1984); (4) Sheldon and Dickinson (1986); (5) Tanaka (1991); (6) Hogan (1987); (7) Brown and Lemmon (1990); (8–10) Brown and Lemmon (1991a,b,c); (11) Traas *et al.* (1989); (12) Tiwari (1989); (13) Perez-Munoz and Webster (1990); (14) Staiger and Cande (1990); (15) Zhou and Liu (1990); (16) Heslop-Harrison *et al.* (1988); (17) Derksen *et al.* (1985); (18) Tiwari and Polito (1988c); (19) Tiwari and Polito (1990b); (20) Heslop-Harrison and Heslop-Harrison (1988b); (21) Heslop-Harrison and Heslop-Harrison (1988c); (22) Pierson (1989); (23) Pierson *et al.* (1985); (24) Pierson *et al.* (1986a); (25) Åstrom *et al.* (1991); (26) Derksen and Traas (1985); (27) Pierson *et al.* (1989); (28,29) Raudaskoski *et al.* (1986,1987); (30) Tiezzi *et al.* (1986; (31) Palevitz and Cresti (1989); (32) Zhou (1990b); (33) Zee and Aziz-un-Nisa (1991); (34) Terasaka and Niitsu (1989b); (35) Zhou (1987); (36) Zhou *et al.* (1988); (37) Cresti and Tiezzi (1990); (38) Tiezzi *et al.* (1989); (39) Tiezzi *et al.* (1989); (40) Zhu *et al.* (1990); (41) Tiezzi *et al.* (1988c); (42) Bartalesi *et al.* (1991); (43) Tiezzi *et al.* (1989); (44) Taylor *et al.* (1989); (45) Theunis and Wilms (1988); (46) Terasaka and Niitsu (1989b); (47) Liu and Palevitz (1991); (48) Palevitz (1990); (49) Palevitz and Cresti (1988); (50) Murgia *et al.* (1986); (51) Sanger and Jackson (1971a); (52) Cresti *et al.* (1986); (53) Tiwari and Polito (1988d); (54) Franke *et al.* (1972); (55) Jensen and Fisher (1970); (56) Yen *et al.* (1986); (57) Cresti *et al.* (1984); (58) Lancelle and Hepler (1988); (59) Lancelle *et al.* (1987); (60) Cresti *et al.* (1986a); (61) Cresti *et al.* (1976); (62) Ciampolini *et al.* (1988); (63) Tiezzi *et al.* (1988b); (64) Hoefert (1969); (65) Cresti *et al.* (1990a); (66) Dumas *et al.* (1985); (67) Chen *et al.* (1989); (68) Hu and Yu (1988); (69) Burgess (1970); (70) Sanger and Jackson (1971b); (71) Cass (1973); (72) Cass and Karas (1975); (73) Charzynska *et al.* (1988); (74) Tanaka *et al.* (1989); (75) Cresti *et al.* (1988); (76) Cresti *et al.* (1985); (77) Linskens *et al.* (1989); (78) Yu *et al.* (1989); (79) Herrero and Dickinson (1981); (80) Russell and Cass (1981); (81) Kaul *et al.* (1987); (82) Karas and Cass (1976); (83) Brighigna *et al.* (1980); (84) Zhou (1980); (85) McConchie *et al.* (1987); (86) Xu *et al.* (1990).

bundles in *Clivia* and *Lilium* (Franke *et al.*, 1972) or as short solitary microtubules in *Nicotiana* (Lancelle *et al.*, 1987). However, these data concern the zone adjacent to the extreme tip (about 2 μm from the tip, Franke *et al.*, 1972) where a distinct cell wall is already apparent. Basically, no labeling of microtubules has been detected in the extreme apex of mature pollen tubes in all of the published immunofluorescence investigations, which use antibodies against animal or yeast tubulin. After FITC–antitubulin staining, an intense but diffuse fluorescence was observed only in the dome of young pollen tube tips in *Nicotiana tabacum* (Derksen *et al.*, 1985) and in dried pollen tubes of *Lilium* (Thompson and Brower, 1985). This type of labeling may represent tubulin monomers or oligomers (Derksen *et al.*, 1985). Recent immunofluorescence results on pollen tubes of *Zea mays,* for which several antibodies were directed against angiosperm tubulin isoforms, show that at least in some pollen tubes microtubules are also present in the zone between 6 and 12 μm immediately behind the tip (Soon-Ok Cho, unpublished observations). We are inclined to conclude that, as a general trend, microtubules are absent from the youngest, outermost part of the pollen tube and that the site of origin of microtubules, therefore, would be located in the subapical part of the pollen tube, although no clear sites of nucleation could be observed there.

b. Orientation of Microtubules   Cortical microtubules may run in interconnected or single parallel arrays of axial (Derksen *et al.*, 1985; Tiezzi *et al.*, 1986; Raudaskoski *et al.*, 1987; Heslop-Harrison and Heslop-Harrison, 1988b, 1988c; Heslop-Harrison *et al.*, 1988; Zhou *et al.*, 1990; Åström *et al.*, 1991), (Fig. 3a) or helical (Derksen *et al.*, 1985; Heslop-Harrison *et al.*, 1988; Raudaskoski *et al.*, 1987; Åström *et al.*, 1991) orientation with respect to the tube axis. The organization of microtubules in the pollen tube depends on many factors. The first determining parameter is the species. For example, arrays of microtubules in the pollen tubes of *Nicotiana tabacum* are more widely spaced and steeper than those in the pollen tube of *Lilium longiflorum* (Derksen *et al.*, 1985; P. K. Hepler, personal communication). The second variable is the location in the tube, which is directly related to the age of that particular portion of the cell; crisscross patterns have been seen essentially behind the tip in *Nicotiana tabacum* (Derksen *et al.*, 1985) (Fig. 3b), whereas in *Lilium auratum* (Heslop-Harrison and Heslop-Harrison, 1988c) discrete, longitudinally oriented bands have been visualized in the subapical zone only, and in *Zea mays* a loose band of circumferentially aligned microtubules has sometimes been observed in the zone between 18 and 60 μm from the tip (Soon-Ok Cho and S. M. Wick, unpublished results). The third factor is the condition of the pollen tubes in culture. Cold shock (change in temperature from 22°C

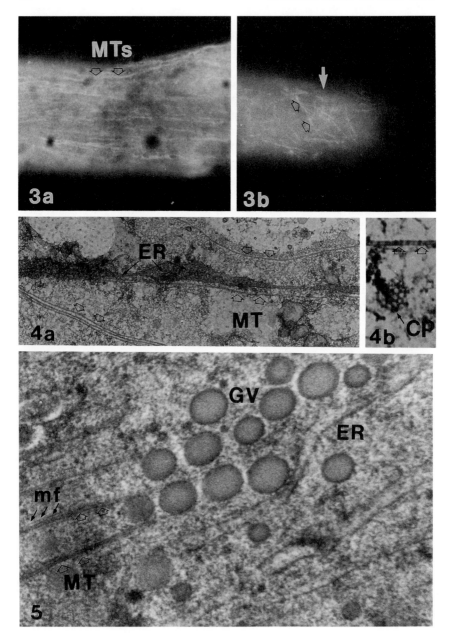

to 4°C) induced fragmentation of the microtubules in *Nicotiana* pollen tubes (Åström *et al.*, 1991), whereas the presence of 0.2% DMSO or 1 μg taxol/ml medium resulted in more labeled tubulin fibers near the tip dome of the tube or better stabilization of microtubules with respect to control pollen (Raudaskoski *et al.*, 1987). The fourth factor, which requires more investigation, is the possible occurrence of different pools of microtubules, corresponding to various tubulin isoforms. Using antibodies against specific plant tubulins, Soon-Ok Cho *et al.* (personal communication) observed in some pollen tubes patterns that were unusual in respect to other antitubulin labelings.

It is clear that in pollen tubes, microtubules tend to be oriented in orderly, mostly axial arrays, yet the microfibrils of the cell wall have a random organization (Sassen, 1964; Kroh and Knuiman, unpublished results). Apparently in pollen tubes and root hairs, in contrast to many other cell types (Seagull, 1991), microtubules are not directly involved in the deposition of organized arrays of wall microfibrils.

c. Intermicrotubular Connections and Associations between Microtubules and Other Elements   Various types of intermicrotubular connections and associations have been observed between microtubules, microfilaments, and other cytoplasmic components in pollen tubes using the microscope, but their nature and function(s) are still not well understood.

In one of the most intensively studied species, *Lilium longiflorum*, microtubules have repeatedly been seen to be interlinked by bridges

---

FIG. 3   Conventional indirect fluorescence microscopy labeling of pollen tubes grown *in vitro* using MAS 077 (Sera Lab. Ltd) antitubulin. (a) Axially oriented microtubules (MTs; open arrows) in the cortical region of the middle part of a pollen tube of *Lilium longiflorum*. Micrograph courtesy of E. S. Pierson and J. Derksen. ×1000, (b) Crisscross pattern of microtubule bundles (open arrows) often found in the subapical zone (white arrow) in pollen tubes of *Nicotiana tabacum*. Micrograph courtesy of J. Derksen and E. S. Pierson. ×1600.

FIG. 4   Dry-cleaved preparations of pollen tubes visualized by means of transmission electron microscopy. (a) Portion of the cortical cytoplasm of a pollen tube of *Lilium longiflorum* cv. White Europe, grown in the style and prepared from the stylar canal 24 hours after self-pollination. The method allows one to observe microtubules (MT, open arrows) over a long distance. One microtubule co-aligns with the tubular cisternae of endoplasmic reticulum (ER). ×19,300, (b) Detail of a coated pit (Cp) and a microtubule (open arrows) in the cortical cytoplasm of a pollen tube of *Lilium longiflorum* grown *in vitro*. Micrographs courtesy of J. A. Traas and E. S. Pierson. ×180,000.

FIG. 5   Glancing thin section of a cryofixed and freeze-substituted pollen tube of *Nicotiana tabacum* grown *in vitro*. Single microtubules (MT, open arrows) are often accompanied by one or more microfilaments (mf, closed arrows). Golgi vesicles (GV) seem to align with the microtubules. Putative endoplasmic reticulum (ER) is found between the vesicles. Micrograph courtesy of J. Derksen and B. Knuiman. ×55,000.

or zigzag-like structures in cross and longitudinal sections (periodicity 22.5 nm, Franke *et al.*, 1972; RF–FS, Lancelle and Hepler, submitted) and in dry-cleaved preparations (Pierson *et al.*, 1986a). Is it possible that these interconnections give mechanical support to microtubules so that the microtubules contribute, like a scaffold, to the position of internal structures in the cell? Are these intermicrotubular connections related perhaps to the acentric cross-links sometimes encountered between microtubules in angiosperm male gametic cells (Burgess, 1970; Lancelle *et al.*, 1987; Ciampolini *et al.*, 1988; Tiezzi *et al.*, 1988b; Cresti *et al.*, 1990a) and in the axonemes of lower plants and animal sperms (Tiezzi and Cresti, 1990)?

Whether microtubules are closely associated with fine filaments composed of actin in pollen tubes has not been proven (immunogold labeling at the electron-microscopic level could provide the final answer), but there already are strong arguments in this direction. Well-preserved cortical or cytoplasmic microtubules appear to be invariably associated, sometimes even cross-linked at 14-nm intervals, with one or more microfilaments of approximately 7 nm coinciding with the size of actin filaments (Franke *et al.*, 1972; Lancelle *et al.*, 1987; Raudaskoski *et al.*, 1987; Tiwari and Polito, 1988c). Microtubules from polypeptide extracts of tobacco pollen reconstituted in the presence of taxol ($5-20 \ \mu M$) are associated with fine filaments, and extracts from the taxol-induced aggregates cross-react with antiactin, also after MAP extraction by salt treatment (Tiezzi *et al.*, 1987, 1988a, 1988b; Moscatelli *et al.*, in press). Finally, labeling with antitubulin and rhodamine–phalloidin has demonstrated the partial co-localization of bundles of microtubules with bundles of actin filaments in pollen tubes of *Nicotiana tabacum* (Pierson *et al.*, 1989). In pollen tubes of *Pyrus communis,* associations between cytoplasmic microtubules and microfilaments are often found in the vicinity of vesicles and vacuoles (Tiwari and Polito, 1988c). One of the most attractive hypotheses suggests that microtubules form the relatively stiff support on which a network of more dynamic actin filaments rests.

Other associations between microtubules and noncytoskeletal elements include the following: (1) microtubule–plasma membrane connections (Franke *et al.*, 1972; Lancelle *et al.*, 1987; Zhu and Liu, 1990); (2) coalignment of microtubules and cytoplasmic membranes of putative ER tubules (Franke *et al.*, 1972; Pierson *et al.*, 1986a; Lancelle *et al.*, 1987), (Fig. 4a); (3) close proximity of microtubules and coated pits, mitochondria, or Golgi vesicles (Franke *et al.*, 1972; Lancelle *et al.*, 1987; Raudaskoski *et al.*, 1987), (Figs. 4b, 5); and (4) association of microtubules with the envelope of the vegetative nucleus (in *Nicotiana alata,* Lancelle *et al.*, 1987 and *Nicotiana tabacum,* Raudaskoski *et al.*, 1987, but not in *Pyrus communis* where associations with microfilaments are found, Tiwari

and Polito, 1988c). In some preparations of *Nicotiana tabacum* a bright ring or spot attached to the vegetative nucleus, ahead of the generative cell, is stained by antitubulin (Derksen *et al.*, 1985; Åström *et al.*, 1991). We can only speculate that these tubulin-rich structures are perhaps remnants of a tail-like connection between the generative cell and the vegetative nucleus (Mogensen, this volume).

C. Myosin and Kinesin

1. Pollen Tubes

a. Myosin Monoclonal antimyosin antibodies (anti-LMM and anti-S1), as described in Section III,C,2, appear to label organelles and/or vesicles as fluorescent spots in pollen tubes of *Nicotiana alata* (Tang *et al.*, 1989a). The staining was more accentuated in the tip region. Similar mottled patterns and ubiquitous staining have been shown on the surface of isolated organelles of *Alopecurus pratensis*, *Secale cereale*, and *Hyacinthus orientalis* by Heslop-Harrison and Heslop-Harrison (1989b, 1989d), using a commercial antibody reported to recognize both the heavy and light chains of muscle myosin. All of the standard controls were carried out and showed negative results.

Moreover, anti-S1, but not anti-LMM, stains the generative cell and the envelope of the vegetative nucleus of *Nicotiana alata* (Tang *et al.*, 1989a). Heslop-Harrison and Heslop-Harrison (1989b, 1989d) also found staining at the surface of the vegetative nucleus or the generative cell in *Hyacinthus orientalis* and *Helleborus foetidus*, after extraction from the pollen grain. These observations need to be combined with immunolabeling at the electron-microscopic level. In this view it would be interesting to establish whether the lateral projections (about 35 nm high and 40 nm space) shown at the surface of the outer membrane of the generative cell in *Capparis spinosa* (Van Went and Gori, 1989) and *Amaryllis belladonna* (Cresti *et al.*, 1991) contain myosin.

b. Kinesin-like Protein K71S23, which is a monoclonal antibody specifically directed against the heavy chain of kinesin (Moscatelli *et al.*, 1988; Cai *et al.*, in press; Tiezzi *et al.*, 1992; Section III,C,2) stains punctate structures, possibly Golgi vesicles, in the tip of the pollen tube of *Nicotiana tabacum*, but it does not label the generative cell (Moscatelli *et al.*, 1988b). The similarity between the pattern and location of myosin and kinesin staining is striking. This location-dependent superficial staining of organelles is difficult to reconcile with the fact that all large particles undergo cyclosis in the pollen tube. Do preparative barriers perhaps mask some

cytoplasmic domains of myosin and kinesin, which may be even more widespread in the vegetative cytoplasm than the method indicates?

V. Effects of Inhibitors on the Cytoskeleton

Cytochalasins are fungal metabolites that reversibly bind to the barbed end of actin filaments and inhibit both association and dissociation at that end. They can promote nucleation and sometimes shorten actin filaments (Cooper, 1987; Bershadsky and Vasiliev, 1988). Cytochalasins are therefore widely used to investigate the physiological effects of deactivation of the actin cytoskeleton. Cytochalasin D is probably the component with the highest specificity for actin, being about 10 times more effective than cytochalasin B (Cooper, 1987). Another reason that cytochalasin D appears to be more appropriate than cytochalasin B for assaying cell motility is that the latter may also bind to glucose transporters (Cooper, 1987).

The presence of cytochalasin in the culture medium of pollen tubes has been reported to induce fragmentation or decrease rhodamine–phalloidin-labeling of actin filaments (Perdue and Parthasarathy, 1985; Heslop-Harrison and Heslop-Harrison, 1989c, 1991a; Tiwari and Polito, 1990a). However, contradictory results have been found by RF–FS and immunofluorescence, which show that microfilaments aggregate into massive bundles as a consequence of cytochalasin treatment (Lancelle and Hepler, 1988; Tang et al., 1989b). Addition of a micromolar concentration of cytochalasin B (Franke et al., 1972; Herth et al., 1972; Mascarenhas and Lafountain, 1972; Derksen and Traas, 1985; Parthasarathy et al., 1985; Perdue and Parthasarathy, 1985; Lancelle and Hepler, 1988; Ma and Yen, 1988; Heslop-Harrison and Heslop-Harrison, 1989a; Tang et al., 1989b) or cytochalasin D (Picton and Steer, 1981, 1983b; Lancelle and Hepler, 1988; Heslop-Harrison and Heslop-Harrison, 1989a, 1989c, 1991b; Tang et al., 1989b; Tiwari and Polito, 1990a; Heslop-Harrison et al., 1991) to a culture of pollen leads to multiple physiological responses: (1) inhibition of pollen germination, (2) cessation of organelle movement, (3) loss of organelle zonation, (4) accumulation of vesicles in the tip, (5) decrease or even complete elimination of pollen tube growth, and (6) contraction of the vegetative nucleus.

The morphological and physiological effects of cytochalasins on growing pollen tubes agree well with the models asserting that actin filaments modulate organelle movement, transport vesicles to the tip, and control tip growth (Steer, 1990). Actin filaments also seem to be involved in a sorting-out mechanism that governs the characteristic zonation of organelles in the tube (e.g., Heslop-Harrison et al., 1991) and in controlling the

elongated shape of the vegetative nucleus (Heslop-Harrison and Heslop-Harrison, 1989c, 1991b).

Colchicine, a secondary metabolite of the two plants in the *Liliaceae*, *Colchicium autumnale* and *Colchicium speciosum*, is the chemical most widely used to cause depolymerization of microtubules. Colchicine inhibits the polymerization of tubulin *in vitro* and causes rapid disassembly of most types of microtubules *in vivo* (Bershadsky and Vasiliev, 1988). To be affected, plant tubulins and microtubules must be treated with a millimolar concentration of colchicine, instead of micromolar quantities, which are effective for members of the animal kingdom (Dustin, 1984; Seagull, 1989; Fosket, 1989). Colchicine has a limited effect on inhibiting pollen germination, pollen tube, growth, and organelle movement (Mascarenhas, 1966; Franke *et al.*, 1972; Mascarenhas and Lafountain, 1972; Derksen and Traas, 1985; Parthasarathy *et al.*, 1985; Heslop-Harrison *et al.*, 1988; Ma and Yen, 1988), unless it is used in extremely high (5 g/l) concentrations (Dexheimer, 1966). If 0.02% colchicine is included in the pollen culture medium in *Tradescantia*, a large number of the generative nuclei are observed in metaphase after 6–7 hours of growth (Mascarenhas, 1966). However, colchicine (Sanger and Jackson, 1971b; Palevitz and Cresti, 1988, 1989; Heslop-Harrison *et al.*, 1988) and isopropyl *N*-phenylcarbamate (Sanger and Jackson, 1971b) cause the spindle-shaped generative cell to become spheroidal (Palevitz and Tiezzi, this volume). In *Endymion nonscriptus* the rate of movement of the vegetative nucleus was halved and its length reduced to two-thirds by 5 μM of the antimicrotubule agent nocodazole (structure in Bershadsky and Vasiliev, 1988), but no effect of this kind was observed in the generative cell (Heslop-Harrison, 1988).

Three classes of herbicides are also known to act on plant microtubules in the micromolar range and could be used to further explore microtubule function in the male gametophyte. These include phosphoric amide compounds, dinitroaniline derivatives (e.g., trifluralin and oryzalin), and carbamates (Fedtke, 1982). Triethyl lead at a concentration of 50 μM causes complete selective destruction of the cortical microtubules in pollen tubes of *Lilium longiflorum*, but 60 μM lead did not impair the structure of these cytoskeletal elements (Roederer and Reiss, 1988).

VI. Conclusions

The presence of an actomyosin driven system generating organelle movement in pollen tubes is strongly indicated by the biochemical and morphological data on the presence of actin in a proteinaceous form or arranged in filamentous arrays (Sections III,A and IV,A), a myosin coating on the

surface of organelles (Sections III,C,2 and IV,C,1), and the effects of cytochalasin (Section V). The involvement of ER with myosin may also be possible, because it is evident that ER cisternae are often observed to be associated with microtubules and microfilaments and appear to be mobile. Video microscopy has clearly demonstrated that lipid globuli also show unidirectional movement in the pollen tube, but it appears improbable that their surface is coated by myosin, in contrast to the membrane-bearing globular organelles. The close association of lipid globuli and ER, as occasionally shown in pollen grains, for example, *Arabidopsis* (Cresti *et al.,* unpublished data), is compatible with the idea that the ER acts as a carrier for the globuli in their movement through the tube. From the results shown in Sections II,D and III,C,2, it appears furthermore that the mechanism of locomotion for cyclosis in pollen tubes is calcium sensitive. There are two candidates to explain calcium sensitivity of actomyosin translocation: fragmentation of actin filaments and myosin inactivation. The second option has become more plausible with the discovery of a protein kinase that co-localizes with actin filaments in pollen tubes and can (in)activate a myosin-like peptide in a Ca^{2+}-dependent manner (Section II,D). The confirmed existence of a tip-to-base gradient of cytosolic free calcium along the pollen tube (Section II,D) may result in a spatial differentiation of the activity of the cytoskeleton. It is perhaps premature, but attractive, to postulate that the general concentration of calcium in the pollen tube is regulated by calmodulin-like proteins and calcium channels, whereas local control is achieved by calcium sequestration through ER and mitochondria. The pH gradient observed along the length of the pollen tube is interesting because it may be a crucial factor in the regulation of ATPase activity of locomotor molecules, such as myosin, in the vegetative cytoplasm.

Other important aspects regarding the cytoskeleton are the transport of the nuclei and the GC–SC and the maintenance of their elongated shape. Although the exact mechanisms are not well understood, data from structural, physiological, and morphological responses to drugs disturbing the cytoskeleton suggest that the microtubular system of the GC–SC and the actin cytoskeleton of the vegetative cytoplasm surrounding the GC–SC and vegetative nucleus collaborate in these two functions. The role of the cortical microtubules of the pollen tube remains more obscure, especially because pollen tubes continue to exhibit normal growth in the presence of high colchicine concentrations. Taking the fact that microtubules and actin–microfilaments co-localize in the cortex, it seems possible that microtubules function as an auxiliary support for the more fragile actin filaments. Microtubules could also be utilized as a scaffold for the positioning of fine structures related to the plasma membrane and as guide elements for the internalization of coated vesicles.

The idea of the widespread occurrence of kinesin has already been raised by Williamson (1986) in a discussion about the possible existence of a microtubule-based motility system for the transport of Golgi vesicles in the phragmoplast of plants. In this part of the review, we would like to emphasize again the presence of kinesin-like molecules described in pollen tubes of tobacco (Sections III,C,2 and IV,C,1), which appear to be located in the vesicular region of the pollen tube apex. We do not yet know whether the kinesin-like molecule is involved in translocation, nor do we understand how it may interact with microtubules, because these elements seem to be almost absent from the pollen tube tips where the kinesin-like protein is found. However, at this moment we can assert that *Nicotiana* pollen contains a molecule with great biochemical homology to animal (calf brain) kinesin.

Besides being involved in the generation of forces for the movement of organelles, the actin cytoskeleton may also regulate the rigidity and extension of the apical plasma membrane (Picton and Steer, 1982; Steer, 1990), again utilizing calcium as a pivotal element. However, not only the actin cytoskeleton but also the apical cell wall must substantially contribute to the consolidation of the pollen tube tip to resist the force of osmotic pressure. This can be easily exemplified by adding a cell wall-degrading enzyme to a culture of growing pollen tubes (e.g., 1% cellulysin to *Lilium* pollen tubes in 10% sucrose borate medium). The pollen tubes will burst at the tip within a few seconds. The effect can be overcome by raising the concentration of sucrose in the medium. Besides, we should not forget the results reported by Kwack (1967) (also, Weisenseel *et al.*, 1975) that exogenous calcium binds to pectins of the pollen tube wall with the consequence that wall rigidity increases. The pH may also affect the viscosity of the cytosol of the pollen tube tip. Thus, we favor a model of pollen tube tip growth that integrates both the actin cytoskeleton and the cell wall and that is not restrictive toward other potential regulatory factors. Because of the near-absence of microtubules in the tip zone of pollen tubes, it is more logical to regard their possible contribution to growth as indirect.

VII. Considerations for the Near Future

Important advances were made during the last decade in describing four major elements of the cytoskeleton in pollen: tubulin, actin, myosin, profilin and a kinesin-like protein. It is expected that in pollen tubes, as in certain other plant cells (Derksen *et al.*, 1990; Staiger and Lloyd, 1991; Wang and Yen, 1991), other cytoskeletal proteins will be discovered soon: perhaps MAPs, membrane skeleton proteins, such as spectrin, microtubule, and actin

filament cross-linking proteins, intermediate filaments, and dynein analogs. We envisage that research will continue to focus on understanding the function of the cytoskeleton and rapidly extend into molecular biology. The mechanisms of genetic regulation leading to the expression of various isoforms of cytoskeletal proteins will certainly receive much attention. The study of defective mutants is a promising approach in this context. The significance of calcium as a second messenger will be further questioned. It will be interesting to investigate whether the disruption of the calcium gradient, for example, after treatment with shuttle buffers or after local supply of free calcium with caged calcium probes, has effect on the zonation of the cytoplasm and the organization of the cytoskeleton. The role of calcium-dependent regulatory proteins on the activity of the cytoskeleton is in progress. The dynamics of the cytoskeleton and its impact on tip growth and the movement of organelles and nuclei will be further studied in living pollen tubes, integrating modern techniques, such as microinjection, analog cytochemistry, *in vivo* labeling with fluorescent phalloidin, and application of powerful light microscopic methods and image deconvolution processing. We imagine that RF–FS and immunogold labeling (Beesley, 1989) will soon become the standard method for electron microscopy and that the application of freeze immobilization will be enlarged to light microscopy as well. The time is ripe to ask whether changes in cytoskeletal organization occurring during pollen maturation, activation, and pollen tube emergence and growth are related to reorientation or to depolymerization–repolymerization events. Pollen is becoming a model system for the research on the plant cytoskeleton. In the next few years many exciting advances can be expected in this rapidly expanding field of modern sciences.

Acknowledgments

The authors thank Prof. P. K. Hepler and Dr. U.K. Tirlapur for their critical comments on the manuscript and all persons who provided unpublished results. The encouragement of Prof. H. F. Linskens, Prof. J. Heslop-Harrison, Dr. C. T. Theunis, and colleagues of the Dipartimento di Biologie Ambientale (Siena) was much appreciated during the preparation of this chapter. This work was supported by grants from the European Communities awarded to Elisabeth S. Pierson (grants BAP-0587-I-CH and BIOT 0078-I-CH).

Note added: Three papers report on ratiometric measurement of free calcium gradients in pollen tubes of *Lilium longiflorum*. In these studies, pollen tubes were incubated in medium containing methyl-ester forms of Fura-2 and Quin-2 (Obermeyer, G., and Weisenseel,M. H. (1991). *Eur. J. Cell. Biol.* **56,** 319–327), or microinjected with the fluorophores Indo-1 (Rathore, K. S., Cork, R. J., and Robinson, K. R. (1990). *Devel. Biol.* **148,** 612–619) or Fura-2 covalently bound to a dextran molecule in order to prevent sequestration of the dye in cellular

compartments (Miller, D. D., Callaham, D. A., Gross, D. J., and Hepler, P. K. (1992) *J. Cell Sci*.**101**, 7–12). The values reported in the three papers significantly differ between each other, but they all point out that the concentration of free Ca^{2+} ions in the tip region of the pollen tube is more elevated than was previously published. The high Ca^{2+} ion concentration in the tip are interpreted in view of pollen tube growth, cytoplasmic streaming, vesicle fusion and fragmentation of actin filaments.

References

Adams, R. J., and Pollard, T. D. (1989). *Cell Motil.Cytoskeleton* **14**, 178–182.

Amici, G. B. (1824). *Ann. Sci. Nat.* **1**, 41–70.

Åström, M., Virtanen, I., and Raudaskoski, M. (1991). *Protoplasma* **160**, 99—107.

Baldi, B. G., Franceschi, V. R., and Loewus, F. A. (1987). *Plant Physiol.* **83**, 1018–1021.

Bartalesi, A., Del Casino, C., Moscatelli, A., Cai, G., and Tiezzi, A. (1991). *Giorn. Bot. Ital.* **125**, 21–28.

Bednarska, E. (1989). *Sex. Plant Reprod.* **2**, 53–58.

Bednarska, E. (1991). *Sex. Plant Reprod.* **4**, 36–38.

Beesley, J. E. (1989). "Colloidal Gold: A New Perspective for Cytochemical Marking." Oxford University Press, Oxford.

Bershadsky, A. D., and Vasiliev, J. M. (1988). *In* "Cytoskeleton" (P.H. Siekievitz, ed.), Series on Cellular Organelles. Plenum Press, New York.

Brewbaker, J. L., and Kwack, B. H. (1963). *Am. J. Bot.* **50**, 859–865.

Brighigna, L., Fiordi, A. C., and Palandri, M. R. (1980). *Am. J. Bot.* **67**, 1483–1494.

Brown, R. C., and Lemmon, B. E. (1988). *Am. J. Bot.* **75**, 1848–1856.

Brown, R. C., and Lemmon, B. E. (1990). *Am. J. Bot.* (Suppl.) **77**, 32.

Brown, R. C., and Lemmon, B. E. (1991a). *J. Cell Sci.* **99**, 273–281.

Brown, R. C., and Lemmon, B. E. (1991b). *Protoplasma* **163**, 9–18.

Brown, R. C., and Lemmon, B. E. (1991c). *Protoplasma* **165**, 155–166.

Burgess, J. (1970). *Planta* **95**, 72–85.

Cai, G. P., Bartalesi, A., Moscatelli, A., and Cresti, M. (1991). *Proc. Int. Symp. on Angiosperm Pollen and Ovules: Basic and Applied Aspects, Como, June 23–27* (in press).

Callaham, D. A., and Hepler, P. K. (1991). *In* "Cellular Calcium: A Practical Approach" (J. G. McCormack and P. H. Cobbald, eds.), pp. 383–410. Oxford University Press, Oxford.

Cass, D. D. (1973). *Can. J. Bot.* **51**, 601–605.

Cass, D. D., and Karas, I. (1975). *Can. J. Bot.* **53**, 1051–1062.

Charzynska, M., Ciampolini, F., and Cresti, M. (1988). *Sex. Plant Reprod.* **1**, 240–247.

Charzynska, M., Murgia, M., and Cresti, M. (1989). *Protoplasma* **152**, 22–28.

Chen, F., Ciampolini, F., Tiezzi, A., and Cresti, M. (1989). *Sex. Plant Reprod.* **2**, 193–198.

Ciampolini, F., Moscatelli, A., and Cresti, M. (1988). *Sex. Plant Reprod.* **1**, 88–96.

Clapham, D. H., and Östergren, G. (1984). *Hereditas* **101**, 137–142.

Clarke, E., and Steer, M. W. (1983). *Caryologia* **36**, 299–305.

Condeelis, J. S. (1974). *Exp. Cell Res.* **88**, 435–439.

Cooke, R. (1989). *Cell Motil. Cytoskeleton* **14**, 183–186.

Cooper, J. A. (1987). *J. Cell Biol.* **105**, 1473–1478.

Cresti, M., and Keijzer, C. J. (1985). *J. Submicrosc. Cytol.* **17**, 615–620.

Cresti, M., and Tiezzi, A. (1990). *In* "Microspores: Evolution and Ontogeny" (S. Blackmore, and R. B. Knox, eds.), pp. 239–263. Academic Press, San Diego.

Cresti, M., Pacini, E., Sarfatti, G., and Simoncioli, C. (1975). In "Gamete Competition in Plants and Animals" (D. L. Mulcahy, ed.), pp. 19–28. North Holland, Amsterdam.

Cresti, M., Van Went, J. L., Willemse, M. T. M., and Pacini, E. (1976). Acta Bot. Neerl. 25, 381–383.

Cresti, M., Pacini, E., Ciampolini, F., and Sarfatti, G. (1977). Planta 136, 239–247.

Cresti, M., Ciampolini, F., and Sarfatti, G. (1980). Planta 150, 211–217.

Cresti, M., Ciampolini, F., and Kapil, R. N. (1983a). Acta Bot. Neerl. 32, 177–183.

Cresti, M. Ciampolini, F., and Sarfatti, G. (1983b). In "Pollen Biology and Implications for Plant Breeding" (D. L. Mulcahy and E, Ottaviano, eds.), pp. 165–172. Elsevier, Amsterdam.

Cresti, M., Ciampolini, F., and Kapil, R. N. (1984). J. Submicrosc. Cytol. 16, 317–326.

Cresti, M., Ciampolini, F., Mulcahy, D. L., and Mulcahy, G. (1985). Am. J. Bot. 72, 719–727.

Cresti, M., Ciampolini, F., and Tiezzi, A. (1986a). Acta Bot. Neerl. 35, 285–292.

Cresti, M., Hepler, P. K., Tiezzi, A., and Ciampolini, F. (1986b). In "Biotechnology and Ecology of Pollen" (D. L. Mulcahy, G. B. Mulcahy, and E. Ottaviano, eds.), pp. 283–288. Springer-Verlag, New York.

Cresti, M., Lancelle, S. A., and Hepler, P. K. (1987). J. Cell Sci. 88, 373–378.

Cresti, M., Milanesi, C., Tiezzi, A., Ciampolini, F., and Moscatelli, A. (1988). Acta Bot. Neerl. 37, 379–386.

Cresti, M., Murgia, M., and Theunis, C. H. (1990a). Protoplasma 154, 151–156.

Cresti, M., Milanesi, C., Salvatici, P., and Van Aelst, A. C. (1990b). Bot. Acta 103, 349–354.

Cresti, M., Ciampolini, F., and Van Went, J. L. (1991). Ann. Bot. 68, 105–107.

Cross, R. A., and Kendrick-Jones, J. (eds.) (1991). Motor Proteins. J. Cell. Sci. (Suppl. 14).

Cyr, R. J., and Palevitz, B. A. (1989). Planta 177, 245–260.

De Brabander, M., Geuens, G., Nuyens, R., Willebrords, R., Aerts, F., De Mey, J., and McIntosh, J. R. (1986). Int. Rev. Cytol. 101, 215–274.

De Nettancourt, D. (1977). "Incompatibility in Angiosperms. Springer-Verlag, New York.

Derksen, J., and Emons, A. M. (1990). In "Tip Growth in Plant and Fungal Cells" (I. B. Heath, ed.), pp. 147–181. Academic Press, San Diego.

Derksen, J., and Traas, J. A. (1985). In "Sexual Reproduction in Seed Plants, Ferns and Mosses" (M. T. M. Willemse and J. L. Van Went, eds.), pp. 64–67. Pudoc, Wageningen.

Derksen, J., Pierson, E. S., and Traas, J. A. (1985). Eur. J. Cell Biol. 38, 142–148.

Derksen, J., Wilms, F. H. A., and Pierson, E. S. (1990). Acta Bot. Neerl. 39, 1–18.

Dexheimer, J. (1966). C. R. Acad. Sci. Paris 263, 1703–1705.

Dewey, M., Evans D., Coleman J., Priestley, R., Hull, R., Horsley, D., and Hawes, C. (1991). Acta Bot. Neerl. 40, 1–27.

Dickinson, D. B. (1968). Plant Physiol. 43, 1–8.

Dickinson, H. G., and Sheldon, J. (1984). Planta 161, 86–90.

Dixon, G. K., Brownlee, C., and Merrett, M. J. (1989). Planta 178, 443–449.

Dumas, C., Knox, R. B., and Gaude, T. (1985). Protoplasma 124, 168–174.

Dustin, P. (1984). "Microtubules," 2nd ed. Springer-Verlag, New York.

Elleman, C. J., Willson, C. E., and Dickinson, H. G. (1987). Pollen Spores 29, 273–290.

Emons, A. M. C., Pierson, E. S., and Derksen, J. (1991). In "Biotechnology: Current Progress" (P. N. Cheremisimoff and L. Ferrante, eds.), pp. 311–335. Technomic, Basel.

Ender, C., Li, M. Q., Martin, B., Povh, B., Nobiling, R., Reiss, H. D., and Traxel, K. (1983). Protoplasma 116, 201–203.

Fähnrich, P. (1964). In "Pollen Physiology and Fertilization" (H. F. Linskens, ed.), pp. 120–127. North Holland, Amsterdam.

Fedtke, C. (1982). "Biochemistry and Physiology of Herbicide Action." Springer Verlag, New York.

Fosket, D. E. (1989). In "The Biochemistry of Plants: A Comprehensive Treatise" (A. Marcus, ed.), Vol. 15: Molecular Biology, pp. 393–455. Academic Press, San Diego.

Franke, W. W., Herth, W., Van Der Woude, W. J., and Morré, D. J. (1972). *Planta* **105**, 317–341.

Gilissen, L. J. W. (1977). *Planta* **137**, 299–301.

Gilroy, S., Blowers, D. P., and Trewavas, A. J. (1987). *Development* **100**, 181–184.

Gunning, B. E. S., and Hardham, A. R. (1982). *Annu. Rev. Plant Physiol.* **33**, 651–698.

Haughland R. P. (1989). "Handbook of Fluorescent Probes and Research Chemicals." Molecular Probes, Eugene.

Hausser, I., Herth, W., and Reiss, H. D. (1984). *Planta* **162**, 33–39.

Heath, I. B. (1990). *Int. Rev. Cytol.* **123**, 95–127.

Hepler, P. K. (1988). *In* "Proc. of the XIV Int. Bot. Congress" (W. Greuter and B. Zimmer, eds.), pp. 225–240. Koeltz, Koenigstein and Taunus.

Hepler, P. K., and Palevitz, B. A. (1974). *Annu. Rev. Plant Physiol.* **25**, 309–362.

Hepler, P. K., and Wayne, R. O. (1985). *Annu. Rev. Plant Physiol.* **36**, 397–439.

Herrero, M., and Dickinson, H. G. (1981). *J. Cell Sci.* **47**, 365–383.

Herth, W. (1978). *Protoplasma* **96**, 275–282.

Herth, W. (1989). *Eur. J. Cell Biol.* **48**(Suppl. 26), 82.

Herth, W., Franke, W. W., and Van Der Woude, W. J. (1972). *Naturwissenschaften* **59**, 38–39.

Herth, W., Reiss, H. D., and Hartmann, E. (1990). *In* "Tip Growth in Plant and Fungal Cells" (I. B. Heath, ed.), pp. 91–118. Academic Press, San Diego.

Hesemann, C. U. (1972). *Flora* **161**, 509–526.

Heslop-Harrison, J. (1979a). *Ann. Bot.* **44**(Suppl. 1), 1–47.

Heslop-Harrison, J. (1979b). *Ann. Missouri Bot. Gard.* **66**, 813–829.

Heslop-Harrison, J. (1983). *Proc. R. Soc. Lond. [Biol.]* **218**, 371–395.

Heslop-Harrison, J. (1987). *Int. Rev. Cytol.* **107**, 1–78.

Heslop-Harrison, J. (1988). *In* "Sexual Reproduction in Higher Plants" (M. Cresti, P. Gori, and E. Pacini, eds.), pp. 195–203. Springer-Verlag, New York.

Heslop-Harrison, J., and Heslop-Harrison, Y. (1982). *Protoplasma* **112**, 71–80.

Heslop-Harrison, J., and Heslop-Harrison, Y. (1984). *Acta Bot. Neerl.* **33**, 131–134.

Heslop-Harrison, J., and Heslop-Harrison, Y. (1987). *Plant Sci.* **51**, 203–213.

Heslop-Harrison, J., and Heslop-Harrison, Y. (1988a). *Sex. Plant Reprod.* **1**, 16–24.

Heslop-Harrison, J., and Heslop-Harrison, Y.(1988b). *Ann. Bot.* **61**, 249–254.

Heslop-Harrison, J., and Heslop-Harrison, Y. (1988c). *Ann. Bot.* **62**, 455–461.

Heslop-Harrison, J., and Heslop-Harrison, Y. (1989a). *Sex. Plant Reprod.* **2**, 27–37.

Heslop-Harrison, J., and Heslop-Harrison, Y. (1989b). *Sex. Plant Reprod.* **2**, 199–207.

Heslop-Harrison, J., and Heslop-Harrison, Y. (1989c). *J. Cell Sci.* **93**, 299–308.

Heslop-Harrison, J., and Heslop-Harrison, Y. (1989d). *J. Cell Sci.* **94**, 319–325.

Heslop-Harrison, J., and Heslop-Harrison, Y. (1990). *Sex. Plant Reprod.* **3**, 187–194.

Heslop-Harrison, J., and Heslop-Harrison, Y. (1991a). *Sex. Plant Reprod.* **4**, 6–11.

Heslop-Harrison, J., and Heslop-Harrison Y. (1991b). *Philos. Trans. R. Soc. Lond. [Biol.]* **331**, 225–235.

Heslop-Harrison, J., Heslop-Harrison, J. S., and Heslop-Harrison, Y. (1986a). *Ann. Bot.* **58**, 1–12.

Heslop-Harrison, J., Heslop-Harrison, Y., Cresti, M., Tiezzi, A., and Ciampolini, F. (1986b). *J. Cell Sci.* **86**, 1–8.

Heslop-Harrison, Y., Heslop-Harrison, J. S., and Heslop-Harrison, J. (1986c). *Acta Bot. Neerl.* **35**, 265–284.

Heslop-Harrison, J., Heslop-Harrison, Y., Cresti, M., Tiezzi, A., and Moscatelli, A. (1988). *J. Cell Sci.* **91**, 49–60.

Heslop-Harrison, J., Heslop-Harrison, Y., Cresti, M., and Ciampolini, F. (1991). *Sex. Plant Reprod.* **4**, 73–80.

Heslop-Harrison, J. S., Heslop-Harrison, J., Heslop-Harrison, Y., and Reger, B. J. (1985). *Proc. R. Soc. Lond.* [*Biol.*] **225**, 315–327.
Hightower, R. C., and Meagher, R. B. (1986). *Genetics* **114**, 315–332.
Hoefert, L. L. (1969). *Protoplasma* **68**, 237–240.
Hogan, C. J. (1987). *Protoplasma* **138**, 126–136.
Horak, A., and Hill, R. D. (1972). *Plant Physiol.* **49**, 365–370.
Hu, S. Y., and Yu, H. S. (1988). *Protoplasma* **147**, 55–63.
Inoué, S. (1986). "Video Microscopy." Plenum Press, New York.
Iwanami, Y. (1956). *Phytomorphology* **6**, 288–295.
Jaffe, L. A., Weisenseel, M. H., and Jaffe, L. F. (1975). *J. Cell Biol.* **67**, 488–492.
Jensen, W. A., Ashton, M., and Heckard, L. R. (1974). *Bot. Gaz.* **135**, 210–218.
Jensen, W. A., and Fisher, D. B. (1970). *Protoplasma* **69**, 215–235.
Jokusch, B. M., Fürchtbauer, A., Wiegand, C., and Höner, B. (1986). *In* "Cell and Molecular Biology of the Cytoskeleton" (J. W. Shay, ed.), pp. 1–40. Plenum Press, New York.
Kakimoto, T., and Shibaoka, H. (1987). *Plant Cell Physiol.* **28**, 1581–1585.
Karas, I., and Cass, D. D. (1976). *Phytomorphology* **26**, 36–45.
Kaul, V., Theunis, C. H., Palser, B. F., Knox, R. B., and Williams, E. G. (1987). *Ann. Bot.* **59**, 227–235.
Kengen, H. M. P., and Derksen, J. (1991). *Acta Bot. Neerl.* **40**, 29–40.
Kilmartin, J. V., Wright, B., and Milstein, C. (1982). *J. Cell Biol.* **93**, 576–582.
Kirkeby, S., and Moe, D. (1983). *Electrophoresis* **4**, 236–237.
Knebel, W., and Reiss, H. D. (1990). *Leitz-Mitt. Wiss. Tech.* **9**, 279–283.
Knox, R. B. (1984). *Encycl. Plant Physiol. New Ser.* **17**, 508–608.
Knox, R. B., and Singh M. B. (1987). *Ann. Bot.* **60**(Suppl. 4), 15–37.
Kohno, T., and Shimmen, T. (1987). *Protoplasma* **141**, 177–179.
Kohno, T., and Shimmen, T. (1988a). *J. Cell Sci.* **91**, 501–509.
Kohno, T., and Shimmen, T. (1988b). *J. Cell Biol.* **106**, 1539–1543.
Kohno, T., Chaen, S., and Shimmen, T. (1990). *Protoplasma* **154**, 179–183.
Kohno, T., Okagaki, T., Kohama, K., and Shimmen, T. (1991). *Protoplasma* **161**, 75–77.
Kristen, U. (1986). *In* "Progress in Botany," Vol. 48, pp. 1–20. Springer-Verlag, Berlin.
Kristen, U. (1987). *In* "Progress in Botany," Vol. 49, pp. 1–12. Springer-Verlag, Berlin.
Kuroda, K. (1990). *Int. Rev. Cytol.* **121**, 267–307.
Kwack, B. H. (1967). *Physiol. Plant.* **20**, 825–833.
Lackie, J. M. (1986). "Cell Movement and Cell Behavior." Allen and Unwin, London.
Laemmli, U. (1970). *Nature* **227**, 680–685.
Lancelle, S. A., and Hepler, P. K. (1988). *Protoplasma* (Suppl. 2), 65–75.
Lancelle, S. A., and Hepler, P. K. (1989). *Protoplasma* **150**, 72–74.
Lancelle, S. A., Callaham, D. A., and Hepler, P. K. (1986). *Protoplasma* **131**, 153–165.
Lancelle, S. A., Cresti, M., and Hepler, P. K. (1987). *Protoplasma* **140**, 141–150.
Lehrer, S. S. (1981). *J. Cell Biol.* **90**, 459–466.
Lewandowska, E., and Charzynska, M. (1977). *Acta Soc. Bot. Polon.* **46**, 587–597.
Lichtscheidl, I. K., and Url, W. G. (1987). *Eur. J. Cell Biol.* **43**, 93–97.
Lichtscheidl, I. K., and Weiss, D. G. (1988). *Eur. J. Cell Biol.* **46**, 376–382.
Linskens, H. F., Ciampolini, F., and Cresti, M. (1989). *Proc. Kon. Ned. Acad. Wet.* **C92**, 465–475.
Liu, B., and Palevitz, B. A. (1991). *J. Cell Sci.* **98**, 475–482.
Liu, G. Q., and Yen, L. F. (1987). *Acta Biol. Exp. Sin.* **20**, 267–273.
Liu, G. Q., and Yen, L. F. (1991). *Acta Bot. Sin.* **33**, 214–218.
Liu, G. Q., Teng, X. Y., and Yen, L. F. (1990). *Acta Bot. Sin.* **32**, 187–190.
Lloyd, C. W. (1987). *Annu. Rev. Plant Physiol.* **38**, 119–139.
Lloyd, C. (1988). *J. Cell Sci.* **90**, 185–188.

Lloyd, C. (1989). *Curr. Opinion Cell Biol.* **1**, 30–35.

Ludwig, S. R., Oppenheimer, D. G., Silflow, C. D., and Snustad, D. P. (1987). *Proc. Natl. Acad. Sci. U.S.A.* **84**, 5833–5837.

Ludwig, S. R., Oppenheimer, D. G., Silflow, C. D., and Snustad, D. P. (1988). *Plant Mol. Biol.* **10**, 311–321.

Ma, Y. Z., and Yen, L. F. (1988). *Acta Bot. Sin.* **30**, 285–291.

Ma, Y. Z., and Yen, L. F. (1989). *Chin. Biochem. J.* **5**, 320–325.

Maheshwari, P. (1949). *Bot. Rev.* **15**, 1–75.

Mascarenhas, J. P. (1966). *Am. J. Bot.* **53**, 563–569.

Mascarenhas, J. P. (1975). *Bot. Rev.* **41**, 259–314.

Mascarenhas, J. P., and Lafountain, J. (1972). *Tissue Cell* **4**, 11–14.

Maupin-Szamier, P., and Pollard, T. D. (1978). *J. Cell Biol.* **77**, 837–852.

McConchie, C. H., Hough, T., and Knox, R. B. (1987). *Protoplasma* **139**, 9–19.

McLean, B., and Juniper, B. E. (1988). *Cell Biol. Int. Rep.* **12**, 509–517.

Mesland, D. A. M., and Spiele, H. (1984). *J. Cell Sci.* **68**, 113–137.

Miki-Hirosige, H., and Nakamura, S. (1982). *J. Electron Microsc.* **31**, 51–62.

Morejohn, L. C., and Fosket, D. E. (1986). *In* "Cell Molecular Biology of the Cytoskeleton" (J. W. Shay, ed.), pp. 257–330. Plenum Press, New York.

Moscatelli, A., Tiezzi, A., Cai, G., Ciampolini, F., and Cresti, M. (1988a). *Giorn. Bot. Ital.* **122**, 90–91.

Moscatelli, A., Tiezzi, A., Vignani, R., Bartalesi, A., and Cresti, M. (1988b). *In* "Sexual Reproduction in Higher Plants" (M. Cresti, P. Gori, and E. Pacini, eds.), pp. 205–209. Springer-Verlag, New York.

Moscatelli, A., Tiezzi, A., Cai, G., and Cresti, M. (in press). *Phytomorphology*

Murgia, M., Wilms, H. J., Cresti, M., and Cesca, G. (1986). *Acta Bot. Neerl.* **35**, 405–424.

Nagle, R. B. (1988). *Am. J. Surg. Pathol.* **12**(Suppl. 1), 4–16.

Nakamura, S. (1979). *J. Electron Microsc.* **28**, 275–284.

Nobiling, R., and Reiss, H. D. (1987). *Protoplasma* **139**, 20–24.

Pacini, E. (1990). *Plant Syst. Evol.* (Suppl. 5), 53–69.

Palevitz, B. A. (1990). *Protoplasma* **157**, 120–127.

Palevitz, B. A., and Cresti, M. (1988). *Protoplasma* **146**, 28–34.

Palevitz, B. A., and Cresti, M. (1989). *Protoplasma* **150**, 54–71.

Parthasarathy, M. V. (1987). *Plant Mol. Biol. Rep.* **5**, 251–259.

Parthasarathy, M. V., Perdue, T. D., Witztum, A., and Alvernaz, J. (1985). *Am. J. Bot.* **72**, 1318–1323.

Pawley, J. B. (1989). "Handbook of Biological Confocal Microscopy." Plenum Press, New York.

Perdue, T. D., and Parthasarathy, M. V. (1985). *Eur. J. Cell Biol.* **39**, 13–20.

Perez-Munoz, C. A., and Webster, B. D. (1990). *Am. J. Bot.* **77**(Suppl.), 21.

Peter, J. K., and Stanley, R. G. (1974). *In* "Fertilization in Higher Plants" (H. F. Linskens, ed.), pp. 131–136. North Holland, Amsterdam.

Picton, J. M., and Steer, M. W. (1981). *J. Cell Sci.* **49**, 261–272.

Picton, J., and Steer, M. W. (1982). *J. Theor. Biol.* **98**, 15–20.

Picton, J. M., and Steer, M. W. (1983a). *Protoplasma* **115**, 11–17.

Picton, J., and Steer, M. W. (1983b). *Eur. J. Cell Biol.* **29**, 133–138.

Picton, J. M., and Steer, M. W. (1985). *Planta* **163**, 20–26.

Pierson E. S. (1988). *Sex. Plant Reprod.* **1**, 83–87.

Pierson, E. S. (1989). Ph.D. Thesis, University of Nijmegen. Nijmegen, The Netherlands.

Pierson, E. S., Derksen, J., and Traas, J. A. (1985). *Acta Bot. Neerl.* **34**, 437.

Pierson, E. S., Derksen, J., and Traas, J. A. (1986a). *Eur. J. Cell Biol.* **41**, 14–18.

Pierson, E. S., Willekens, P. G. M., Maessen, M., and Helsper, J. P. F. G. (1986b). *Acta Bot. Neerl.* **35**, 249–256.

Pierson, E. S., Kengen, H., and Derksen, J. (1989). *Protoplasma* **150**, 75–77.
Pierson, E. S., Lichtscheidl, I. K., and Derksen, J. (1990). *J. Exp. Bot.* **41**, 1461–1468.
Poddubnaja-Arnoldi, V. (1936). *Planta* **25**, 502–529.
Polito, V. S. (1983). *Protoplasma* **117**, 226–232.
Pollard T. D., and Cooper, J. A. (1986). *Annu. Rev. Biochem.* **55**, 987–1035.
Polunia, N. N., and Sveshnikov, A. I. (1963). *Int. Symp. on Pollen Physiol. and Fertilization, 29–31 August 1963, Leningrad.* (Film from 1953).
Putnam-Evans, C., Harmon, A. C., Palevitz, B. A., Fechheimer, M., and Cormier, M. J. (1989). *Cell Motil. Cytoskeleton* **12**, 12–22.
Putnam-Evans, C., Harmon, A. C., and Cormier, M. J. (1990). *Biochemistry* **29**, 2488–2495.
Que, Q. D., and Tang, X. H. (1988). *Acta Bot. Sin.* **30**, 501–507.
Raudaskoski, M., Åström, H., and Färdig, M. (1986). *J. Ultrastruct. Res.* **94**, 285.
Raudaskoski, M., Åström, H., Perttilä, K., Virtanen, I., and Louhelainen, J. (1987). *Biol. Cell* **61**, 177–188.
Reiss, H. D., and Herth, W. (1978). *Protoplasma* **97**, 373–377.
Reiss, R. D., and Herth, W. (1979). *Planta* **145**, 225–132.
Reiss, R. D., and Herth, W. (1980). *Planta* **147**, 295–301.
Reiss, H. D., and Herth, W. (1985). *J. Cell Sci.* **76**, 247–254.
Reiss, H. D., and Nobiling, R. (1986). *Protoplasma* **131**, 244–246.
Reiss, H. D., and McConchie, C. A. (1988). *Protoplasma* **142**, 25–35.
Reiss, H. D., Herth, W., Schnepf, E., and Nobiling, R. (1983). *Protoplasma* **115**, 218–225.
Reiss, H. D., Herth, W., and Nobiling, R. (1985a). *Planta* **163**, 84–90.
Reiss, H. D., Grime, G. W., Li, M. Q., Takacs, J., and Watt, F. (1985b). *Protoplasma* **126**, 147–152.
Robards, A. W., and Sleytr, U. B. (1985). "Low Temperature Methods in Biological Electron Microscopy," Vol. 10: Practical Methods in Electron Microscopy. Elsevier, Amsterdam.
Robinson, D. G., and Depta, H. (1988). *Annu. Rev. Plant Physiol. Plant Mol. Biol.* **39**, 53–99.
Robinson, D. G., and Hillmer, S. (1990). *In* "The Plant Plasma Membrane: Structure, Function and Molecular Biology" (C. Larsson and I. M. Moller, eds.), pp. 233–255. Springer-Verlag, Berlin.
Roederer, G., and Reiss, H. D. (1988). *Protoplasma* **144**, 101–109.
Rosen, W. G. (1961). *Am. J. Bot.* **48**, 889–895.
Rosen, W. G., Gawlick, S. R., Dashek, W. V., and Siegesmund, K. A. (1964). *Am. J. Bot.* **51**, 61–71.
Russell, S. D. (1990). *Proc. XII Int. Congress Electron Microscopy.* San Francisco Press, San Francisco.
Russell, S. D., and Cass, D. D. (1981). *Protoplasma* **107**, 85–107.
Rutten, T. L. M., and Derksen, J. (1990). *Planta* **180**, 471–479.
Sanger, J. M., and Jackson, W. T. (1971a). *J. Cell Sci.* **8**, 289–301.
Sanger, J. M., and Jackson, W. T. (1971b). *J. Cell Sci.* **8**, 303–315.
Sassen, M. M. A. (1964). *Acta Bot. Neerl.* **13**, 175–181.
Schmit, A. C., and Lambert, A. M. (1990). *Plant Cell* **2**, 129–138.
Schnepf, E. (1986). *Annu. Rev. Plant Physiol.* **37**, 23–47.
Schoch-Bodmer, H. (1932). *Verh. Schweiz. Naturforsch. Ges.* **113**, 368–371.
Schroeder, M. J., Wehland, J., and Weber, K. (1985). *Eur. J. Cell Biol.* **38**, 211–218.
Seagull, R. W. (1989). *Crit. Rev. Plant Sci.* **8**, 131–167.
Seagull, R. W. (1991). *In* "Biosynthesis and Biodegradation of Cellulose and Cellulosic Materials" (C. H. Haigler and P. Weimer, eds.), pp. 143–163. Marcel Dekker, New York.
Seagull, R. W., and Heath, I. B. (1979). *Eur. J. Cell Biol.* **20**, 184–188.
Sheetz, M. P. (1989). *In* "Cell Movement: Kinesin, Dynein, and Microtubule Dynamics" (F. D. Warner, and J. R. McIntosh, eds.), Vol. 2, pp. 277–285. Liss, New York.

Sheldon, J. M., and Dickinson, H. G. (1986). *Planta* **168**, 11–23.
Sheldon, J., and Hawes, C. (1988). *Cell Biol. Int. Rep.* **12**, 471–476.
Shibata-Sekiya, K., and Tonomura, Y. (1975). *J. Biochem.* **77**, 543–557.
Shimmen, T. (1988). *Bot. Mag. Tokyo* **101**, 533–544.
Shivanna, K. R., and Johri, B. M. (1985). "The Angiosperm Pollen." Wiley Eastern, New Delhi.
Shivanna, K. R., and Rangaswamy, N. S. (in press) "Pollen Biology: A Laboratory Manual." Springer-Verlag, Heidelberg.
Shotton, D. M. (1989). *J. Cell Sci.* **94**, 175–206.
Simmonds, D. (1990). *Physiol. Plant.* **79**, A1.
Simmonds, D. (in press). *Int. Symp. on Angiosperm Pollen and Ovules: Basic and Applied Aspects, Como, June 23-27, 1991*. Springer-Verlag, Berlin.
Sonobe, S., and Shibaoka, H. (1989). *Protoplasma* **148**, 80–86.
Southworth, D., and Dickinson, D. B. (1981). *Grana* **20**, 29–35.
Spurr, A. R. (1969). *J. Ultrastruct. Res.* **26**, 31–43.
Staiger, C. J., and Schliwa, M. (1987). *Protoplasma* **141**, 1–12.
Staiger, C. J., and Cande, W. Z. (1990). *Dev. Biol.* **138**, 231–242.
Staiger, C. J., and Lloyd, C. W. (1991). *Curr. Opinion Cell Biol.* **3**, 33–42.
Stanley, R. G., and Linskens, H. F. (1974). "Pollen: Biology, Biochemistry and Management." Springer-Verlag, New York.
Steer, M. W. (1990). *In* "Tip Growth in Plant and Fungal Cells" (I. B. Heath, ed.), pp. 119–145. Academic Press, San Diego.
Steer, M. W., and Steer, J. M. (1989). *New Phytol.* **111**, 323–358.
Steer, M. W., and O'Driscoll, D. (1991). *In* "Endocytosis, Exocytosis, and Vesicle Traffic in Plants" (C. Hawes, J. Coleman, and D. Evans, eds.), pp. 129–142. Cambridge University Press, Cambridge.
Steer, M. W., Picton, J. M., and Earnshaw, J. C. (1984). *J. Microsc.* **134**, 143–149.
Steffen, K. (1953). *Flora* **93**, 140–174.
Stinson, J. R., Eisenberg, A. J., Willing, R. P., Pè, M. E., Hanson, D. D., and Mascarenhas, J. P. (1987). *Plant Physiol.* **83**, 442–447.
Strömgren-Allen, N. (1983). *In* "The Application of Laser Light Scattering to the Study of Biological Motion" (J. C. Earnshaw and M. W. Steer, eds.), pp. 529–543. Plenum, New York.
Tanaka, I. (1991). *J. Cell Sci* **99**, 21–31.
Tanaka, I., Nakamura, S., and Miki-Hirosige, H. (1989). *Gamete Res.* **24**, 361–374.
Tang, P. H. (1973). *Acta Bot. Sin.* **15**, 12–21.
Tang, P. H. (1988). *In* "Sexual Reproduction in Higher Plants" (M. Cresti, P. Gori, and E. Pacini, eds.), pp. 227–232. Springer-Verlag, New York.
Tang, X. J., Hepler, P. K., and Scordilis, S. P. (1988). *J. Cell Biol.* **107**, 259A.
Tang, X. J., Lancelle, S. A., and Hepler, P. K. (1989a). *Cell Motil. Cytoskeleton* **12**, 216–224.
Tang, X. J., Hepler, P. K., and Scordilis, S. P. (1989b). *J. Cell Sci.* **92**, 569–574.
Taylor, P., Kenrick, J., Li, Y., Kaul, V., Gunning, B. E. S., and Knox, R. B. (1989). *Sex. Plant Reprod.* **2**, 254–264.
Terasaka, O., and Niitsu, T. (1989a). *J. Jpn. Pollen Soc.* **182**, 7–12.
Terasaka, O., and Niitsu, T. (1989b). *Bot. Mag. Tokyo* **102**, 143–147.
Tewinkel, M., Kruse, S., Quader, H., Volkmann, D., and Sievers, A. (1989). *Protoplasma* **149**, 178–182.
Theunis, C. H., and Wilms H. J. (1988). *In* "Plant Sperm Cells as Tools for Biotechnology" (H. J. Wilms, and C. J. Keijzer, eds.), pp. 81–84. Pudoc, Wageningen.
Theunis, C. H., Pierson, E. S., and Cresti, M. (1991). *Sex. Plant Reprod.* **4**, 145–154.
Thompson, M. P., and Brower, D. P. (1985). *Fed. Proc.* **44**, 1450.

Tiezzi, A. (1991). *Electron Microsc. Rev.* **4**, 205–219.

Tiezzi, A., and Cresti, M. (1990). *In* "Mechanism of Fertilization: Plants to Humans" (B. Dale, ed.), NATO ASI Ser. Vol. H45, pp. 17–34. Springer-Verlag, Berlin.

Tiezzi, A., Cresti, A., and Ciampolini, F. (1986). *In* "Biology of Reproduction and Cell Motility in Plants and Animals" (M. Cresti and R. Dallai, eds.), pp. 87–94. University of Siena, Siena.

Tiezzi, A., Moscatelli, A., Milanesi, C., Ciampolini, F., and Cresti, M. (1987). *J. Cell Sci.* **88**, 657–661.

Tiezzi, A., Moscatelli, A., and Cresti, M. (1988a). *J. Submicrosc. Cytol. Pathol.* **20**, 613–617.

Tiezzi, A., Moscatelli, A., Cai, G., Ciampolini, F., and Cresti, M. (1988b). *Giorn. Bot. Ital.*, **122**, 88–89.

Tiezzi, A., Moscatelli, A., Ciampolini, F., Milanesi, C., Murgia, M., and Cresti, M. (1988c). *In* "Sexual Reproduction in Higher Plants" (M. Cresti, P. Gori, and E. Pacini, eds.), pp. 215–220. Springer-Verlag, New York.

Tiezzi, A., Moscatelli, A., Murgia, M., Russell, S. D., Del Casino, C., Bartalesi, A., and Cresti, M. (1989). *In* "Characterization of Male Transmission Units in Higher Plants" (B. Barnabás and K. Liszt, eds.), pp. 17–21. MTA Copy Budapest, Budapest.

Tiezzi, A., Bednara, J., and Del Casino, C. (in press). *Int. Symp. on Angiosperm Pollen and Ovules: Basic and Applied Aspects, Como, June 23-27, 1991.* Springer-Verlag, New York.

Tiezzi, A., Moscatelli, A., Cai, G. P., Bartalesi, A., and Cresti, M. (1992). *Cell Motil. Cytoskeleton* **21**, 132–137.

Tirlapur, U.K. (1987). *Fifth All India Symp. on Palynol., Nagpur, October 7–9, 1987,* p. 37.

Tirlapur, U. K. (1988). *J. Plant Sci. Res.* **4**, 77–80.

Tirlapur, U. K., and Shiggaon, S. V. (1987). *Jpn. J. Palynol.* **33**, 1–5.

Tiwari, S. C. (1989). *Can J. Bot.* **67**, 1244–1253.

Tiwari, S. C., and Polito, V. S. (1988a). *Am. J. Bot.* **75**, 51.

Tiwari, S. C., and Polito, V. S. (1988b). *Protoplasma* **147**, 5–15.

Tiwari, S. C., and Polito, V. S. (1988c). *Protoplasma* **147**, 100–112.

Tiwari, S. C., and Polito, V. S. (1988d). *In* "Sexual Reproduction in Higher Plants" (M. Cresti, P. Gori, and E. Pacini, eds.), pp. 211–214. Springer-Verlag, New York.

Tiwari, S. C., and Polito, V. S. (1990a). *Sex. Plant Reprod.* **3**, 121–129.

Tiwari, S. C., and Polito, V. S. (1990b). *Eur. J. Cell Biol.* **53**, 384–389.

Tiwari, S. C., Wick, S. M., Williamson, R. E., and Gunning, B. E. S. (1984). *J. Cell Biol.* **99**,(Suppl.) 63–69.

Tiwari, S. C., Polito, V. S., and Webster, B. D. (1990). *Protoplasma* **153**, 157–168.

Traas, J. A. (1984). *Protoplasma* **119**, 212–218.

Traas, J. A. (1990). *In* "The Plant Plasma Membrane: Structure, Function and Molecular Biology" (C. Larsson and I. M. Moller, eds.), pp. 269–292. Springer-Verlag, Berlin.

Traas, J. A., and Kengen, H. M. P. (1986). *J. Histochem. Cytochem.* **34**, 1501–1504.

Traas, J. A., Doonan, J. H., Rawlins, D., Shaw, P. J., Watts, J., and Lloyd, C. W. (1987). *J. Cell Biol.* **105**, 387–395.

Traas, J. A., Burgain, S., and Dumas De Vaulx, R. (1989). *J. Cell Sci.* **92**, 541–550.

Turian, G. (1981). *Bot. Helv.* **91**, 161–167.

Vale, R. D. (1987). *Annu. Rev. Cell Biol.* **3**, 347–378.

Vale, R. D., Rees, T. S., and Sheetz, M. P. (1985). *Cell* **42**, 39–50.

Valenta, R., Duchêne, M., Pettenburger, K., Sillaber, C., Valent, P., Bettelheim, P., Breitenbach, M., Rumpold, H., Kraft, D., and Scheiner, O. (1991). *Science* **253**, 557–560.

Valenta, R., Duchêne, M., Ebner, C., Valent, P., Sillaber, C., Deviller, P., Ferreira, F., Tejkl, M., Edelmann, H., Kraft, D., and Scheiner, O. (1992). *J. Exp. Med.* **175**, in press.

Van Lammeren, A. A. M., Keijzer, C. J., Willemse, M. T. M., and Kieft, H. (1985). *Planta* **165**, 1–11.

Van Lammeren, A. A. M., Bednara, J., and Willemse, M. T. M. (1988). *In* "Sexual Reproduction in Higher Plants" (M. Cresti, P. Gori, and E. Pacini, eds.), p. 476. Springer-Verlag, New York.

Van Lammeren, A. A. M., Bednara, J., and Willemse, M. T. M. (1989). *Planta* **178,** 531–539.

Vantard, M., Levilliers, N., Hill, A. M., Adoutte, A., and Lambert, A. M. (1990). *Proc. Natl. Acad. Sci. U.S.A.* **87,** 8825–8829.

Van Went, J., and Gori, P. (1989). *J. Submicrosc. Cytol. Pathol.* **21,** 149–156.

Vasil, I. K. (1964). *In* "Pollen Physiology and Fertilization," pp. 107–119. North Holland, Amsterdam.

Venema, G., and Koopmans, A. (1962). *Cytologia* **27,** 11–24.

Wagner, V. T., Cresti, M., Salvatici, P., and Tiezzi, A. (1990). *Planta* **181,** 304–309.

Wang, Y. D., and Yen, L. F. (1991). *Sin. Sci. Bull.* **36,** 862–866.

Warner, F. D., and McIntosh, J. R. (eds.) (1989). "Cell Movement: Kinesin, Dynein, and Microtubules Dynamics," Vol. II, p. 478 Liss, New York.

Warner, F. D., Satir, P., and Gibbons, I. R. (eds.) (1989). "Cell Movement: The Dynein ATPases," Vol. I, p. 337 Liss, New York.

Weber, M. (1989). *Protoplasma* **152,** 69–76.

Weisenseel, M. H., Nuccitelli, R., and Jaffe, L. F. (1975). *J. Cell Biol.* **66,** 556–567.

Weiss, D. (1986). *J. Cell Sci.* (Suppl. 5), 1–15.

Wick, S. M., Seagull, R. W., Osborn, M., Weber, K., and Gunning, B. E. S. (1981). *J. Cell Biol.* **89,** 685–690.

Wieland, T., and Faulstich, H. (1978). *Crit. Rev. Biochem.* **5,** 185–260.

Williamson, R. E. (1986). *Plant Physiol.* **82,** 631–634.

Xu, S. X., Zhu, C., and Hu, S. Y. (1990). *Acta Bot. Sin.* **32,** 821–826.

Yen, L. F., Wang, X. Z., Teng, X. Y., Ma, Y. Z., and Liu, G. Q. (1986). *Chin. Sci. Bull.* **31,** 267–272.

Yen, L. F., Liu, X., and Liu, G. Q. (in press). *Proc. Int. Symp. on Angiosperm Pollen and Ovules: Basic and Applied Aspects, Como, 1991.*

Yu, H. S., Hu, S. Y., and Zhu, C. (1989). *Protoplasma* **152,** 29–36.

Zee, S. Y., and Aziz-Un-Nisa (1991). *Sex. Plant Reprod.* **4,** 132–137.

Zhang, D. H., Wadsworth, P., and Hepler, P. K. (1990). *Proc. Natl. Acad. Sci. U.S.A.* **87,** 8820–8824.

Zhou, C. (1987). *Acta Bot. Sin.* **29,** 117–122.

Zhou, C., Orndorff, K., Daghlian, C. P., and De Maggio, A. E. (1988). *Sex. Plant Reprod.* **1,** 97–102.

Zhou, C., Yang, H. Y., and Xu, S. X. (Zee, S. Y.) (1990a). *Acta Bot. Sin.* **32,** 657–662.

Zhou, C., Zee, S. C., and Yang, H. Y. (1990b). *Sex. Plant Reprod.* **3,** 213–218.

Zhu, C., and Liu, B. (1990). *Chin. J. Bot.* **2,** 1–6.

Zhu, C., Hu, S. Y., Xu, L. Y., Li, X. R., and Shen, J. H. (1980). *Sci. Sin.* **23,** 371–376.

Zhu, C., Li, C. G., and Hu, S. Y. (1991). *Acta Bot. Sin.* **33,** 1–6.

Part II
Gametes

The Male Germ Unit: Concept, Composition, and Significance

H. Lloyd Mogensen

Department of Biological Sciences, Northern Arizona University
Flagstaff, Arizona 86011

I. Introduction

The pollen grain of flowering plants represents an extreme reduction of the male gametophyte down to the bare essentials for gamete formation and transport. The typical pollen grain at the time of its release from the anther consists of two haploid cells: a larger vegetative cell that will elongate to form the pollen tube and a smaller generative cell that will divide to produce two male sex cells, each of which is destined to be involved in a separate fusion event (Maheshwari, 1950).

The evolution of the pollen tube has allowed the flowering plants (and gymnosperms) to become true land plants because the pollen tube provides a means of sperm transport without the requirement of free water (Foster and Gifford, 1974). Because pairs of male gametes are confined to an individual pollen tube throughout all but the final few micrometers of their journey to the female gametes, unlike the situation in animal systems, direct sperm competition is nearly eliminated. Instead, opportunity for competition occurs primarily among microspores (the haploid precursors of pollen grains), the pollen grains, or the pollen tubes (D. Mulcahy, personal communication; Russell, 1991).

The focus of this chapter is on angiosperm generative cells and sperm cells, and their interaction with the vegetative cell nucleus to form a functional assemblage, termed the male germ unit. The formulation of the male germ unit is believed to be an essential prerequisite for the successful completion of the unique process of double fertilization, that is, the fusion of one sperm with the egg to form the zygote and the fusion of the other sperm with the central cell to form the precursor of the nutritive endosperm tissue (Friedman, this volume; Russell, this volume).

Here, data from quantitative three-dimensional transmission electron microscopy are emphasized. Other features of the male germ unit are addressed elsewhere in this volume.

II. Concepts and Misconceptions

A. Are the Sperms True Cells Or Just Naked Nuclei?

The sperm cells of flowering plants (and higher gymnosperms) are unique from other members of the plant kingdom in that they are considerably less complex structurally and completely lack flagella (although flagellated sperms are retained in *Ginkgo* and the cycads; Foster and Gifford, 1974). Just how reduced the sperms have become has been a topic of considerable controversy, and whether or not flowering plant sperms are complete cells with a plasma membrane enclosing the usual organelles was a viable question up to the late 1930s. However, by the early 1940s it was well established in the scientific literature that the typical sperms are true cells with their own cytoplasmic sheath containing numerous inclusions (Maheshwari, 1949, 1950). It is now known, however, that many sperm cells lack plastids, typically due to their exclusion during generative cell formation at the time of microspore division (Hagemann and Schröder, 1989), but all of the other expected cytoplasmic components are usually present, including mitochondria, dictyosomes, ribosomes, endoplasmic reticulum, vesicles, and microtubules, with the possible exception of microfilaments (Palevitz and Tiezzi, this volume). Nevertheless, it is not uncommon even in current introductory textbooks to see the sperms depicted as naked nuclei. Consequently, the notion that the sperms have no cytoplasm is often used by those unfamiliar with the literature as the basis for explaining the predominance of strict maternal cytoplasmic inheritance in flowering plants.

Early electron-microscopic observations of male gametes, such as in corn (Larson, 1965), cotton (Jensen and Fisher, 1968, 1970), and beet (Hoefert, 1969, 1971), corroborated light microscopic studies by showing the sperms to be intact cells; however, they were still viewed as being highly simplified, "reduced to little more than a package for the nucleus" (Jensen and Fisher, 1968). Based upon electron microscopy, only two species to date, to the author's knowledge, have been reported to possess sperm cells that apparently lack any double membrane organelles other than the nucleus: *Epidendrum scutella* (Cocucci, 1988) and *Orlaya daucoides* (Weber, 1988a).

B. Are the Two Sperms of a Pair Identical?

Although some early light microscopic investigations indicated that the two sperms of a given pollen tube might be morphologically distinct from one another, Maheshwari (1950), in his extensive review of sexual reproduction in flowering plants, pointed out that the apparent differences were likely due to the sperms being viewed in different planes of section, or that they may take on varying shapes at different stages of development or fertilization. The general belief that the two sperms are isomorphic prevailed up through the 1970s because even electron-microscopic observations did not indicate any differences between the sperms of a pair (Jensen and Fisher, 1968, 1970; Hoefert, 1969, 1971; Cass and Karas, 1975; Karas and Cass, 1976).

III. Results from Quantitative, Three-dimensional Studies and the Concept of the Male Germ Unit

Since Jensen and subsequent workers (Jensen, 1973, 1974) had established that the pollen tube consistently discharges the sperms into one of the two synergids (cells that typically flank the egg), a prominent question during the 1970s was: What is the pollen tube pathway in plants that lack synergids? The selection of *Plumbago zeylanica,* therefore, as the study organism for a detailed investigation of embryogenesis was a logical one because of the absence of synergids in this species (Cass and Karas, 1974). As it turned out, in addition to providing details on fundamental aspects of the actual fertilization process in a flowering plant (Russell, this volume), results from studies on *Plumbago* have stimulated renewed interest worldwide in the field of plant sperm cell biology (Dumas *et al.,* 1984; Wilms and Keijzer, 1988; Barnabas and Liszt, 1990; Hu, 1990; Russell, 1991).

A. Connections—Associations between the DNA-Containing Components of Male Heredity

One of the first significant findings to emerge from work on the male reproductive system in *Plumbago* involves the intimate associations that exist between the two sperms, and between one sperm and the vegetative cell nucleus (Figs. 1 and 2). In one of their initial reports, Russell and Cass (1981) demonstrated that the two sperm cells are directly linked to each other by a common transverse cell wall containing plasmodesmata. The sperms are also encompassed by the inner vegetative cell plasma mem-

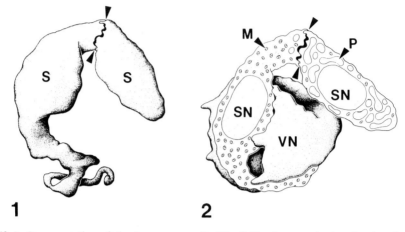

FIG. 1   Representation of the two sperm cells (S) of *Plumbago zeylanica* showing their connection to each other (unlabeled arrowheads) and external features. The larger of the sperms is consistently associated with the vegetative nucleus (not shown). Reproduced with permission from Russell (1984).

FIG. 2   Internal view of the two sperms of *Plumbago zeylanica* showing the predominance of plastids (P) in the smaller sperm cell and most of the mitochondria (M) in the larger sperm cell which is consistently associated with the vegetative nucleus (VN). The sperm cross-wall is indicated by the unlabeled arrowheads. SN, sperm nucleus. Reproduced with permission from Russell(1984).

brane, and one sperm is consistently associated with the vegetative nucleus through a sperm cell extension that embraces the vegetative nucleus over a large surface area and is partially surrounded by lobes of the vegetative nucleus. These associations exist within mature pollen (in this species the sperms are formed before pollen is shed from the anther and, thus, the pollen is termed tricellular) and are maintained throughout pollen tube growth. The sperms were also shown to be highly cytoplasmic, containing mitochondria, endoplasmic reticulum, ribosomes, vesicles, microtubules, possible microfilaments, and plastids, and it was noted that the male gametes of a pair "appear to differ in cell volume" (Russell and Cass, 1981).

   Although previous reports based upon light microscopic observations had noted persistently connected sperm cells (Maheshwari, 1949, 1950), and Jensen and Fisher (1968, 1970), using electron microscopy, had demonstrated a long-lasting association of the connected sperm pair with the vegetative nucleus in cotton pollen tubes, the condition in *Plumbago* caught the attention of the international scientific community because it was viewed in the context of a functional unit of male heredity that may be of fundamental importance to certain critical steps leading up to gametic fusion (Dumas *et al.*, 1984; Heslop-Harrison and Heslop-Harrison, 1984).

B. Sperm Dimorphism in *Plumbago*

The aspect of sperm differences in *Plumbago* was the focus of a subsequent study, utilizing the techniques of serial ultrathin sectioning and computer-assisted three-dimensional reconstruction (Russell, 1984). Not only were the sperms of a pair found to differ in cell size and shape, but also in the size of their nucleus and the number of cytoplasmic organelles (Figs. 1 and 2). The smaller sperm, which is unassociated with the vegetative nucleus, contains an average of 24 plastids and 40 mitochondria. The larger sperm, which is closely associated with the vegetative nucleus through a narrow extension, usually contains no plastids and has an average of 256 mitochondria (Russell, 1984).

C. Prevalence and Implications of the Male Germ Unit

It was not clear initially whether the occurrence of sperm cell–vegetative nucleus packaging and sperm dimorphism may be somehow related to the fact that *Plumbago* is synergidless and has plastids in one of its sperm cells. However, similar sperm cell connections and vegetative nucleus–sperm cell associations were soon found in spinach (Wilms and Van Aelst, 1983) and two species of *Brassica* (Dumas *et al.*, 1985), plants that have synergids, and the sperms of which lack plastids. Sperm dimorphism also appeared to be present in these plants, and this was later confirmed to be the case through quantitative studies (Wilms, 1986; McConchie *et al.*, 1987a), which showed that the sperms of a pair differ in size, shape, and number of mitochondria. These observations were immediately recognized as being of considerable significance because the presence of a "polarised fertilisation-unit" (Heslop-Harrison and Heslop-Harrison, 1984) might mean that double fertilization is not random, as previously thought, and, thus, a complete reassessment of the process would be needed (Heslop-Harrison and Heslop-Harrison, 1984). Moreover, differences in organelle content between the sperms of a pair may be of some importance to the field of cytoplasmic inheritance (Heslop-Harrison and Heslop-Harrison, 1984).

The summary article by Dumas *et al.* (1984) probably did more to stimulate new interest and research in plant sperm cell biology than any other. Here the concept of the male germ unit was introduced, suggesting that the nuclear and cytoplasmic DNA-containing compartments of male heredity are linked so that the male cells and the vegetative nucleus appear to function as a vehicle for transmission, recognition, and fusion with the female target cells during double fertilization. Several exciting possibilities became evident. If the sperms of a pair are different, perhaps they are

predestined to fuse with either the egg or central cell. Ultrastructural tracking of male organelles in *Plumbago* has shown that the smaller sperm, the one containing the plastids, united with the egg cell in 16 of 17 cases (Russell, 1985). Using different techniques, supporting evidence for directed or preferential fertilization was actually demonstrated much earlier by Roman (1947, 1948) who showed that, in certain lines of corn *(Zea mays)*, the sperm cell containing B chromosomes most often fuses with the egg cell.

If the individual sperms of a pair are recognizable, and it can be determined which one will combine with a given female target cell, it may be possible to utilize certain techniques, such as sperm cell isolation and subsequent microinjection, electroporation, or biolistic particle gun to effect specific genetic transformations of a given sperm type. Genetically transformed sperms, then, conceivably could be utilized in *in vitro* fertilizations (Kranz *et al.*, 1991) followed by regeneration to produce specific desirable traits in important crop plants (Wilms and Keijzer, 1988; Russell *et al.*, 1990; Russell, 1991; Chaboud and Perez, Huang and Russell, and Kranz, *et al.*, this volume).

Because the original concept of the male germ unit was based upon a limited number of species, all of which are dicots with tricellular pollen, some immediate questions were as follows: How widespread is the male germ unit and sperm dimorphism? Do sperm cell connections function primarily in keeping the dissimilar sperms together while they exit the pollen grain and enter the pollen tube? In plants with bicellular pollen, the sperms are not formed until the generative cell is well on its way down the pollen tube. If the sperms are not dimorphic, perhaps they would have a natural tendency to stay near one another during their passage through the pollen tube. However, an earlier study in cotton, a dicot plant with bicellular pollen, had shown the presence of a male germ unit without any apparent differences between the sperm cells (Jensen and Fisher, 1968, 1970). Subsequent studies soon demonstrated the existence of a male germ unit in other plants with bicellular pollen, such as *Hippeastrum vitatum,* a monocot (Mogensen, 1986a), and *Petunia hybrida,* a dicot (Wagner and Mogensen, 1987). However, in *Petunia,* which was investigated using quantitative three-dimensional reconstruction and morphometric analysis, the sperms of a pair are not significantly different morphologically (Wagner and Mogensen, 1987). So, it was appearing that the male germ unit, at least from the standpoint of connected sperms in close association with the vegetative nucleus, is a common feature. More recent studies have reported male germ units in additional species with bicellular pollen, and if the concept is expanded to include an intimate association between the generative cell and the vegetative nucleus (Fig. 3), the list is becoming long, including, for example, bicellular dicots, *Rhododendron* spp. (Kaul

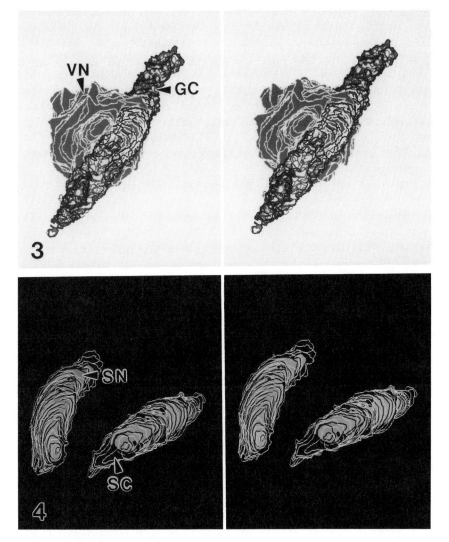

FIG. 3   Computer-generated stereoscopic pair of the generative cell (GC) and encompassing vegetative nucleus (VN) within the mature pollen grain of *Medicago sativa*. ~×4600. Reproduced with permission from Zhu *et al.* (1990a).

FIG. 4   Sperm cells of *Hordeum vulgare* reconstructed in their exact arrangement within the mature pollen grain. Prominent is the similarity between the two sperms, the lack of any connections between them, and the absence of the vegetative nucleus, which was not close enough to be included in the reconstruction. SC, sperm cytoplasm; SN, sperm nucleus. ~×3600. Reproduced with permission from Mogensen and Rusche (1985).

et al., 1987), *Acacia retinoides* (McCoy and Knox, 1988), *Cyphomandra betacea* (Hu and Yu, 1988), *Nicotiana tabacum* (Yu *et al.*, 1989), and *Medicago sativa* (Zhu *et al.*, 1990a); and bicellular monocots, *Aloe ciliaris* (Ciampolini *et al.*, 1988), *Galanthus nivalis* (Wilms *et al.*, 1988), and *Gladiolus gandavensis* (Shivanna *et al.*, 1988). Additional tricellular pollen species have also been reported to have a male germ unit, for example, tricellular dicots, *Cichorium intybus* (Levieil, 1986), *Catananche caerulea* (Barnes and Blackmore, 1987), *Gerbera jamesonii* (Provoost *et al.*, 1988), and *Sambucus nigra* (Charzynska and Lewandowska, 1990). Hu (1990) provided a more extensive list of species and described the techniques employed.

D. Exceptions to the Rule?

In the grasses, which are monocots with tricellular pollen, the literature before 1985 suggested that the sperms are not connected to each other, that neither sperm is in close association with the vegetative nucleus, and that the sperm cells are not dimorphic (barley, Cass, 1973, Cass and Karas, 1975; rye, Karas and Cass, 1976; wheat, Zhu *et al.* 1980, Mogensen, 1986c; *Triticale,* Schröder, 1983). A quantitative, three-dimensional study on barley confirmed that a male germ unit does not exist in this species at pollen maturity and that the sperms are essentially isomorphic (Mogensen and Rusche, 1985; Fig. 4). Yet, it was already known that the sperms arrive at the embryo sac at the same time and that double fertilization occurs normally (Pope, 1937; Cass and Jensen, 1970; Mogensen, 1982).

1. Postpollination Connections–Associations

The question then became: How important is the male germ unit after all? Perhaps some other mechanism(s) is operating to effect tandem movement of the sperms within the pollen tube and sperm recognition within the embryo sac in grasses. A subsequent study, which concentrated on post-pollination stages, helped to clarify the situation in barley (Mogensen and Wagner, 1987). Interestingly, it was found that within 5 minutes after *in vitro* pollination, close associations begin to form between the vegetative nucleus and the sperm cells. Within 15 minutes, while still within the pollen grain, the two sperms form cellular extensions and closely align with each other (to a minimum distance of 60 nm with no intervening cytoplasm) in various regions (Fig. 5). By this time, the vegetative nucleus was seen to completely surround one sperm cell extension over a distance of 3 μm (Figs. 6 and 7). Upon entering the pollen tube, the sperm cells appear to have fibrillar cell walls, at least in the areas where the sperm

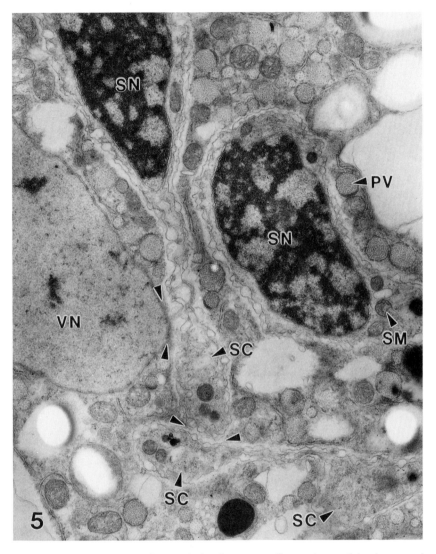

FIG. 5   Transmission electron micrograph showing sperm cell extensions and the close associations (unlabeled arrowheads) between the two sperms and between one sperm and the vegetative nucleus (VN) of *Hordeum vulgare* approximately 15 minutes after *in vivo* pollination. PV, pollen tube wall precursor vesicle; SC, sperm cytoplasm; SM, sperm mitochondrion; SN, sperm nucleus. ×12,500. Reproduced with permission from Mogensen and Wagner (1987).

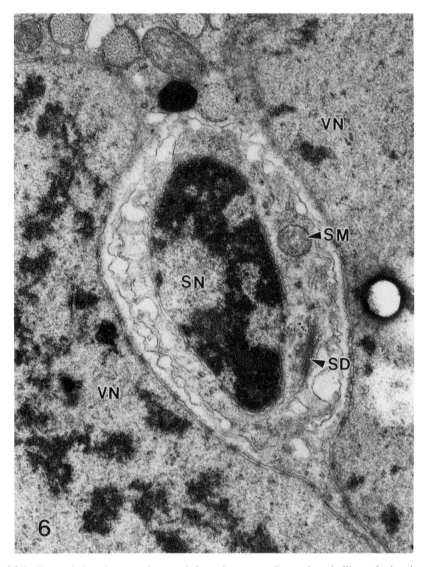

FIG. 6   Transmission electron micrograph from the same pollen grain as in Figure 5, showing the vegetative nucleus (VN) nearly surrounding one end of a sperm cell. SD, sperm dictyosome; SM, sperm mitochondrion; SN, sperm nucleus. ×20,500. Reproduced with permission from Mogensen and Wagner (1987).

cells are now physically connected. The sperms remain connected as the pollen tube grows through the stigma branches; as the pollen tube approaches the ovule, three-dimensional reconstruction shows the sperms to be connected at both ends (Mogensen and Wagner, 1987; Fig. 8). Results from this study provide an explanation for barley, and perhaps other grasses, as to how the sperm cells remain together as they travel through the pollen tube, that is, they become physically connected after pollination but before their exit from the pollen grain into the pollen tube. In a sense, then, barley is the exception that proves the rule, in that not only does a male germ unit form but it appears to be formed as the result of an active process, occurring only after pollen germination, rather than as a developmental consequence of generative cell division. This would seem to underscore the significance of the male germ unit. Why the sperms separate after their formation from generative cell division is not understood. Also curious is the short duration of the vegetative nucleus–sperm cell associations in barley; in no case was the vegetative nucleus seen to accompany the sperm cells within the pollen tube (Mogensen and Wagner, 1987). A wide separation between the sperm pair and the vegetative nucleus has also been reported in *Helleborus foetidus,* a dicot with bicellular pollen (Heslop-Harrison *et al.,* 1986) and in the grass *Alopecurus pratensis* (Heslop-Harrison and Heslop-Harrison, 1984). In wheat the sperms migrate into the pollen tube 5–15 minutes after pollination, whereas the vegetative nucleus does so only after 15–30 minutes (Chandra and Bhatnagar, 1974). In *Euphorbia dulcis* mature pollen contains two connected sperm cells, but they have no close associations with the vegetative nucleus (Murgia and Wilms, 1988). Even though the vegetative nucleus–sperm cell association is ephemeral in barley, the fact that it takes place, and to such an intimate degree, would appear to be significant when taken together with observations of similar, although more persistent, contacts in other species. It is apparent that observations of the lack of vegetative nucleus–sperm cell associations should take into account the possibility of brief associations that are more difficult to detect and the possibility that these associations may occur at later stages. Postpollination production of sperm cell extensions and the establishment of an intimate association between the vegetative nucleus and the leading sperm cell have been reported in *Hippeastrum* (Mogensen, 1986a, 1986b), and the formation of a complex association between the vegetative nucleus and the generative cell has been described to take place within 24 hours after pollination in *Rhododendron* (Kaul *et al.,* 1987). A similar course of postpollination events, as seen in barley, has been reported in corn (Rusche, 1988; Rusche and Mogensen, 1988). In corn it was found that the sperms and vegetative nucleus become closely associated only after pollen activation, but in this case the vegetative nucleus remains in close association with the connected sperms as they travel within the pollen tube.

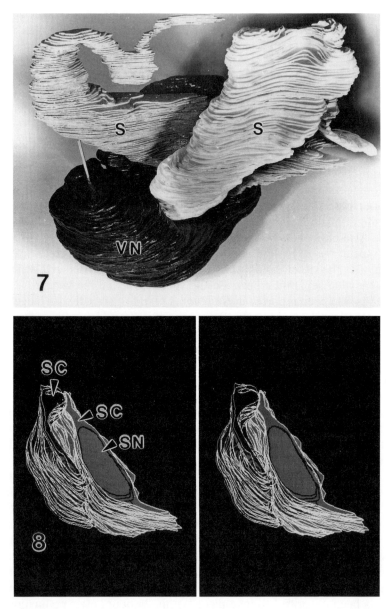

FIG. 7   Physical reconstruction of the complete male germ unit from a pollen grain of *Hordeum vulgare* fixed approximately 15 minutes after *in vivo* pollination, showing the close association of sperm cells (S) and the vegetative nucleus (VN). ~×5300. Reproduced with permission from Mogensen and Wagner (1987).

Thus, the formation of a maintained linkage between sperms of common origin, whether it exists from the time of sperm formation or is established later, appears to be a rather universal feature among flowering plants. The apparent exceptions, such as in *Secale cereale* (Heslop-Harrison and Heslop-Harrison, 1987), where living sperms have been observed in real-time to pass each other and to move in opposite directions within the pollen tube, must be viewed with caution until additional studies are made at the electron-microscopic level. It is not inconceivable that contacts between the sperm cells could shift, or slide, and that sperm cell extensions could stretch while the sperms move down the pollen tube. The disconnected sperm pairs observed at the electron-microscopic level (Weber, 1988b) in the tricellular pollen of *Avena sativa* (*Poaceae*), *Galium mollugo* (*Rubiaceae*), and *Trichodiadema setuliferum* (*Aizoaceae*) will predictably be found to become connected after pollen germination.

Intimate associations between the vegetative nucleus and generative–sperm cells also appear to be the rule, however brief these associations may be. Again, the apparent exceptions, such as in *Helleborus foetidus* (Heslop-Harrison *et al.*, 1986) and in *Alopecurus pratensis* (Heslop-Harrison and Heslop-Harrison, 1984), should be confirmed with electron microscopy, and the notion of short-lived, but intimate, associations should be kept in mind.

2. The Scale of Sperm Dimorphism

The high degree of sperm dimorphism that occurs in *Plumbago zeylanica* (Russell, 1984) has not yet been found in any other species. Considerable morphological differences do occur between the sperms of a pair in *Brassica* (McConchie *et al.*, 1987a) and spinach (Wilms, 1986), but the organellar differences are restricted to mitochondrial number, because neither sperm contains any plastids. Sperm dimorphism, based upon cell and/or nuclear size and shape, as well as on mitochondrial number and/or morphology, has been reported to occur to some degree in a number of additional species to date, including a monocot with bicellular pollen, *Gladiolus gandavensis* (Shivanna *et al.*, 1988); dicots with bicellular pollen, *Rhododendron* spp. (Shivanna *et al.*, 1988); dicots with tricellular pollen, *Euphorbia dulcis* (Murgia and Wilms, 1988) and *Gerbera jamesonii*

FIG. 8   Computer-generated stereoscopic pair of a partial reconstruction of the sperm cells of *Hordeum vulgare* within a pollen tube that was approaching the ovule. The sperms are still attached at both ends. The vegetative nucleus is no longer associated with the sperm pair. SC, sperm cytoplasm; SN, sperm nucleus. ~×3600. Reproduced with permission from Mogensen and Wagner (1987).

(Provoost *et al.*, 1988); and a monocot with tricellular pollen, *Zea mays* (McConchie *et al.*, 1987b; Rusche, 1988, 1990; Rusche and Mogensen, 1988). In the case of *Zea mays*, it should be noted that the extreme differences found between the reconstructed sperm cells of a single pair figured by McConchie *et al.* (1987b) were not observed in computer-generated reconstructions of three sperm pairs from a different variety of corn, black Mexican, which was determined to carry B chromosomes (Rusche, 1988, 1990; Rusche and Mogensen, 1988).

In a few cases so far, evidence for isomorphic sperm cells has been described using the techniques of serial ultrathin sectioning, computer reconstruction, and quantitation, *Hordeum vulgare*, a monocot with bicellular pollen (Mogensen and Rusche, 1985), and three dicots with bicellular pollen, *Petunia hybrida* (Wagner and Mogensen, 1987), *Nicotiana tabacum* (Yu, as cited in Russell, 1991), and *Medicago sativa* (Zhu *et al.*, 1990b; Zhu *et al.*, unpublished). In the case of *Medicago sativa*, which is known from genetic studies to have biparental plastid inheritance (Smith *et al.*, 1986), both sperms contain large numbers of plastids (about 60–80 per cell) and very few mitochondria (about 8–10 per cell).

IV. Functional Significance of the Male Germ Unit

All features of the male germ unit, namely, sperm-to-sperm connections, sperm dimorphism, and vegetative nucleus associations, appear to be related in a precisely controlled, integrated series of steps leading to the successful completion of double fertilization.

A. Sperm-to-Sperm Connections

Clearly, the two fertilization events that initiate embryo and endosperm development must be coordinated such that they occur essentially at the same time. Sperm-to-sperm connections during pollen tube transport would help to ensure this. Moreover, the available space within the embryo sac (usually within the degenerate synergid) for pollen tube discharge is limited, necessitating that the sperms be in close proximity and that they are near the pollen tube tip at the time of discharge. Because the two sperms have been seen to be in contact simultaneously with their respective female target cells, the egg and central cell, before actual fusion occurs (Mogensen, 1982; Russell, 1983), the order of arrival of the sperms at the fertilization site may not be critical. Likely, specific recognition events at

the level of the gamete plasma membranes presumably determine which cells will fuse (Russell, 1985, this volume).

Sperm-to-sperm connections may also play a role in reducing the occurrence of heterofertilization, namely, double fertilization involving sperms from different pollen tubes, which is known to occur in corn (Sprague, 1932), and barley (R.T. Ramage, personal communication).

The nature of the sperm-to-sperm connections in those cases where the sperms remain attached to each other after generative cell division is easily understood as a typical adhesion between two cell walls, possibly involving a middle lamella-like system. However, in those cases in which the sperms are initially separated and then rejoined, the mechanism of rejoining and the basis of the connections are not so easily explained. In barley after the sperms have rejoined, what appear to be fibrillar cell walls are present particularly at the points of attachment between the two sperms as they enter the pollen tube; however, at later stages sperm cell walls are indiscernible, yet the cells remain attached (Mogensen and Wagner, 1987).

B. Sperm Dimorphism

We have seen that morphological differences between the sperms of a pair can vary from the extreme in *Plumbago zeylanica* (Figs. 1 and 2) to little or none in *Hordeum vulgare* (Fig. 4). The high degree of sperm dimorphism in *Plumbago* has been correlated with preferential fertilization (Russell, 1985); yet, preferential fertilization is also known to occur in *Zea mays* (Roman, 1947, 1948), in which the degree of dimorphism may be rather slight (Rusche, 1988, 1990; Rusche and Mogensen, 1988). Future studies, utilizing immunological techniques to recognize sperm differences (Pennell *et al.*, 1987; Chaboud and Perez, this volume) may show that even in those cases in which the sperms are isomorphic, recognition mechanisms are operating that are not correlated with gross morphological differences. Also, it should be emphasized that postpollination modifications, whether morphological or biochemical, may be important in effecting differences between sperms of a pair, as well as the ability of the sperms to participate in normal fertilization. It would not be surprising if something similar to capacitation in animal sperms (Sidhu and Guraya, 1989) is also found to occur in flowering plants. Thus, investigations involving characterizations of isolated plant sperm cells, with the ultimate goal of *in vitro* fertilization (Russell, 1991; Chaboud and Perez and Kranz, *et al.*, this volume), should take this into account.

Differential packaging of organelles between the two sperms may be related to cytoplasmic inheritance in some cases, or to the initial formation of a certain organellar ratio, male-to-female, within the embryo and/or

endosperm in those plants where male cytoplasm is transferred during syngamy (Heslop-Harrison *et al.*, 1986; Russell, 1987).

C. The Vegetative Nucleus

As is the case with the other features of the male germ unit, the functional significance of the association of the vegetative nucleus with the generative and sperm cells is highly speculative at this time. The author is aware of only two studies that may shed some light on this important question. Tang (1988), using histochemical techniques at the light microscope level, followed ATPase activity in pollen and pollen tubes of *Amaryllis vittata* and *Clivia nobilis* and concluded that ATPase activity reached its highest level at the vegetative nucleus–generative cell interface at the time when these two structures first come together. Shi *et al.* (1991) employed the techniques of quantitative ultrastructure to study nuclear pore densities during pollen development in *Medicago sativa*. They found that, at pollen maturity, mean pore density is 69% higher on that portion of the vegetative nucleus which is in intimate association with the generative cell, compared with the rest of the nuclear envelope. Before the association is established, nuclear pore density is equal around the entire vegetative nucleus; the surface area of the vegetative nucleus did not change during the stages examined. Because numerous studies have shown a positive correlation between nuclear pore density and nuclear synthetic activity (Shi *et al.*, 1991), and it is known that macromolecular traffic into and out of the nucleus occurs through the pores and is regulated by an active transport system requiring ATP (Shi *et al.*, 1991), the studies of Tang (1988) and Shi *et al.* (1991) suggested that there may be a physiological relationship between the vegetative nucleus and the generative cell. Perhaps this association functions in facilitating the positioning of mRNA from the vegetative nucleus specifically in the vicinity of the generative cell. Subsequent translation products may influence generative cell differentiation and, later, sperm cell morphogenesis (Shi *et al.*, 1991).

Exactly how the vegetative nucleus becomes initially associated with the generative and sperm cells, and in most cases, remains so during pollen tube growth is not entirely understood because no actual physical connections have been observed between the vegetative nucleus and either the generative or sperm cells. In many cases the vegetative nucleus appears to embrace the associated cell(s), or extensions thereof, and this configuration may be sufficient to keep the entities together. Heslop-Harrison and Heslop-Harrison (1989a, 1989b) viewed the vegetative nucleus and generative cell (and the sperm cells by implication) as remaining together as the result of being located within the nonvacuolated portion

of the cytoplasm near the pollen tube tip and being near a position of dynamic equilibrium "between acropetal and basipetal forces applied to their (vegetative nucleus and generative cell) surfaces at points or zones of contact with actin fibrils of different polarities."

V. Conclusions and Prospects

Renewed interest and participation in plant sperm cell research have spread worldwide within recent years as exemplified by international congresses on this subject held in The Netherlands (Wilms and Keijzer, 1988), Hungary (Barnabas and Liszt, 1990), and the United States (published abstracts of the symposium, Plant Gametes and Fertilization, Richmond, Virginia, *Am. J. Bot.*, **77** (6) [Suppl.], 1990). Results from quantitative, three-dimensional ultrastructure, the focus of this chapter, have laid the groundwork for other exciting and promising approaches to studies in plant sperm cell biology, including sperm isolation and subsequent characterization from the standpoint of physiology, molecular biology, and immunochemistry (Russell *et al.*, 1990; Russell, 1991; Russell, this volume).

An ultimate goal of plant gamete research is the achievement of true *in vitro* fertilization in a flowering plant. The recent success of Kranz *et al.* (1991) with electrofusion of the isolated egg and sperm of *Zea mays* represents an important first step toward this goal (Kranz *et al.*, this volume).

The concept of the male germ unit has served as a catalyst for much of the research presented in this chapter, as well as for that of other chapters of this volume. Clearly, additional observations and experimental approaches are badly needed before we can hope to fully understand the functional significance of the male germ unit. The years ahead promise to provide interesting answers as well as intriguing new questions.

Acknowledgments

The author thanks Tong Zhu for technical help with the preparation of the manuscript, and Tong Zhu, Liang Shi, and Jean-Emmanuel Faure for their critical reading of the manuscript. The original work mentioned was supported in part by the National Science Foundation under grant DCB-8501995, by the United States Department of Agriculture under grants 83-CRCR-1-1270 and 88-37234-3876, and by the Organized Research Fund, Northern Arizona University.

References

Barnabas, B., and Liszt, K. (eds.) (1990). "Characterization of Male Transmission Units in Higher Plants." Hungarian Acad. Sci., Budapest.

Barnes, S. H., and Blackmore, S. (1987). *Protoplasma* **138**, 187–189.
Cass, D. D. (1973). *Can. J. Bot.* **51**, 601–605.
Cass, D. D., and Jensen, W. A. (1970). *Am. J. Bot.* **57**, 62–70.
Cass, D. D., and Karas, I. (1974). *Protoplasma* **81**, 49–62.
Cass, D. D., and Karas, I. (1975). *Can. J. Bot.* **53**, 1051–1062.
Chandra, S., and Bhatnagar, S. P. (1974). *Phytomorphology* **24**, 211–217.
Charzynska, M., and Lewandowska, E. (1990). *Ann. Bot.* **65**, 685–689.
Ciampolini, F., Moscatelli, A., and Cresti, M. (1988). *Sex. Plant Reprod.* **1**, 88-96.
Cocucci, A. E. (1988). *In* "Sexual Reproduction in Higher Plants" (M. Cresti, P. Gori, and
    E. Pacini, eds.), pp. 251–256. Springer, New York.
Dumas, C., Knox, R. B., McConchie, C. A., and Russell, S. D. (1984). *What's New Plant
    Physiol.* **15**, 17–20.
Dumas, C., Knox, R. B., and Gaude, T. (1985). *Protoplasma* **124**, 168–174.
Foster, A. S., and Gifford, E. M., Jr. (1974). "Comparative Morphology of Vascular Plants,"
    2nd ed. W. H. Freeman, San Francisco.
Hagemann, R., and Schröder, M. B. (1989). *Protoplasma* **152**, 57–64.
Heslop-Harrison, J., and Heslop-Harrison, Y. (1984). *Acta Bot. Neerl.* **33**, 131–134.
Heslop-Harrison, J., and Heslop-Harrison, Y. (1987). *Plant Sci.* **51**, 203–213.
Heslop-Harrison, J., and Heslop-Harrison, Y. (1989a). *J. Cell Sci.* **94**, 319–325.
Heslop-Harrison, J., and Heslop-Harrison, Y. (1989b). *Sex. Plant Reprod.* **2**, 199–207.
Heslop-Harrison, J., Heslop-Harrison, J. S., and Heslop-Harrison, Y. (1986). *Ann. Bot.* **58**,
    1–12.
Hoefert, L. L. (1969). *Protoplasma* **68**, 237–240.
Hoefert, L. L. (1971). *In* "Pollen: Development and Physiology" (J. Heslop-Harrison, ed.),
    pp. 68–69. Butterworths, London.
Hu, S. Y. (1990). *Acta Bot. Sin.* **32**, 230–240.
Hu, S. Y., and Yu, H. S. (1988). *Protoplasma* **147**, 55–63.
Jensen, W. A. (1973). *Bioscience* **23**, 21–27.
Jensen, W. A. (1974). *In* "Dynamic Aspects of Plant Ultrastructure" (A. W. Robards, ed.),
    pp. 481–503. McGraw-Hill, New York.
Jensen, W. A., and Fisher, D. B. (1968). *Protoplasma* **65**, 277–286.
Jensen, W. A., and Fisher, D. B. (1970). *Protoplasma* **69**, 215–235.
Karas, I., and Cass, D. D. (1976). *Phytomorphology* **26**, 36–45.
Kaul, V., Theunis, C. H., Palser, B. F., and Knox, R. B. (1987). *Ann. Bot.* **59**, 227–235.
Kranz, E., Bautor, J., and Lörz, H. (1991). *Sex. Plant Reprod.* **4**, 12–16.
Larson, D. A. (1965). *Am. J. Bot.* **52**, 139–154.
Levieil, C. (1986). *C. R. Acad. Sci. Paris Ser. D* **303**, 769–774.
Maheshwari, P. (1949). *Bot. Rev.* **15**, 1–75.
Maheshwari, P. (1950). "An Introduction to the Embryology of Angiosperms. McGraw-Hill,
    New York.
McConchie, C. A., Russell, S. D., Dumas, C., Tuohy, M., and Knox, R. B. (1987a). *Planta*
    170, 446–452.
McConchie, C. A., Hough, T., and Knox, R. B. (1987b). *Protoplasma* **139**, 9–19.
McCoy, K., and Knox, R. B. (1988). *Protoplasma* **143**, 85–92.
Mogensen, H. L. (1982). *Carlsberg Res. Commun.* **47**, 313–354.
Mogensen, H. L. (1986a). *In* "Biotechnology and Ecology of Pollen" (D. L. Mulcahy,
    G. B. Mulcahy, and E. Ottaviano, eds.), pp. 297–305. Springer-Verlag, New York.
Mogensen, H. L. (1986b). *Protoplasma* **134**, 67–42.
Mogensen, H. L. (1986c). *In* "Pollination '86" (E. G. Williams, R. B. Knox, D. Irvine, eds.),
    pp. 166–171. University of Melbourne, Melbourne.
Mogensen, H. L., and Rusche, M. L. (1985). *Protoplasma* **128**, 1–13.

Mogensen, H. L., and Wagner, V. T. (1987). *Protoplasma* **138**, 161–172.
Murgia, M., and Wilms, H. J. (1988). *In* "Plant Sperm Cells as Tools for Biotechnology" (H. J. Wilms and C. J. Keijzer, eds.), pp. 75–79. Pudoc, Wageningen.
Pennell, R. I., Geltz, N. R., Koren, E., and Russell, S. D. (1987). *Bot. Gaz.* **148**, 401–406.
Pope, M. (1937). *J. Agricult. Res.* **54**, 525–529.
Provoost, E., Southworth, D. A., and Knox, R. B. (1988). *In* "Plant Sperm Cells as Tools for Biotechnology" (H. J. Wilms and C. J. Keijzer, eds.), pp. 69–73. Pudoc, Wageningen.
Roman, H. (1947). *Genetics* **32**, 391–409.
Roman, H. (1948). *Proc. Natl. Acad. Sci. U.S.A.* **34**, 36–42.
Rusche, M. L. (1988). *In* "Plant Sperm Cells as Tools for Biotechnology" (H. J. Wilms and C. J. Keijzer, eds.), pp. 61–68. Pudoc, Wageningen.
Rusche, M. L. (1990). Ph.D. dissertation, Northern Arizona University, Flagstaff, Arizona.
Rusche, M. L., and Mogensen, H. L. (1988). *In* "Sexual Reproduction in Higher Plants" (M. Cresti, P. Gori, and E. Pacini, eds.), pp. 221–226. Springer-Verlag, Berlin.
Russell, S. D. (1983). *Am. J. Bot.* **70**, 416–434.
Russell, S. D. (1984). *Planta* **162**, 385–391.
Russell, S. D. (1985). *Proc. Natl. Acad. Sci. U.S.A.* **82**, 6129–6132.
Russell, S. D. (1987). *Theor. Appl. Genet.* **74**, 693–699.
Russell, S. D. (1991). *Annu. Rev. Plant Physiol. Plant Molec. Biol.* **42**, 189–204.
Russell, S. D., and Cass, D. D. (1981). *Protoplasma* **107**, 85–107.
Russell, S. D., Cresti, M., and Dumas, C. (1990). *Physiol. Plant.* **80**, 669–676.
Schröder, M. B. (1983). *In* "Fertilization and Embryogenesis in Ovulated Plants." (O. Erdelská, ed.), pp. 101–104. VEDA, Bratislava, Czechoslovakia.
Shi, L., Mogensen, H. L., Zhu, T., and Smith, S. E. (1991). *J. Cell Sci.* **99**, 115–120.
Shivanna, K. R., Xu, H., Taylor, P., and Knox, R. B. (1988). *Plant Physiol.* **87**, 647–650.
Sidhu, K. S., and Guraya, S. S. (1989). *Int. Rev. Cytol.* **118**, 231–280.
Smith, S. E., Bingham, E. T., and Fulton, R. W. (1986). *J. Hered.* **80**, 214–217.
Sprague, G. F. (1932). *Genetics* **17**, 358–368.
Tang, P. H. (1988). *In* "Sexual Reproduction in Higher Plants" (M. Cresti, P. Gori, and E. Pacini, eds.), pp. 227–232. Springer-Verlag, New York.
Wagner, V. T., and Mogensen, H. L. (1987). *Protoplasma* **143**, 101–110.
Weber, M. (1988a). *In* "Sexual Reproduction in Higher Plants" (M. Cresti, P. Gori, and E. Pacini, eds.), p. 491. Springer-Verlag, New York.
Weber, M. (1988b). *Plant System. Evol.* **161**, 53–64.
Wilms, H. J. (1986). *In* "Biology of Reproduction and Cell Motility in Plants and Animals" (M. Cresti and R. Dallai, eds.), pp. 193–198. University of Siena, Siena.
Wilms, H. J., and Van Aelst, A. C. (1983). *In* "Fertilization and Embryogenesis in Ovulated Plants" (O. Erdelská, ed.), pp. 105–112. VEDA, Bratislava, Czechoslovakia.
Wilms, H. J., and Keijzer, C. J. (1988). "Plant Sperm Cells as Tools for Biotechnology." Pudoc, Wageningen.
Wilms, H. J., Murgia, M., and Van Spronsen, E. A. (1988). *In* "Plant Sperm Cells as Tools for Biotechnology." (H. J. Wilms and C. J. Keijzer, eds.), pp. 35–40. Pudoc, Wageningen.
Yu, H. S., Hu, S. Y., and Zhu, C. (1989). *Protoplasma* **152**, 29–36.
Zhu, C., Hu, S. Y., Xu, L. Y., Li, X. R., and Shen, J. H. (1980). *Sci. Sin.* **23**, 371–376.
Zhu, T., Mogensen, H. L., and Smith, S. E. (1990a). *Protoplasma* **158**, 66–72.
Zhu, T., Mogensen, H. L., and Smith, S. E. (1990b). *Am. J. Bot.* **77**(Suppl.), 42.

Organization, Composition, and Function of the Generative Cell and Sperm Cytoskeleton

Barry A. Palevitz* and Antonio Tiezzi†

* Department of Botany, University of Georgia, Athens, Georgia 30602
† Dipartimento di Biologia Ambientale, Università di Siena, Siena, Italy

I. Introduction

Division of the generative cell in angiosperms produces the two sperm needed for double fertilization. Thus, the properties and behavior of generative cells and sperm have long interested plant biologists. Until recently, however, knowledge about these cells has been fragmentary, in large part due to their relative inaccessibility within pollen grains and tubes, and because of our own inability to obtain isolated cells in numbers and purity sufficient for analysis. Strangely, even when discoveries have been forthcoming, misconceptions have been difficult to dispel. Forty years after Maheshwari (1950) stated in his classic volume on angiosperm reproduction that "recent work has shown . . . that in all cases the male gametes are definite cells" and not just nuclei, a widely-used biology text recently maintained that " . . . the generative nucleus divides to form two sperm nuclei, which function as male gametes" (Campbell, 1990, p. 741). Modern electron-microscopic and morphometric analyses have unequivocally established that sperm consist of more than just a nucleus, and detection of the microtubule cytoskeleton by immunofluorescence makes this principle obvious. Before embarking on our discussion of the generative cell and sperm cytoskeleton, a number of introductory remarks are appropriate. In addition to their ontogenetic relationship, the two cell types share a number of properties, including an elongate, lenticular, or spindle shape that usually tapers into tail-like extensions. Because similarities also extend into characteristics of their cytoskeletal arrays, distinctions between generative cells and sperm are not made in this review. Instead, for the sake of convenience generative and sperm cells will be referred to in aggregate as GSPs, except where reference to either is specifically appro-

priate. Reports of possible differences in generative cell versus sperm microtubules (Mts) are also noted, however.

Among the aspects of GSP that have concerned researchers are as follows: (1) the nature of the interface between the generative cell and surrounding vegetative cell, including the nature of the intervening wall; (2) interactions between GSP and the vegetative nucleus in those species characterized by a distinct male germ unit; (3) the factors responsible for the elongate shape and flexible nature of GSP; (4) the manner in which GSPs move down the pollen tube, and in particular, whether they actively locomote; and (5) the phenomenology and mechanisms governing generative cell division. The cytoskeleton is of direct importance in the last three topics and perhaps the second as well.

Some general remarks about the cytoskeleton are also in order at this time. The cytoskeleton in plants consists of at least two classes of filamentous elements, tubulin-containing MTs, and actin-containing microfilaments (MFs), which serve a variety of structural and motility-related functions (Staiger and Schliwa, 1987; Lloyd, 1987, 1989; Fosket, 1989; Seagull, 1989; Derksen *et al.*, 1990). For example, MTs comprise the spindle fibers necessary for mitosis and meiosis, whereas MFs are responsible for the vigorous cytoplasmic streaming seen in most plant cells. In some instances, the two systems interact. These interactions may operate at the level of a specific function (e.g., organelle movement), or they may be involved in the organization of one system or the other. A close association between MTs and MFs has been reported in the vegetative cortex of pollen tubes (Lancelle *et al.*, 1987; Raudaskoski *et al.*, 1987; Tiwari and Polito, 1988; Pierson *et al.*, 1989; Lancelle and Hepler, 1991; Pierson and Cresti, this volume).

A variety of associated proteins regulate the organization and function of MTs and MFs. Some of these proteins belong to families of mechanochemical motors that govern motility (e.g., dynein, kinesin, dynamin, myosin; Staiger and Schliwa, 1987; Kiehart, 1990; Vallee and Shpetner, 1990), whereas others affect filament assembly, stability, or breakdown (Vallee and Bloom, 1984; Stossel *et al.*, 1985; Pollard and Cooper, 1986; Schliwa, 1986; Fosket, 1989; Wiche *et al.*, 1991). A third class of cytoskeletal proteins belonging to the intermediate filament family has been reported in plants (Ross *et al.*, 1991) but is far less extensively characterized than that in animal cells (Schliwa, 1986). MTs dominate the GSP cytoskeleton.

As in many other fields in biology, progress in our understanding of the cell biology of plant reproduction has been driven by technology. With regard to the cytoskeleton, fluorescence localization techniques have had a particularly important impact. The use of specific probes such as rhodamine–phalloidin (for MFs) and antibodies to tubulin, actin, and kinetochores, in conjunction with nuclear stains, such as Hoechst 33258 and

DAPI, have provided global views of cytoskeletal organization and distribution unobtainable with electron microscopy (Lloyd, 1987). The advent of image processing and confocal laser scanning microscopy should potentiate even further progress at the light microscope level (Taylor *et al.*, 1989; Palevitz, 1990; Bartalesi *et al.*, 1991; Theunis *et al.*, 1991; Liu and Palevitz, 1992), especially when combined with microinjection of fluorescent analogs of tubulin and actin (Zhang *et al.*, 1990). However, new fixation regimens, such as freeze substitution, have allowed more precise, definitive determinations of cell detail at the electron-microscopic level than heretofore possible (Russell and Cass, 1981; Cresti *et al.*, 1987; Lancelle *et al.*, 1987; Tiwari and Polito, 1988; Noguchi and Ueda, 1990). Hopefully, use of freeze substitution will expand, perhaps in conjunction with high pressure freezing, because GSPs can be difficult to reliably preserve for electron microscopy using conventional chemical means.

Lastly, this chapter does not claim to exhaustively cite the increasingly hefty literature on the subject. Instead, we wish to elaborate on various facts and concepts germane to the GSP cytoskeleton, with examples of relevant articles cited where appropriate. In the case of literature prior to 1970, key publications or reviews are used. We have tried to include more of the literature since 1970 but again acknowledge that our citations are incomplete.

II. Microtubules of the Generative Cell and Sperm

A. Structural Organization

1. General Characteristics

Examination of a variety of species has yielded a surprisingly uniform image of MT organization in GSP. The MTs of generative cells and sperm (at least for a time after generative cell division) are arranged in prominent bundles that are aligned helically or longitudinally relative to the long axis of the cell (Figs. 1A–D, 2A, and 3A; for other global immunofluorescence images, Derksen *et al.*, 1985; Pierson *et al.*, 1986; Tiezzi *et al.*, 1986, 1988; Raudaskoski *et al.*, 1987; Heslop-Harrison and Heslop-Harrison *et al.*, 1988; Zhou *et al.*, 1988; Palevitz and Cresti, 1988, 1989; Pierson, 1989; Taylor *et al.*, 1989; Terasaka and Niitsu, 1989b, 1990; Zhou *et al.*, 1990; Zhu and Liu, 1990; Bartalesi *et al.*, 1991; Theunis *et al.*, 1991). The bundles often appear flexuous as well (Lancelle *et al.*, 1987). Because the helically disposed bundles are often arranged in a crisscross pattern (Fig. 1C; Derksen *et al.*, 1985; Terasaka and Niitsu, 1990; Zhou *et al.*, 1990), descrip-

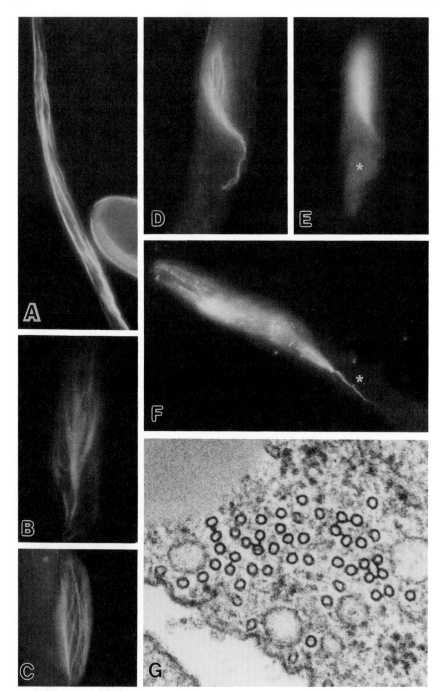

tions of the array as a basket or cage ensheathing the nucleus seem appropriate (Tiezzi *et al.*, 1986, 1988; Heslop-Harrison *et al.*, 1988; Tanaka *et al.*, 1989; Taylor *et al.*, 1989; Terasaka and Niitsu, 1990; Tiezzi and Cresti, 1990; Zhu and Liu, 1990; Theunis *et al.*, 1991).

Electron microscopy shows that the bundles consist of tightly packed MTs interlinked by prominent cross-bridges (Fig. 1G; Burgess, 1970a,b; Sanger and Jackson, 1971; Cass, 1973; Karas and Cass, 1976; Cresti *et al.*, 1984, 1988a, 1990; Tiezzi *et al.*, 1986, 1988; Lancelle *et al.*, 1987; Raudaskoski *et al.*, 1987; Tiezzi, 1991). Bridges also connect bundles near the cell surface with the plasma membrane. Based on antitubulin localizations following cold treatments, Åström *et al.* (1991) proposed that generative cell MTs associate with the plasma membrane and nuclear envelope.

The number of bundles per cross section varies; Heslop-Harrison *et al.* (1988) reported 14 major bundles in the generative cell of *Endymion*, while values for *Nicotiana* vary from 4 to 15 (Cresti *et al.*, 1984; Derksen *et al.*, 1985; Raudaskoski *et al.*, 1987; Theunis *et al.*, 1991). Values in this range have also been reported for other species (Cresti *et al.*, 1984). Variations presumably depend on preparation procedures and species-specific factors, including the average size of the generative cell. The bundles branch profusely, and consequently their dimensions vary greatly, from rather thin elements composed of only one or two MTs, to thick fibers (Figs. 1A,B,D, and 2A) containing as many as 35 MTs (e.g., Cresti *et al.*, 1984; Theunis *et al.*, 1991). The bundles are usually lengthy; Raudaskoski *et al.* (1987) followed bundles for distances up to two thirds the length of the *Nicotiana* generative cell (Fig. 1C) .

The bundles can be preferentially located near the plasma membrane in some species (Burgess, 1970b; Cass, 1973; Raudaskoski *et al.*, 1987; Hu and Yu, 1988; Zhou *et al.*, 1988; Tanaka *et al.*, 1989; Noguchi and Ueda, 1990; Bartalesi *et al.*, 1991) but seem to occupy all regions of the cytoplasm in others (Tiezzi *et al.*, 1986; Lancelle *et al.*, 1987; Palevitz and

FIG. 1 (A) Longitudinally disposed MTs in a generative cell of *Tradescantia virginiana* revealed by antitubulin immunocytochemistry. From Palevitz and Cresti (1989). (B) MT bundles in a sperm cell of *Tradescantia virginana*. From Palevitz and Cresti (1988). (C) Basket-like configuration of MTs in a generative cell of *Nicotiana tabacum*. One bundle can be followed most of the length of the cell. Micrograph courtesy of B. Liu. (D, E) A long MT-containing extension at one end of a tobacco generative cell (D) is associated with the vegetative nucleus (*) in (E), as revealed by Hoechst fluorescence. (F) A MT-containing extension at one end of a sperm cell. The position of the vegetative nucleus (*) was determined with coordinate DAPI fluorescence. From Palevitz and Cresti (1988). (G) A MT bundle in a sperm cell of *Brassica oleracea* viewed in the electron microscope. There are cross-bridges between MTs. From Cresti *et al.* (1990).

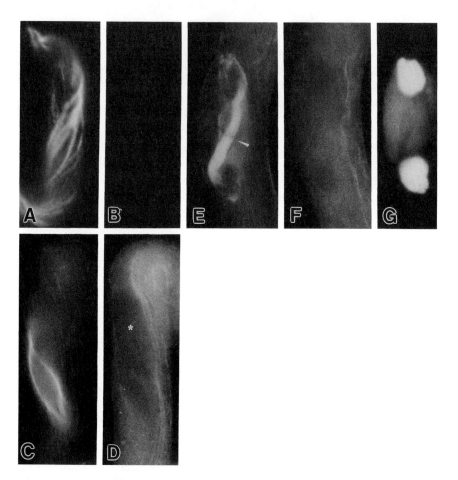

FIG. 2 (A, B) MT bundles in *Tradescantia virginiana* are revealed by tubulin immunofluorescence (A) but dual staining with rhodamine–phalloidin for F-actin (B) yields negative results. From Palevitz and Cresti (1988). (C, D) MTs are revealed in a *Nicotiana tabacum* generative cell by tubulin immunofluorescence (C), but the complimentary localization with antiactin only shows MFs in the vegetative cytoplasm. The vegetative nucleus is located at (*). (E–G) Dividing *Nicotiana* generative cells have a prominent phragmoplast that contains MTs (E) but not F-actin (F). The complimentary Hoechst image (G) reveals the two prominent sperm nuclei and the intervening, less fluorescent vegetative nucleus. Micrographs C–G courtesy of B. Liu.

Cresti, 1988, 1989; Yu *et al.,* 1989). It is noteworthy in this regard that only a thin layer of cytoplasm often separates the nucleus from the plasma membrane in these cells, so definitive determinations may be difficult. Thus, while the bundles in *Nicotiana* may appear to be peripheral in some views (e.g., Bartalesi *et al.,* 1991; Theunis *et al.,* 1991), others located close to the nucleus can also be found (Fig. 4 in Tiezzi *et al.,* 1986; Yu *et*

al., 1989). Cresti *et al.* (1990) noted that single MTs and bundles are present in *Brassica* sperm, with the former located mainly near the nuclear envelope. Where generative cells have a lobed or scalloped outline, the bundles may be situated in the lobes (Tanaka *et al.*, 1989; Tiezzi, 1991; but also see Tiezzi *et al.*, 1988, and Noguchi and Ueda, 1990).

The MT bundles may merge into one or two main superbundles that taper into the cytoplasmic extensions commonly seen at the ends of the cells (Fig. 1D and F; Derksen *et al.*, 1985; Pierson *et al.*, 1986, 1989; Tiezzi *et al.*, 1986, 1988; Raudaskoski *et al.*, 1987; Heslop-Harrison *et al.*, 1988; Palevitz and Cresti, 1988, 1989; Taylor *et al.*, 1989; Zhou *et al.*, 1990; Zhu and Liu, 1990; Bartalesi *et al.*, 1991; Tiezzi, 1991). Raudaskoski *et al.* (1987) also reported V-shaped MT configurations in this region. The prominent superbundles in young *Tradescantia* sperm (Figs. 1B, 4C; Palevitz and Cresti, 1988) are derived from prior division events (Section IV,D).

2. Possible Variations

As already noted, MT organization is relatively uniform between the species so far examined. The GSPs surveyed vary from short lenticular or spindle-shaped cells to those that are highly elongate. They belong to both monocot and dicot taxa. Differences between species or individual cells are largely a matter of degree. However, there are indications that in some cases the organization of MTs changes in sperm as they mature following generative cell division. Moreover, the origin of sperm MTs varies as well. Young *Tradescantia* sperm contain prominent MT bundles and branches inherited as a result of MT rearrangements during division, but the array may change later on (Fig. 4C; Palevitz and Cresti, 1989; Palevitz, 1990; Liu and Palevitz, 1992). In contrast, available evidence indicates that sperm MTs are not inherited from the previous generation in *Nicotiana* but are assembled at the end of division (Palevitz and Liu, in preparation). Furthermore, Yu *et al.* (1989) pictured many single or small groups of MTs in *Nicotiana* sperm, whereas the preceding generative cell contains prominent bundles (Cresti *et al.*, 1984; Lancelle *et al.*, 1987; Raudaskoski *et al.*, 1987). Taylor *et al.* (1989) reported that MTs in *Rhododendron* sperm are fewer in number and less prominent than those in generative cells. Thus, sperm MTs in these species appear to be less bundled than those in generative cells. However, highly bundled MTs are evident in the sperm of *Hordeum* (Cass, 1973) and *Brassica* (Fig. 1G; Cresti *et al.*, 1990).

3. Comparison with MTs in the Vegetative Cytoplasm and Somatic Cells

MT organization in GSP contrasts sharply with that in the surrounding vegetative cytoplasm (Pierson and Cresti, this volume). Vegetative MTs

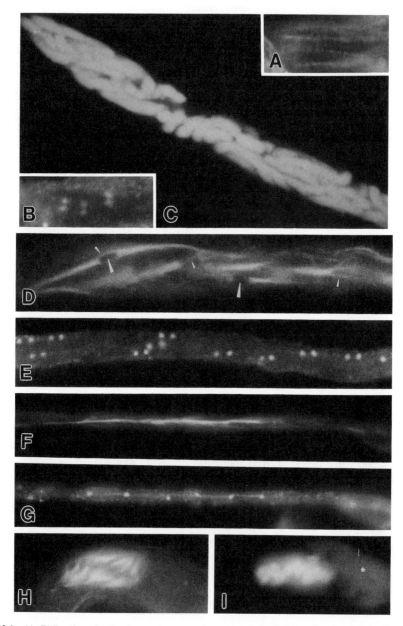

FIG. 3 (A, B) Portion of a *Tradescantia* generative cell early in division showing longitudinally disposed MTs (A, antitubulin image) and transverse pairs of kinetochores (B, CREST serum image) positioned on the MTs like the rungs of a ladder. From Liu and Palevitz (1991). (C) Confocal scanning micrograph of chromosomes in a metaphase *Tradescantia* generative cell. The chromosomes are typically arranged in complex patterns, and a metaphase plate

in hydrated pollen grains (Sheldon and Dickinson, 1986; Tiwari and Polito, 1988, 1990a) and pollen tubes (Franke *et al.*, 1972; Derksen *et al.*, 1985; Pierson *et al.*, 1986; Lancelle *et al.*, 1987; Lancelle and Hepler, 1991; Raudaskoski *et al.*, 1987; Heslop-Harrison *et al.*, 1988; Tiwari and Polito, 1988; Tiezzi, 1991) do not occur in similar thick bundles but instead comprise much thinner elements organized in various patterns. In pollen tubes MTs are arranged primarily in helical to longitudinal arrays in the outer cortex. The cortical elements consist of single MTs or planar groups that are cross-bridged to each other and to the plasma membrane. Generative cell MTs seem to be more stable to the disruptive effects of fixation and cold treatments than those in the vegetative cytoplasm (Lancelle *et al.*, 1987; Raudaskoski *et al.*, 1987; Åström *et al.*, 1991), which could result from any of several biochemical and molecular characteristics of these arrays, including the numerous cross-bridging moieties. Nevertheless, generative cell MTs do break down during cold (Åström *et al.*, 1991), colchicine, and oryzalin treatments (Heslop-Harrison *et al.*, 1988; Palevitz and Cresti, 1989); they also seem to disorganize after the cells are isolated (Tanaka *et al.*, 1989; Zhou *et al.*, 1990; Zee and Aziz-Un-Nisa, 1991; Theunis *et al.*, 1992 Section IV,B,2).

The MT bundles in GSP are also unlike somatic cell arrays. While cortical MTs are linked to the plasma membrane and do associate with each other in somatic cells (Hardham and Gunning, 1978; Palevitz, 1982; Derksen *et al.*, 1990), the kind of large, flexuous bundles that ramify through the cytoplasm of GSP are not seen. Instead, the MTs are largely limited to a single layer next to the plasma membrane. Although the preprophase band (Gunning, 1982) can be thought of as a bundle, it is also limited to the cortex.

---

typical of somatic cells is not evident. From Palevitz (1990). (D) Metaphase arrangement of kinetochore fibers in another *Tradescantia* generative cell. MT branches (small arrowheads) link filial and nonfilial kinetochore fibers. Kinetochores (large arrowheads) are located at the dark, unstained patches. (E) Another metaphase *Tradescantia* generative cell stained with CREST serum and showing that the kinetochores are now longitudinally aligned, compared with the transverse arrangement seen earlier (B). From Liu and Palevitz (1991). (F, G) During anaphase in *Tradescantia,* MTs are rearranged as two superbundles carrying attached sets of nonfilial kinetochores. (F) Shows one MT superbundle and the companion CREST image. (G) show the now single kinetochores. From Liu and Palevitz (1992). (H, I) A dividing *Nicotiana* generative cell. Unlike *Tradescantia,* the mitotic apparatus consists of a distinct spindle containing two obvious halves bisected by a metaphase plate (H). The spindle is oblique, however. The vegetative nucleus, marked by (*) in the Hoechst image in (I), is positioned distal to the generative cell, toward the pollen tube tip. Micrographs courtesy of B. Liu.

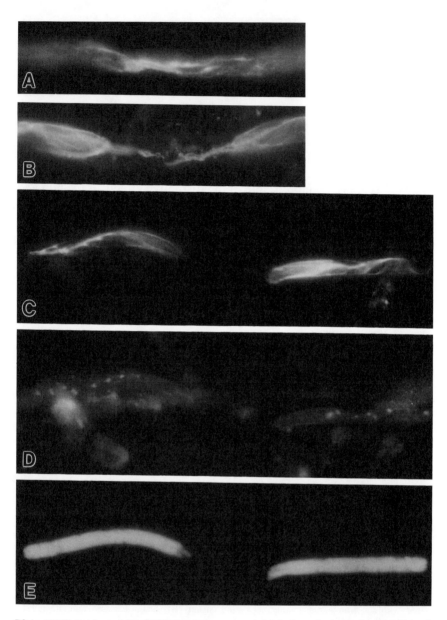

FIG. 4 (A) Zone of constricted MTs between sperm nuclei during telophase of generative cell division in *Tradescantia virginiana*. From Palevitz and Cresti (1989). (B) The constriction zone has fragmented, leaving opposing tail-like extensions on the sperm. From Palevitz and Cresti, (1989). (C, D) Two recently formed sperm cells of *Tradescantia virginiana*. Each sperm contains a superbundle derived from division, plus attached MT branches (C). The kinetochores are located at notches on the superbundles (D). (E) The Hoechst fluorescence image of the sperm nuclei. From Liu and Palevitz (1992).

B. Biochemical and Molecular Considerations

The characteristics of GSP MT arrays are undoubtedly based on molecular and/or biochemical properties peculiar to these cells. Relevant factors typically invoked to explain differences in MT organization include the differential expression of tubulin isotypes, posttranslational modifications of tubulin, and/or the presence or alteration of MT-associated proteins. Relatively little evidence is available so far on any of these factors in plants.

A MT is composed of heterodimers of α- and β-tubulin, which share sequence homology at the DNA and protein levels with counterparts in animals (Schliwa, 1986; Fosket, 1989). Each of the tubulin classes consists of a multigene family, and multiple tubulin genes have been detected in the limited number of plants so far examined (Silflow *et al.*, 1987; Fosket, 1989). *Arabidopsis thaliana* contains at least four α-tubulin genes and seven β-tubulin genes (Silflow *et al.*, 1987; Ludwig *et al.*, 1988). At least two β-tubulin genes have been reported in soybean (Fosket, 1989; Han *et al.*, 1991). These numbers are increasing at this writing, and the genes can now be organized into distinct subfamilies. However, the number of distinct protein isotypes coded by these genes may be more limited.

Differential expression of tubulin isotypes occurs in various tissues and organs, and partial but interesting data on the subject are now available in plants (Hussey *et al.*, 1988; Ludwig *et al.*, 1988; Han *et al.*, 1991). Most appropriate to the present discussion is the work of Ludwig *et al.* (1988) and Hussey *et al.* (1988), who have identified tubulins enriched in anthers of *Arabidopsis* and carrot, respectively. In the case of the β-tubulins of carrot, the $\beta4$ isotype is specific for stamens and is expressed overwhelmingly in pollen, while others are notably absent from pollen compared with other locations. Silflow and co-workers have directed their attention to the α subunits in *Arabidopsis thaliana,* where expression of the $\alpha1$ (*Tua1*) gene predominates in pollen (Ludwig *et al.*, 1988; C. Silflow, personal communication). Preferential enrichment and exclusion of tubulins have also been detected in maize pollen (C. Silflow, personal communication). So far, although tubulin has been identified in electrophoretic analyses of pollen and pollen tubes (Tiezzi *et al.*, 1986; Raudaskoski *et al.*, 1987; Hussey *et al.*, 1988; Åström *et al.*, 1991), comparisons between GSPs and vegetative components have not been reported. The limited biochemical studies of isolated GSPs performed so far have not been directed at cytoskeletal proteins (Russell *et al.*, 1990).

The significance of differential tubulin expression is still largely unknown. Microinjection and genetic manipulation experiments with both animals and plants show that cells can utilize heterologous tubulin in all MT arrays (Sullivan, 1988; Joshi and Cleveland, 1990; Zhang *et al.*, 1990).

Moreover, studies using isotype-specific antibodies show that MTs can be composed of all available subunits. For example, the interphase cortical array of carrot cells appears to be composed of a variety of tubulin isotypes (Hussey *et al.*, 1987). However, evidence also favors specific functions for certain isotypes in some cases. The testis-specific β2-tubulin isotype identified in *Drosophila* seems to be critical for a number of developmental events in spermatogenesis, even though it is present in all MT arrays in these cells (Sullivan, 1988; Joshi and Cleveland, 1990). Thus, more definitive information on GSP tubulins is now needed. Such information may be forthcoming with successful efforts to isolate GSPs and subject them to polypeptide and nucleic acid analyses, and with immunolocalizations using isotype-specific antibodies.

Tubulin is also subject to posttranslational modification, including the reversible removal of a terminal tyrosine residue, acetylation, and phosphorylation. These modifications effectively increase the diversity of tubulin subunits beyond that specified by the multigene families. Considerable debate has centered on the role of these modifications in properties, such as MT stability. While modifications seem to accompany increased stability, their role as a causative factor is not settled (Sullivan, 1988; Prescott *et al.*, 1989). Progress in animal cells has been potentiated by antibodies specific for various modified forms of tubulin. Little is known about posttranslational modifications in plant tubulins.

Finally, a variety of accessory proteins influence MT assembly and association in higher order arrays (Vallee and Bloom, 1984; Schliwa, 1986; Fosket, 1989; Wiche *et al.*, 1991), as well as serve as mechanochemical transducers during MT-related motility (Vallee and Shpetner, 1990). The activity of these proteins is subject to close regulation (e.g., via phosphorylation). Unfortunately, little is known about MT-associated proteins in plant cells. That they exist is made evident by the electron-microscopic visualization of bridges (such as those in GSP; Fig. 1G) between adjacent MTs and linking MTs to nearby membranes. Recently, polypeptides were identified in carrot cells that promote the bundling and enhanced stability of both brain and carrot MTs *in vitro* (Cyr and Palevitz, 1989). The activity of similar polypeptides may be responsible for the pronounced bundling of MTs characteristic of GSP.

III. Actin in the Generative Cell and Sperm (?)

Actin is clearly evident in pollen and pollen tubes based on biochemical criteria and specific visual probes (Pierson and Cresti, this volume). However, although actin and associated proteins are major components of the

vegetative cytoskeleton, their presence in GSPs as manifested by discernible MFs has been the subject of some controversy. The point is important because the absence of MFs would have substantial significance and repercussions relative to a number of questions concerning these cells.

A. Fluorescence and Electron Microscope Evidence

Initial reports using rhodamine–phalloidin (RP) as a fluorescent reporter probe indicated that GSPs do not contain MFs visible at the light microscope level. Thus, Pierson *et al.* (1986) and Palevitz and Cresti (1988, 1989) found no evidence of MFs in *Lilium, Nicotiana, Petunia,* and *Tradescantia* (Fig. 2A and B). The probe clearly labels MFs in the vegetative cytoplasm (e.g., Perdue and Parthasarathy, 1985; Pierson *et al.,* 1986, 1989; Pierson, 1988; Tang *et al.,* 1989b; Heslop-Harrison and Heslop-Harrison, 1991a,b; Pierson and Cresti, this volume), including those that invest the generative cell (e.g., Pierson, 1989). Furthermore, Lancelle *et al.* (1987) maintained that MFs could not be seen in GSPs of *Nicotiana* at the electron-microscopic level. Similar negative findings were reported for *Nicotiana, Aloe,* and *Linaria* by Tiezzi *et al.* (1988). Most ultrastructural articles on GSPs are conspicuous in the absence of any mention of MFs, despite considerable attention to MTs. Two exceptions are articles by Zhu *et al.* (1980) and Russell and Cass (1981), which reported MFs in the sperm of wheat and *Plumbago,* respectively. However, the micrographs in both cases are not convincing. Russell and Cass (1981) also claimed that MTs could not be detected in their material. Because MT bundles have been universally observed in the GSPs of other species, their absence in this case injects a degree of caution. Thus, the filamentous material could have been the remnants of degraded MTs.

   The initial reports of negative findings at the light microscope level utilized RP to detect F-actin. Thus, it was of great interest when a subsequent article using antiactin immunofluorescence as a specific probe reported impressive localizations in the generative cells of *Rhododendron* (Taylor *et al.,* 1989). Although fluorescence was present prior to division, it was particularly striking in the phragmoplast. What could be the basis for the difference between these reports? First, it could be due to differences in the efficacy of antiactin versus RP as a probe for GSP MFs. Actin, like tubulin, comprises a multigene family. In *Petunia,* for example, more than 100 actin genes have been detected (Meagher, 1991). Meagher and colleagues reported sequence differences between members of this family and animal actins. Surprisingly, however, significant differences in amino acid sequence separate the various plant actins, which also exhibit differential expression in root tissues (McLean *et al.,* 1990; Meagher, 1991).

Furthermore, some of these sequence differences occur around the phalloi-din binding site. It is therefore possible that GSPs contain differentially expressed actin(s) with reduced ability to bind phalloidin, which would account for the negative results with RP. Regions of homology that are recognized by antiactin might remain, however, allowing for detection with at least some antibodies. Another possibility is that *Rhododendron* has retained GSP MFs while other taxa have not.

To address this issue, our laboratory (Palevitz and Liu, 1992) embarked on a systematic search for F-actin in GSP using both RP and antiactin. We examined *Tradescantia virginiana, Nicotiana tabacum,* and the same species used by Taylor *et al.* (1989), *Rhododendron laetum.* Several differ-ent antiactins were employed, including the one used by Taylor *et al.* (1989). Coordinate localizations with Hoechst and antitubulin were also performed to firmly identify division stages and MT arrays. The results were unequivocal. We found no MF signal in the GSPs of any species, although MFs could be seen in the vegetative cytoplasm (Figs. 2C and D). With all three species, negative results were obtained with GSPs in intact pollen tubes as well as those liberated during processing; the latter re-moved the problem of confusing MFs in the surrounding vegetative cyto-plasm as a complicating factor.

Because Taylor *et al.* (1989) reported MFs in the midzone of dividing generative cells, we specifically focused our attention on this region during anaphase and telophase in each species. We found that filamentous fluores-cence is often evident in the vicinity of the cytokinetic constriction zone in intact pollen tubes of *Tradescantia virginiana,* but close scrutiny of succes-sive optical sections shows that the MFs again are in the vegetative cyto-plasm. Previous findings demonstrated that vegetative MFs cluster around the generative cell (Pierson *et al.,* 1986; Pierson, 1989). A key difference between *Rhododendron* and *Tradescantia* lies in the mechanism of cytoki-nesis. The former utilizes a phragmoplast, whereas the latter does not (Sec-tion IV,D). Because actin is typically found in phragmoplasts (e.g., Palevitz, 1987; Kakimoto and Shibaoka, 1989; Schmit and Lambert, 1990), one could reason that the presence of MFs in *Rhododendron* is somehow related to the use of this structure for cytokinesis. Thus, we sampled *Rhododendron* and *Nicotiana,* which both utilize a prominent phragmoplast. Again, no sign of MFs could be seen at any stage, including phragmoplasts identified by dual antitubulin staining (Figs. 2E–G). As in our earlier work, none of the GSPs stained with RP (Palevitz and Liu, 1992).

B. Correlative Studies

Our results conform to other studies using antiactin localizations. Immuno-gold detection of actin at the electron-microscopic level demonstrated

distinct labeling of MFs in the vegetative cytoplasm of *Nicotiana* (Lancelle and Hepler, 1989, 1991), but none could be seen in adjacent generative cells (Lancelle and Hepler, 1991). Because actin usually functions in cell motility together with some form of myosin, if actin were present in GSP, it is reasonable to suppose that myosin would be there as well. Strong evidence indicates that myosin linked to organelles is present in the vegetative cytoplasm (e.g., Kohno *et al.,* 1990). However, Tang *et al.* (1989b) and Heslop-Harrison and Heslop-Harrison (1989d) have shown that three different antibodies to myosin, although yielding a distinctly punctate pattern (which would be expected if myosin is linked to organelles) in the vegetative cytoplasm, produce no signal within the generative cell. Punctate fluorescence is concentrated at the periphery of the generative cell (where vegetative MFs also congregate) but likely is associated with the cytoplasmic surface of the vegetative plasma membrane. Our own studies with antimyosin confirm these findings (Palevitz and Liu, 1992). The myosin could be related to projections recently identified on the surface of the vegetative membrane at this location (Cresti *et al.,* 1991).

These results are also consistent with studies on the effects of cytochalasins on the pollen tube cytoskeleton. These drugs have drastic effects on F-actin in the vegetative cytoplasm that are clearly visible with the fluorescence and electron microscopes as fragmentation and/or abnormal thickening of MF arrays (Lancelle and Hepler, 1988; Tang *et al.,* 1989a; Tiwari and Polito, 1990b; Heslop-Harrison and Heslop-Harrison, 1991a; Pierson and Cresti, this volume). The effects are accompanied by the cessation of movement of organelles, vegetative nucleus, and generative cell (Heslop-Harrison and Heslop-Harrison, 1989a, 1991a). However, no alteration is seen in the shape or ultrastructure of the generative cell in response to treatment (Lancelle and Hepler, 1988; Heslop-Harrison and Heslop-Harrison, 1989a). In the fluorescence microscope, cytochalasin-induced rods are plentiful in the vegetative cytoplasm but are conspicuously absent from the generative cell (Palevitz and Liu, 1992).

In summary, we cannot account for the positive results on actin reported by Taylor *et al.* (1989). However, the balance of evidence points to the lack of F-actin in GSPs.

C. Implications

If F-actin is indeed absent, it seems to disappear during the microspore division creating the generative cell. Recently, Brown and Lemmon (1989 and personal communication) have shown that the cytoplasm around the generative cell pole becomes depauperate in RP-binding material as the

new generative cell detaches from the parental wall at the end of this highly asymmetrical division.

The reasons for this "housecleaning" of generative cell MFs are unclear, but its implications are significant. First, GSPs may be the only plant cells so far examined that are devoid of MFs, with the possible exception of cells in the quiescent zone of root meristems (Ding *et al.,* 1991). Second, if F-actin is absent, it cannot take part in motility-related processes, such as organelle movements and cell locomotion. Thus, the lack of MFs may be correlated with an absence of intracellular saltations and streaming in GSPs (J. Heslop-Harrison, personal communication). However, because MTs also play a role in intracellular transport (e.g., Vallee and Shpetner, 1990), more information is needed on the presence or absence of such movements inside GSPs. Much speculation has surrounded the transport of GSPs down the pollen tube and whether GSPs are self-propelled (i.e., capable of locomotion). An extended discussion of this subject occurs herein (Section IV,C).

The negative data on MFs also indicate that F-actin cannot participate in cytokinesis, via either phragmoplast or cleavage-mediated mechanisms. Indeed, treatment of pollen tubes with cytochalasins does not seem to disrupt cytokinesis per se in *Nicotiana* or *Tradescantia* (Palevitz and Cresti, 1989; Palevitz and Liu, 1992), although it may delay division. The drugs alter normal tip morphology and fragment vegetative MFs (e.g., Lancelle and Hepler, 1988; Heslop-Harrison and Heslop-Harrison, 1989a; Steer and Steer, 1989; Tang *et al.,* 1989a; Tiwari and Polito, 1990b). The anti-MT agents colchicine and oryzalin have severe effects on generative cell division (Palevitz and Cresti, 1989). The presence of F-actin in the phragmoplast of somatic cells has engendered considerable speculation about its possible function in cytokinesis (e.g., Palevitz, 1987). However, the negative observations on F-actin in generative cells that divide with the aid of a phragmoplast (e.g., in *Nicotiana*) may limit the range of its possible functions at this location. For example, MFs associated with the phragmoplast could aid in the consolidation of the new cell wall. Thus, the absence of F-actin in cytokinetic generative cells and the production of a seemingly fragile partition between the sperm (e.g., Hu and Yu, 1988; Charzynska *et al.,* 1989; Taylor *et al.,* 1989) may not be coincidental.

The mechanism responsible for constriction in generative cells of such species as *Tradescantia* is also unclear. The negative data in this situation, however, seem to preclude the operation of an actin-myosin–based purse-string mechanism analogous to that responsible for cleavage in animal cells (Cao and Wang, 1990).

Despite the absence of MFs in GSPs, actin in its monomeric or G form may still be present. Indeed, two different studies using protein and nucleic acid methods reported that actin increases in pollen after the generative

cell forms (Stinson *et al.*, 1987; Bedinger and Edgerton, 1990). If G-actin is present, an accompanying actin-binding protein may prevent its assembly into filaments. Thus, a recent article on the presence of significant quantities of profilin in pollen is noteworthy, because this protein binds G-actin and prevents its assembly (Valenta *et al.*, 1991; Pollard and Cooper, 1986). Given these unknowns, experiments with isolated GSPs assume increased significance. Isolated cells can be used to search for actin polypeptides. Moreover, polymerase chain reaction (PCR) would be valuable in assessing whether actin mRNAs are present.

IV. Function(s) of the Generative Cell and Sperm Cytoskeleton

A. Evolutionary Considerations

The high degree of order displayed by the bundled MTs in GSPs is sufficiently novel for plant cells to warrant comparisons with male reproductive cells in nonflowering taxa. Specifically, are analogous structures found in the sperm of other species? At least one interesting observation stands out. Terrestrial species with flagellated, motile sperm contain a distinctive multilayered structure (MLS) with which the flagellar basal bodies and mitochondria are associated. The MLS consists of a layer of MTs (the MT ribbon or spline) and associated fibers and partitions that participate in nuclear shaping and/or other aspects of cell differentiation and organization (Hepler and Myles, 1977; Carothers and Duckett, 1980; Li *et al.*, 1989). It is tempting to speculate that the MT bundles in angiosperm GSPs represent the partial retention of the MLS or a modified form of its expression. Alternatively, the bundles may bear homologies with the force producing axonemes of progenitor flagella (Lancelle *et al.*, 1987; Cresti *et al.*, 1990). In either case, the architecture of the array is altered, perhaps in relation to a change in function away from cell locomotion. Nevertheless, GSP MT bundles undoubtedly govern specific functions, which we will now address.

B. GSP MTs and Cell Shape

1. Evidence in Favor

The most commonly mentioned function of GSP MTs is the establishment and/or maintenance of the elongate, lenticular, and spindle shapes charac-

teristic of these cells. Because the MTs are arranged in cross-bridged bundles aligned primarily parallel to the long axis of the cell, it is reasonable to suppose that they are involved in cell shaping (Hoefert, 1969; Burgess, 1970b; Cass, 1973; Cresti *et al.*, 1984; Pierson *et al.*, 1986; Raudaskoski *et al.*, 1987; Tiezzi *et al.*, 1988; Pierson, 1989). That the MT bundles often occupy longitudinal ridges or scallops also supports this idea. Further, Burgess (1970b) noted that the generative cell of *Endymion* changes shape as its MT complement reorganizes during prophase.

A shape-determining role of MTs may have considerable functional significance. For example, the elongate shape of GSPs may aid in their transport through the pollen tube (Cass, 1973; Heslop-Harrison *et al.*, 1988). Moreover, GSP shape is not static; instead, the cells are pleomorphic (Steffen, 1953; Heslop-Harrison *et al.*, 1988; Cresti *et al.*, 1990; Fig. 1D), which may further aid their entry into and progress through the tube (Cass, 1973; Heslop-Harrison *et al.*, 1988). Typically, the end of GSPs toward the tip of the pollen tube is characterized by a particularly long extension (Venema and Koopmans, 1962; Navashin, 1969; Heslop-Harrison *et al.*, 1988), which may be a response to or a causative factor in cell movement. The elaboration of pointed, strap-like, or lamellar appendages (Figs. 1D and F) also appears to mediate the association of GSPs with the vegetative nucleus (e.g., Figs. 1D and E; Russell, 1984; Dumas *et al.*, 1985; Mogensen and Wagner, 1987; Hu and Yu, 1988; Knox *et al.*, 1988; Rusche and Mogensen, 1988; Wagner and Mogensen, 1988; Taylor *et al.*, 1989; Yu *et al.*, 1989; Russell *et al.*, 1990). Again, such appendages may be generated with the aid of MTs. The function of MTs in GSP shaping would be analogous to the direct skeletal function that they play in various animal and protist cells (Dustin, 1978). Furthermore, it is not unexpected, given the fact that GSPs are surrounded by a much more flexible extracellular matrix than the reinforced cellulosic walls of somatic plant cells (e.g., Cresti *et al.*, 1987, 1988b; Charzynska *et al.*, 1988). The latter are also thought to utilize MTs to determine cell shape, but the action is indirect, via the alignment of new cellulose microfibrils (Seagull, 1989).

MTs may regulate changes in cell shape via the participation of the many cross-bridges seen in the electron microscope. Whereas many of these bridges may be structural in nature, others may consist of mechanochemical motor proteins that could promote MT-to-MT sliding. Cross-linking proteins have been identified in other cells that facilitate the sliding of bundled MTs (Vallee and Shpetner, 1990).

In a pioneering study on the generative cell cytoskeleton, Sanger and Jackson (1971b) found that MTs in newly formed generative cells of *Haemanthus,* which are still spherical in shape, are randomly arranged, but increase in number and become longitudinally aligned as the cells elongate. If the MTs in elongated cells are disrupted by colchicine or

isopropyl *N*-phenylcarbamate, the cells revert to a spherical shape. These results support a role for MTs in generative cell morphogenesis. The Heslop-Harrisons have provided the most intensive, recent efforts on this subject. While repeating Sanger and Jackson's results with colchicine (but on *Galanthus*), they also provided extensive observations on the pleomorphic behavior of the generative cell as it emerges from the grain and moves through the pollen tube (Heslop-Harrison *et al.*, 1988). They concluded that generative cells actively control their shape in response to environmental signals via the mechanochemical activity of cross-bridges in the MT bundles. The resulting changes in MT distribution–organization produce the cell contractions and extensions seen *in vivo*. In a somewhat different interpretation, they allowed in a later article that the generative cell, like the vegetative nucleus, is deformed by forces produced by actin–myosin interactions in the vegetative cytoplasm, but the former is less malleable, due in part to its MT cytoskeleton (Heslop-Harrison and Heslop-Harrison, 1989b). In this scenario, the MT bundles would presumably act as an internal skeleton to restrain excessive generative cell deformation.

2. Are MTs Sufficient to Maintain Cell Shape?

While these observations stress a direct skeletal role for generative cell MTs, the matter is more complicated. It seems that although MTs may be necessary for cell shape generation–maintenance *in situ*, they are not sufficient to maintain shape when GSPs are placed outside the confines of the pollen grain–tube. Most studies with isolated GSPs (e.g., Cass, 1973; Dupuis *et al.*, 1987; Zhou *et al.*, 1988, 1990; Cass and Fabi, 1988; Tanaka *et al.*, 1989; Theunis and Van Went, 1989; Russell *et al.*, 1990; Theunis *et al.*, 1991, 1992; Chaboud and Perez, this volume) clearly show that the cells become spherical soon after liberation (although reports of cells retaining an elongate shape are available; Knox *et al.*, 1988). Clearly, if MTs are still present, they are not maintaining cell shape under these drastically altered conditions. Thus, MTs may operate in shape maintenance along with other environmentally sensitive determinants near the cell surface. Interestingly, the cortical MTs of somatic cells cannot maintain cylindrical or specialized shapes either, despite extensive cross-bridging to the plasma membrane (Doohan and Palevitz, 1980; Lloyd *et al.*, 1980; Palevitz, 1982; Wang *et al.* 1989; Melan, 1990); these cells also become spherical during protoplast formation. Examination of somatic protoplasts shows that MTs are still present but are randomly arranged (e.g., Doohan and Palevitz, 1980; Lloyd *et al.*, 1980; Wang *et al.*, 1989; Akashi *et al.*, 1990; Melan, 1990).

Thus, the fate of GSP MTs after cell isolation is of interest. This issue has now been addressed in the generative cells of *Lilium* (Tanaka *et al.*, 1989), *Zephyranthes* (Zhou *et al.*, 1990), *Nicotiana* (Theunis *et al.*, 1992), and *Allemanda* (Zee and Aziz-Un-Nisa, 1991). Although the cells retain a positive reaction to antitubulin as they round up, the signal assumes diffuse, random, or net-like patterns. Although rearrangement of the MT array is indicated in the last two patterns, a diffuse fluorescence signal could arise from MT depolymerization. This conclusion is supported by the results of Cass and Fabi (1988), who could not detect MTs in the sphericalized sperm isolated from *Zea mays*. The generative cells may be rounding up in response to changes in MTs (which may be due to factors attending cell isolation), in which case the MTs could have maintained cell shape had their organization been retained. Conversely, the MTs may be responding to the change in cell shape, in which case they are insufficient as direct skeletal determinants. Again, analogous work with somatic cell protoplasts shows that MTs do not disappear during sphericalization. Studies have indicated that interactions with cell wall proteins mediated by protease-sensitive plasma membrane moieties stabilize MTs in tobacco protoplasts (Akashi *et al.*, 1990; Akashi and Shibaoka, 1991). These results strengthen the possibility of potential relationships among MTs, cell shape, and environmentally sensitive elements in the thin cell wall surrounding GSPs.

C. GSP Locomotion

1. Background

The movement of GSPs out of the pollen grain and through the advancing tube has engendered a lot of attention. That the movement often occurs in association with the vegetative nucleus adds further interest in the underlying motility mechanism. GSPs generally follow the advancing tube tip, but movement can be discontinuous, with the cells slowing spontaneously or even reversing direction temporarily. The GSPs can also trail, overtake, lead, and retreat relative to the vegetative nucleus (Venema and Koopmans, 1962; Heslop-Harrison and Heslop-Harrison, 1987, 1989a,b). Two hypotheses have dominated thinking in this area (e.g., Navashin, 1969; Heslop-Harrison *et al.*, 1988): (1) the power of movement resides within the GSP, as it does in flagellated sperm, and (2) GSP transport is governed by forces exerted in the vegetative cytoplasm. Interest in the first possibility is fueled by the presence of highly cross-bridged MT bundles in GSPs (Hoefert, 1969; Lancelle *et al.*, 1987; Cresti *et al.*, 1990). The analogy is drawn to the complex architectural arrangement of flagellar axonemes,

in which dynein bridges provide the ATPase activity necessary for active MT sliding and resulting flagellar motion. Thus, any undulatory or ameboid movements of GSPs would be governed by MTs and associated proteins (Steffen, 1953; Pierson, 1989).

2. GSPs Do Not Appear to Be Self-Propelled

Despite the attraction of this hypothesis, the bulk of evidence now favors motion of GSPs determined by the vegetative cytoskeleton (Heslop-Harrison and Heslop-Harrison, 1987, 1989b; Heslop-Harrison *et al.*, 1988). For example, isolated GSPs generally do not exhibit directed movements (e.g., Cass and Fabi, 1988; but see Steffen, 1953). Colchicine, which disrupts both the MTs and shape of generative cells, does not interfere with their emergence into or movement through the pollen tube (Heslop-Harrison *et al.*, 1988). The vegetative nucleus undoubtedly does not have self-propelling faculties, yet it moves in close association with the GSPs. Extensive light microscopic observations of live and fixed pollen tubes strongly point to interactions between the vegetative nucleus and streaming vegetative cytoplasm containing MFs (Navashin, 1969; Pierson, 1989; Heslop-Harrison and Heslop-Harrison, 1988, 1989b,c). Cytochalasin, unlike colchicine, stops movement in the pollen tube, although the timing of the response varies between the generative cell and vegetative organelles (Heslop-Harrison and Heslop-Harrison, 1989a). Because generative cells probably lack MFs whereas the vegetative cytoplasm contains many, the conclusion seems inescapable that movement of the GSPs and vegetative nucleus is governed by the vegetative cytoplasm. The conclusion is reinforced by the concentration of vegetative MFs in the vicinity of the GSPs (Pierson, 1989; Heslop-Harrison and Heslop-Harrison, 1989c). Indeed, recent observations show that the generative cell is closely ensheathed in MFs (J. Derksen, personal communication), although it is still unclear whether the MFs represent a specialized component of the vegetative actin network. The localization of myosin on the vegetative cytoplasmic surface bordering the GSPs (Heslop-Harrison and Heslop-Harrison, 1989d; Tang *et al.*, 1989b; Palevitz and Liu, 1992) is also relevant. Myosin would likely provide the enzymatic basis for GSP movement along MFs. The discontinuous nature of the movement may be due to successive encounters with MFs or cytoplasmic streams of opposite polarity and direction.

   In summary, although it is possible that the complex arrangement of MTs and bridging moieties in GSPs is evolutionarily derived from the axonemes of flagellated sperm antecedents and may retain mechanochemical activities (Lancelle *et al.*, 1987; Cresti *et al.*, 1990), the bulk of evidence

indicates that the bundles now function in cell morphogenesis rather than cell locomotion.

D. Generative Cell Division

1. A Bit of History

The premier function of the generative cell is to divide and thereby produce two functional sperm. Consequently, events attending this division have engendered a lot of research over the years. Many of the studies appeared in the first half of the century and were followed by a relative hiatus that lasted until the 1980s. There has been renewed attention to the subject.

It is perhaps an understatement to say that the phenomenology and mechanisms governing generative cell division have been the subject of considerable debate. Seemingly conflicting observations, sometimes on the same genus or species, have provided fodder for numerous reviews and reexaminations. In 1936 Cooper published a reinvestigation of generative cell division in *Lilium* because "Differing observations have been made regarding the history of the male gametes in the *Liliaceae,* and especially in the genus *Lilium.* The present investigation was undertaken because the findings of certain recent investigators are almost diametrically opposed to those of earlier workers." Five years later the situation had not changed much because Johnston (1941) noted that "Numerous investigations have been made of the development of male gametes in angiosperms, but cytologists are still not in agreement concerning many phases of the subject. This is true especially with regard to the mechanism by which division of the generative cell is accomplished. The chief points at issue are whether a regular metaphase plate is present, whether a spindle is formed, [and] whether cytokinesis is by furrowing, constriction, or cell-plate formation . . . " That same year, Sax and O'Mara published observations diametrically opposite to those of Johnston.

Later reports have revived the discussion. However, it now seems clear that while events similar to those in other cells are the rule in many dividing generative cells, seemingly unconventional arrangements of the mitotic apparatus govern this division in other species, such as *Tradescantia virginiana.* Instead of trying to resolve which observations or investigators are correct, efforts have now shifted to explain the reasons for the unconventional arrangements and how they operate to achieve accurate chromosome separation. As a result, we may also clarify the reasons behind the dichotomy.

2. Conventional Spindles and Phragmoplast–Cell Plates

Many reports have documented spindles and phragmoplast–cell plate complexes in dividing generative cells. The reader is referred to Cooper (1936),

Johnston (1941), Maheshwari (1950), and Ota (1957) for coverage of the older literature. In recent years, complimentary reports have come from bicellular and tricellular species, such as barley (Charzynska *et al.*, 1988), tobacco (Raudaskoski *et al.*, 1987), *Rhododendron* (Taylor *et al.*, 1989), rye (Karas and Cass, 1976), *Sambucus* (Charzynska and Lewandowska, 1990), and *Brassica* (Charzynska *et al.*, 1989). Metaphase plates are easily recognizable and may be transverse or obliquely aligned relative to the long axis of the cell and spindle (Figs. 3H and I). Kinetochore fibers are also clearly distinguishable in some cases. Karyokinesis is followed by the appearance of a phragmoplast rich in MTs (Fig. 2E). As already noted, evidence is split on F-actin in the phragmoplast, but the balance of data indicates that it is absent (e.g., Fig. 2F).

The origin of the phragmoplast is still not completely clear, which mirrors the situation in somatic cells and endosperm (Bajer and Molé-Bajer, 1986; Vantard *et al.*, 1990; Zhang *et al.*, 1990; Asada *et al.*, 1991). That the phragmoplast is at least superficially similar to that in other cells is attested to by the presence of interdigitating MTs and clustered vesicles in the midzone (Karas and Cass, 1976; Charzynska *et al.*, 1988, 1989). At the immunofluorescence level, the midzone is clearly marked by a thin, dark line or cell plate (Fig. 2E; Raudaskoski *et al.*, 1987; Taylor *et al.*, 1989).

The final result of phragmoplast activity and the behavior and fate of the cell plate are somewhat unclear. The cell plate is reported to fade or degenerate in some cases (Eigsti, 1940). However, it clearly persists in many others and leads to a partition or wall that can be distinctly fibrillar in nature (e.g., Cresti *et al.*, 1987; Hu and Yu, 1988). Recent results show that the new wall also contains callose (Yu *et al.*, in press), as does the cell plate of somatic cells (Fulcher *et al.*, 1976). The plate often seems fragile, however, and localized regions of incomplete cytokinesis could be responsible for sustained associations between sperm cells in some species (e.g., Hu and Yu, 1988).

As already noted, the origin of the MT cytoskeleton in *Nicotiana* sperm may differ from that in *Tradescantia*. Preliminary evidence (Palevitz and Liu, in preparation) and examination of published micrographs (e.g., Raudaskoski *et al.*, 1987) indicate that the MTs are assembled *de novo* at the end of division.

Because the generative cell is contained within an additional membrane derived from the vegetative plasma membrane, whether the two sperms completely separate is of interest. The cell plate would be expected to bisect only the generative cell proper, leaving the vegetative membrane unaffected. However, reports vary as to whether the sperm cells plus surrounding membrane completely separate (e.g., Cass and Karas, 1975; Russell and Cass, 1981; Mogensen and Rusche, 1985; Charzynska *et al.*, 1988, 1989; Weber, 1988; Yu *et al.*, 1989, in press). Potential mechanisms

by which the vegetative membrane could be cleaved remain largely speculative (Weber, 1988).

3. Unconventional Arrangement of the Division Apparatus as Typified by the Generative Cell of Tradescantia

a. Background   Reports of unusual arrangements of the division apparatus in generative cells of various species have appeared repeatedly over the years and have been challenged about as often. Debate has most notably surrounded *Tradescantia* and members of the *Liliaceae* (e.g., *Lilium, Tulipa*). For example, whereas O'Mara (1933) and Sax and O'Mara (1941) reported atypical arrangements in lily, Cooper (1936) and Johnston (1941) forcefully disagreed, all in the same time period. The Sax and O'Mara work also covered *Tradescantia virginiana*. Whereas Lewandowska and Charzynska (1977) and Terasaka and Niitsu (1989b) obtained similar findings with *Tradescantia bracteata* and *Tradescantia paludosa,* respectively, Lafleur *et al.* (1981), and J. P. Mascarenhas (personal communication, 1981) reported conventional spindles and metaphase plates in *Tradescantia paludosa.* Our own results with *Tradescantia virginiana* are in line with those of Sax and O'Mara, Lewandowska and Charzynska, and Terasaka and Niitsu.

Part of the problem behind these conflicting results may be misinterpretations of chromosome arrangements in material prepared with traditional histochemical stains (Cooper, 1936; Maheshwari, 1950), as well as newer fluorescence probes, such as Hoechst and DAPI. For example, Eigsti (1940) reported conventional metaphase plates in *Tradescantia,* but an examination in that article, which is supposed to show chromosome arms extending away from a rather compact metaphase plate, actually represents a generative cell in late anaphase–telophase, in which two new sets of sperm chromosomes are in close apposition and probably linked by a chromosome bridge. The image is equivalent to Figs. 2B and 13B in Palevitz and Cresti (1989) in which the mitotic stage was confirmed using antitubulin localizations. Precise determinations of division stages, given the complex, convoluted arrangements of chromosomes in these cells, often require a separate assay, with the visualization of MTs being the most efficacious.

b. Establishment of the Mitotic Apparatus   Generative cell division in *Tradescantia* begins approximately 3 hours after pollen germination (Lewandowska and Charzynska, 1977; Palevitz and Cresti, 1989). Until this time, the generative cell nucleus appears uniformly bright after DAPI or Hoechst staining, with no sign of condensation of individual chromosomes. The nucleus is highly elongate and usually situated behind or astride the vegetative nucleus relative to the tip of the pollen tube. The

vegetative nucleus is less fluorescent than that of the generative cell and often appears mottled or nonhomogenous. This image changes little during the course of generative cell division. Initial chromosome condensation is witness to the first interesting manifestation of division. Although the interphase MT arrays of somatic cells typically disassemble prior to division, prophase in *Tradescantia* generative cells occurs in the presence of the MT bundles so prominent to this point.

From the earliest signs of condensation, the chromosomes are arranged in complex twisted patterns (Fig. 3C). Sax and O'Mara (1941) described the chromosome distribution as "scattered." To clarify this arrangement further, kinetochores were visualized using a human autoimmune serum obtained from a patient with the CREST form of scleroderma (Palevitz, 1990; Liu and Palevitz, 1991). Immunolocalizations with this preparation reveal kinetochores as pairs of spherical dots distributed along the length and depth of the cell (Liu and Palevitz, 1991). At around 4 hours after germination, the dot pairs are predominantly oriented transverse to the long axis of the cell (Figs. 3A and B). Dual localizations with antitubulin show that the kinetochores are laterally associated with the MT bundles (Figs. 3A and B). Early lateral interactions between kinetochores and spindle MTs have been the subject of intense interest in the mitosis community (Liu and Palevitz, 1991). Examination of a range of preparations of different age indicates that the orientation of the kinetochore pairs shifts from transverse to oblique and longitudinal (Figs. 3D and E) and that the shift is associated with a switch from lateral associations to end-on kinetochore fibers. At the same time, the distance between opposing filial kinetochores increases, probably due to axially directed forces exerted by the fibers. The new kinetochore fibers are heterogenous in length, with some short and others long. They're also linked to each other and to surrounding MT bundles by a variety of connecting fibers (Fig. 3D; Liu and Palevitz, 1991). The kinetochore fibers appear to increase in complexity and assume a number of characteristics similar to kinetochore fibers of somatic cells (Palevitz and Cresti, 1989; Liu and Palevitz, 1991). In particular, MTs branch from the trunk of the fibers in fir-tree fashion (Palevitz, 1988; Bajer, 1990). All these observations indicate that kinetochore fibers are derived from the preexisting MT bundles and then progressively modified while maintaining connections with each other and with surrounding bundles. The establishment of the mitotic apparatus may therefore differ from that in species such as *Nicotiana,* in which Raudaskoski *et al.* (1987) maintained that kinetochore fibers form as or after the interphase bundles disassemble, a situation similar to that in somatic divisions. The *Tradescantia* generative cell arrives at the full-blown metaphase state with its kinetochore pairs still strung out, often in tandem fashion, along the length and depth of the cell, and chromosome arms still

entwined in complex twisted patterns (Figs. 3C,E; Lewandowska and Charzynska, 1977; Palevitz and Cresti, 1989; Terasaka and Niitsu, 1989b; Liu and Palevitz, 1991).

These observations indicate that generative cell division is mediated by a system of fibers, although to this point they are not organized in a manner typical of somatic spindles. In some cells, where many of the kinetochore pairs and kinetochore fibers are obliquely aligned, some hint of bipolar order can be seen. In most instances, however, an organization consisting of discrete half spindles typical of somatic spindles is not obvious. Nevertheless, old reports that generative cells lack spindle fibers (Maheshwari, 1950) are inaccurate in the strict sense. Indeed, the kinetochore fibers bear important similarities to their counterparts in other cell types. Still, it is difficult to visualize how two groups of brother kinetochores will become sorted toward opposite ends of the cell given such a metaphase arrangement. Kinetochores must pass each other going in opposite directions starting from distant points. The chromosomes do not engage obvious half spindles, so the mitotic apparatus seemingly lacks the bipolarity that is a hallmark of typical spindles and provides the framework for orderly chromosome movement. But sorted they are, with apparent fidelity. The division apparatus works; the question is, how?

c. Anaphase–Telophase   The onset of anaphase occurs around 7 hours after germination. Lewandowska and Charzynska (1977) indicated that anaphase begins in a wave from one part of the *Tradescantia bracteata* generative cell to the other. We could not confirm this finding in *Tradescantia virginiana* (Liu and Palevitz, 1992). Early anaphase is characterized by a reorganization of the kinetochore fibers and surrounding MT bundles such that nonfilial kinetochores become grouped into two obvious sets as kinetochore fibers are linked on two large MT superbundles (Liu and Palevitz, 1992; Figs. 3F and G). The immunofluorescence images indicate that chromosome motion occurs in two stages. The kinetochore fibers proper shorten, linking the kinetochores directly to the superbundles in what appears to be a form of anaphase A. The superbundles also separate, elongating the mitotic apparatus and carrying the attached kinetochores further apart. This second component seems analogous to anaphase B. The *Tradescantia* generative cell differs from many somatic cells in this regard, which lack significant anaphase B motion (Hepler and Palevitz, 1986; Baskin and Cande, 1990). Because the interkinetochore distance on each superbundle shortens, the kinetochores further cluster at the ends of the cell (Fig. 4D).

The events in early anaphase in effect transform the mitotic apparatus into a bipolar array or make a preexisting but cryptic bipolarity more obvious. They also indicate that the motion of chromosomes in each set is probably coordinated, an interesting conclusion given that, in an effort

to explain the unusual events in this cell, investigators have speculated that individual chromosome pairs organize their own independent spindles (Sax and O'Mara, 1941; Lewandowska and Charzynska, 1977).

Cytokinesis begins as a localized constriction of interzonal MTs trailing away from attachment sites on the superbundles (Fig. 4A). The constriction zone spreads, finally cleaving the two new sperm cells (Fig. 4B; Palevitz and Cresti, 1989; Palevitz, 1990; Liu and Palevitz, 1992). Each sperm inherits a now shortened superbundle and MT branches as its initial cytoskeleton. The superbundles occupy the distal end of each cell (Fig. 4C) and may help create the cell extension at that location. The proximal end of each sperm adjacent to the cleavage site may also contain one or more thick bundles, which appear to be derived from constriction of the interzonal MTs (Fig. 4B). The MT cytoskeleton may then be modified as the sperm cells mature.

The mechanism responsible for cytokinesis is still unclear. As already noted, actin MFs are not detectable inside the generative cell at any time during division. Moreover, sperm cells are formed in the presence of cytochalasin D at concentrations that affect vegetative MFs (Palevitz and Cresti, 1989; Palevitz and Liu, 1992). Thus, an actin-myosin–based system analogous to that responsible for animal cell cleavage (Cao and Wang, 1990) seems unlikely. It is possible that interzonal MTs exert the force necessary for constriction. Colchicine and oryzalin interfere with division (Palevitz and Cresti, 1989), but such results are not terribly illuminating viz. cytokinetic mechanisms because the effects of the drugs are massive and nonselective (they disrupt the entire division apparatus).

Ultrastructural data are needed to further clarify a number of aspects of this division. For example, the interaction between MT bundles and the kinetochores should be more intensively investigated, as should the nature of the constriction zone and the organization of its MTs. It is of interest whether the compressed MTs bear any resemblance to a phragmoplast that is not evident at the light microscope level. The fate of the vegetative membrane surrounding the dividing generative cell is also unclear, as is whether the two sperm cells maintain connections or a degree of continuity. Antitubulin immunofluorescence consistently fails to detect MTs between the sperm cells proper (Palevitz and Cresti, 1988) but does not rule out tenuous cytoplasmic connections devoid of MTs. Finally, superbundles with attached single kinetochores remain evident in sperm cells for some time, an observation that raises interesting questions about the status of the nuclear envelope.

4. Further Considerations

a. Architecture of the Mitotic Apparatus   The factors responsible for differences in the architecture of the mitotic apparatus deserve further

attention. Sax and O'Mara (1941) believed that space constraints determined by the narrow confines of the pollen tube and/or the size–volume of the chromosomes or spindle lead to the extended metaphase arrangements typified by *Tradescantia*. Thus, the mitotic apparatus is constrained by the rigid wall of the pollen tube rather than the flexible matrix around the generative cell. This hypothesis was reiterated by Lewandowska and Charzynska (1977), and Terasaka and Niitsu (1989a,b) took up the argument to explain mitotic arrangements in *Calanthe* and *Tradescantia paludosa*. Thus, comparisons with species that utilize spindles of more conventional appearance are in order. Inspection of metaphase spindles and chromosome assemblages in *Nicotiana* and *Rhododendron* generative cells shows that they are indeed smaller than the massive chromosome volumes taken up in *Tradescantia*. Nevertheless, even here the metaphase plates are usually distinctly oblique and kinetochore fibers may be linked in higher order aggregates (Figs. 3H and I; Raudaskoski *et al.*, 1987; Taylor *et al.*, 1989; Palevitz and Liu, in preparation; Terasaka and Niitsu, 1989a,b). In the generative cells of *Calanthe,* Terasaka and Niitsu (1989a) maintained that a spindle is present but skewed to such an extent that the metaphase plate is fully longitudinal. We may be seeing in these diverse observations a gradation in arrangement between the relatively orthodox spindles typified by *Nicotiana* and the seemingly chaotic condition of *Tradescantia*. It is worth stressing again in this context that the relative lack of organization in *Tradescantia* during metaphase is probably only apparent. It seems unlikely that the bipolarity manifested at the start of anaphase arises *de novo;* instead, anaphase reorganization is probably visualizing a bipolar architecture present in metaphase but not immediately obvious (Liu and Palevitz, 1992).

Oblique spindles are commonly encountered in various locations including guard mother cells, stamen hairs, and rib meristems (Palevitz and Hepler, 1974; Palevitz and Cresti, 1989; Palevitz, 1986, 1990). These arrangements have also been attributed to space constraints. Wada (1965) reported that somatic spindles become increasingly oblique as the ploidy level increases in various *Triticum* species. Somatic cells have reorientation movements in anaphase and telophase that correct the oblique alignment, in response to the presence of morphogenetic information on the division plane in the cell cortex (Gunning, 1982; Palevitz, 1986; Zhang *et al.*, 1990). The movements are driven in part by phragmoplast–interzone interactions with the preprophase band (PPB) site. The anaphase *Allium* root cell shown in Fig. 17 of Palevitz and Cresti (1989) appears to have a skewed spindle not totally unlike the mitotic apparatus of *Tradescantia* generative cells, with kinetochore fibers seemingly fused into large aggregates. Similar arrangements are also evident in the spindles of *Calanthe* (Terasaka and Niitsu, 1989a) and *Nicotiana* (Palevitz and Liu, in prepara-

tion). In the latter, the spindle also appears distorted in the same manner noted in somatic cells with oblique metaphase plates (e.g., Palevitz and Hepler, 1974). Thus, the arrangement of the mitotic apparatus in *Tradescantia* generative cells may reflect a general tendency seen in all tissues.

We have tried to test the applicability of the space hypothesis to generative cells by examining *Tradescantia* species of different ploidy levels. So far, we have found little or no difference in generative cell division in *Tradescantia hirsuticaulis* and *Tradescantia paludosa*, both diploids with N = 6, and *Tradescantia virginiana* and *Tradescantia ohiensis*, both tetraploids at N = 12 (Liu and Palevitz, in preparation). These results are preliminary and await a more rigorous examination of chromosome volume versus pollen tube width. Nevertheless, similar results were obtained for *Tradescantia paludosa* by Terasaka and Niitsu (1989b); the results are also similar to those with *Tradescantia bracteata*, another diploid, reported by Lewandowska and Charzynska (1977). They are seemingly at odds with those on *Tradescantia paludosa* reported by Lafleur *et al.* (1981). A strict dependency of spindle arrangement on space constraints also runs counter to the situation in *Nicotiana*. Although the spindle is almost always oblique, the small pleomorphic generative cell seems to have a considerable amount of room within the pollen tube and is free to assume various orientations from longitudinal to nearly transverse (Palevitz and Liu, in preparation). Whatever the cause of skewed or scattered mitotic arrangements, they attest to a remarkable degree of architectural plasticity, which nevertheless results in accurate chromosome sorting.

b. Absence of a PPB   Another comparison between typical and nonconventional division systems in generative cells is also noteworthy: both types of division are not preceded by a PPB (Burgess, 1970b; Raudaskoski *et al.*, 1987; Palevitz and Cresti, 1989; Taylor *et al.*, 1989; Terasaka and Niitsu, 1989a,b; Liu and Palevitz, 1991). Although the PPB marks the division site in most tissue cells, it is notably absent in dividing endosperm, microspore mother cells, and microspores (Gunning, 1982; Sanger and Jackson, 1971a; Terasaka and Niitsu, 1990; Brown and Lemmon, 1991). It has been proposed that the absence of a PPB in such cases may be a function of the isolation of these cells from somatic tissues, a less stringent requirement for precise division planes, the uncoupling of cytokinesis from karyokinesis, or the lack of continuity of cell plates with the parental wall (Gunning, 1982). Any of these arguments could be invoked in the case of generative cells. The absence of a PPB could also reflect specialized properties of the generative cell and/or resulting sperm, such as the flexible nature of the wall or the propensity for sperm cells to remain in close contact via a unique, perhaps incomplete partition. In *Tradescantia* generative cells, where the cytokinetic constriction zone is rather broad, a

precise positioning of the division plane may not be required. Yet in *Nicotiana* and *Rhododendron*, where the presence of a phragmoplast–cell plate would seem to define a more limited division plane, a PPB is absent as well (Raudaskoski *et al.*, 1987; Taylor *et al.*, 1989). In somatic cells the phragmoplast interacts with the PPB site to achieve the final fusion location of the cell plate on the parental plasma membrane (Palevitz and Hepler, 1974; Gunning, 1982; Palevitz, 1986). Interestingly, a mechanism for adjusting the final division plane may also occur in *Nicotiana* generative cells, because the great majority of metaphase plates are oblique but the cell plate is most commonly transverse (Palevitz and Liu, in preparation). Thus, the presence and alignment of a phragmoplast–cell plate is not related to a PPB.

Another intriguing hypothesis is that during division of reproductive cells, cytoplasmically inherited morphogenetic determinants are eliminated in preparation for the next generation. The lack of a PPB, which appears to represent such determinants in somatic tissues (Gunning, 1982), might not be unexpected in this context in generative cells.

c. Interactions with the Vegetative Cell   Interactions between the generative cell and surrounding vegetative structures during division have been addressed a number of times over the years. For example, Raudaskoski *et al.* (1987) reported that vegetative MTs break down in the vicinity of the dividing generative cell in *Nicotiana*. Venema and Koopmans (1962) noted that streaming of the vegetative cytoplasm halts during division, so an effect on the actin–myosin system is indicated. Attention has also been directed at the behavior of the vegetative nucleus (Bishop and McGowan, 1953; Venema and Koopmans, 1962). The vegetative nucleus commonly is positioned distal (toward the tip of the tube) relative to the generative cell by the time division begins. However, division also occurs in the minority of cases in which the vegetative nucleus lags behind the generative cell. Whereas these observations seem to negate a relationship between the two, additional observations suggest otherwise. For example, the vegetative nucleus in *Nicotiana* is primarily located around the distal end of the generative cell at the onset of division (Fig. 3I) but seems to shift to a position straddling the phragmoplast and sperm by telophase (Fig. 2G). The nucleus is often wrapped around the distal sperm and extends along the surface of the phragmoplast, as if appressed to it. In *Tradescantia* the vegetative nucleus passes the generative cell in that significant number of cases where it emerges second during germination, so that in almost all instances it leads the generative cell by the time division ensues. However, the vegetative nucleus then tends to shift proximally after division (Palevitz and Cresti, 1989). These observations could add an additional dimension to emerging concepts on interactions between

GSPs and the vegetative nucleus. As already noted, generative cell MTs are likely causal agents in the formation of cell extensions that interact with the vegetative nucleus (Figs. 1D and E). However, induced changes in MTs and/or MFs in the nearby vegetative cytoplasm could easily modify the relative positions of the generative cell and vegetative nucleus during division, given their flexible nature. The nature of these interactions and their expression via the GSPs and vegetative cytoskeletons deserve more intensive investigation.

5. The Generative Cell Cycle

Key information that might help to clarify some of the problems discussed herein also resides in a more complete knowledge of the generative cell cycle. In addition, other aspects of division relative to the cell cycle deserve attention.

a. Prophase   It has been proposed that the generative cell enters prophase in binucleate species prior to pollen germination (Cooper, 1936; Johnston, 1941; Maheshwari, 1950; Venema and Koopmans, 1962; Lewandowska and Charzynska, 1977). In *Nicotiana*, for example, chromosome condensation is evident well before division and the appearance of a spindle. If the same were the case in *Tradescantia*, the MT bundles in the generative cell might be viewed as a preset mitotic system rather than a specialized interphase array. As already noted, however, while the generative cell nucleus is homogenously bright in the ungerminated grain, indicative of dense chromatin packing, individual chromosomes do not become evident until 3 hours after germination. Prophase, as defined by traditional cytological criteria, is thus delayed until this time. Lewandowska and Charzynska (1977) maintained that the chromosomes actually do condense early in pollen development but then despiralize as the grains mature. More information is clearly needed on cell cycle stages in *Tradescantia* and their relation to the MT array.

b. MT Dynamics   Another aspect of generative cell division in *Tradescantia* may also be worth emphasizing at this point. The progression from generative cells to sperm in this group seems to involve reutilization of intact MTs (Palevitz and Cresti, 1989; Liu and Palevitz, 1991, 1992; Palevitz, 1991). The MT bundles are not broken down prior to division and instead appear to be used in the construction of the division apparatus. Moreover, each sperm cell inherits half the division apparatus as its initial MT cytoskeleton. This characteristic sets these cells apart from somatic cells, where there appears to be no continuity in MTs between generations, although reutilization may occur within and between arrays of a given

generation. *Tradescantia* also appears to differ from *Nicotiana,* in which a phragmoplast replaces the spindle apparatus and appears in turn to be succeeded by a new population of cytoplasmic MTs (Palevitz and Liu, in preparation). It would be interesting to learn more about the relationship between MT dynamics and the cell cycle in these systems.

c. Heterochrony   A third aspect of the cell cycle also deserves closer attention, namely, the factors that control the timing of division. Angiosperms are classified as tricellular or bicellular species, based on whether the division occurs before or after pollen germination (Brewbaker, 1967). The significance of this difference is largely unknown, but correlations with pollen viability, metabolic activity, germination–growth rates, and patterns of self-incompatibility have been made (Brewbaker, 1967; Heslop-Harrison, 1987; J. Heslop-Harrison, personal communication). Although there is considerable variation in division timing within each category, it may be proper to classify the division as broadly heterochronic, that is, a division, the timing of which differs relative to a developmental event (in this case, germination). Thus, generative cells could provide useful information on the control of this class of division. Hopefully, more information about the biochemical and molecular characteristics of generative cells, perhaps obtained from isolated material, will soon shed more light on this subject.

V. Concluding Remarks

Generative cells and sperm possess fascinating and in some ways unique properties of interest not only to those concerned with plant reproduction but to cell biologists as well. As experimental objects, they provide alternative, excellent material for investigating the organization, behavior, and function of plant MTs. Furthermore, our ideas about the cytoskeleton may be significantly augmented by some of the surprising characteristics found in GSPs (e.g., utilization and rearrangement of MTs in *Tradescantia;* lack of MFs). Although the isolation, purification, and characterization of GSPs are crucial for elucidating the nature of cell-to-cell interactions involved in fertilization, more definitive information on the molecular properties of the cytoskeleton should also be sought. This new information will be important in further clarifying the role of the cytoskeleton in the development and behavior of generative cells and sperm.

Acknowledgments

The authors thank Bo Liu for his efforts in much of the unpublished work described in this review. Barry A. Palevitz wishes to particularly thank Professore Mauro Cresti for providing

the stimulus and opportunity to get into this fascinating field. Supported by grants DCB-8703292 and DCB-9019285 to Barry A. Palevitz from the U.S. National Science Foundation, and Biot-CT900172 to Antonio Tiezzi from the E.C.C. Bridge Program.

References

Akashi, T., and Shibaoka, H. (1991). *J. Cell Sci.* **98**, 169–174.
Akashi, T., Kawasaki, S., and Shibaoka, H. (1990). *Planta* **182**, 363–369.
Asada, T., Sonobe, S., and Shibaoka, H. (1991). *Nature* **350**, 238–241.
Åström, H., Virtanen, I., and Raudaskoski, M. (1991). *Protoplasma* **160**, 99–107.
Bajer, A. S. (1990). *Adv. Cell Biol.* **3**, 65–93.
Bajer, A. S., and Molé-Bajer, J. (1986). *J. Cell Biol.* **102**, 263–281.
Bartalesi, A., Del Casino, C., Moscatelli, A., Cai, G., and Tiezzi, A. (1991). *Gior. Bot. Ital.* **125**, 21–28.
Baskin, T. I., and Cande, W. Z. (1990). *Annu. Rev. Plant Physiol. Plant Mol. Biol.* **41**, 277–315.
Bedinger, P. A., and Edgerton, M. D. (1990). *Plant Physiol.* **92**, 474–479.
Bishop, C. J., and McGowan, L. J. (1953). *Am. J. Bot.* **40**, 658–659.
Brewbaker, J. L. (1967). *Am. J. Bot.* **54**, 1069–1083.
Brown, R. C., and Lemmon, B. E. (1989). *J. Cell Biol.* **109**, 89a.
Brown, R. C., and Lemmon, B. E. (1991). *J. Cell Sci.* **99**, 273–281.
Burgess, J. (1970a). *Planta* **92**, 25–28.
Burgess, J. (1970b). *Planta* **95**, 72–85.
Campbell, N. A. (1990). "Biology," 2nd edition. Benjamin/Cummings, Redwood City, California.
Cao, L. G., and Wang, Y. L. (1990). *J. Cell Biol.* **110**, 1089–1096.
Carothers, Z. B., and Duckett, J. G. (1980). *Bull. Torrey Bot. Club* **107**, 281–297.
Cass, D. D. (1973). *Can. J. Bot.* **51**, 601–605.
Cass, D. D., and Fabi, G. C. (1988). *Can. J. Bot.* **66**, 819–825.
Cass, D. D., and Karas, I. (1975). *Can. J. Bot.* **53**, 1051–1062.
Charzynska, M., and Lewandowska, E. (1990). *Ann. Bot.* **65**, 685–689.
Charzynska, M., Ciampolini, F., and Cresti, M. (1988). *Sex. Plant Reprod.* **1**, 240–247.
Charzynska, M., Murgia, M., Milanesi, C., and Cresti, M. (1989). *Protoplasma* **149**, 1–4.
Cooper, D. C. (1936). *Bot. Gaz.* **98**, 169–177.
Cresti, M., Ciampolini, F., and Kapil, R. N. (1984). *J. Submicrosc. Cytol.* **16**, 317–326.
Cresti, M., Lancelle, S. A., and Hepler, P. K. (1987). *J. Cell Sci.* **88**, 373–388.
Cresti, M., Milanesi, C., Tiezzi, A., Ciampolini, F., and Moscatelli, A. (1988a). *Acta Bot. Neerl.* **37**, 379–386.
Cresti, M., Murgia, M., Milanesi, C., Ciampolini, F., and Tiezzi, A. (1988b). *In* "Plant Sperm Cells as Tools for Biotechnology" (H. J. Wilms and C. J. Keijzer, eds.), pp. 27–33. Pudoc, Wageningen.
Cresti, M., Murgia, M., and Theunis, C. H. (1990). *Protoplasma* **154**, 151–156.
Cresti, M., Ciampolini, F., and Van Went, J. L. (1991). *Ann. Bot.* **68**, 105–107.
Cyr, R. J., and Palevitz, B. A. (1989). *Planta* **177**, 245–260.
Derksen, J., Pierson, E. S., and Traas, J. A. (1985). *Eur. J. Cell Biol.* **38**, 142–148.
Derksen, J., Wilms, H. A., and Pierson, E. S. (1990). *Acta Bot. Neerl.* **39**, 1–18.
Ding, B., Turgeon, R., and Parthasarathy, M. V. (1991). *Protoplasma* **165**, 96–105.
Doohan, M. E., and Palevitz, B. A. (1980). *Planta* **149**, 389–341.
Dumas, C., Knox, R. B., and Gaude, T. (1985). *Protoplasma* **124**, 168–174.

Dupuis, I., Roeckel, P., Matthys-Rochon, E., and Dumas, C. (1987). *Plant Physiol.* **85,** 876–878.

Dustin, P. (1978). "Microtubules." Springer-Verlag, Berlin.

Eigsti, O. J. (1940). *Am. J. Bot.* **27,** 512–524.

Fosket, D. E. (1989). *In* "The Biochemistry of Plants" (A. D. Marcus, ed.), Vol. 15, pp. 393–484. Academic Press, San Diego.

Franke, W. W., Herth, W., Van Der Woude, W. J., and Morre, D. J. (1972). *Planta* **105,** 317–341.

Fulcher, R. G., McCully, M. E., Setterfield, G., and Sutherland, J. (1976). *Can. J. Bot.* **54,** 539–542.

Gunning, B. E. S. (1982). *In* "The Cytoskeleton in Plant Growth and Development" (C. W. Lloyd, ed.), pp. 229–292. Academic Press, London.

Han, I. S., Jongewaard, I., and Fosket, D. E. (1991). *Plant Mol. Biol.* **16,** 225–234.

Hardham, A. R., and Gunning, B. E. S. (1978). *J. Cell Biol.* **77,** 14–34.

Hepler, P. K., and Myles, D. G. (1977). *In* "International Cell Biology 1976–1977" (B. R. Brinkley and K. R. Porter, eds.), pp. 569–579. Rockefeller University Press, New York.

Hepler, P. K., and Palevitz, B. A. (1986). *J. Cell Biol.* **102,** 1995–2005.

Heslop-Harrison, J. (1987). *Int. Rev. Cytol.* **107,** 1–78.

Heslop-Harrison, J., and Heslop-Harrison, Y. (1987). *Plant Sci.* **51,** 203–213.

Heslop-Harrison, J., and Heslop-Harrison, Y. (1988). *Sex. Plant Reprod.* **1,** 16–24.

Heslop-Harrison, J., and Heslop-Harrison, Y. (1989a). *Sex. Plant Reprod.* **2,** 27–37.

Heslop-Harrison, J., and Heslop-Harrison, Y. (1989b). *Sex. Plant Reprod.* **2,** 199–207.

Heslop-Harrison, J., and Heslop-Harrison, Y. (1989c). *J. Cell Sci.* **93,** 299–308.

Heslop-Harrison, J., and Heslop-Harrison, Y. (1989d). *J. Cell Sci.* **94,** 319–325.

Heslop-Harrison, J., and Heslop-Harrison, Y. (1991a). *Philos. Trans. R. Soc. Lond. [Biol]* **331,** 225–235.

Heslop-Harrison, J., and Heslop-Harrison, Y. (1991b). *Sex. Plant Reprod.* **4,** 6–11.

Heslop-Harrison, J., Heslop-Harrison, Y., Cresti, M., Tiezzi, A., and Moscatelli, A. (1988). *J. Cell Sci.* **91,** 49–60.

Hoefert, L. L. (1969). *Protoplasma* **68,** 237–240.

Hu, S. Y., and Yu, H. S. (1988). *Protoplasma* **147,** 55–63.

Hussey, P. J., Traas, J. A., Gull, K., and Lloyd, C. W. (1987). *J. Cell Sci.* **88,** 225–230.

Hussey, P. J., Lloyd, C. W., and Gull, K. (1988). *J. Biol. Chem.* **263,** 5474–5479.

Johnston, G. W. (1941). *Am. J. Bot.* **28,** 306–319.

Joshi, H. C., and Cleveland, D. W. (1990). *Cell Motil. Cytoskeleton* **16,** 159–163.

Kakimoto, T., and Shibaoka, H. (1989). *Protoplasma* (Suppl. 2), 95–103.

Karas, I., and Cass, D. D. (1976). *Phytomorphology* **26,** 36–45.

Kiehart, D. P. (1990). *Cell* **60,** 347–350.

Knox, R. B., Southworth, D., and Singh, M. B. (1988). *In* "Eukaryotic Cell Recognition. Concepts and Model Systems" (G. P. Chapman, C. C. Ainsworth, and C. J. Chatham, eds.), pp. 175–193. Cambridge University Press, Cambridge.

Kohno, T., Chaen, S., and Shimmen, T. (1990). *Protoplasma* **154,** 179–183.

Lafleur, G. J., Gross, A. E., and Mascarenhas, J. P. (1981). *Gam. Res.* **4,** 35–40.

Lancelle, S. A., and Hepler, P. K. (1988). *Protoplasma* (Suppl. 2), 65–75.

Lancelle, S. A., and Hepler, P. K. (1989). *Protoplasma* **150,** 72–74.

Lancelle, S. A., and Hepler, P. K. (1991). *Protoplasma* **165,** 167–172.

Lancelle, S. A., Cresti, M., and Hepler, P. K. (1987). *Protoplasma* **140,** 141–150.

Lewandowska, E., and Charzynska, M. (1977). *Acta Soc. Bot. Pol.* **46,** 587–598.

Li, Y., Wang, F. H., and Knox, R. B. (1989). *Protoplasma* **149,** 57–63.

Liu, B., and Palevitz, B. A. (1991). *J. Cell Sci.* **98,** 475–482.

Liu, B., and Palevitz, B. A. (1992). **166,** 122–133. *Protoplasma*

Lloyd, C. W. (1987). *Annu. Rev. Plant Physiol. Plant Mol. Biol.* **38**, 119–139.

Lloyd, C. W. (1989). *Curr. Opin. Cell Biol.* **1**, 30–35.

Lloyd, C. W., Slabas, A. R., Powell, A. J., and Lowe, S. B. (1980). *Planta* **147**, 500–506.

Ludwig, S. R., Oppenheimer, D. G., Silflow, C. D., and Snustad, D. P. (1988). *Plant Mol. Biol.* **10**, 311–321.

Maheshwari, P. (1950). "An Introduction to the Embryology of Angiosperms." McGraw-Hill, New York.

McLean, B. G., Eubanks, S., and Meagher, R. B. (1990). *Plant Cell* **2**, 335–344.

Meagher, R. B. (1991). *Int. Rev. Cytol.* **125**, 139–162.

Melan, M. (1990). *Protoplasma* **153**, 169–177.

Mogensen, H. L., and Rusche, M. L. (1985). *Protoplasma* **128**, 1–13.

Mogensen, H. L., and Wagner, V. T. (1987). *Protoplasma* **138**, 161–172.

Navashin, M. S. (1969). *Rev. Cytol. Biol. Veg.* **32**, 141–148.

Noguchi, T., and Ueda, K. (1990). *Cell Struct. Funct.* **15**, 379–384.

O'Mara, J. (1933). *Bot. Gaz.* **94**, 567–578.

Ota, T. (1957). *Cytologia* **22**, 15–27.

Palevitz, B. A. (1982). *In* "The Cytoskeleton in Plant Growth and Development" (C. W. Lloyd, ed.), pp. 345–376. Academic Press, London.

Palevitz, B. A. (1986). *Dev. Biol.* **117**, 644–654.

Palevitz, B. A. (1987). *Protoplasma* **141**, 24–32.

Palevitz, B. A. (1988). *Protoplasma* **142**, 74–78.

Palevitz, B. A. (1990). *Protoplasma* **157**, 120–127.

Palevitz, B. A. (1991). *In* "The Cytoskeletal Basis of Plant Growth and Form" (C. W. Lloyd, ed.). Academic Press, London.

Palevitz, B. A., and Cresti, M. (1988). *Protoplasma* **146**, 28–34.

Palevitz, B. A., and Cresti, M. (1989). *Protoplasma* **150**, 54–71.

Palevitz, B. A., and Hepler, P. K. (1974). *Chromosoma* **46**, 297–326.

Palevitz, B. A., and Liu, B. (1992). *Sex. Plant Reprod.* **5**, 89–100.

Perdue, T., and Parthasarathy, M. V. (1985). *Eur. J. Cell Biol.* **39**, 13–20.

Pierson, E. S. (1988). *Sex. Plant Reprod.* **1**, 83–87.

Pierson, E. S. (1989). Ph.D. Dissertation, Catholic University of Nijmegen, Nijmegen, The Netherlands.

Pierson, E. S., Derksen, J., and Traas, J. A. (1986). *Eur. J. Cell Biol.* **41**, 14–18.

Pierson, E. S., Kengen, H. M. P., and Derksen, J. (1989). *Protoplasma* **150**, 75–76.

Pollard, T. D., and Cooper, J. A. (1986). *Annu. Rev. Biochem.* **55**, 987–1035.

Prescott, A. R., Vestberg, M., and Warn, R. M. (1989). *J. Cell Sci.* **94**, 227–236.

Raudaskoski, M., Åström, H., Pertilla, K., Virtanen, I., and Louhelainen, J. (1987). *Biol. Cell.* **61**, 177–188.

Ross, J. H. E., Hutchings, A., Butcher, G. W., Lane, E. B., and Lloyd, C. W. (1991). *J. Cell Sci.* **99**, 91–98.

Rusche, M. L., and Mogensen, H. L. (1988). *In* "Sexual Reproduction in Higher Plants" (M. Cresti, P. Gori, and E. Pacini, eds.), pp. 221–226. Springer-Verlag, Berlin.

Russell, S. D. (1984). *Planta* **162**, 385–391.

Russell, S. D., and Cass, D. D. (1981). *Protoplasma* **107**, 85–107.

Russell, S. D., Cresti, M., and Dumas, C. (1990). *Physiol. Plant* **80**, 669–676.

Sanger, J. M., and Jackson, W. T. (1971a). *J. Cell Sci.* **8**, 175–181.

Sanger, J. M., and Jackson, W. T. (1971b). *J. Cell Sci.* **8**, 303–315.

Sax, K., and O'Mara, J. G. (1941). *Bot. Gaz.* **102**, 629–636.

Schliwa, M. (1986). "The Cytoskeleton." Springer-Verlag, New York.

Schmit, A. C., and Lambert, A. M. (1990). *Plant Cell* **2**, 129–148.

Seagull, R. W. (1989). *C.R.C. Crit. Rev. Plant Sci.* **8**, 131–167.

Sheldon, J. M., and Dickinson, H. G. (1986). *Planta* **168**, 11–23.

Silflow, C. D., Oppenheimer, D. G., Kopczak, S. D., Ploense, S. E., Ludwig, S. R., Haas, N., and Snustad, D. P. (1987). *Dev. Gen.* **8**, 435–460.

Staiger, C. J., and Schliwa, M. (1987). *Protoplasma* **141**, 1–12.

Steer, M. W., and Steer, J. M. (1989). *New Phytol.* **111**, 323–358.

Steffen, K. (1953). *Flora* **140**, 140–174.

Stinson, J. R., Eisenberg, A. J., Willing, R. P., Pe, M. E., Hanson, D. D., and Mascarenhas, J. P. (1987). *Plant Physiol.* **83**, 442–447.

Stossel, T. P., Chaponnier, C., Ezzell, R. M., Hartwig, J. H., Janmey, P. A., Kwiatkowski, D. J., Lind, S. E., Smith, D. B., Southwick, F. S., Yin, H. L., and Zaner, K. S. (1985). *Annu. Rev. Cell Biol.* **1**, 353–402.

Sullivan, K. F. (1988). *Annu. Rev. Cell Biol.* **4**, 687–716.

Tanaka, I., Nakamura, S., and Miki-Hirosige, H. (1989). *Gam. Res.* **24**, 361–374.

Tang, X., Hepler, P. K., and Scordillis, S. P. (1989a). *J. Cell Sci.* **92**, 569–574.

Tang, X., Lancelle, S. A., and Hepler, P. K. (1989b). *Cell Motil. Cytoskeleton* **12**, 216–224.

Taylor, P., Kenrick, J., Li, Y., Kaul, V., Gunning, B. E. S., and Knox, R. B. (1989). *Sex. Plant Reprod.* **2**, 254–264.

Terasaka, O., and Niitsu, T. (1989a). *Bot. Mag. Tokyo* **102**, 143–147.

Terasaka, O., and Niitsu, T. (1989b). *J. Jpn. Pollen Soc.* **35**, 7–12.

Terasaka, O., and Niitsu, T. (1990). *Bot. Mag. Tokyo* **103**, 133–142.

Theunis, C. H., and Van Went, J. L. (1989). *Sex. Plant Reprod.* **2**, 97–102.

Theunis, C. H., Pierson, E. S., and Cresti, M. (1991). *Sex. Plant Reprod.* **4**, 145–154.

Theunis, C. H., Pierson, E. S., and Cresti, M. (1992). *Sex. Plant Reprod.* **5**, 64–71.

Tiezzi, A. (1991). *Elec. Microsc. Rev.* **4**, 205–219.

Tiezzi, A., and Cresti, M. (1990). *In* "Mechanism of Fertilization," NATO ASI Series, Vol. H45, pp. 17–34.

Tiezzi, A., Cresti, M., and Ciampolini, F. (1986). *In* "Biology of Reproduction and Cell Motility in Plants and Animals" (M. Cresti and R. Dallai, eds.), pp. 87–94. University of Siena, Siena, Italy.

Tiezzi, A., Moscatelli, A., Ciampolini, F., Milanesi, C., Murgia, M., and Cresti, M. (1988). *In* "Sexual Reproduction in Higher Plants" (M. Cresti, P. Gori, and E. Pacini, eds.), pp. 215–220. Springer-Verlag, Berlin

Tiwari, S. C., and Polito, V. S. (1988). *Protoplasma* **147**, 100–112.

Tiwari, S. C., and Polito, V. S. (1990a). *Eur. J. Cell Biol.* **53**, 384–389.

Tiwari, S. C., and Polito, V. S. (1990b). *Sex. Plant Reprod.* **3**, 121–129.

Valenta, R., Duchene, M., Pettenburger, K., Sillaber, C., Valent, P., Bettelheim, P., Breitenbach, M., Rumpold, H., Kraft, D., and Scheiner, O. (1991). *Science* **253**, 557–560.

Vallee, R. B., and Bloom, G. S. (1984). *Mod. Cell Biol.* **3**, 21–75.

Vallee, R. B., and Shpetner, H. S. (1990). *Annu. Rev. Biochem.* **59**, 909–932.

Vantard, M., Levilliers, N., Hill, A. M., Adoutte, A., and Lambert, A. M. (1990). *Proc. Natl. Acad. Sci. U.S.A.* **87**, 8825–8829.

Venema, G., and Koopmans, A. (1962). *Cytologia* **27**, 11–24.

Wada, B. (1965). *Cytologia* **30**(Suppl.), 92–149.

Wagner, V. T., and Mogensen, H. L. (1988). *Protoplasma* **143**, 101–110.

Wang, H., Cutler, A. J., Saleem, M., and Fowke, L. C. (1989). *Protoplasma* **150**, 48–53.

Weber, M. (1988). *Plant Syst. Evol.* **161**, 53–64.

Wiche, G., Oberkanins, C., and Himmler, A. (1991). *Int. Rev. Cytol.* **124**, 217–273.

Yu, H. S., Hu, S. Y., and Russell, S. D. (1992). *Protoplasma*.

Yu, H. S., Hu, S. Y., and Zhu, C. (1989). *Protoplasma* **152**, 29–36.

Zee, S. Y., and Aziz-Un-Nisa. (1991). *Sex. Plant Reprod.* **4**, 132–137.

Zhang, D., Wadsworth, P., and Hepler, P. K. (1990). *Proc. Natl. Acad. Sci. U.S.A.* **87**, 8820–8824.

Zhou, C., Orndorff, K., Daghlian, C. P., and DeMaggio, A. E. (1988). *Sex. Plant Reprod.* **1**, 97–102.

Zhou, C., Zee, S. Y., and Yang, H. Y. (1990). *Sex. Plant Reprod.* **3**, 213–218.

Zhu, C., and Liu, B. (1990). *Chin. J. Bot.* **2**, 1–6.

Zhu, C., Hu, S. Y., Xu, L., Li, X., and Shen, J. (1980). *Sci. Sin.* **23**, 371–376.

Freeze Fracture of Male Reproductive Cells

Darlene Southworth

Department of Biology, Southern Oregon State College, Ashland, Oregon 97520

I. Introduction

A precise sequence of cell divisions and differentiation distinguishes the formation of male reproductive cells in flowering plants. Between meiosis and fertilization, haploid cells differentiate into three cell types that form the male germ line: vegetative cell, generative cell, and sperm. These three, formed in a particular relationship to each other, have distinct functions related to membrane structure.

Following meiosis in anthers, four haploid cells are formed. These are not sperm, as they would be in animals; they are immature pollen grains or microspores. Each haploid microspore divides once to form a vegetative cell and a generative cell. Vegetative cells and generative cells form as unequal sibling cells in pollen grains. The fate of these two cells is distinctly different. When pollen lands on a stigma, the vegetative cell elongates to form a pollen tube that grows toward the egg. The vegetative cell does not divide; its primary function is growth. The generative cell divides once more, in either the pollen grain or pollen tube, to form two sperm cells. If generative cell division occurs in the anther before pollen is discharged, the pollen will be tricellular (one vegetative cell plus two sperm cells) at anthesis. If generative cell division is delayed so that its division occurs after pollen is discharged, pollen will be bicellular (one vegetative cell and one generative cell) at anthesis. The old terms "trinucleate pollen" and "binucleate pollen" are inaccurate in that each cell within a pollen grain is uninucleate.

The generative cell, itself, functions strictly as a developmental stage between the haploid microspore and two sperm cells. The two sperm have related functions in double fertilization: one fuses with the egg forming a zygote, the other with the central cell forming endosperm.

The cells of the pollen grain develop in ways consistent with their functions. The differentiation of the vegetative cell involves acquiring the ability to recognize the correct stigma and to activate the growth process. The pattern of growth of the vegetative cell into a pollen tube is conferred by cellular properties, including the cytoskeleton and the cell wall. *In vivo* the pollen tube exhibits directed growth as it responds to external signals that include the substratum, gradients of minerals and nutrients, and possibly other signals that result in the arrival of the pollen tube at the micropyle of an ovule, which serves as gateway to the egg.

Membrane changes might be associated with several differentiation events. The vegetative cell changes from a quiescent, mature cell of the pollen grain at anthesis to a rapidly elongating pollen tube displaying a major increase in surface area. I would predict at least four configurations for membranes in the vegetative cell: (1) a region of pollen grain membrane that does not elongate, (2) a region of pollen tube tip actively elongating, (3) a completed, postelongation region of pollen tube membrane, and (4) an inner vegetative cell membrane that encloses the generative cell or sperm.

Meanwhile, the generative cell responds to signals from the vegetative cell. The decision to divide at a given time is precisely controlled, although the regulatory factors are unknown. The generative cell differentiates from its inception at cytokinesis through its own cell cycle to the cell division forming two sperm. At first it is lens-shaped and attached to the intine. Then the cell wall loosens; the generative cell detaches from the pollen wall and, as a spheroidal cell, becomes wholly embedded in the vegetative cell. In some plants the cell wall is entirely removed, and the generative cell remains embedded in a periplasmic pocket of the vegetative cell. Finally, the generative cell becomes convoluted in shape and reduced in volume. As the generative cell moves through its cell cycle and the cell wall changes, the generative cell plasma membrane may also change. Cell division of the generative cell to form two sperm cells occurs within a cavity of the vegetative cell.

Sperm cells also differentiate with the result that they can recognize and fuse with the egg and central cell. Sperm are embedded in the vegetative cell and separated from it by two plasma membranes. Following cytokinesis sperm cells mature. They elongate, and cytoplasmic vesicles may be eliminated. Differentiation of sperm plasma membrane may occur either in the pollen tube or in the embryo sac when sperm are released from the pollen tube. Sperm experience two major environmental changes: (1) a decrease in osmotic potential of the surrounding vegetative cell with uptake of water during pollen tube growth, and (2) contact with synergid cytoplasm or other fluids in the embryo sac during its release from the pollen tube.

After sperm cells are released into the embryo sac near the egg and central cell, they recognize the cells with which they will fuse: either the egg or the central cell. In some species two sperm from one pollen grain are dimorphic, differing in relative numbers of mitochondria and plastids and also differing in shape, volume, and surface area. If double fertilization depends on dimorphic sperm (one to recognize the egg and one the central cell), the sperm plasma membranes might differ in composition and structure.

Many of the activities of the germ cells and changes in their structure are related to events at the cell surface. Responses to signals, cell-to-cell communication, recognition, and fusion are basically membrane events. The technique of freeze fracture provides a unique opportunity to observe membrane substructure. Fracture of frozen material appears to split membranes along their hydrophobic interior so that arrangements of intramembrane particles can be observed.

The relationship between vegetative cell and generative cell and later between vegetative cell and sperm is one of close membrane contact. All cells retain their separate plasma membranes; they do not fuse. Raw materials needed by generative cells and by sperm must pass through vegetative cell membranes, which can be compared in structure. Membrane contacts or bridges, if any, should be visible with freeze fracture.

In animal sperm there are structural domains on the plasma membrane that correlate with sperm function. If specialized fusion areas exist on plant sperm, they might have a distinctive membrane structure.

Freeze fracture has been useful in visualizing membrane asymmetries correlated with lipid content and lipoprotein arrangement. Cell membranes are fundamentally asymmetric. The plasma membrane, for example, has one side closer to the cytoplasm (the protoplasmic or P-face) and the other side away from the cytoplasm and toward the extracellular space (the E-face). This concept can be extended to all cell membranes. One half (P) is closer to the soluble cytoplasm; the other half (E) is closer to an extracellular or an endoplasmic space, such as the interior of endoplasmic reticulum channels or of vesicle contents. The fracture faces of membranes can therefore be identified as the P-face, adjacent to the protoplasm, or the E-face, adjacent to the exoplasm or endoplasm (Branton et al., 1975). The arrangement of particles on membrane fracture faces has been used to characterize membranes. The precise chemical identity of the molecules in intramembrane particles, as distinct from the rest of the membrane, has not been determined in most cases. However, there are correlations of membrane particle density with protein content and sometimes with specific membrane proteins or with enzyme complexes. The size and density of intramembrane particles can be determined, although the accuracy of such counts is limited by factors such as the time between fracture and

replication and contamination rates at the surface (Rash and Giddings, 1989). If the fracture is a result of successive cuts, some membrane faces may be exposed earlier than others as a result of a localized deep fracture. In the additional time between the exposure of the fracture face and replication, water vapor or other volatiles may condense on the particles adding to their apparent size.

In addition to membrane structure, the association and arrangement of membranous organelles and their distribution in cells can be determined (Platt-Aloia and Thomson, 1989). Partial three-dimensional configurations can be observed without serial thin sections and with rapid freezing. Substructural differences among membranes of various organelles can be detected. Freeze fracture is potentially a useful way of examining membranes of male reproductive cells.

II. Vegetative Cells in Pollen and Pollen Tubes

A. External Plasma Membranes

The outer plasma membrane of the vegetative cell in the pollen grain and pollen tube surrounds the entire male germ line. It has two specialized features: distinctive patterns of intramembrane particles and membranous vesicles external to the plasma membrane.

Patterns of intramembrane particles on fracture faces include hexagonal arrays, rosettes, compound rosettes (honeycombs), and sinuous ridges. Hexagonal arrays observed in nongerminating pollen of *Lilium humboldtii* and *Artemisia pycnocephala* and in pollen tubes of *Nicotiana tabacum* occur as irregular patches 0.2–0.5 μm across with uniform particles of center-to-center spacing of 20–25 nm (Southworth and Branton, 1971; Kroh and Knuiman, 1985). Rosettes of six equispaced particles arranged in a ring are found on pollen tubes of *Lilium longiflorum* but not on tubes of *Nicotiana tabacum* or on pollen of *Lilium humboldtii* (Southworth and Branton, 1971; Kroh and Knuiman, 1985, 1986; Reiss *et al.*, 1985). Spacing of intramembrane particles in rosettes is similar to intramembrane particle spacing in hexagonal arrays.

Both hexagonal arrays and rosettes of intramembrane particles have been observed on fracture faces of plasma membranes of other plant cell types. Although it has been suggested that rosettes correlate with cellulose synthesis, no definitive evidence exists to postulate a function for either pattern. The external vegetative cell plasma membrane lacks unique features that distinguish it from that of other plant cells.

A third fracture pattern, a meandering network of ridges, has been observed on plasma membranes of *Lilium longiflorum* pollen tubes (Kroh and Knuiman, 1986). This pattern differs from the parallel ridges observed on fracture faces of the inner vegetative cell plasma membrane surrounding the generative cell in *Phoenix dactylifera* (Figs. 1 and 2; Southworth *et al.*, 1989).

The plasma membrane and organelle membranes of dry pollen produce fracture faces similar to those of hydrated pollen (Platt-Aloia *et al.*, 1986). Membranes of *Zea mays* pollen dried to 3% water and *Cucurbita pepo* pollen with 25% water produce fracture faces similar to those of more hydrated pollen (Kerhoas *et al.*, 1987; Digonnet-Kerhoas *et al.*, 1989). These results indicate that the plasma membrane of the vegetative cell is arranged in a bilayer even under dehydrated conditions.

The presence of membranous vesicles located between cell wall and plasma membrane or embedded in the cell wall is unusual. In frozen pollen tubes of *Nicotiana tabacum*, membranous vesicles occur external to, or fused with, plasma membrane both in plasmolyzed and nonplasmolyzed conditions. In fixed pollen of *Zea mays*, vesicles occur in the intine (Kerhoas *et al.*, 1987). In a tangential fracture in pollen of *Artemisia pycnocephala*, some vesicles occur outside the plasma membrane. Vesicles outside the plasma membrane in aldehyde-fixed, sectioned pollen grains or tubes are not uncommon, but they have been difficult to interpret because of the possibility of fixation artifacts. Observation of external vesicles in freeze-fractured pollen tubes suggests that they are not induced by fixation, an interpretation supported by the finding of inclusions in the intine of freeze-substituted pollen (Martinez and Wick, 1991). Numerous vesicles are present in cortical cytoplasm. The relationship of external to internal vesicles is unknown.

B. Vegetative Nucleus

The vegetative nucleus is generally larger and more hydrated than the generative nucleus with less condensed chromatin, indicating active transcription (Southworth *et al.*, 1988, 1989; Wagner *et al.*, 1990). In *Zea mays* the nucleus is lobed; nuclear pores are regularly dispersed over the nuclear envelope (Fig. 3). Serial reconstructions of *Zea mays* pollen show that one sperm cell contacts an evagination of the vegetative nucleus (McConchie *et al.*, 1987). However, no observed fracture could be interpreted as a sperm extension associated with the vegetative nucleus or the vegetative cell plasma membrane (Southworth *et al.*, 1988). In *Phoenix dactylifera* chromatin is somewhat condensed, the nucleus is unlobed, and some areas of the nuclear envelope lack pores (Fig. 4). In thin sections of *Medicago sativa*, nuclear pore density is reduced on the side of the nucleus away

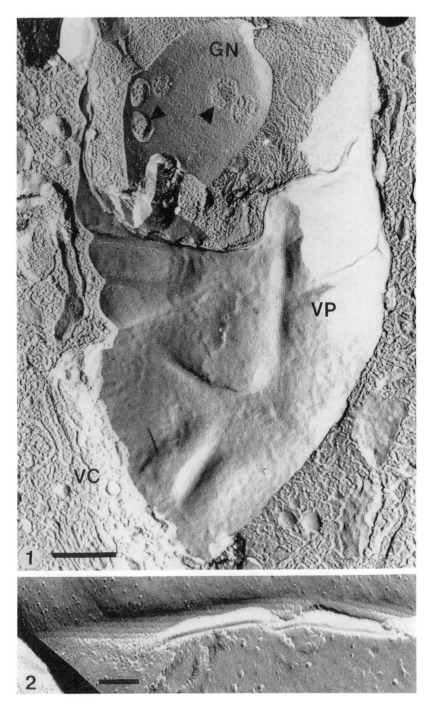

from the generative cell (Shi *et al.*, 1991). The patterns of pores on vegetative nuclei are diverse.

C. Organelles

The organelles of vegetative cells are variable and include plastids, mitochondria, and an endomembrane system with dictyosomes, endoplasmic reticulum, and numerous storage vesicles (Figs. 1, 3–10). These reflect the major functions of the vegetative cell including synthesizing vesicles and cell wall for pollen tube growth and providing an energy source, new plasma membrane, and cell wall material.

D. Cell Walls: Intine and Exine

The intine, a heterogeneous inner wall layer consisting of hydrated polysaccharides and proteins, gives no distinctive fracture pattern (Southworth and Branton, 1971; Kroh and Knuiman, 1985). Variations occur in step fractures across the intine, but these have not been correlated with chemical composition.

Fractures of exine are difficult to interpret. In *Lilium humboldtii* and in *Brassica* sp., a fracture propagates along the surface of the exine (Southworth and Branton, 1971; Gaude and Dumas, 1984). In other species and in some fractures of these species, the exine did not direct the fracture along a preferred surface. The exine either cross-fractured smoothly or deformed to create a rough surface suggestive of a fibrillar substructure (Southworth and Branton, 1971).

III. Generative Cells

A. Interface Between Vegetative and Generative Cells

In many species the separation between the vegetative cell inner plasma membrane and the generative cell plasma membrane is irregular in

---

FIG. 1 Generative cell of *Phoenix dactylifera*. Generative nucleus (GN) with fracture face of outer nuclear membrane with large, irregularly spaced pores (arrowheads). Vegetative cell (VC) with P-face (VP) of vegetative cell inner plasma membrane showing parallel ridges on indentations. ×35,000. Reproduced with permission from Southworth *et al.* (1989). Bar, 0.5 μm.

FIG. 2 *Phoenix dactylifera*. Parallel ridges in the E-face of the vegetative cell inner plasma membrane. ×109,000. Reproduced with permission from Southworth *et al.* (1989). Bar, 0.1 μm.

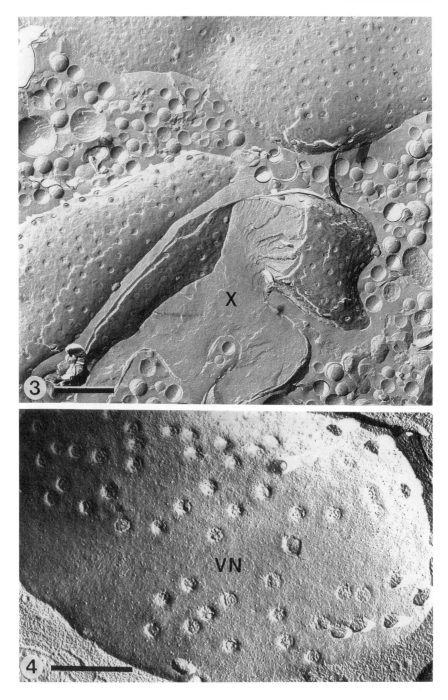

aldehyde-fixed pollen grains viewed in thin sections. The space around the generative cell is electron-translucent with fine fibrils and is interpreted as a cell wall traversed by membrane bridges called trabeculae. Thin sections prepared by rapid freezing followed by freeze-substitution and replicas from freeze fracture (Figs. 1 and 5) show that these membranes are closely parallel but without bridges (Platt-Aloia *et al.*, 1986; Cresti *et al.*, 1987; Emons *et al.*, 1988; Southworth *et al.*, 1989; Van Aelst *et al.*, 1989; Wagner *et al.*, 1990). This leads to the conclusion that the irregular separation in chemically fixed thin sections is a fixation artifact.

In *Phoenix dactylifera* the generative cell surface was convoluted with irregular narrow indentations into the generative cell (Figs. 1, 2, and 5; Platt-Aloia *et al.*, 1986; Southworth *et al.*, 1989). In many indentations 4–10 parallel ridges occur in both P- and E-faces of the vegetative cell plasma membrane (Fig. 2). Plasma membranes of vegetative cells and generative cells are appressed, evenly separated at a uniform distance, with no cell wall material occurring between them and no membrane bridges linking the two cells. By contrast, in *Collomia grandiflora* plasma membranes of vegetative cells and generative cells are neither parallel nor appressed (Platt-Aloia *et al.*, 1986). An exoplasmic space between them has been interpreted as a cell wall.

Pollen of *Papaver dubium* was cryofractured and examined by scanning electron microscopy at −130°C (Van Aelst *et al.*, 1989). The generative cell is elongated in this plant with membrane ridges along the axis of the cell. No cell wall was observed nor were there cross bridges linking plasma membranes of vegetative and generative cells. The technique did not allow determination of the presence of parallel ridges in the fracture face of the vegetative cell membrane as in *Phoenix dactylifera*. The generative cell of *Nicotiana tabacum* was also highly convoluted (Wagner *et al.*, 1990).

A honeycomb pattern, similar to a hexagonal array, but with one particle absent from the center of each hexagon, was found on the generative cell plasma membrane in pollen tubes of *Lilium longiflorum* (Emons *et al.*, 1988).

Surface convolutions are typical in generative cells. The immediate effect of a lobed surface is to increase the surface area-to-volume ratio. This facilitates transport across membranes and provides the capacity to

---

FIG. 3 *Zea mays.* Vegetative nucleus with fractures of inner and outer membranes of the nuclear envelope and cross fracture (X) of nucleoplasm. Channels between lobes of the nucleus are filled with granular cytoplasm or vesicles. ×16,600. Reproduced with permission from Southworth *et al.* (1988). Bar, 1 μm.

FIG. 4 *Phoenix dactylifera.* Vegetative nucleus (VN). ×45,000. Reproduced with permission from Southworth *et al.* (1989). Bar, 0.5 μm.

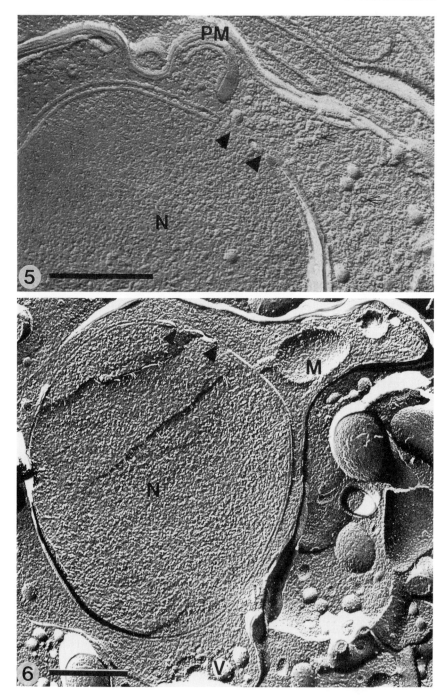

accommodate increases in volume and surface area without synthesizing new membrane. The generative cells of *Phoenix dactylifera* have a much lower osmotic potential than vegetative cells (Southworth, unpublished results). As the vegetative cell swells and the pollen tube emerges, the generative cell remains small and convoluted. At the time of generative cell cytokinesis, the total cell surface area of the two sperm will exceed that of the generative cell. The extra plasma membrane material from the generative cell convolutions may be used as sperm plasma membrane. Cytokinesis in pollen tubes generally produces a small cell plate with few vesicles and proceeds by furrowing.

B. Generative Nucleus

Although condensed chromatin has been observed in generative nuclei by fluorescence microscopy and in thin sections, no evidence of chromatin condensation has been seen in fractured generative nuclei (Southworth *et al.*, 1989; Wagner *et al.*, 1990). Apparently condensed chromatin does not alter the fracture. Pores in the generative nucleus appear larger (95–160 μm) and more variable than those of the vegetative nucleus (Figs. 1, 4, and 5). A large portion of the nuclear envelope is devoid of pores; those present are irregularly spaced and often touching (Fig. 1). Pore-free areas of the nuclear envelope have been correlated with underlying regions of condensed chromatin that probably have reduced transcriptional activity (Myles *et al.*, 1978).

C. Organelles

Endomembranes of the generative cell consist of endoplasmic reticulum, dictyosomes, and vesicles. Mitochondria, but not plastids, have been identified in fractures of *Phoenix dactylifera* generative cells (Southworth *et al.*, 1989). Microtubules, observed in fractures and in thin sections of generative cells, are associated with neither the indentation nor with the

---

FIG. 5 *Phoenix dactylifera*. Cross fracture of generative cell with microtubules (arrows) and nucleus (N) with pores (arrowheads). Plasma membranes (PM) of vegetative cell and generative cell are parallel. ×57,000. Reproduced with permission from Southworth *et al.* (1989). Bar, 0.5 μm.

FIG. 6 *Gerbera jamesonii*. Cross fracture of sperm cell and nucleus (N) with pores (arrowheads) in the nuclear envelope. Sperm organelles include a mitochondrion (M) and vesicles (V). ×43,300. Reproduced with permission from Southworth (1990). Bar, 0.5 μm.

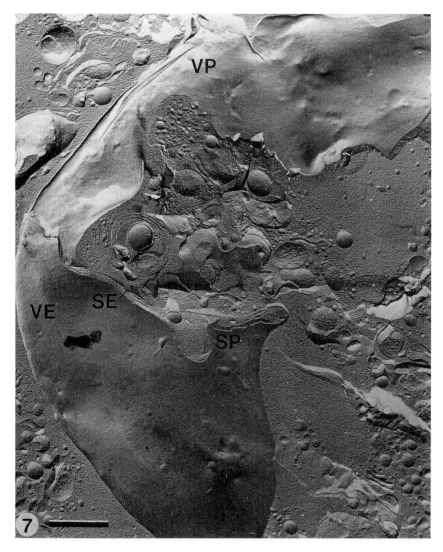

FIG. 7   *Gerbera jamesonii.* Sperm cell with narrow patches of E- and P-faces of sperm plasma membrane (SE, SP) surrounded by broad regions of vegetative cell inner plasma membrane with both E- and P-faces (VE, VP). ×16,300. Reproduced with permission from Southworth (1990). Bar, 1 μm.

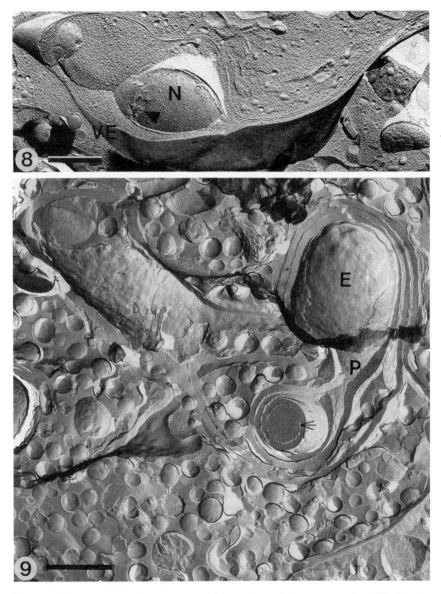

FIG. 8 *Gerbera jamesonii.* Fracture across the envelope of the sperm nucleus (N) showing nuclear pores (arrowhead). Inner vegetative cell plasma membrane (VE) surrounds sperm. ×29,400. Reproduced with permission from Southworth (1990). Bar, 0.5 μm.

FIG. 9 *Zea mays.* Sperm cell (possibly two cells) with endoplasmic lamellae (arrow). Intramembrane particles are more dense on the P-face (P) of endoplasmic lamellae than on the E-face (E). ×18,100. Reproduced with permission from Southworth *et al.* (1988). Bar, 1 μm.

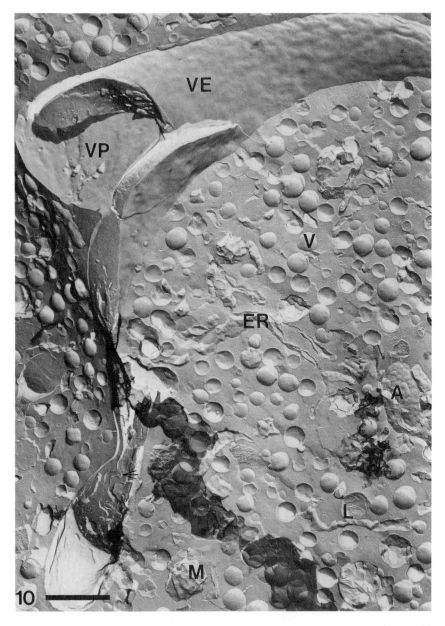

FIG. 10 *Zea mays*. Sperm cell surrounded by vegetative cell inner plasma membrane with both E- and P-faces (VE, VP). Sperm cytoplasm contains endoplasmic lamellae (arrow). Vegetative cell cytoplasm contains endoplasmic reticulum (ER), an amyloplast (A), mitochondria (M), lipid droplets (L), and vesicles (V). ×17,500. Reproduced with permission from Southworth *et al*. Bar, 1 μm.

bulge of generative cell and vegetative cell membranes (Fig. 5). Arms link cross-fractured microtubules but were not observed to link microtubules to plasma membrane.

IV. Sperm Cells

A. Plasma Membrane

Sperm cells from only three species of flowering plants have been freeze-fractured: two *in vivo* and one *in vitro*.

In *Gerbera jamesonii* plasma membranes of vegetative cell and sperm were uniformly appressed with no cell wall material between them and no membranous connections linking them (Figs. 6-8; Southworth, 1990). Lack of cytoplasmic connections was consistent with the ease of separating sperm cells from vegetative cells (Southworth and Knox, 1989). The irregular spacing and apparent bridges between sperm and vegetative cell in thin sections must be interpreted as artifacts of chemical fixation. The vegetative cell inner plasma membrane fractured along extensive faces, whereas the sperm plasma membrane fractured over much more restricted areas. The fracture faces of the sperm plasma membrane appear pebbly to particulate. No distinctive pattern of intramembrane particle distribution was observed on either the vegetative cell or sperm plasma membrane.

In *Zea mays* sperm plasma membranes produced only cross fractures (Figs. 9, 10). The fracture faces associated with sperm were those of the vegetative cell inner plasma membrane or of endoplasmic reticulum layers (endoplasmic lamellae) inside sperm.

In isolated sperm cells of *Spinacea oleracea,* the fracture produced extensive faces of the sperm plasma membrane and rarely crossed through the cytoplasm (Van Aelst *et al.,* 1990). Both faces of the sperm plasma membrane were pebbly with dispersed intramembrane particles. No distinctive intramembrane particle patterns were observed.

In animal sperm fracture faces showed specialized regions of intramembrane particle distribution. In the golden hamster extensive hexagonal intramembrane particle arrangements occur on the sperm head. Linear patterns of intramembrane particles occur on guinea pig and golden hamster sperm near the posterior ring or annulus (Friend and Heuser, 1981; Friend, 1986; Suzuki, 1988). A double row of particles, the zipper, was observed on the flagellar membrane of guinea pig sperm (Friend and Heuser, 1981). The lack of plasma membrane fracture faces in *Zea mays* and limited sperm plasma membrane fracture faces in *Gerbera jamesonii* made the presence of these features on plant sperm impossible to deter-

mine; however, extensive surfaces of isolated sperm of *Spinacea oleracea* showed none of these distinctive patterns.

B. Sperm Nucleus

Sperm nuclei also contain condensed chromatin as observed by fluorescence microscopy or by thin section transmission electron microscopy (Southworth and Knox, 1989). No evidence of chromatin condensation could be recognized by freeze-fracture (Fig. 6).

The nuclear envelope is readily identified in *Gerbera jamesonii* (Figs. 6, 8) and in *Spinacea oleracea*. Nuclear pores are sparse and irregularly arranged. In *Zea mays* no nuclear pores are present, a condition that rendered the sperm nucleus difficult to identify (Southworth *et al.*, 1988). Numerous membranes, including endoplasmic reticulum and possibly the nuclear envelope, are present in sperm cytoplasm of *Zea mays*. As in generative cells of *Phoenix dactylifera*, the paucity of nuclear pores correlates with condensed chromatin and reduced transcriptional activity.

C. Organelles

Endoplasmic reticulum in *Zea mays* sperm occurs as paired concentric membranes of 3–7 lamellae under the plasma membrane (Fig. 9). The paired membranes are appressed at a uniform distance with no evidence of annulate lamellae. E-faces have a lower particle density than P-faces. Vesicles, but not dictyosomes, are present. In *Gerbera jamesonii* endoplasmic reticulum is more limited; dictyosomes and vesicles are present. Mitochondria are recognized in sperm of both species.

V. Conclusions and Prospects

The most useful discoveries made by freeze-fracture of male reproductive cells are that: (1) no membrane bridges connect vegetative cells to either generative cells or to sperm cells; (2) a distinctive fracture pattern of parallel ridges is associated with the vegetative cell's inner plasma membrane at indentations of the generative cell surface; (3) nuclear pores on the generative nucleus are reduced in number, larger in size, and irregular in distribution; (4) nuclear pores on sperm nuclei are similarly reduced in number, in some cases to zero, and are more irregular in size and distribution; and (5) the outer vegetative cell plasma membrane contains patterns of intramembrane particles as hexagonal arrays or rosettes.

Important questions remain concerning sperm plasma membranes. The limited area of fracture faces of sperm cells in pollen does not yet permit a conclusion about dimorphism of sperm membranes nor about membrane domains associated with recognition and fusion. The dilemma is that fractures *in vivo* propagate preferentially along the vegetative cell membrane and rarely along the sperm plasma membrane. Sperm cells *in vitro* do fracture; however, their structure has changed from that in pollen. The alterations of sperm structure during the isolation process may be similar to those occurring at pollen tube rupture *in vivo*, in which case, membranes of isolated sperm could still be used to investigate sperm membrane dimorphism.

Other questions concern developmental changes in the generative cell plasma membrane and in the relationship between the vegetative cell and either generative or sperm cells. No studies are known to exist on transfer of material from vegetative cell to germ cell nor on modulation of material by vegetative cells for germ cells.

It is almost essential that freeze-fracture studies of male germ cells be accompanied by thin sections of the same material, even if fixed with aldehydes. Interpretation of replicas is sufficiently confusing that additional information is needed to identify sperm cells in particular and the organelles of vegetative cells.

The most difficult part of freeze-fracture studies of male reproductive cells in pollen is removing the exine during the cleaning of replicas. Replicas of small, thin-walled pollen, such as *Phoenix dactylifera,* were in pieces containing fractures across many pollen grains. However, replicas of *Zea mays* pollen ruptured when exines swelled during cleaning. Each replica consisted of a single pollen grain without exine.

Sperm and generative cells from few species have been freeze-fractured. There is considerable variability in sperm and generative cell organization among species. Future studies will show whether features observed thus far are widely shared or are unique and distinctive of particular species.

Acknowledgments

This work was supported by National Science Foundation Grant DCB-8801902 through the Research at Undergraduate Institutions Program.

References

Branton, D., Bullivant, S., Gilula, N. B., Karnovsky, M. J., Moor, H., Muhlethaler, K., Northcote, D. H., Packer, L., Satir, B., Speth, V., Staehelin, L. A., Steere, R. L., and Weinstein, R. S. (1975). *Science* **190,** 54–56.

Cresti, M., Lancelle, S. A., and Hepler, P. K. (1987). *J. Cell Sci.* **88**, 373–378.

Digonnet-Kerhoas, C., Gilles, G., Duplan, J. C., and Dumas, C. (1989). *Planta* **179**, 165–170.

Emons, A. M. C., Kroh, M., Knuiman, B., and Platel, T. (1988). *In* "Plant Sperm Cells as Tools for Biotechnology" (H. J. Wilms and C. J. Keijzer, eds.), pp. 41–48. Pudoc, Wageningen.

Friend, D. S. (1986). *Proc. Int. Cong. Electron Microsc. Kyoto.* **11**, 1877–1880.

Friend, D. S., and Heuser, J. E. (1981). *Anat. Rec.* **199**, 159–175.

Gaude, T., and Dumas, C. (1984). *Ann. Bot.* **54**, 821–825.

Kerhoas, C., Gay, G., and Dumas, C. (1987). *Planta* **171**, 1–10.

Kroh, M., and Knuiman, B. (1985). *Planta* **166**, 287–299.

Kroh, M., and Knuiman, B. (1986). *Acta Bot. Neerl.* **35**, 361–365.

Martinez, L., and Wick, S. M. (1991). *J. Electron Microsc. Tech.* **18**, 305–314.

McConchie, C. A., Hough, T., and Knox, R. B. (1987). *Protoplasma* **139**, 9–19.

Myles, D. G., Southworth, D., and Hepler, P. K. (1978). *Protoplasma* **93**, 419–431.

Platt-Aloia, K. A., and Thomson, W. W. (1989). *J. Electron Microsc. Tech.* **13**, 288–299.

Platt-Aloia, K. A., Lord, E. M., DeMason, D. A., and Thomson, W. W. (1986). *Planta* **168**, 291–298.

Rash, J. E., and Giddings, F. D. (1989). *J. Electron Microsc. Tech.* **13**, 204–215.

Reiss, H. D., Herth, W., and Schnepf, E. (1985). *Naturwissenschaften* **72**, 276.

Shi, L., Mogensen, H. L., Zhu, T., and Smith, S. E. (1991). *J. Cell Sci.* **99**, 115–120.

Southworth, D. (1990). *J. Struct. Biol.* **103**, 97–103.

Southworth, D., and Branton, D. (1971). *J. Cell Sci.* **9**, 193–207.

Southworth, D., and Knox, R. B. (1989). *Plant Sci.* **60**, 273–277.

Southworth, D., Platt-Aloia, K. A., and Thomson, W. W. (1988). *J. Ultrastruct. Mol. Struct. Res.* **101**, 165–172.

Southworth, D., Platt-Aloia, K. A., DeMason, D. A., and Thomson, W. W. (1989). *Sex. Plant Reprod.* **2**, 270–276.

Suzuki, F. (1988). *J. Ultrastruct. Mol. Struct. Res.* **100**, 39–54.

Van Aelst, A. C., Muller, T., Dueggelin, M., and Guggenheim, R. (1989). *Acta Bot. Neerl.* **38**, 25–30.

Van Aelst, A. C., Theunis, C. H., and van Went, J. L. (1990). *Protoplasma* **153**, 204–207.

Wagner, V. T., Cresti, M., Salvatici, P., and Tiezzi, A. (1990). *Protoplasma* **181**, 304–309.

Generative Cells and Male Gametes: Isolation, Physiology, and Biochemistry

Annie Chaboud and Réjane Perez
Reconnaissance Cellulaire et Amélioration des Plantes,
Université Claude Bernard-Lyon 1
Villeurbanne, France

I. Introduction

In angiosperms the male gametes originate from the division of the generative cell, which occurs either in the pollen grain before anthesis or in the pollen tube after pollination. These cells are devoted to the transport of the genetic information of the male to the embryo sac for fertilization. For this reason male gametes arc interesting cells for use in fundamental research and biotechnological manipulation. The aim of this chapter is to review isolation procedures developed for generative cells and male gametes with an emphasis on the physiology and biochemistry of these isolated cells. Most of the research reviewed has been conducted since 1985.

II. Formation of Generative Cells and Male Gametes during Gametophyte Development

Sperm cell formation and maturation occurs either during pollen tube growth through the style in bicellular pollen grains usually several hours after pollen germination (about 70% of angiosperms) or during pollen development within the anther in tricellular pollen grains (about 30% of angiosperms) (Knox *et al.*, 1986). In either case pollen development involves a dramatic process of differentiation with numerous distinctive steps (Willemse, 1988) and characterized by significant changes in cytomorphology, biochemistry, and gene expression.

The developmental program of pollen formation has been studied mostly by classical cytomorphological techniques (Giles and Prakash, 1987). An

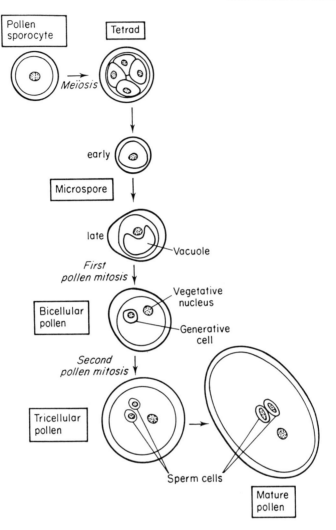

FIG. 1 Diagrammatic representation of cytological stages of tricellular pollen development. Reproduced with permission from Roeckel *et al.* (1990a).

example of the different cytological stages of tricellular pollen development is illustrated in Figure 1. These cytological stages can be assessed by evaluating the nuclear state of fresh anthers (occurrence of division, number of nuclei) using the DNA-specific fluorochrome DAPI (4,6-diamidino-2-phenylindole) in association with membrane permeabilization by Triton X-100 (Vergne *et al.*, 1987). This improved procedure allows a rapid and effective description of the entire male pattern of development

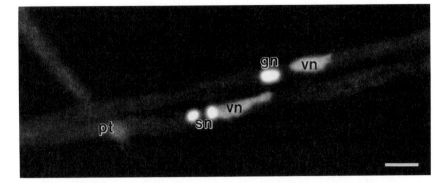

FIG. 2   Fluorescence photomicrography of DAPI-stained pollen tubes (pt) of *Populus del-toides* after 7 hours of growth *in vitro*. Vegetative nucleus (vn) displays less fluorescence and is elongated compared with the generative nucleus (gn) and sperm nuclei (sn). In one pollen tube the vegetative nucleus is ahead of the generative nucleus and separate from it. The vegetative nucleus is closely associated with one of the two sperm cells in the other pollen tube, forming an MGU. ×1200. Bar, 10 μm. Reproduced with permission from Rougier *et al.* (1991).

(Delvallée and Dumas, 1988; Vergne and Dumas, 1988; Detchepare *et al.*, 1989). It can also be particularly useful in visualizing the onset of the generative cell division within the pollen tube in bicellular pollen species (Rougier *et al.*, 1991), as illustrated in Figure 2.

III. Characterization of Generative Cells and Male Gametes *in Situ*

A. Structural Characteristics

The structural characteristics of generative cells and male gametes *in situ* have been examined extensively by transmission electron microscopy (TEM), freeze-fracture, and immunocytochemistry to better understand the relationship between the different cells of the male gametophyte and the mechanism of their transmission in the pollen tube.

In numerous species the two male gametes remain together in close association with the vegetative nucleus in the pollen grain, in the pollen tube, or both. This association, in which all the DNA of male heredity is contained, has been termed the male germ unit (MGU; Dumas *et al.*, 1984; Matthys-Rochon and Dumas, 1988). Extensive three-dimensional reconstructions have been made to provide insights about its organization

in vivo (Mogensen, this volume). In all species the two male gametes are enclosed together within the inner plasma membrane of the vegetative cell (Mogensen and Rusche, 1985; McConchie *et al.*, 1987a,b; Wagner and Mogensen, 1987; Yu *et al.*, 1989). They generally appear elongated or spindle-shaped and contain a nucleus that occupies a major volume of the cell (Knox *et al.*, 1988). In most cases one of the two male gametes displays an evagination, which is usually physically associated with the vegetative nucleus (Russell, 1984; McConchie *et al.*, 1985, 1987a,b; Dumas *et al.*, 1985; Yu *et al.*, 1989). In barley the vegetative nucleus and the male gametes are separated in mature pollen, but after pollination they approach one another and form an association that is maintained for a considerable time thereafter (Mogensen and Wagner, 1987). In *Nicotiana tabacum* and *Cyphomandra* the relationship between the two male gametes is a real structural connection (Yu *et al.*, 1989; Hu and Yu, 1988). Similar patterns of linkage are described in other species, such as *Plumbago* (Russell and Cass, 1981), *Spinacia* (Wilms and Van Aelst, 1983), *Euphorbia* (Murgia and Wilms, 1988), *Petunia* (Wagner and Mogensen, 1987), and barley (Mogensen and Wagner, 1987). In contrast, corn male gametes have been shown to be long and discrete cells that are not physically associated in the pollen grain (McConchie *et al.*, 1987b). Apart from a few exceptional cases (*Plumbago zeylanica*, Russell, 1984; *Spinacea oleracea*, Wilms, 1986), no cell wall has been reported to occur in the male gametes.

Male gamete nuclei often possess densely stained heterochromatic granular chromatin, suggesting a transcriptionally inactive state (Knox *et al.*, 1988). The cytoplasm of the male gametes generally contains mitochondria, endoplasmic reticulum, ribosomes, dictyosomes, and vesicles (McConchie *et al.*, 1987a,b; Yu *et al.*, 1989). The heritable organelles (mitochondria and plastids) have a characteristic morphology and are mostly organized in close association with the nucleus. In *Nicotiana, Zea mays*, and *Brassica* no plastids were seen in the sperm cell (McConchie *et al.*, 1987a,b; Yu *et al.*, 1989), whereas some plastids were found in *Plumbago* (Russell, 1984) and *Rhododendron* (Taylor *et al.*, 1989). The organization of the chondriome of the male gametes is variable from one species to the other. In *Zea mays* three-dimensional reconstruction of the cell indicates the existence of filamentous mitochondrial complexes in the male gametes of mature pollen (McConchie *et al.*, 1987b). This type of mitochondrial morphology has not been reported in any other male gametes, in which the mitochondria are generally small and ellipsoidal with a simple morphology, as in *Brassica oleracea* and *Brassica campestris* (Dumas *et al.*, 1985; McConchie *et al.*, 1987a). In the pollen tube of *Nicotiana tabacum*, both mitochondria and small vacuoles often aggregate into groups within the male gametes (Yu *et al.*, 1989).

Cytoplasmic heterospermy was first described by Russell (1984, 1985) in *Plumbago zeylanica*, in which one sperm cell is mitochondria-rich and

the other is plastid-rich. The two male gametes in pollen grains of *Brassica oleracea* also possess differences in size, shape, morphology, and association with the vegetative nucleus (Dumas *et al.,* 1985; McConchie *et al.,* 1985). Evidence from quantitative microscopic observations indicates that the pair of male gametes can be dimorphic in terms of cell volume, surface area, and organelle content. In dicots this is the case in *Plumbago* (Russell, 1984), *Brassica* (McConchie *et al.,* 1985, 1987a), *Spinacia* (Wilms, 1986), *Euphorbia* (Murgia and Wilms, 1988), and *Rhododendron* (Taylor *et al.,* 1989), whereas sperm cells are isomorphic in *Petunia* (Wagner and Mogensen, 1987). In monocots slight dimorphism has been reported for male gametes of *Zea mays* (McConchie *et al.,* 1987b; Rusche and Mogensen, 1988), whereas male gametes appear to be isomorphic in *Hordeum vulgare* (Mogensen and Rusche, 1985).

Freeze-fracture data on the components of the MGU are available for a limited number of species (Southworth, this volume). The analysis of pollen grains of *Zea mays* (Southworth *et al.,* 1988) and *Gerbera* (Southworth, 1990) by this technique shows that each sperm cell is surrounded by two tightly appressed membranes: the inner plasma membrane of the vegetative cell and the membrane of the sperm cell. No connection between these membranes was observed in these two species in contrast to the appearance using conventional TEM. This indicates that chemical fixation can induce artifacts in the cell surface and that freezing gives a more accurate picture of the relationship of the two membranes. In freeze-fractured pollen grains of *Phoenix dactylifera,* structural differences between the vegetative and generative cells were found within the plasma membranes but also at the nuclear envelope level. The nucleus of the generative cell harbors fewer and larger nuclear pores than the vegetative nucleus (Southworth *et al.,* 1989). Nuclear pores also seem to be reduced in number in *Zea mays* and *Gerbera* male gametes, suggesting low nuclear activity (Southworth *et al.,* 1988; Southworth, 1990). Moreover, no distinct pattern of particle distribution was observed in membranes of the two male gametes in either *Zea mays* (Southworth *et al.,* 1988) or *Gerbera* (Southworth, 1990), providing no clues to possible membrane dimorphism.

Cell shaping and the movement of sperm or generative cells in pollen tubes are believed to be closely related to the microtubular cytoskeleton (Heslop-Harrison and Heslop-Harrison, 1988a). This has led to numerous studies of the organization of the microtubular cytoskeleton using immunofluorescence visualization (review of pollen grains and tubes: Pierson and Cresti, this volume; review of gametes: Palevitz and Tiezzi, this volume). Using this technique the organization of microtubules in generative cells is characteristically basket-like and often helically twisted in the angiosperm species examined *in situ* (Derksen *et al.,* 1985; Pierson *et al.,* 1986; Lancelle *et al.,* 1987; Palevitz and Cresti, 1988; Tiezzi *et al.,* 1988; Taylor *et al.,* 1989; Zhou *et al.,* 1990). In *Brassica oleracea* sperm cells traveling

within the pollen tube, microtubule bundles are preferentially located in tails and lobes of the cell (Cresti *et al.*, 1990). However, no longitudinally oriented microtubule patterns of this type were observed in the male gametes of *Alopecurus pratensis* (Heslop-Harrison and Heslop-Harrison, 1988b). In this species the two male gametes appear to remain interconnected by tubulin-containing cytoplasmic processes expressing varying degrees of attenuation during their passage through the pollen tube.

The microtubule cytoskeleton is presumed to play an important role in determining cell shape of the generative cell and sperm cells. Immunofluorescence and ultrastructural studies of the generative cell cytoskeleton reveal that bundles of microtubules are interconnected by electron-dense structures in pollen tubes of *Hyacinthus orientalis* (V. T. Wagner *et al.*, personal communication) and *Nicotiana* (Cresti *et al.*, 1987; Lancelle *et al.*, 1987). These linked microtubule bundles could give added strength to the cytoskeleton to maintain the cellular shape. The microtubules also cooperate in the movement of chromosomes, mitochondria, or vesicles inside the male gamete (V. T. Wagner *et al.*, personal communication). Another potential role of the cytoskeleton is its possible participation in the movement of male gametes, forming an active system for translocation that propels the male gametes in the tube (Cresti *et al.*, 1990). In other species movement seems more the result of translocation along cytoskeletal elements in the tube cytoplasm (Palevitz and Cresti, 1988). The presence of myosin at the surface of the vegetative nuclei and generative cells in pollen grains and tubes of *Hyacinthus orientalis* and *Helleborus foetidus,* taken in conjunction with the observation of longitudinally oriented microfilament bundles in the pollen tube (Heslop-Harrison and Heslop-Harrison, 1989a), suggests that an actomyosin system may be involved in the movement of the MGU in the pollen tube (Heslop-Harrison and Heslop-Harrison, 1989b).

B. Gene Expression during Formation of Generative Cells and Male Gametes

Male gametophyte development is characterized by two distinct phases of gene activity during the formation of pollen grains (Mascarenhas, 1990, this volume). These include an early phase lasting throughout meiosis to the end of microspore mitosis and a later phase encompassing the maturation of the microspore to the time of anthesis (Mascarenhas, 1990). Although there is an extensive overlap (60–90%) between genes expressed in pollen and those in vegetative tissues, a specific set of genes appears to be expressed just in pollen (Willing and Mascarenhas, 1984; Willing *et al.*, 1988).

The male developmental program has been analyzed at the translational level in several recent biochemical studies. Increases in protein content and in protein synthesis were observed at different stages in pollen development of *Nicotiana tabacum* (Tupý *et al.*, 1983; Villanueva *et al.*, 1985), *Datura innoxia* (Villanueva *et al.*, 1985), and *Hyoscyamus niger* (Raghavan, 1984). Qualitative developmental changes in protein or isoenzyme patterns were reported during male gametogenesis in *Lilium henryi* (Linskens, 1966), *Nicotiana tabacum* (Zarsky *et al.*, 1985), *Zea mays* (Frova *et al.*, 1987; Delvallée and Dumas, 1988; Frova, 1990), and *Brassica oleracea* (Detchepare *et al.*, 1989). Gametophyte development can apparently be divided into two main periods in tricellular pollen species, such as *Zea mays* (Delvallée and Dumas, 1988) and *Brassica oleracea*

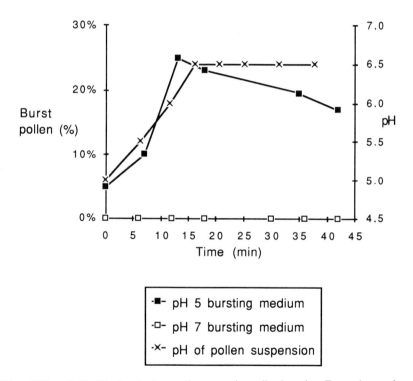

FIG. 3   Effect of pH of the incubation medium on maize pollen bursting. For each experiment 250 mg maize pollen was incubated either in 5 ml pH 5 bursting medium (Brewbaker and Kwack salts, 15% sucrose) or in 5 ml pH 7 bursting medium (Brewbacker and Kwack salts, 15% sucrose, 10 mM MOPS, pH 7). Aliquots of the pollen suspension were sampled at different times during incubation for burst pollen determinations. Pollen grains that have liberated all or part of their cytoplasm were counted as burst. Variation in pH was measured in the pH 5 pollen suspension using pH indicator strips.

(Detchepare *et al.*, 1989), as well as in bicellular pollen, such as *Lilium henryi* (Linskens, 1966). The first period of pollen development corresponds to the microspore and young vacuolate pollen stages. After generative cell division a major transition in protein content occurs, resulting in a second major period of development corresponding to pollen maturation (i.e., storage product accumulation, sperm cell differentiation and the acquisition of the ability to germinate). Using [^{35}S]methionine incorporation in *Brassica* anthers, Detchepare *et al.* (1989) obtained evidence that this second set of specific proteins corresponds to newly synthesized proteins. In contrast, in *Nicotiana tabacum,* a bicellular pollen species, most of the stage-specific variations consist of a progressive increase in the amount of preexisting proteins but not in the appearance of new ones (Zarsky *et al.*, 1985). Moreover, when proteins were extracted from isolated male gametophytes of different developmental stages, electrophoretic analyses demonstrated that a new set of proteins was also detected at the onset of sperm cell formation in wheat (Vergne and Dumas, 1988) and in corn (Bedinger and Edgerton, 1990; Mandaron *et al.*, 1990). Presumably, some of these proteins are related to the formation and differentiation of sperm cells, which occurs in this late stage of development.

Gene expression during gametophyte development has also been extensively studied at the transcriptional level. To characterize mRNAs corresponding to pollen-specific genes, recombinant cDNA libraries have been constructed to poly(A)RNA isolated from the mature pollen of corn and *Tradescantia paludosa* (Stinson *et al.*, 1987), the mature anther of tomato (Twell *et al.*, 1989), and the mature pollen of *Oenothera* (Brown and Crouch, 1990). Pollen-specific cDNAs thus isolated are representative of the late phase of gene expression and have been the subject of detailed molecular characterization (corn, Hanson *et al.*, 1989; tomato, Ursin *et al.*, 1989, Wing *et al.*, 1989). These genes code only for mRNAs synthesized after microspore mitosis that increase in concentration up to maturity. *In situ* hybridization has shown that they are expressed in the cytoplasm of the pollen vegetative cell, and after germination, throughout the cytoplasm of the pollen tube. Therefore, these pollen-specific late genes appear to have roles in pollen tube germination and growth but not during microspore development.

A new set of genes expressed specifically in developing microspores and not in mature pollens have been identified using cDNA libraries of transcripts of defined stages of anther development in *Brassica napus* (Albani *et al.*, 1990; Scott *et al.*, 1991; Roberts *et al.*, 1991). The identified genes were temporally expressed, indicating that some are activated during the phase between tetrad formation and the completion of generative cell division (Albani *et al.*, 1990). Some other genes cover more discrete periods, for example, generative cell formation or first to second pollen mitosis (Scott *et al.*, 1991; Roberts *et al.*, 1991). Therefore, these genes

are presumably involved in processes unique to microspore development and gametogenesis.

Similar conclusions have been drawn from the analysis of stage-related expression of mRNAs during bicellular pollen development in lily and tobacco using 2D-electrophoresis of *in vitro* translated products (Schrauwen *et al.*, 1990). These authors showed that there is a steep increase in transcription immediately after microspore mitosis (300% increase of mRNA content in young binucleate cells), representing a dynamic period of pollen development; presumably, this involves the maturation of the generative cell. In addition, the transient appearance of some RNAs only detectable at a certain stage of development strongly supports their involvement in specific developmental processes.

These studies suggest that some of the stage-related, pollen-specific genes are involved in the formation and maturation of generative cells and male gametes and that they will be further characterized in the near future.

In contrast, gene activity in generative cells or male gametes themselves is not well documented. At present, limited and conflicting information is available regarding the synthesis of RNA in male gametes. In the tricellular pollen of *Secale cereale,* [5-^{3}H]uridine was incorporated into both the sperm and vegetative nuclei during *in vitro* pollen tube growth (Haskell and Rogers, 1985), but the types of RNA synthesized by the sperm nuclei were not determined. In the bicellular pollen of *Hyoscyamus niger,* however, exposure of male gametes formed in the pollen tube to short pulses of [5-^{3}H]uridine fail to reveal RNA synthesis (Reynolds and Raghavan, 1982). These apparently contradictory data could also reflect physiological differences between tricellular and bicellular pollen grains. Possibly, the sperm nuclei in tricellular pollen are preactivated in terms of transcription, whereas the newly formed sperm of bicellular pollen may require further maturation before activation occurs.

IV. Isolation of Generative Cells and Male Gametes

Numerous reports have appeared concerning the isolation of generative cells and male gametes from flowering plants. Because generative cells and male gametes are enclosed in entirety within the vegetative cell of the pollen grain and tube, isolation requires first the liberation of these cells and then their purification from the components of the vegetative cell.

A. Release of Generative Cells and Male Gametes

The first direct observations of free generative cells or male gametes were made on microscope slides just after their release from a few pollen grains

obtained by osmotic shock (Cass, 1973; Russell and Cass, 1981; Dumas *et al.*, 1986; Matthys-Rochon et al., 1987) or after squashing (Zhou *et al.*, 1986; Matthys-Rochon *et al.*, 1987; Zhou *et al.*, 1988). The cited authors observed morphological changes of male gametes and generative cells after their release from pollen grains. *In situ* the cells were usually spindle-shaped, but after release from the pollen the cells commonly become rounded and finally assume a spherical shape. However, for further characterization, it is necessary to obtain isolated generative cells or male gametes in large numbers, and numerous isolation protocols have been reported (Theunis *et al.*, 1991).

The release of sperm or generative cells can be obtained by three general techniques: osmotic shock, direct physical separation by grinding or applied pressure, or wall degrading enzymes (Roeckel *et al.*, 1990a; Russell, 1991). The methods used must be adapted to the species studied to allow pollen grains to break without damaging male gametes.

Osmotic shock is widely used to burst pollen grains or tubes (Roeckel *et al.*, 1990a; Russell, 1991). This method relies on the greater sensitivity of certain types of pollen to rapid changes in water flux. Osmotic shock has also been used to burst pollen tubes of bicellular (*Lilium, Rhododendron*, Shivanna *et al.*, 1988) or tricellular (*Brassica*, Taylor *et al.*, 1991) species. A prehydration step of the pollen before osmotic shock may be necessary for some species (Southworth and Knox, 1988, 1989). In *Zea mays* osmotic shock is coupled with a pH shock to release the male gametes through the germinal aperture (Fig. 3). At the same osmotic pressure, no bursting occurs if the pH is not acidic. Moreover, as more pollen cytoplasm is released, the pH of the pollen suspension becomes neutral, apparently inhibiting the bursting of additional pollen. This phenomenon is possibly the result of modifications of the aperture cell wall and the plasma membrane at the Zwischenkörper, as described by Heslop-Harrison and Heslop-Harrison (1980). The neutral pH achieved by the release of cytoplasmic elements is convenient in maintaining male gametes in a lifelike state (Roeckel *et al.*, 1990b).

The release of generative cells or male gametes can also be accomplished by using physical stress to break the pollen wall. Grinding methods, employing tissue homogenizers or glass beads, take advantage of the smaller size of the male gametes, which renders them more difficult to crush. Grinding has been used successfully in releasing generative cells of *Lilium* (Tanaka, 1988) and male gametes of *Brassica* (Hough *et al.*, 1986; Roeckel *et al.*, 1988; Southworth and Knox, 1988) and *Gerbera* (Southworth and Knox, 1989). The release of *Spinacia* sperm cells was conducted using the pressure of a glass roller applied to a smooth glass surface to press the pollen grains open (Theunis and Van Went, 1989).

Wall-degrading enzymes have also been used for pretreating pollen grains before liberating generative cells by osmotic shock in *Vicia faba*

(Zhou, 1988) and *Zephyranthes* (Zhou *et al.*, 1990), and for the release of male gametes from pollen tubes in *Rhododendron* and *Gladiolus* (Shivanna *et al.*, 1988).

The medium most frequently used for generative and sperm cell isolation is a modification of the pollen germination medium developed by Brewbaker and Kwack (1963), including only mineral salts and sucrose. More complex media derived from protoplast regeneration nutrients have also been successfully used (Van der Maas and Zaal, 1990). It is likely that medium composition could be further improved to optimize the release and recovery of isolated generative and sperm cells.

B. Purification of Released Cells

The second step of the isolation procedure is to separate released generative cells or male gametes from free organelles and cytoplasmic debris of the vegetative cell. The larger contaminants can be easily removed by filtration using stainless sieves, nylon filters, or Nucleopore membranes. Subsequent purifications of isolated cells from smaller contaminants are usually achieved by sucrose or Percoll gradient centrifugation (Table I, isolation method), as exemplified by the two first described procedures. Russell (1986) centrifuged released male gametes over a single pad of 30% sucrose using rate-zonal purification; the heavier sperms pelleted, whereas lighter organelles floated. Dupuis *et al.*, (1987) separated male gametes from pollen contaminants by discontinuous Percoll gradient centrifugation, which is effective in partially eliminating starch grains, the most important storage compound in corn pollen grains. Male gametes are concentrated in the 15% Percoll layer, whereas starch grains form a pellet. The sperm cell-enriched fraction obtained in this manner is still contaminated by some amyloplasts (Fig. 4a), but these are eliminated by another gradient centrifugation (Fig. 4b). In this case a large part of isolated male gametes is lost, but purity is significantly improved (Chaboud and Perez, unpublished results).

The yield of the described isolation procedures has been estimated by several different methods. In most reports the concentration of isolated cells per milliliter of medium or the percentage of isolated cells recovered from the initial number of male gametes within treated pollen grains is given. Starting pollen amounts and final isolated cell volume, however, vary considerably from one species to the other, so for biochemical work yield is perhaps more appropriately evaluated by the total number of isolated cells recovered in a single isolation procedure (Table I, yield). On this basis of comparison, corn appears as the best suited species for obtaining quantities of isolated male gametes *en masse*.

TABLE I

En Masse Purification of Isolated Generative and Sperm Cells

Species	Cell type[a]	Isolation method[b]	Yield[c]	Viability[d]	References
Dicotyledons					
Plumbago zeylanica	S	OS, filtration, sucrose pad	1.7×10^5 cells, 60–75%	95%, (EvB$^-$)	Russell, 1986
Rhododendron spp.	S	OS or enzymatic digestion, filtration	90–270 pairs per style[e]	ND	Shivanna et al., 1988
Brassica oleracea	S	Grinding, filtration	2%	ND	Roeckel et al. 1988
Beta vulgaris	S	OS, filtration, Percoll gradient	7×10^4 cells	30%, (FCR$^+$)	Nielsen and Olesen. 1988
Vicia faba	G	OS, filtration	ND	FCR$^+$	Zhou, 1988
Gerbera jamesonii	S	Grinding, filtration	6×10^4 cells	ND	Southworth and Knox, 1989
Spinacia oleracea	S	Squash, filtration, Percoll gradient	10^5 cells, 5–10%	90%, (FCR$^+$)	Theunis and Van Went, 1989
Brassica napus	S	OS, filtration	2.7×10^4 cells	90%, (TEM)	Taylor et al., 1991
Monocotyledons					
Zea mays	S	pHS-OS, filtration, Percoll gradient	3×10^6 cells, 20–30%	80–90%, (FCR$^+$)	Dupuis et al., 1987; Roeckel et al., 1988
Zea mays	S	OS, filtration, sucrose gradient	1.5×10^6 cells	50%, (EbB$^-$)	Cass and Fabi, 1988
Lolium perenne	S	Squash, filtration, Percoll gradient	2%	ND	Van der Maas and Zaal, 1990
Gladiolus gandavensis	S	OS or enzymatic digestion, filtration	65–84 pairs per style[e]	ND	Shivanna et al., 1988
Lilium longiflorum	G	Grinding, Percoll gradient	10^5 cells, 33%	ND	Tanaka, 1988

[a] S, sperm cells; G, generative cells.

[b] OS, osmotic shock; pHS, pH shock.

[c] Yield was calculated as the total number of isolated cells obtained from a single isolation procedure, followed by the percentage of isolated cells recovered, as available. ND, not determined.

[d] Percentage of viable cells at the end of the isolation procedure. FCR$^+$, positive fluorochromatic reaction using fluorescein diacetate; EvB$^-$, Evans blue excluded; TEM, transmission electron microscopy; ND, not determined.

[e] In this study male gametes were isolated from pollen tubes growing in stylar segments using the *semi vivo* method; yield has been estimated by the number of pairs of male gametes isolated per style.

V. Physiology and Biochemistry of Isolated Generative Cells and Male Gametes

The ability to isolate free, living sperm or generative cells from the vegetative cell provides the opportunity to examine the physiology and biochemistry of these isolated cells, as well as their structural characteristics.

A. Structural Characteristics of the Isolated Cells

Ultrastructural studies of the isolated male cells have been undertaken to examine the presence of a cell wall in the generative and sperm cells, the complement of cellular organelles, and also shape changes resulting from the isolation procedure.

1. Generative Cells

The presence of a cell wall around the isolated generative cells has been shown in *Lilium longiflorum* (Tanaka, 1988), *Lilium* hybrid cv., *Tulipa gesneriana,* and *Trillium kamtschaticum* by Tanaka *et al.* (1989). When treated with cell wall digestion enzymes, they lose their wall and give rise to intact protoplasts, which are surrounded by the single generative cell membrane. In contrast, isolated generative cells from mature *Haemanthus katherinae* pollen grains are unlikely to possess a typical cellulosic wall (Zhou *et al.*, 1986). It must be kept in mind that the isolation process itself may alter the cell wall.

Isolated generative cells usually progressively lose their spindle shape and become ellipsoidal or spherical over time (Tanaka *et al.*, 1989; Zhou *et al.*, 1988, 1990; Zee and Azis-Un-Nisa, 1991). The shape of the isolated cells also varies considerably in media with different concentrations of sucrose (Zhou *et al.*, 1986). These observations prompted further investigations of the organization of the cytoskeleton because of the role of microtubules in determining cellular shape. No microtubules were observed in the generative cells of *Trillium kamtschaticum* and their protoplasts (Tanaka *et al.*, 1989). In *Zephyranthes grandiflora* Zhou *et al.* (1990) showed that the spindle shape of generative cells is correlated with longitudinally oriented microtubules, whereas spherical cells have predominantly a mesh-like cytoskeletal structure. In the ellipsoidal cells a transitional form consisting of a mixture of microtubule bundles and meshes can be found. These results indicate that the structure of the microtubular cytoskeleton depends on environmental conditions and plays an important role in determining cell shape.

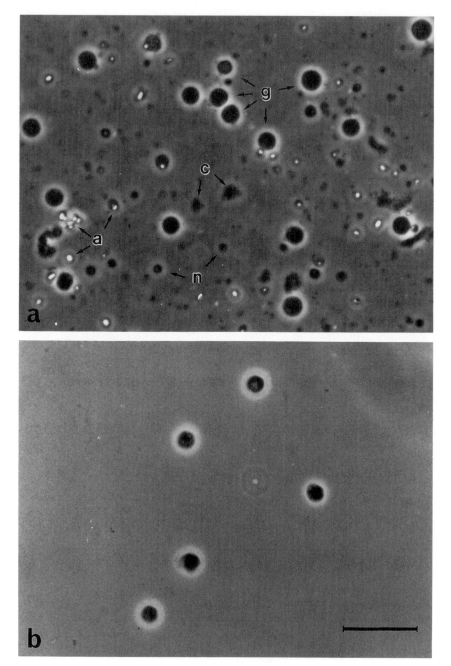

An immunofluorescence study by Zee and Azis-Un-Nisa (1991) gives more detail about changes in microtubule organization during mitosis in *Allemanda neriifolia* generative cells. In this species most of the newly isolated cells remain either in early or late prophase. In early prophase the nuclear membrane of the cells appears intact and the cytoplasm is full of reticulate microtubules. Later, the nuclear membrane breaks up and the chromosomes condense, migrate into the cytoplasm, and combine with microtubules. When generative cells enter metaphase, the spindle forms. In anaphase sister chromatids separate and the spindle disappears. A new array of longitudinally oriented "cage" microtubules can then be seen that disappear at telophase, as interzonal microtubules begin to form. Later, in some telophase cells the interzonal microtubules become highly elongated or are soon replaced by thick bands of microtubules in the midplane between the two clusters of chromosomes. The cell shape then reverts back to spheroidal.

2. Male Gametes

Upon release from the pollen grain, male gametes rapidly lose their characteristic shape and become spherical in *Zea mays, Brassica napus, Spinacia oleracea,* and *Lolium perenne* (Dupuis *et al.,* 1987; Matthys-Rochon *et al.,* 1988; Cass and Fabi, 1988; Theunis and Van Went, 1989; Van der Maas and Zaal, 1990), whereas they remain ellipsoidal in *Rhododendron* (Taylor *et al.,* 1989). In *Beta vulgaris* the male gametes assume a banana or S-shape but become round with time in the isolation medium, becoming spherical as well (Nielsen and Olesen, 1988). These cells typically have a small round appendicular structure on their surface at the point where the two ends of the horseshoe-shaped nucleus approach each other (Nielsen and Olesen, 1988). Similar surface structures have been shown in *Hordeum* (Cass, 1973) and *Plumbago zeylanica* (Russell, 1986). In *Gerbera jamesonii* the isolated male gametes keep their elongated shape as in the pollen, displaying extensions at one or both ends of the gamete (Southworth and Knox, 1989). The morphology of released male gametes, however, often

---

FIG. 4 Phase-contrast microscopy of sperm cell-enriched fraction during two isolation steps using maize pollen. ×500. Bar, 45 μm. (a) First one-step gradient purification in isolating sperm cell-enriched fraction. Numerous viable male gametes (g) appear as individual spherical cells with an average diameter of 10 μm (isolation yield, ~20%). Frequent pollen contaminants such as amyloplasts (a), cytoplasmic debris (c), or dead sperm cell remnants such as nuclei (n) are still present. (b) Sperm cell-enriched fraction after second purification step obtained by two successive gradient purifications. Contaminants have been removed, but isolation yield is considerably reduced (~4%).

depends on the constitution and osmolarity of the isolation medium. In *Triticum aestivum* Szakács and Barnabas (1990) observed that media with 30% sucrose provided paired spindle-shaped male gametes in a native configuration, whereas lower sucrose concentrations resulted in spherical separated cells. In all of these studies no sign of autonomous sperm cell motility was observed.

Fixation of isolated plant male gametes has proven difficult for TEM; however, the ultrastructure of isolated male gametes has been reported in *Zea mays* (Dupuis *et al.*, 1987; Matthys-Rochon *et al.*, 1988; Cass and Fabi, 1988; Wagner *et al.*, 1989), *Spinacia oleracea* (Theunis, 1990), and *Brassica napus* (Taylor *et al.*, 1991). In the two first species ultrastructural observations in TEM revealed the absence of the vegetative plasma membrane around the isolated male gametes. No cell wall can be seen around their intact single plasma membrane. These isolated male gametes can therefore be considered as true protoplasts (Dupuis *et al.*, 1987; Cass and Fabi, 1988; Matthys-Rochon *et al.*, 1988; Wagner *et al.*, 1989; Theunis, 1990). In contrast, ultrastructural observations of male gametes isolated from *Brassica* pollen tubes grown *in vitro* reveal that many isolated sperm pairs were still enclosed within the pollen tube inner plasma membrane. The periplasmic space between the pollen tube and sperm plasma membranes contains fibrillar components often appearing to form bridges between the two membranes; a remnant of this connective region may still be apparent on the sperm surface after their release from the tube inner plasma membrane (Taylor *et al.*, 1991).

To date, freeze-fracture analysis of plasma membrane of isolated male gametes is limited to *Spinacia oleracea* (Van Aelst *et al.*, 1990). No distinctive particle domains were observed on fracture faces of the isolated male gametes, but the ectoplasmic fracture face is reported to contain three times more intramembrane particles than the protoplasmatic fracture face (Southworth, this volume).

The nucleus of the male gametes is surrounded by a well-structured envelope that exhibits different chromatin conformations. In *Spinacia oleracea* male gametes, the nucleus is euchromatic or heterochromatic (Theunis, 1990), as in *Zea mays* where chromatin is either dispersed or not (Dupuis *et al.*, 1987; Wagner *et al.*, 1989). In *Brassica* male gametes the temperature of isolation seems to affect the state of the nuclei, with euchromatic sperm nuclei obtained at 4°C reportedly becoming heterochromatic at room temperature (Taylor *et al.*, 1991).

The cytoplasm of isolated male gametes appears to contain all normal components except plastids and microtubules (Cass and Fabi, 1988; Wagner *et al.*, 1989; Theunis, 1990; Taylor *et al.*, 1991). This result has been confirmed in *Zea mays* by morphometric analysis of numerous sperm cells (Wagner *et al.*, 1989). Numerous protein bodies have been observed

in isolated male gametes of *Zea mays,* some of which are in contact with rough endoplasmic reticulum (ER) (Cass and Fabi, 1988). This observation suggests the possibility of translational involvement of ER in protein body formation. In *Spinacia oleracea,* however, ER typically has few ribosomes in contact with it, suggesting low protein synthetic activity (Theunis, 1990). Presumably, sperm cell metabolism may be affected by culture conditions and must be examined by biochemical approaches to better answer this question.

Detailed information on isolated sperm cell structure has also been obtained by morphometric analysis (Wagner *et al.,* 1989; Theunis, 1990) and three-dimensional reconstruction (Mogensen *et al.,* 1990). These techniques were used to examine the possible dimorphism between the two gametes of the same pollen grain. In *Zea mays* morphometric analysis of 400 isolated male gametes confirms the presence of heterochromatic and nonheterochromatic nuclei (Wagner *et al.,* 1989). Although statistical analysis of cell components shows some significant differences between the two nuclear types, they do not occur in a 1 : 1 ratio and therefore are not believed to represent dimorphism between the paired sperm cells occurring within each pollen grain. It seems more likely that the variability within the sperm population is the result of different developmental stages or metabolic activity. The existence of a high degree of populational variability among isolated *Zea mays* male gametes has been confirmed by three-dimensional reconstruction (Mogensen *et al.,* 1990). In *Spinacia oleracea* morphometric analysis of the isolated male gametes similarly reveals only single peak in the distribution of cell diameter and organelle volume, indicating no morphological dimorphism in the sperm cell population (Theunis, 1990). Because no data about the number of organelles per cell were determined by morphometric analysis, the mitochondrial dimorphism previously described *in situ* (Wilms, 1986) cannot be confirmed using isolated *Spinacia* sperm cells (Theunis, 1990).

B. Viability and Maintenance of Isolated Cells

Before physiological or biochemical characterization, *en masse* isolated male gametes or generative cells must be evaluated to establish their quality. Numerous methods have been used for this assessment (Table I, viability).

Some researchers have estimated sperm cell quality by morphological study using scanning electron microscopy (SEM) (Shivanna *et al.,* 1988; Southworth and Knox, 1988; Taylor *et al.,* 1991). These studies use the ability of the SEM to visualize surfaces and show the exterior of male gametes to be examined for defects. A more sensitive visualization tech-

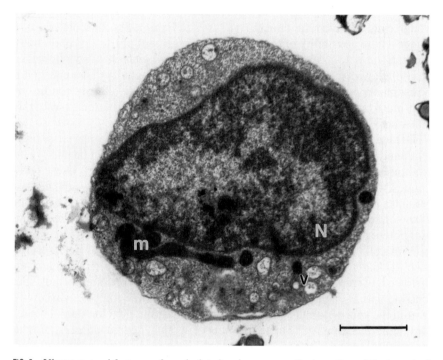

FIG. 5   Ultrastructural features of one isolated maize sperm cell. A portion of the branched mitochondrial complex (m) is found in close association with the heterochromatic sperm nucleus (N). Small cytoplasmic vacuoles (v) are also visible. ×10,000. Bar, 2 μm.

nique is the use of TEM (Fig. 5), as used by Dupuis *et al.* (1987) and Cass and Fabi (1988), which allows direct visualization of both plasma membrane configuration and intactness, and also the observable condition of cellular organelles (Wagner *et al.*, 1989; Theunis, 1990).

The method most commonly used to test viability is the fluorochromatic reaction (FCR), which is based on the evaluation of both membrane integrity and enzyme activity. This test (Heslop-Harrison and Heslop-Harrison, 1970) is based on the penetration of fluorescein diacetate (nonfluorescent and apolar) into the pollen grain and its hydrolysis by an esterase to yield fluorescein (fluorescent and polar). Viable male gametes accumulate the fluorescein, fluoresce brightly, and are designated FCR$^+$ (Fig. 6). This method is usually used to score the percentage of viable cells just after isolation (Table I, viability). In maize cell viability has been demonstrated using both FCR (Fig. 6) and the functional state of the plasma membrane, demonstrated by measuring transmembrane currents recorded using patch–clamp experiments in attached cell configuration (Roeckel *et al.*, unpublished data).

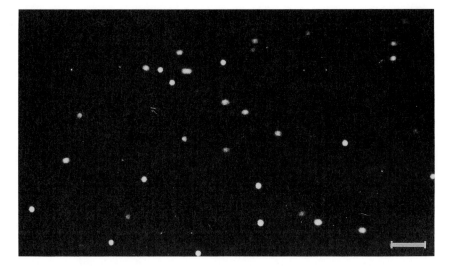

FIG. 6   Fluorescence microscopy of a sperm cell-enriched fraction stained by FCR test. Numerous brightly fluorescent cells are visible. ×100. Bar, 100 μm.

Another cytological stain can also be used. Evans blue assays the ability of the membrane to prevent entry of a polar molecule into the cell (Gaff and Okang'O-Ogola, 1971). Russell (1986) obtained exclusion of Evans blue for 20 hours in male gametes of *Plumbago zeylanica,* whereas these cells remained FCR$^+$ for only 5 minutes. This enigmatic result indicates that further studies are needed to understand the physiology of isolated male gametes.

Cass and Fabi (1988) suggested that for future fertilization studies, a more useful criterion for cell viability in sperm protoplasts is whether they will fuse with other protoplasts. Preliminary experiments by Tanaka (1988) indicated that generative cell protoplasts of *Lilium longiflorum* are suitable for successful fusion with other protoplasts. No direct assay of sperm cell function is, however, currently available. Whether isolated male gametes remain able to fertilize the egg or central cell will be the ultimate biotechnological assay (Southworth and Knox, 1988).

Long-term viability and maintenance of the isolated generative cells or male gametes are key points if one intends to further investigate their functional and biochemical characteristics. However, only a few reports mention the life expectancy of the isolated cells and, when mentioned, it varies considerably from one species to the other: 15 minutes for wheat sperm cells (Szakács and Barnabas, 1990) to 50 hours for corn sperm cells at room temperature (Roeckel *et al.,* 1988).

The addition of either organic supplements (nicotinic acid, pyridoxine-HCl, thiamine-HCl, and glycine; Roeckel *et al.,* 1988) or vitamins C and E (Van der Maas and Zaal, 1990; Theunis and Van Went, 1990) has been shown to significantly increase the life span of isolated male gametes at room temperature. Maintenance of isolated corn sperm cells in a living state can be achieved by storage in ice for 1 or 2 days without much decrease in cell viability (Roeckel, 1990). Moreover, long-term storage of corn sperm cells by freezing at −80°C for several months results in the recovery of 50–70% viable cells upon thawing (Roeckel, 1990). Nevertheless, further improvements of storage media are required to extend the useful life span of isolated generative cells and male gametes.

C. Functional Characteristics of the Isolated Cells

At the present time, the functional characteristics of the isolated generative cells and male gametes remain largely unknown. This kind of investigation has been hampered by the requirement to develop larger-scale isolation procedures and significantly improve the long-term viability of the isolated cells. Nevertheless, the first trials at functionally characterizing the isolated male gametes of corn have begun.

1. Metabolic Activity

The presence of intracellular adenosine triphosphate (ATP), a nucleotide either consumed or regenerated in metabolic pathways (Leach, 1981), can be an indicator of the energy potential of a cell. Thus, the metabolic activity of isolated male gametes of corn has been estimated by measuring the ATP level of cell suspensions using the luciferin–luciferase assay (Roeckel *et al.,* 1990b). At the end of the isolation procedure, a linear correlation was found between ATP content and the number of FCR$^+$ cells in 47 independent cell suspensions, indicating that a stable nonnegligible ATP level can be determined for freshly isolated male gametes. During natural aging of isolated cells, the concentration of FCR$^+$ cells and ATP content decrease as a function of time, but ATP is still evident even when the cells are no longer viable according to the FCR test. This finding suggests that during aging energy-producing organelles, such as mitochondria, may remain capable of ATP production even after the breakdown of the cell membrane (Roeckel *et al.,* 1990b). Nevertheless, from these data we can conclude that isolated male gametes have potent energy supplies and active metabolic pathways.

2. Transcriptional Activity

Although conflicting information is available regarding RNA synthesis in the male gametes of pollen grains and tubes (Section III,B), the use of isolated male gametes provides an opportunity to clarify this problem. To examine transcriptional activity in corn gametes, isolated cells were incubated under sterile conditions in the presence of ^{32}P-labeled uridine triphosphate and analyzed quantitatively to determine the amount of radioactivity incorporated into the nucleic acids (Perez and Chaboud, unpublished data). Problems of cell survival and sterility have hampered this research, however. The long incubation times at room temperature in a 15% sucrose medium cause the few remaining contaminants to actively divide, invalidating the results. These results indicate two problems that must be overcome in this area of research, in general: (1) adequate pollen sterilization must precede the isolation of cells, and (2) a suitable medium must be developed to meet the conditions of the assay and maintain the living cells for long periods of time. Once these technical problems are resolved, it will be possible to investigate whether RNA synthesis is turned off in the nucleus of the sperm cell at the end of pollen development and whether it is reactivated only in the cytoplasm of the egg cell after fertilization and prior to karyogamy, as described for sea urchin spermatozoa (Poccia, 1989).

3. Translational Activity

To provide evidence of translational activity in isolated male gametes, corn sperm cells isolated under sterile conditions were incubated for 2–36 hours in the presence of [^{35}S]methionine, and newly synthesized proteins were analyzed by SDS-PAGE and fluorography (Roeckel, 1990; Roeckel *et al.*, unpublished data). A large range of [^{35}S]methionine-labeled polypeptides from 15 to 90 kDa was found in isolated sperm cell fractions. [^{35}S]Methionine incorporation into proteins was maximal after a 18 hours of incubation when 50% of the initial cell population was still viable. No incorporation was detected in the control containing cytoplasmic fragments from pollen and male gametes without intact cells. These results indicate that: (1) the observed protein synthesis was dependent on intact, isolated male gametes, (2) translatable mRNA pools exist for a number of polypeptides, and (3) the translational machinery of the isolated sperm cells is fully functional. Thus, even after pollen maturation, protein synthesis may still occur in male gametes in vivo during pollen tube growth. These newly synthesized sperm cell proteins could be relevant for further successful fertilization and may be related to possible "plant sperm capacitation" similar to animals (Kranz *et al.*, this volume). Further research is needed to answer this question.

D. Biochemical Characteristics of the Isolated Cells

To date, biochemical characterization has been essentially devoted to cellular determinants on male gametes; such studies of generative cells are still lacking.

1. Immunological Characterization

The potential of studying cell surface determinants of isolated male gametes is offered by monoclonal antibody technology. Initial attempts were performed by immunizing mice with intact male gametes to elicit production of antibodies to surface epitopes during the time when the cells retain their integrity.

Monoclonal antibodies were first raised to isolated male gametes of *Brassica campestris* (Hough *et al.*, 1986). The authors developed a rapid immunofluorescent screening method in which the male gametes are immobilized on polycarbonate membranes without fixation. By this procedure 5 of more than 100 supernatants were shown to bind to sperm antigens, but neither the site of binding on the sperm nor cross-reactivity to other pollen fractions was determined. Moreover, this method is not suited to all male gametes; Pennell *et al.* (1987) reported that male gametes of *Plumbago zeylanica* are apparently destroyed by this technique.

A more detailed investigation on the potential use of monoclonal antibodies for male gamete characterization was reported by Pennell *et al.* (1987). In this study male gametes of *Plumbago zeylanica* appeared to be strongly immunogenic, as 43% of all reactive antibodies bound to the sperm cell fraction. Moreover, a large proportion of reactive lines were sperm-specific (23%), indicating that pollen wall contaminants present in the antigen preparation were not immunodominant as had been suspected, based on the ability of pollen wall components to elicit allergenic reactions. When pollen cytoplasm (immobilized on poly-L-lysine–coated microscope slides) was used for immunofluorescence screening, the strongly reactive lines appeared to be directed against nuclear or cytoplasmic components of the sperm cell; however, many more reacted with distinctly particulate or soluble components of pollen cytoplasm. Our own first trials of mice immunizations with isolated corn male gametes pointed out the same limiting factor. Polyclonal sera obtained were strongly reactive with pollen contaminants, such as starch grains. This result is in accordance with the recent findings of Singh *et al.*, (1991), showing that the major allergen of another grass, *Lolium*, is located within amyloplasts (Davies *et al.*, this volume).

These data clearly suggest that the sensitivity of screening procedures will need to be significantly increased before it is possible to produce a

useful monoclonal antibody library directed against gamete-specific cell surface determinants. Consequently, improving the large-scale purification of isolated male gametes has become a major objective with the goal of membrane isolation. However, because the pollen grain of corn is essentially filled with amyloplasts and the volume of the sperm cells represents only 0.02–0.2% of the volume of the grain, the purification of sperm cells may require considerable effort to obtain a fraction exclusive of contaminants.

2. Polypeptide–Protein Characterization

Another method to characterize male gametes is the analysis of their protein content. For this purpose homogenates of isolated male gametes and corresponding pollen preparations can be compared by electrophoretic methods.

Preliminary data have been reported for *Brassica campestris* and *Gerbera jamesonii* (Knox *et al.*, 1988), suggesting that several proteins and glycoproteins separated by SDS-PAGE may be present only in the male gametes. These results have been confirmed and extended by 2D-PAGE for *Plumbago* (Geltz and Russell, 1988) and *Zea* (Roeckel, 1990). Geltz and Russell (1988) have shown significant heterogeneity between polypeptides isolated from an MGU-rich fraction, a cytoplasmic-particulate fraction, and a water-soluble fraction of *Plumbago zeylanica* mature pollen. A high degree of overlap between MGU-rich and cytoplasmic-particulate fractions was noticed and interpreted as partially the result of common proteins required for cellular metabolism in vegetative and male gametes. The possibility of some contamination of the MGU-rich fraction by cytoplasmic pollen components also cannot be eliminated. Nevertheless, approximately 13.5% of the most conspicuous polypeptide spots were unique to the MGU-rich fraction and may include cell surface determinants of the male gametes. For *Zea mays* polypeptide heterogeneity was compared for a sperm-enriched fraction, the last supernatant of sperm purification (containing cytoplasmic contaminants from pollen and bursted male gametes but no intact cells), and pollen (Roeckel, 1990). Although some sperm-specific spots were evident, a rather high pollen protein overlap was once again noticed.

These different data clearly indicate that for further characterization of the cell surface protein, isolated male gamete preparations need to be scaled up to get plasma membrane fractions suitable for analysis. This may require considerable effort to reach a satisfactory conclusion since Geltz and Russell (1988) calculated that MGU proteins represent only 0.14% of the total protein of the mature pollen in *Plumbago* (about 14.8 pg/pollen grain). Similarly, in corn, based on average diameter of

male gametes (5–10 μm) and pollen grain (100 μm), we can estimate that the ratio of male gametes versus pollen plasma membranes ranges from 0.5% to 2% of the grain.

VI. Concluding Remarks

In conclusion, sperm cells are now available as isolated cells in many species, and they are well characterized at the structural level. Similar progress has been made for the female target cells: the egg and central cells (Huang and Russell, this volume). Angiosperm gametes can now be considered as experimental tools for further functional investigations. In particular, the intimate process of the double fertilization must still be examined in detail. Genetic studies of the Black Mexican Sweet (BMS) maize lines have long indicated that the male gamete bearing the nondis-junct B-chromosomes preferentially fuses with the egg cell (Roman, 1948). The resulting hypothesis of a preferential fusion occurring between one predetermined sperm cell of the pollen grain and the egg cell has recently been supported by cytological data from *Plumbago*. In this species one of the two sperm cells is plastid-rich and preferentially fertilizes the egg cell (Russell, 1985). At the present time, ultrastructural analyses of the isolated male gametes provide valuable information for use in studying the fertiliza-tion process but do not resolve the dimorphism question, which must be considered as a functional characteristic at the biochemical and molecular level.

Our success in isolating intact, viable, and functional corn gametes offers the possibility of direct access to the interaction between gamete membranes that are presumed to harbor the specific determinants of inter-est in double fertilization. Thus, we are currently designing an experimen-tal strategy based on the construction of a monoclonal antibody library directed against the cell surface of isolated viable male gametes that will be used for cell-sorting the population of isolated gametes (Chaboud *et al.*, 1992). This would allow rapid assessment of cellular dimorphism at the membrane level to discriminate between the two male gametes of the pollen grain. The development of an *in vitro* model of intergametic fusion without any artificial inducers, currently in progress (Kranz *et al.*, this volume), will be used for experiments of fusion inhibition by monoclonal antibodies. This would allow specific cell surface determinants involved in gametic recognition in corn to be sorted and examined individually. This approach may lead in the future to a detailed examination of the ontogenesis and role of gamete-specific determinants in the double fertil-ization process.

References

Albani, D., Robert, L. S., Donaldson, P. A., Altosaar, I., Arnison, P. G., and Fabijanski, S. F. (1990). *Plant Mol. Biol.* **15,** 605–622.

Bedinger, P. A., and Edgerton, M. D. (1990). *Plant Physiol.* **92,** 474–479.

Brewbaker, J. L., and Kwack, B. H. (1963). *Am. J. Bot.* **50,** 859–865.

Brown, S. M., and Crouch, M. L. (1990). *Plant Cell* **2,** 263–274.

Cass, D. D. (1973). *Can. J. Bot.* **51,** 601–605.

Cass, D. D., and Fabi, G. C. (1988). *Can. J. Bot.* **66,** 819–825.

Chaboud, A., Perez, R., Digonnet, C., and Dumas, C. (1992). In "Perspectives in Plant Cell Recognition" (J. A. Callow and J. R. Green, eds.). Cambridge University Press, Cambridge, Massachusetts.

Cresti, M., Lancelle, S. A., and Hepler, P. K. (1987). *J. Cell Sci.* **88,** 373–378.

Cresti, M., Murgia, M., and Theunis, C. H. (1990). *Protoplasma* **154,** 151–156.

Delvallée, I., and Dumas, C. (1988). *J. Plant Physiol.* **132,** 210–217.

Derksen, J., Pierson, E. S., and Traas, J. A. (1985). *Eur. J. Cell Biol.* **38,** 142–148.

Detchepare, S., Heizmann, P., and Dumas, C. (1989). *J. Plant Physiol.* **135,** 129–137.

Dumas, C., Knox, R. B., McConchie, C. A., and Russell, S. D. (1984). *What's New Plant Physiol.* **15,** 17–20.

Dumas, C., Knox, R. B., and Gaude, T. (1985). *Protoplasma* **124,** 168–174.

Dumas, C., Kerhoas, C., Matthys-Rochon, E., Vergne, P., Gaude, T., Detchepare, S., and Gay, G. (1986). In "Biology of Reproduction and Cell Motility in Plants and Animals" (M. Cresti and R. Dallai, eds.), pp. 155–162. University of Siena, Siena, Italy.

Dupuis, I., Roeckel, P., Matthys-Rochon, E., and Dumas, C. (1987). *Plant Physiol.* **85,** 876–878.

Frova, C. (1990). *Sex. Plant Reprod.* **3,** 200–206.

Frova, C., Binelli, G., and Ottaviano, E. (1987). In "Isozymes: Current Topics in Biological and Medical Research" (M. C. Rattazi, J. G. Scandalios, and G. S. Whitt, eds.), pp. 97–120, Vol. 15: Genetics, Development and Evolution. Liss, New York.

Gaff, D. F., and Okang'O-Ogola, O. (1971). *J. Exp. Bot.* **22,** 756–758.

Geltz, N. R., and Russell, S. D. (1988). *Plant Physiol.* **88,** 764–769.

Giles, K. H., and Prakash, J. (eds.) (1987). *Int. Rev. Cytol.* **107.**

Hanson, D. D., Hamilton, D. A., Travis, J. L., Bashe, D. M., and Mascarenhas, J. P. (1989). *Plant Cell* **1,** 173–179.

Haskell, D. W., and Rogers, O. M. (1985). *Cytologia* **50,** 805–809.

Heslop-Harrison, J., and Heslop-Harrison, Y. (1970). *Stain Tech.* **45,** 115–120.

Heslop-Harrison, J., and Heslop-Harrison, Y. (1980). *Pollen Spores* **22,** 5–10.

Heslop-Harrison, J., and Heslop-Harrison, Y. (1988a). *Sex. Plant Reprod.* **1,** 16–24.

Heslop-Harrison, J., and Heslop-Harrison, Y. (1988b). *Ann. Bot.* **61,** 249–254.

Heslop-Harrison, J., and Heslop-Harrison, Y. (1989a). *J. Cell Sci.* **94,** 319–325.

Heslop-Harrison, J., and Heslop-Harrison, Y. (1989b). *J. Cell Sci.* **93,** 299–308.

Hough, T., Singh, M. B., Smart, I. J., and Knox, R. B. (1986). *J. Immunol. Methods* **92,** 103–107.

Hu, S. Y., and Yu, H. S. (1988). *Protoplasma* **147,** 55–63.

Knox, R. B., Williams, E. G., and Dumas, C. (1986). *Plant Breeding Rev.* **4,** 8–79.

Knox, R. B., Southworth, D., and Singh, M. B. (1988). In "Eucaryote Cell Recognition: Concepts and Model Systems" (G. P. Chapman, C. C. Ainsworth, and C. J. Chatham, eds.), pp. 175–193. Cambridge University Press, Cambridge, Massachusetts.

Lancelle, S. A., Cresti, M., and Hepler, P. K. (1987). *Protoplasma* **140,** 141–150.

Leach, F. R. (1981). *J. Appl. Biochem.* **3,** 473–517.

Linskens, H. F. (1966). *Planta* **69,** 79–91.

Mandaron, P., Niogret, M. F., Mache, R., and Moneger, F. (1990). *Theor. Appl. Genet.* **80,** 134–138.

Mascarenhas, J. P. (1990). *Annu. Rev. Plant Physiol. Plant Mol. Biol.* **41,** 317–338.

Matthys-Rochon, E., and Dumas, C. (1988). *In* "Plant Sperm Cells as Tools for Biotechnology" (H. J. Wilms and C. J. Keijzer, eds.), pp 51–60. Pudoc, Wageningen.

Matthys-Rochon, E., Vergne, P., Detchepare, S., and Dumas, C. (1987). *Plant Physiol.* **83,** 464–466.

Matthys-Rochon, E., Detchepare, S., Wagner, V., Roeckel, P., and Dumas, C. (1988). *In* "Sexual Reproduction in Higher Plants" (M. Cresti, P. Gori, and E. Pacini, eds.), pp. 245–250. Springer-Verlag, Berlin.

McConchie, C. A., Jobson, S., and Knox, R. B. (1985). *Protoplasma* **127,** 57–63.

McConchi, C. A., Russell, S. D., Dumas, C., Tuohy, M., and Knox, R. B. (1987a). *Planta* **140,** 7–13.

McConchie, C. A., Hough, T., and Knox, R. B. (1987b). *Protoplasma* **139,** 9–19.

Mogensen, H. L., and Rusche, M. L. (1985). *Protoplasma* **128,** 1–13.

Mogensen, H. L., and Wagner, V. T. (1987). *Protoplasma* **138,** 161–172.

Mogensen, H. L., Wagner, V. T., and Dumas, C. (1990). *Protoplasma* **153,** 136–140.

Murgia, M., and Wilms, H. J. (1988). *In* "Plant Sperm Cells as Tools for Biotechnology" (H. J. Wilms and C. J. Keijzer, eds.), pp. 75–79. Pudoc, Wageningen.

Nielsen, J. E., and Olesen, P. (1988). *In* "Plant Sperm Cells as Tools for Biotechnology" (H. J. Wilms and C. J. Keijzer, eds.), pp. 111–112 Pudoc, Wageningen.

Palevitz, B. A., and Cresti, M. (1988). *Protoplasma* **146,** 28–34.

Pennell, R. I., Geltz, N. R., Koren, E., and Russell, S. D. (1987). *Bot. Gaz.* **148,** 401–406.

Pierson, E. S., Derksen, J., and Traas, J. A. (1986). *Eur. J. Cell Biol.* **41,** 14–18.

Poccia, D. (1989). *In* "The Molecular Biology of Fertilization" (H. Schatten and G. Schatten, eds.), pp. 115–135. Academic Press, San Diego.

Raghavan, V. (1984). *Can. J. Bot.* **62,** 2493–2513.

Reynolds, T. L., and Raghavan, V. (1982). *Protoplasma* **111,** 177–182.

Roberts, M. R., Robson, F., Foster, G. D., Draper, J., and Scott, R. J. (1991). *Plant Mol. Biol.* **17,** 295–299.

Roeckel, P. (1990). Ph.D. Dissertation, Université de Lyon, Lyon, France.

Roeckel, P., Dupuis, I., Detchepare, S., Matthys-Rochon, E., and Dumas, C. (1988). *In* "Plant Sperm Cells as Tools for Biotechnology" (H. J. Wilms and C. J. Keijzer, eds.), pp. 105–110. Pudoc, Wageningen.

Roeckel, P., Chaboud, A., Matthys-Rochon, E., Russell, S., and Dumas, C. (1990a). *In* "Microspores: Evolution and Ontogeny" (S. Blackmore and R. B. Knox, eds.), pp. 281–307. Academic Press, London.

Roeckel, P., Matthys-Rochon, E., and Dumas, C. (1990b). *In* "Characterization of Male Transmission Units in Higher Plants" (B. Barnabas and K. Liszt, eds.), pp. 41–48. Agricultural Research Institute, Hungarian Academy of Sciences, Budapest.

Roman, H. (1948). Directed fertilization in maize. *Proc. Natl. Acad. Sci. U.S.A.* **34,** 36–42.

Rougier, M., Jnoud, N., Said, C., Russell, S. D., and Dumas, C. (1991). *Protoplasma* **162,** 140–150.

Rusche, M. L., and Mogensen, H. L. (1988). In "Sexual Reproduction in Higher Plants" (M. Cresti, P. Gori, and E. Pacini, eds.), pp. 221–226. Springer-Verlag, Berlin.

Russell, S. D. (1984). *Planta* **162,** 385–391.

Russell, S. D. (1985). *Proc. Natl. Acad. Sci. U.S.A.* **82,** 6129–6132.

Russell, S. D. (1986). *Plant Physiol.* **81,** 317–319.

Russell, S. D. (1991). *Annu. Rev. Plant Physiol. Plant Mol. Biol.* **42,** 189–204.

Russell, S. D., and Cass, D. D. (1981). *Protoplasma* **107,** 85–107.

Schrauwen, J. A. M., De Groot, P. F. M., Van Herpen, M. M. A., van der Lee, T., Reynen, W. H., Weterings, K. A. P., and Wullems, G. J. (1990). *Planta* **182,** 298–304.

Scott, R. J., Dagless, E., Hodge, R., Paul, W., Soufleri, I., and Draper, J. (1991). *Plant Mol. Biol.* **17**, 195–207.

Shivanna, K. R., Xu, H., Taylor, P., and Knox, R. B. (1988). *Plant Physiol.* **87**, 647–650.

Singh, M. B., Hough, T., Theerakulpisut, P., Avjioglu, A., Davies, S., Smith, P. M., Taylor, P., Simpson, R. J., Ward, L. D., McCluskey, J., Puy, R., and Knox, R. B. (1991). *Proc. Natl. Acad. Sci. U.S.A.* **88**, 1384–1388.

Southworth, D. (1990). *J. Struct. Biol.* **103**, 97–103.

Southworth, D., and Knox, R. B. (1988). *In* "Plant Sperm Cells as Tools for Biotechnology" (H. J. Wilms and C. J. Keijzer, eds.), pp. 87–95. Pudoc, Wageningen.

Southworth, D., and Knox, R. B. (1989). *Plant Sci.* **60**, 273–277.

Southworth, D., Platt-Aloia, K. A., and Thomson, W. W. (1988). *J. Ultrastruct. Mol. Struct. Res.* **101**, 165–172.

Southworth, D., Platt-Aloia, K. A., DeMason, D. A., and Thomson, W. W. (1989). *Sex. Plant Reprod.* **2**, 270–276.

Stinson, J. R., Eisenberg, A. J., Willing, R. P., Pe, M. E., Hanson, D. D., and Mascarenhas, J. P. (1987). *Plant Physiol.* **83**, 442–447.

Szakács, É., and Barnabas, B. (1990). *In* "Characterization of Male Transmission Units in Higher Plants" (B. Barnabas and K. Liszt, eds.), pp. 37–40. Agricultural Research Institute, Hungarian Academy of Sciences, Budapest.

Tanaka, I. (1988). *Protoplasma* **142**, 68–73.

Tanaka, I., Nakamura, S., and Miki-Hirosige, H. (1989). *Gamete Res.* **24**, 361–374.

Taylor, P., Kenrick, J., Li, Y., Kaul, V., Gunning, B. E. S., and Knox, R. B. (1989). *Sex. Plant Reprod.* **2**, 254–264.

Taylor, P., Kenrick, J., Blomsted, C. K., and Knox, R. B. (1991). *Sex. Plant Reprod.* **4**, 226–234.

Theunis, C. H. (1990). *Protoplasma* **158**, 176–181.

Theunis, C. H., and Van Went, J. L. (1989). *Sex. Plant Reprod.* **2**, 97–102.

Theunis, C. H., and Van Went, J. L. (1990). *In* "Characterization of Male Transmission Units in Higher Plants" (B. Barnabas and K. Liszt, eds.), pp. 25–29. Agricultural Research Institute, Hungarian Academy of Sciences, Budapest.

Theunis, C. H., Pierson, E. S., and Cresti, M. (1991). *Sex. Plant Reprod.* **4**, 145–154.

Tiezzi, A., Moscatelli, A., Ciampolini, F., Milanesi, C., Murgia, M., and Cresti, M. (1988). *In* "Sexual Reproduction in Higher Plants" (M. Cresti, P. Gori, and E. Pacini, eds.), pp. 215–220. Springer-Verlag, Berlin.

Tupý, J., Suss, J., Hrabetova, E., and Rihova, L. (1983). *Biol. Plant.* **25**, 231–237.

Twell, D., Wing, R., Yamaguchi, J., and McCormick, S. (1989). *Mol. Gen. Genet.* **247**, 240–245.

Ursin, V. M., Yamaguchi, J., and McCormick, S. (1989). *Plant Cell* **1**, 727–736.

Van Aelst, A., Theunis, C. H., and van Went, J. L. (1990). *Protoplasma* **153**, 204–207.

Van der Maas, H. M., and Zaal, M. A. C. M. (1990). *In* "Characterization of Male Transmission Units in Higher Plants" (B. Barnabas and K. Liszt, eds.), pp. 31–36. Agricultural Research Institute, Hungarian Academy of Sciences, Budapest.

Vergne, P., and Dumas, C. (1988). *Plant Physiol.* **88**, 969–972.

Vergne, P., Delvalée, I., and Dumas, C. (1987). *Stain Tech.* **62**, 299–304.

Villanueva, V. R., Mathivet, V., and Sangwan, R. S. (1985). *Plant Growth Regulation* **3**, 293–307.

Wagner, V. T., and Mogensen, H. L. (1987). *Protoplasma* **143**, 93–100.

Wagner, V. T., Dumas, C., and Mogensen, H. L. (1989). *J. Cell Sci.* **93**, 179–184.

Willemse, M. T. M. (1988). *In* "Plant Sperm Cells as Tools for Biotechnology" (H. J. Wilms and C. J. Keijzer, eds.), pp. 11–16. Pudoc, Wageningen.

Willing, R. P., and Mascarenhas, J. P. (1984). *Plant Physiol.* **75**, 865–868.

Willing, R. P., Basche, D., and Mascarenhas, J. P. (1988). *Theor. Appl. Genet.* **75,** 751–753.

Wilms, H. J. (1986). *In* "Biology of Reproduction, and Cell Motility in Plants and Animals" (M. Cresti and R. Dallai, eds.), pp. 193–198. University of Siena, Siena.

Wilms, H. J., and Van Aelst, A. C. (1983). *In* "Fertilization and Embryogenesis in Ovulated Plants", (O. Erdelská, ed.), pp. 105–112. VEDA, Bratislava, Czechoslovakia.

Wing, R. A., Yamaguchi, J., Larabell, S. K., Ursin, V. M., and McCormick, S. (1989). *Plant Mol. Biol.* **14,** 17–28.

Yu, H. S., Hu, S. Y., and Zhu, C. (1989). *Protoplasma* **152,** 29–36.

Zarsky, V., Capkova, V., Hrabetova, E., and Tupý, J. (1985). *Biol. Plant* **27,** 438–444.

Zee, S. Y., and Azis-Un-Nisa, (1991). *Sex. Plant Reprod.* **4,** 132–137.

Zhou, C. (1988). *Plant Cell Rep.* **7,** 107–110.

Zhou, C., Orndorff, K., Allen, R. D., and De Maggio, A. E. (1986). *Plant Cell Rep.* **5,** 306–309.

Zhou, C., Orndorff, K., Daghlian, C. P., and De Maggio, A. E. (1988). *Sex. Plant Reprod.* **1,** 97–102.

Zhou, C., Zee, S. Y., and Yang, H. Y. (1990). *Sex. Plant Reprod.* **3,** 213–218.

Female Germ Unit: Organization, Isolation, and Function

Bing-Quan Huang and Scott D. Russell

Department of Botany and Microbiology, University of Oklahoma, Norman, Oklahoma 73019

I. Introduction

Although the organization of the female gametophyte was first described accurately in the late 1870s (Strasburger, 1878), much literature has since arisen concerning the structure and ontogeny of the embryo sac. This has been extended significantly in the last 30 years by transmission electron microscopy (TEM), quantitative cytochemistry, histochemistry, fluorescence, and phase microscopy (Maheshwari, 1950; Kapil and Bhatnagar, 1981; Willemse and Van Went, 1984). Information from TEM and light microscopy has extended our knowledge about the female gametophyte in angiosperms, our concepts on the diversity of megagametophyte development (Willemse and Van Went, 1984; Haig, 1986, 1990), and the postpollination function of the megagametophyte (Jensen, 1974; Willemse and Van Went, 1984; Van Went and Willemse, 1984; Russell, this volume).

Considerable progress has been made in establishing the angiosperm female gametophyte as a developmental model. In particular, the isolation of living and fixed embryo sacs and female gametophytic cells (Theunis *et al.*, 1991), the isolation of the male gamete (Russell, 1991; Chaboud and Perez, this volume), and the combination of the two in an *in vitro* fertilization system (Kranz *et al.*, 1991a,b; Kranz *et al.*, this volume) represent important accomplishments in exploiting the gametophyte as a source of haploid cells for experimental manipulation. Our understanding of the female gametophyte during megasporogenesis and fertilization has also been furthered by studies using fluorescent probes on intact female gametophytes and their component cells using tubulin immunocytochemistry for microtubules, rhodamine–phalloidin for F-actin (Willemse and Van Lammeren, 1988; Bednara *et al.*, 1988, 1990; Huang *et al.*, 1990; Webb and Gunning, 1990, 1991), and 4′,6-diamidino-2-phenylindole (DAPI) and

Hoechst 33258 for the localization of DNA (Zhou, 1987; Huang and Russell, 1992a). These recent studies provide the methodology and techniques to more fully characterize female gametophytic cells and provide needed insights about their function.

The purpose of this chapter is to review the currently available information about the megagametophyte, emphasizing work during the past two decades, and particularly examining the female germ unit from four new perspectives: (1) the ultrastructure and cytochemistry of megagametophyte development and cellularization, (2) the organization and function of the female germ unit in reproduction, (3) the isolation and characterization of embryo sacs and their component cells, and (4) the involvement of the cytoskeleton in megagametophyte development and function.

II. Development of the Megagametophyte

The development of the megagametophyte is divided into two connected developmental phases: megasporogenesis, which entails the formation and maturation of the initial products of meiosis, followed by megagametogenesis, which begins with the mitotic division of the meiotic products and continues through the cellularization and maturation of the megagametophyte (Fig. 1). Although this review provides an overview of the process of embryo sac development, specialized reviews on the fine structure of the megagametophyte are available elsewhere (Rodkiewicz, 1978; Kapil and Bhatnagar, 1981; Willemse, 1981; Willemse and Van Went, 1984; Fougère-Rifot, 1987). Embryo sac development follows myriad types as described in reviews by Maheshwari (1950) and others (Haig, 1990); however, only about half of the types have been described at the ultrastructural level. The descriptions that follow are a composite of the best described types, principally the *Polygonum* and the *Fritillaria* types.

A. Megasporocyte

Meiotic division in the megasporocyte usually gives rise to four meiotic products known as megaspores, although the arrest of division may sometimes result in dyad cells. Depending on the ensuing pattern of cell wall formation, megasporogenesis is classified into three patterns based on the number of megaspore nuclei participating to form the megagametophyte: monosporic, bisporic, or tetrasporic (Maheshwari, 1950).

The monosporic pattern of megasporogenesis results in the abortion of all but one of the meiotic products, the uninucleate functional megaspore,

Megasporogenesis   **Megagametogenesis**

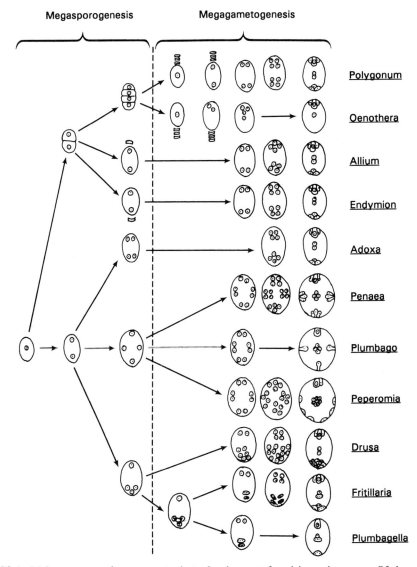

Polygonum

Oenothera

Allium

Endymion

Adoxa

Penaea

Plumbago

Peperomia

Drusa

Fritillaria

Plumbagella

FIG. 1 Major patterns of megagametophyte development found in angiosperms. Of these patterns, the *Polygonum* type accounts for more than 70% of flowering plants. Modified with permission from Gifford and Foster (1988).

which forms the embryo sac. In the *Polygonum* pattern of embryo sac development, the functional megaspore is typically the chalazal one and is present in approximately 70% of the flowering plants studied (Davis, 1966) and in the vast majority of lower seed plants (Gifford and Foster, 1988). The *Polygonum* type is therefore widely regarded as the ancestral type. In rare cases the micropylar megaspore becomes the functional megaspore (*Oenothera* type), and this is principally restricted to the family Onagraceae. In the bisporic pattern of development, the functional megaspore is a 2-nucleate cenomegaspore occupying the cell wall of the non-abortive dyad cell, and abortion occurs in the opposite dyad cell. If the functional cenomegaspore is at the chalazal pole, it is assigned to the *Allium* type of development; the few exceptions with functional cenomegaspores at the micropylar pole are assigned to the *Endymion* type (Fig. 1).

In tetrasporic development all four of the meiotic products are incorporated in the formation of the embryo sac, and this pattern displays the most variation in postmeiotic development (Fig. 1). Three general patterns of polarity become evident during the migration of the megaspore nuclei: (1) a bipolar 2 + 2 pattern of nuclei, in which two megaspore nuclei are at the chalazal pole and two at the micropylar pole, (2) a bipolar 1 + 3 pattern, in which one nucleus migrates to the micropylar pole and three move to the chalazal pole, and (3) a quadripolar 1 + 1 + 1 + 1 pattern, in which the megaspore nuclei become cruciate in placement. The different patterns may also require one, two, or (when some products fuse) three cycles of mitotic divisions to reach their mature complement of nuclei. This contributes to varying degrees of genetic heterogeneity in the embryo sac. In the *Penaea, Drusa,* and *Fritillaria* types the egg and two synergids are derived from the same megaspore nucleus, but in the *Adoxa* type two megaspore nuclei form these cells. The central cell contains the fused derivatives of all four of the megaspore nuclei in most of the tetrasporic patterns, but in the *Adoxa* and *Drusa* types the central cell nucleus is derived from only two of the megaspore nuclei.

Haig (1990) presented a full review of the myriad of patterns of megagametophyte formation and reconciled the apparent degree of variability by suggesting the action of specific algorithms to illustrate the common principles of development. His principal tenet is that by constructing relatively few central rules of organization, the most common patterns of megagametophyte development may be described, along with such major variants as strike (arrest of mitotic division at one pole) or subsequent proliferation. According to theoretically based studies of genetic conflict during megasporogenesis, the most inherently stable form of megasporogenesis should be monosporic development, in which the successful megaspore is positionally determined and megaspores do not directly compete for the ability to form an embryo, which could reduce their vigor (Haig,

1986). He contended that this stabilty is one reason that the *Polygonum* type of development is of such widespread importance. The tetrasporic mode of development in this regard seems the least stable and possesses the most extreme variability in embryo sac organization. The degree of genetic conflict between the ovule and endosperm, however, should be reduced in some tetrasporic plants because the fused central cell nucleus generally is the fusion product of all four of the megaspore nuclei and therefore a genetic equivalent to the megasporocyte.

The cytoplasmic changes accompanying meiosis in the megasporocyte and the transition between the sporophyte and gametophyte generations are in some ways as dramatic as those occurring in the nucleus. The fine structure of the megasporocyte is similar to that of meristematic cells at the initiation of meiosis (Dickinson and Heslop-Harrison, 1977; Eyme, 1965; Schulz and Jensen, 1981; Kapil and Bhatnagar, 1981). Mitochondria and proplastids are not highly differentiated at this stage. Ribosomes are abundant, as is endoplasmic reticulum (ER). The polarity of the cell is variably expressed, although in *Myosurus* all organelles are reported to be present in future cytoplasm of the functional pole (Woodcock and Bell, 1968); in most plants there is a weaker degree of polarization (Rodkiewicz, 1978; Willemse, 1981; Medina *et al.,* 1981; Kennell and Horner, 1985; Schulz and Jensen, 1986; Webb and Gunning, 1990). In most megasporocytes there appears to be an increase in the number of heritable organelles at the functional pole. Polarization of the megasporocyte also includes features of the cell wall. The cell wall of the megasporocyte is intrinsically somewhat thicker than that of surrounding cells and, in many cases, contains callose at the nonfunctional pole in the majority of flowering plants (Kapil and Bhatnagar, 1981). Whether polarity establishes developmental patterns during megasporogenesis (Willemse, 1981) or merely reflects these patterns is unclear.

During prophase I mitochondria and proplastids undergo dramatic structural changes and dedifferentiate, becoming remarkably similar in appearance; this led early cytologists to question whether heritable organelles persisted throughout meiosis or were regenerated from other organelles or the nuclear envelope (Dickinson and Heslop-Harrison, 1977). Ultrastructural and biochemical studies have confirmed that these DNA-containing organelles remain present but are highly modified during prophase. The concentration of ribosomes is also reduced at this stage (Rodkiewicz and Mikulska, 1963, 1965a, 1965b; Dickinson and Heslop-Harrison, 1977; De Boer-de-Jeu, 1978; Willemse and Bednara, 1979). Autophagic vacuoles generated from ER and preexisting vacuoles are frequently found in the cytoplasm and occasionally contain membranes or organelle-like objects (Russell, 1979; Schulz and Jensen, 1981). Two nonspecific reclamation enzymes have been localized in these compartments:

acid phosphatase (Schulz and Jensen, 1981; Willemse and Bednara, 1979) and esterase (Willemse and Bednara, 1979), suggesting that selected cellular materials are degraded in vacuoles during this phase and broken down into their precursor components.

In pollen grains the depression of the mitochondrial population is most extreme during midmeiosis. This, along with the later occurrence of tight associations between mitochondria and nuclear invaginations, supports the possibility of organelle selection and cytoplasmic regeneration in the meiocyte (Dickinson, 1986). In male meiosis decreases in the level of nucleic acids and proteins closely correlate with general increases in hydrolytic enzyme levels. Ultrastructural data confirm that similar changes are likely in the megasporocyte during meiosis (Dickinson and Heslop-Harrison, 1977). Different species of angiosperms appear to express these characteristics to varying degrees. The most extreme case of cytoplasmic modification during female meiosis appears to be in the developing female gametophyte of *Lilium* (Rodkiewicz and Mikulska, 1963, 1965a,b; Dickinson and Potter, 1978), in which the tetrasporic *Fritillaria*-type embryo sac is particularly large. In this plant areas of concentric lamellar ER appear to be retained as reservoirs of protected cytoplasm, while hydrolytic activity is present throughout the cytoplasm. In plants with smaller, monosporic embryo sacs, however, autophagic activity appears to be restricted to specific compartments in which fewer cytoplasmic reservoirs occur, and the elimination of ribosomes is only partial (De Boer-de-Jeu, 1978; Willemse and Franssen-Verheijen, 1978; Jalouzot, 1978; Russell, 1979; Willemse and Bednara, 1979; Medina *et al.*, 1981; Schulz and Jensen, 1981).

B. Meiotic Products and Functional Megaspore Determination

The first cycle of meiosis results in the formation of two dyad nuclei, which typically complete meiosis to form four megaspore nuclei. Cytokinesis occurs at the end of both meiosis I and II in monosporic development, but only at the end of the first phase in bisporic development. In monosporic and bisporic patterns of development, the abortion of specific meiotic products is apparently genetically preprogrammed and is related to the presence of callose (a β-1,3-polyglucan). Callose typically accumulates in walls specifically located at the abortive pole of the megasporocyte. Plasmodesmata occur only in areas of the cell wall that do not accumulate callose. Nonfunctional megaspores and dyad cells are often characterized by a slightly impoverished plastid content, whereas dictyosomes, rough endoplasmic reticulum (RER), and vacuoles may be more abundant than in the functional cell (Willemse, 1981). The deposition of callose in the

tetrads separates these cells from the functional megaspore (Willemse and Van Went, 1984), and frequently aberrant cell walls form between the degenerating megaspores (Schulz and Jensen, 1986; Webb and Gunning, 1990). Abortive cells remain invested in a conspicuous callose-rich cell wall throughout abortion (Rodkiewicz, 1970; Russell, 1979; Schulz and Jensen, 1986) and lack plasmodesmata, unlike the functional megaspore and surrounding cells (De Boer-de-Jeu, 1978; Jalouzot, 1978; Russell, 1979; Schulz and Jensen, 1981, 1986; Willemse, 1981). Haig and Westoby (1986) regarded the callose wall as direct evidence of genetic conflict between megaspores. Whether callose directly obstructs nutrition to degenerate megaspores (or dyad cells) (Dickinson and Heslop-Harrison, 1977) or more subtly interferes with developmental signals to meiotic products is yet unknown, but it is possible that separation from the cellular syncytium alone may be enough to cause eventual degeneration.

Abortive meiotic products are characterized by dense areas of hydrolytic activity, including peroxidases and esterases (Willemse and Bednara, 1979). Acid and alkaline phosphatases are not restricted to the degenerating megaspores, however, and are frequently observed but limited to specific compartments in the functional cell (Schulz and Jensen, 1986; Willemse and Bednara, 1979). The degenerate megaspores are initially similar to the megasporocyte, but eventually the breakdown of materials becomes general within the cell as vital intracellular membrane systems are broken down. Eventually, the plasma membrane is lost and lytic activity becomes general (Cass *et al.*, 1985; Russell, 1979; Schulz and Jensen, 1986). Presumably, the degraded products of these abortive cells are ultimately resorbed, as are those of the surrounding cells of the nucellus (Russell, 1979; Schulz and Jensen, 1986). Unlike the breakdown of nucellar cells, however, the degenerate megaspores and dyad cells rapidly abort prior to the encroachment of the developing megagametophyte (Rodkiewicz, 1978). In tetrasporic development, in which none of the meiotic products abort, callose is absent from the cell wall (Rodkiewicz, 1970) and plasmodesmata are principally restricted to the chalazal cell wall, remaining functional until the completion of meiosis (Rodkiewicz, 1978).

Willemse (1981) proposed that nuclear behavior is one of the principal factors determining polarity during megagametogenesis. In monosporic development nuclear position is typically chalazal to the center, establishing an asymmetric position for dyad cell wall formation. The shape of the megasporocyte and megaspore is likely imposed, to some extent, by the surrounding nucellus cells. That biophysical pressure contributes to the development of polarity (Lintilhac, 1974a,b) is another possible factor that should not be ignored. According to Lintilhac (1974b), cells located in areas of null stress may be developmentally predisposed to depart from the somatic program. Because there is no predetermined orientation by

which cell plate orientation would occur after division, the megaspores undergo free nuclear division, dividing mitotically without accompanying cytokinesis. Thus, megaspores outside of the biophysical focal point may abort, whereas the functional megaspore undergoes cenocytic development forming the megagametophyte.

The organization of the tetrad is subject to considerable variability in flowering plants. In the order Leguminales, Rembert (1971) described 12 general patterns of tetrad organization, based on cell wall orientation (including linear, T-shaped, oblique, or bilateral), arrest of cell division at the dyad stage (triad formation), and the location of the functional megaspore. Interestingly, he found some phylogenetic cogence in the patterns. Whether this may form the basis of phylogenetic conclusions at this time is unclear (Tobe, 1989). Webb and Gunning (1990) have apparently described a new organization of meiotic products in *Arabidopsis* in which the tetrad is decussate, forming tetrahedrally placed meiotic products.

In the resulting functional megaspore of all of the different patterns of development, the mitochondria and plastids redifferentiate, the ultrastructural appearance of the cytoplasm reflects increased synthetic activity, the nucleolus enlarges, cytoplasmic ribosomes accumulate, organelles multiply, and cells enlarge dramatically (Eyme, 1965; De Boer-de-Jeu, 1978; Russell, 1979; Schulz and Jensen, 1986). The cytoplasm becomes dense with ribosomes at this stage (Cass *et al.*, 1985; Schulz and Jensen, 1986). Storage products may also reappear (Willemse and Franssen-Verheijen, 1978), including lipids and starch grains (Willemse and Bednara, 1979). In *Capsella* small microvillus-like projections of the plasma membrane become conspicuous and appear to facilitate the absorption of material from abortive megaspores (Schulz and Jensen, 1986). Small inclusions within the callosic cell walls are conspicuous at this stage and may reflect the mode of cell wall development (Schulz and Jensen, 1986) or an alternative mechanism of absorption (Russell, 1979). Abnormalities in embryo sac development, such as those causing female sterility in soybean, first become evident at this stage (Benavente *et al.*, 1989), suggesting that there are crucial changes in physiology accompanying the redifferentiation of the functional megaspore.

C. Megagametogenesis and Cytoplasmic Maturation

Megasporogenesis ends with the mitotic division of the megaspore nucleus in monosporic plants; however, the megagametophyte in tetrasporic plants contains at least four nuclei at this stage and is therefore two mitotic divisions more advanced than plants with a single functional megaspore

nucleus. Monosporic megaspores require three cycles of mitotic division, bisporic megaspores require two cycles of division, and tetrasporic megaspores may require as few as one cycle to reach their mature nuclear complement (Maheshwari, 1950). Although in monosporic and bisporic types of megagametophyte development, a significant amount of the megasporocyte cytoplasm will abort, tetrasporic megagametophytes retain the entire cytoplasmic volume of the cenomegaspore as they enter megagametogenesis, and appear precocious in terms of size, degree of enlargement, and cytoplasmic maturation (De Boer-de-Jeu, 1978).

Several conspicuous changes occur during cytoplasmic maturation, including the inflation of the central vacuole, the replication of heritable organelles, one or more cycles of mitotic division, cellularization, and the differentiation of individual gametophytic cells (Kapil and Bhatnagar, 1981; Willemse, 1981). Central vacuole formation typically is initiated by the production of small vacuoles from ER and *de novo*. Numerous small vacuoles coalesce to establish the large central vacuole that is characteristic of maturity. In monosporic plants the formation of the central vacuole may become well established at the 2-nucleate stage, but in tetrasporic types the central vacuole usually forms just after the migration of the megaspore nuclei to their characteristic developmental position (i.e., $2 + 2$, $1 + 3$, or $1 + 1 + 1 + 1$, as mentioned; Maheshwari, 1950). Expansion of the megagametophyte coincides with the formation of a large central vacuole and may proceed in all directions around the immature gametophyte, encroaching on surrounding nucellar cells and crushing the remnants of abortive meiotic products (Willemse and Van Went, 1984). In the *Polygonum*-type embryo sac, coalescence of small vacuoles in the 2-nucleate embryo sac occurs after the first mitotic division, and in most types the prominent vacuole is central (Maheshwari, 1950). However, in the *Oenothera* type of development, vacuolization starts on chalazal side and pushes the nuclei of the cenocytic system to the micropylar pole of the young megagametophyte (Jalouzot, 1978); the mature embryo sac contains only four micropylar nuclei in this type: the egg cell, two synergids, and a single polar nucleus. In most plants the embryo sac still contains a prominent vacuole at maturity that is regular in position, but there are exceptions. In *Brassica* the mature central cell is devoid of the large central vacuole that occurs early in megagametophyte development (Sumner and Van Caeseele, 1990). The ultimate position of female gametophytic nuclei depends on the reorganization of the cytoplasm (Willemse and Van Went, 1984).

Although the 2-nucleate embryo sac appears to be involved in intense synthetic activity in *Zea* (Russell, 1979), the 4-nucleate stage appears to be a stage of rapid expansion in which the central vacuole enlarges dramatically and mitochondria undergo a postmeiotic cycle of dedifferenti-

ation and redifferentiation. This coincides with the completion of intense lytic activity in the central vacuole and may indicate the transition from vegetative to reproductive differentiation. In *Crepis* the number of organelles increases at the 2-nucleate stage (Willemse and Van Went, 1984), whereas in *Allium, Lilium,* and *Impatiens* (De Boer-de-Jeu, 1978) the number of organelles does not change until the 4-nucleate stage. In *Capsella* some mitochondria appear to survive in the functional megaspore, whereas others appear to degenerate (Schulz and Jensen, 1986). An association of ER with dictyosomes is the first step in the formation of the provacuole. In general, most of the cell organelles are produced before vacuolation (Willemse and Franssen-Verheijen, 1978). In *Gasteria* (Willemse and Franssen-Verheijen, 1978) an increased number of polysomes is also noted at the 4-nucleate stage. The abundance of RER and the large number of ribosomes in *Calendula* (Plisko, 1971) and *Zea* (Russell, 1979) suggest high levels of protein synthesis. In *Conium* (Dumas, 1978) abundant lipid bodies appear in the 2- and 4-nucleate megagametophytes, and in *Helianthus* strands of RER, more lipid droplets, and starch-containing plastids accumulate (Newcomb, 1973). During the development of the *Oenothera* type of embryo sac (Bednara, 1977), abundant RER appears in the functional megaspore, becoming dense at the 2-nucleate stage, and decreasing to previous levels at the 4-nucleate stage (Kapil and Bhatnagar, 1981). A large number of active dictyosomes are present at the periphery of the embryo sac and presumably are involved in the deposition of the embryo sac cell wall (Newcomb, 1973).

Free nuclear division without the formation of cell walls continues to occur until the mature complement of nuclei is achieved in the embryo sac. In most cases the mature embryo sac contains eight nuclei, but in many cases this is changed by the occurrence of strike or subsequent multiplication of specific cells, typically the antipodals (Maheshwari, 1950).

D. Cellularization of the Embryo Sac

Although no evidence of cell walls is observed at the 4-nucleate stage, cellularization proceeds rapidly enough at the 8-nucleate stage that no free-nucleate embryo sacs are normally evident. The most widely accepted view about the origin of the reproductively functional cells is that the two synergids are derived from the division of one of the two micropylar nuclei in the 4-nucleate embryo sac and that the egg and one polar nucleus are mitotic products of the other micropylar nucleus (Porsch, 1907). This view suggests a separation of reproductive and somatic developmental programs occurring at the 4-nucleate stage, with cellular differentiation

delayed until cytokinesis. There seems to be strong evidence that nuclear positioning is an important factor in cell determination during the maturation of the embryo sac (Herr, 1972; Ehdaie and Russell, 1984), and in *Glycine max* the determination of the egg and synergids is principally attributed to the chalazal movement of one of the two micropylar nuclei (Folsom and Cass, 1990). This chalazal movement has been observed in a number of angiosperms (Folsom and Cass, 1990) and suggests that this is not uncommon. In *Ranunculus scleratus,* however, both nuclei are reported to be at exactly the same level in the broad micropylar pole (Bhandari and Chitralekha, 1989), which complicates the identification of the immature egg and synergids in this plant. The disposition of the nucleus may also coordinate spindle orientation: the more micropylar of the two nuclei forms a transverse spindle, which is perpendicular to the long axis of the megagametophyte, whereas the other nucleus forms an oblique to longitudinal spindle. The cytological explanation for this behavior is unclear, but it may be an indirect effect of the placement of the large central vacuole, which appresses the cytoplasm to the periphery of the embryo sac and might limit microtubule polymerization to the edge of the embryo sac.

Langlet (1927) was among the first to present evidence that the more micropylar of the two nuclei was the source of the two synergids. The transverse spindle formed during its division is regarded as the origin of the synergid common wall. The chalazally displaced micropylar nucleus is regarded as the origin of the egg and micropylar polar nucleus. Following the most widely accepted concept of egg apparatus formation, Cass *et al.* (1986) and Folsom and Cass (1990) regarded the first vertical cell plate as the origin of the synergid common wall and the synergid nuclei as sisters. The longitudinal to oblique spindle is the origin of the cell wall that divides the micropylar end of the embryo sac from the central cell (Cass *et al.,* 1985, 1986; Bhandari and Chitralekha, 1989; Folsom and Cass, 1990). This cell wall is called the egg apparatus top wall (Folsom and Cass, 1990) and forms a single chalazal wall over the unexpanded egg and two synergids (Fig. 2A). Apparently, microtubules remain present between the sister nuclei and directly form the vertical and horizontal phragmoplasts (Cass *et al.,* 1986) (Fig. 2B); however, they found no evidence for microtubules remaining between the nonsister nuclei from the previous cycle of division. Thus, there is no direct evidence for the origin of the phragmoplast between the egg cell and synergids.

The completion of cell wall formation between nonsister nuclei has been the source of conflicting reports. Because there are four nuclei present in the micropylar end of the embryo sac at this stage, as many as four cell plates might be organized to complete cell wall formation. However, at most three cell plates have been described, with one of the cell walls

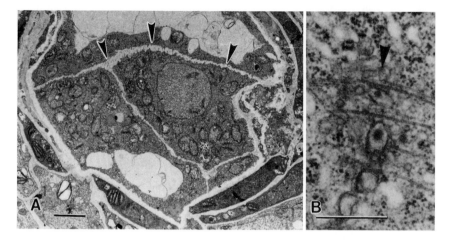

FIG. 2   Cellularization of the megagametophyte. (A) Formation of the egg apparatus of soy-bean, *Glycine max,* soon after cell wall formation. The egg apparatus top wall (arrow-heads) forms a common chalazal wall over all of the cells of the egg apparatus. Bar, 2 μm. Reproduced with permission from Folsom and Cass (1990). (B) Phragmoplast between cells of the egg apparatus during wall formation in barley. Arrowhead indicates vesicle associated with a microtubule in the phragmoplast. Bar, 0.5 μm. Reproduced with permission from Cass *et al.* (1985).

originating between the sister nuclei, curving to completely enclose the cytoplasmic domain of one of the sister nuclei from the nonsister nucleus. Cass *et al.* (1985, 1986) reported that in barley the cell plate forming the egg apparatus top wall bifurcates and extends obliquely toward the micropylar end of the embryo sac to form the lateral walls of the egg cell; in this model a separate third cell plate is not formed. Bhandari and Chitralekha (1989), however, reported that in the egg apparatus of *Ranunculus* a curved vertical cell wall forms between the two sister nuclei and that a separate vertical cell plate forms between the nonsister nuclei just after the formation of the first two cell plates (Fig. 3). In either model the formation of the cell wall around the nonsister nucleus depends on the position of the other nuclei—either directly because of nuclear positioning or indirectly because of the position of cell plates. Whether cell wall formation around the nonsister nuclei is initiated in continuity with the existing horizontal phragmoplast (Cass *et al.,* 1986); or whether a separate third cell plate intervenes between nonsister nuclei (Bhandari and Chitralekha, 1989) needs to be reinvestigated in detail using other approaches and possibly serial TEM reconstruction (Folsom and Cass, 1990). In microsporogenesis cell walls formed in certain monocots without prepro-phase bands appear to have their plane of division determined by the

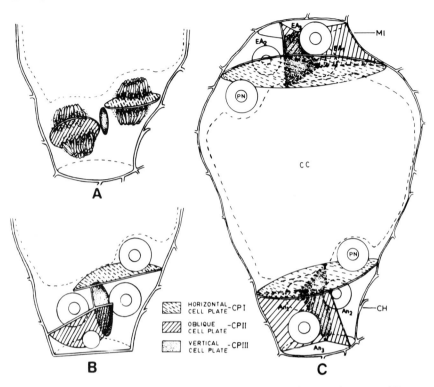

FIG. 3 Diagrammatic reconstruction of phragmoplast formation in the embryo sac of *Ranunculus scleratus*. Three cell plates (CPI–CPIII) are formed in this model. An$_1$, An$_2$, An$_3$, antipodal cells; CC, central cell; CH, chalazal region; EA$_1$, EA$_2$, EA$_3$, egg apparatus cells; MI, micropylar region; PN, polar nucleus. Reproduced with permission from Bhandari and Chitralekha (1989).

formation of cytoplasmic domains that delimit the range of cytoplasm ruled by a specific nucleus; these also appear to determine the future location of the cell wall in this circumstance (Brown and Lemmon, 1991). Cellularization in embryo sacs also appears to occur in the absence of preprophase bands and should be reinvestigated to determine how cell plate orientation is controlled. Approaches including tubulin localization will likely be required to visualize how the phragmoplast is oriented during megagametophyte cellularization and may be necessary to resolve remaining problems in this area.

Cellularization at the chalazal pole is simultaneous with that in the micropylar pole and is mentioned in varying detail in the reports. Bhandari and Chitralekha (1989) reported that three cell plates form the antipodals at the chalazal pole in much the same manner as egg apparatus cells form

at the micropylar pole. The two chalazal nuclei of the 4-nucleate embryo sac divide to form three antipodal cells and the chalazal polar nucleus. Although Folsom and Cass (1990) did not report in detail on the mode of cellularization in the antipodals of *Glycine*, they noted that the occurrence of binucleate antipodal cells could be attributed to the failure of a cell plate to bifurcate during the partitioning of antipodal cells in the *Asteraceae*.

After the completion of cell wall formation, the egg and synergids are small and wedge-shaped, with a flattened (Cass *et al.*, 1986; Folsom and Cass, 1990) or a slightly indented chalazal wall at the egg apparatus common wall where the central vacuole encroaches into the micropylar pole (Bhandari and Chitralekha, 1989). At this phase the egg and synergid cells alike expand toward the chalaza, thereby stretching their component cells until their chalazal cell walls become discontinuous and bead-like in nature, ultimately consisting of islets of cell wall material separated by segments of closely appressed plasma membranes. Initially, the cells of the immature egg apparatus are indistinguishable (Bhandari and Chitralekha, 1989). This similarity in egg apparatus cells is retained until transfer cell walls form in two of these cells, indicating the initiation of the filiform apparatus (FA) and identifying these cells as synergids (Cass *et al.*, 1986). The expansion of the egg apparatus cells changes the form of the cell walls of the egg and both synergids but does not alter their connection with the embryo sac wall, which usually remains perpendicular where the cells are attached. The apparent encroachment of the central cell along the sides of the synergids and egg, which is evident in many embryo sacs, appears to be the result of the differential chalazal expansion of egg apparatus cells with fixed sites of cell wall anchorage rather than by central cell migration. This is believed to form the developmental basis of so-called synergid and egg hooks (Cass *et al.*, 1986) or apical pockets. This dramatic expansion of egg apparatus cells is typical during maturation (Maheshwari, 1950). For example, during the maturation of the egg apparatus in *Spinacia*, Wilms (1981a) noted that the egg, synergids, and central cell expand by up to 10 times during maturation, whereas the antipodals remain essentially the same size.

Surprisingly few published studies have taken a quantitative approach to examining the development of the female gametophyte. Measurements of female gametophyte development were reported in *Glycine* and *Phaseolus* (George *et al.*, 1988), and in Japanese persimmon cv. *Nishmurawase* (Fukui *et al.*, 1989). The greatest increase in mean length occurs between the 2- and 4-nucleate stage in the greenhouse for *Phaseolus* and in the field for *Glycine* (George *et al.*, 1988). In Japanese persimmon rapid enlargement was reported throughout development from the megaspore stage to the completion of nuclear divisions in the embryo sac (Fukui *et al.*, 1989).

Stereological data are also rare. Apparently, the only detailed report is one concerning the relative organellar volumes in the cytoplasm in *Crepis*

tectorum, Epilobium hirsutum, and *Peperomia blanda* (each conforming to different embryo sac cell types) before and after fertilization (Bannikova *et al.,* 1987). This study revealed dramatic increases in the content of organelle levels after fertilization in *Peperomia* (which is relatively depauperate prior to fertilization), but no changes in organelles in *Crepis* and *Epilobium.* This study also indicates that each of the cells of the embryo sac is cytologically distinct and appears to differ from one plant to another among these disparate plants. The stereological approach allows judgments about cell organization to be quantitatively supported. One example of this application could be, for example, trying to identify the characteristics of cells with embryogenetic potential from those with deterministic developmental programs. Such a quantitative approach should be encouraged in future research.

A study of the exact position and variability of the embryo sac within the ovule in *Zea mays* indicates that the X (vertical) and Y (horizontal) axes (relative to the apical silk attachment point) vary by a standard deviation of 51 and 83 μm, respectively (Wagner *et al.,* 1990); this allows the average position of the embryo sac to be reasonably accurately determined. Although this degree of variability appears to be relatively small, the ranges observed in these plants (from a single hybrid line) exceeds the normal size of the embryo sac; this should indicate caution in the selection of external landmarks. The ultimate goal of *in situ* transformation of embryo sacs using microinjection within the ovule (Hepher *et al.,* 1985) will require detailed quantitative information on crop plants that is not yet readily available.

III. Organization of the Female Germ Unit

A. Concept of the Female Germ Unit

The female germ unit is comprised of the egg, two synergids, and the central cell. As such, it constitutes the minimum number of cells required to: (1) receive the pollen tube, (2) cause the discharge of the sperm into the receptive portion of the female gametophyte, and (3) undergo double fertilization. Originally proposed by Dumas *et al.* (1984), the term expands the concept of the egg apparatus; the egg apparatus concept recognizes the integral involvement of the synergids in egg function but does not recognize the essential role of the central cell, which forms the nutritive endosperm. Interestingly, the term egg apparatus (or Eiapparat) was coined at approximately the same time that the synergid was first recognized as a distinct cell type in the micropylar end of the angiosperm female gametophyte. The recognition that the synergids worked in cooperation

with the egg must have been conceived as early as when the cells were first named (Schacht, 1857) because the term "synergid" is derived from the Greek term "synergos" meaning "working together."

The validation of the egg apparatus by electron microscopy (Jensen, 1974; Willemse and Van Went, 1984) brought universal acceptance to the concept that the synergids worked in cooperation with the egg. The work of Jensen (especially, 1964–1974; Jensen and Fisher, 1968; Fisher and Jensen, 1969) developed a strong working model of the functional association of the egg and synergid and clearly stimulated the next generation of structural studies (Fig. 4). The question of how these cells collaborate to accomplish this function is still being actively investigated.

Numerous adaptations that promote successful double fertilization are evident in the female germ unit. These include the following: (1) the

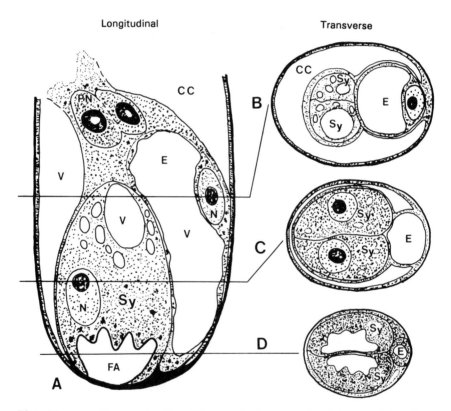

FIG. 4   Diagrammatic reconstruction of the organization of the female germ unit in cotton, *Gossypium hirsutum*, illustrating the relationships among the egg (E), synergids (Sy), and the central cell (CC). (A) Longitudinal view (micropylar region at base). (B–D) Transverse sections at selected levels of the egg apparatus. FA, filiform apparatus; PN, polar nucleus; V, vacuole. Modified with permission from Jensen (1965a).

removal of impediments for the movement of sperm cells within the female gametophyte, which is evident in the poorly developed and frequently absent cell walls near the sites of male gametic transfer; (2) the positioning and frequent proximity of female nuclei near the chalazal end of the egg apparatus; (3) the involvement of subcellular cytoplasmic elements, such as those of the cytoskeleton, in the establishment of polarity, guidance of sperm cells, and migration of male nuclei within the female reproductive cells; and (4) the chalazal migration and establishment of postfertilization patterns within the nascent zygote and endosperm. Similar to the male germ unit, in which the sperm physically associate with vegetative nucleus to form the functional unit of male DNA transmission (Russell, 1990), the nuclei of the female germ unit, also closely associate. After fertilization considerable changes in the female germ unit cause these cells to dissolve their association. New walls regenerate at the site of sperm fusion, where cell walls were originally absent, and the nuclei frequently become distant during their initial stages of development.

B. Organization and Ultrastructure of the Female Germ Unit

In most female germ units the two synergids and the egg cell are pyriform cells of approximately the same size and shape, with the synergids displaced noticeably toward the micropylar end of the embryo sac. The synergids are located at the extreme micropylar end of the embryo sac, which by proximity to the micropyle alone, would be expected to be the first embryo sac cells to encounter an entering pollen tube. The wall of the synergid, which is contiguous with the embryo sac wall at its micropylar end, is attached chalazally by a perpendicular cell wall at its boundary with the central cell—historically called the hook or apical pocket. The egg is attached laterally to the two synergids and frequently has a poorly accentuated point of attachment with the embryo sac wall that is located one-tenth to one-third of its cell length away from the micropyle. Because the egg and both synergids are approximately the same length, the egg is one-tenth to one-third of a cell length more chalazally immersed in the embryo sac than the synergids. In cross sections through the three cells, the egg and synergids are usually arranged in a triangular configuration, frequently with almost equal sectional profiles, despite significant cytoplasmic differences. These cells, as descendants of a single nucleus, are all of the same ploidy level as the megaspore nucleus from which they were obtained.

The central cell, in contrast, occupies most of the embryo sac, bordering with the egg and both synergids at its micropylar end and with the antipodal cells (when present) at its chalazal end. The nuclei that are initially present

within the central cell are termed polar nuclei and are variable in number depending on the developmental pattern of megagametogenesis (Maheshwari, 1950) and whether all nuclei divide predictably or participate in strike, causing some nuclei (typically at the chalazal pole) to remain undivided. The central cell nuclei may or may not fuse prior to fertilization when only two polar nuclei are present. In more complicated embryo sacs, however, with four or more polar nuclei (up to 14 in *Peperomia hispidula;* Johnson, 1914), the nuclei usually fuse prior to fertilization, forming a single so-called secondary nucleus. Frequently, polar nuclei are unfused or only partially fused before fertilization in monosporic embryo sacs, but usually are fused in bisporic and tetrasporic embryo sacs.

1. Synergid

The synergids are highly specialized cells of the egg apparatus that possess a distinctive cytoplasmic organization and cell wall modifications that attract pollen tubes, receive their contents, and promote the passage of male gametes into the egg cell and central cell (Fig. 5). Synergids also appear to be physiologically active cells involved in secretory, chemotropic, and transport functions within the embryo sac.

a. Cell Wall and Filiform Apparatus The function of the synergid is intimately related to the organization of the cell wall because it provides both constraints for the passage of gametes into and out of the cell and forms the external surface used for intercellular communication. At the micropylar end of the embryo sac, the wall is typically thickest and is formed in continuity with the embryo sac wall, providing access to the nucellar apoplast at this region. Typically occupying the extreme micropylar end of the mature synergid is a variably thickened area of proliferated cell wall known as the FA (Figs. 5 and 6). Because the plasma membrane is appressed to the inner surface of the FA, this increases the internal surface area of the synergid. The cell wall becomes thinner toward the chalaza and typically becomes discontinuous or disappears at the chalazal end, exposing the plasma membrane at interfaces with the egg and central cell (Jensen, 1974; Kapil and Bhatnagar, 1981; Willemse and Van Went, 1984).

Despite the apparent commonality in function, the morphology and structure of the FA in angiosperms is highly variable. Figure 6 illustrates the form of the FA in selected species of angiosperms grouped according to family. Dicot families are represented in the first three rows and monocot families on the bottom row. Although few closely related species have been examined, FA organization within a family appears to be similar with regard to mode of insertion of the FA in the embryo sac, continuity (or

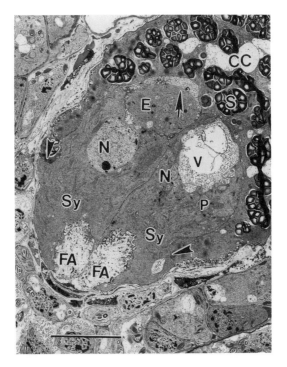

FIG. 5 Organization of the mature synergids (Sy) of soybean, *Glycine max,* prior to fertilization. A prominent filiform apparatus (FA) is evident at the micropylar end of each synergid,
whereas the nucleus (N) and vacuoles (V) are displaced to the chalazal end. Starch (S) is
abundant in the adjacent central cell (CC). Arrowheads indicate the so-called hook region of
the synergids and the arrow indicates the irregularly thickened wall at the apex of the egg
(E). Bar, 10 μm. Reproduced with permission from Dute *et al.* (1989).

lack of continuity) of the FA at the synergid common wall, complexity
and form of cell wall proliferation in the FA, and general extent and
proportion of the cytoplasm occupied by the FA.

The site of attachment of the FA ranges from broad continuity along the
length of the synergid common wall as in *Nicotiana* (Mogensen and Suthar,
1979), *Petunia* (Van Went, 1970a), *Proboscidea* (Mogensen, 1978a,b), and
Helianthus (Newcomb, 1973; Yan *et al.,* 1990, 1991), to one that is formed
at the micropylar edge of the embryo sac and consists of massive proliferations of wall material into the synergid, as in *Brassica* (Sumner and Van
Caeseele, 1989), *Capsella* (Schulz and Jensen, 1968), cotton (Jensen,
1965a), *Populus* (Russell *et al.,* 1990), and *Spinacia* (Wilms, 1981b). Figure
6 illustrates various forms of FA. The emergent FAs of *Jasione, Torenia,*
and *Saintpaulia* are exerted from the integuments and exposed directly to

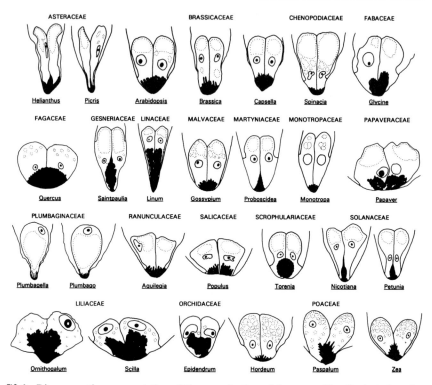

FIG. 6   Diagrammatic representation of the organization of the synergids of selected angiosperms, grouped by family and the eggs of two synergid-lacking angiosperms (*Plumbago* and
Plumbagella). Dicots are located in the top three rows and monocots in the bottom row. The
disposition of the filiform apparatus (black), nucleus (circular lines), and vacuoles (stippled
lines) is indicated. (Different species are not drawn to scale.)

the interior of the ovary; interestingly, these all appear to have essentially
nonfiliform elaborations and tend to be homogenous in structure (Berger
and Erdelská, 1973; Van der Pluijm, 1964; Mogensen, 1981a). In the FA
of *Glycine max* (soybean), different patterns of attachment were reported
in different cultivars (Folsom and Peterson, 1984; Folsom and Cass, 1990);
in var. Gnome the FA proliferates from a restricted region of the synergid
common wall, whereas the FA in var. Bragg is attached to both embryo
sac wall and the synergid common wall. This variability appears to indicate
that the function of the FA can be met by a variety of patterns even at a
cultivar level.

Although Maheshwari (1950) reported variability in the presence of the
FA in angiosperms, it has proven to be present in most species. Successful
detection appears to be technique-dependent. Originally observed as early

as 1856 by Schacht, the existence of the FA was at first contested; only in 1877 when Strasburger used tantalum album did he report that he was able to observe the FA (Habermann, 1906). In light microscopy the synergid appears to be "striated" at the micropylar end as in *Paspalum* (Chao, 1971, 1977) or has "well-developed and multiseriate" striae as in *Acer* (Haskell and Postlethwait, 1971). Paraffin histology and ovule clearings are unreliable techniques for reporting exceptions (Kennell and Horner, 1985); however, ultrastructural observations suggest that a classical FA may be absent from the synergids of some members of the family *Asteraceae,* including *Calendula officinalis, Cichorium intybus, Crepis tectorum, Picris echioides* (Godineau, 1969), and *Crepis capillaris* (Kuroiwa, 1989). The form of these thickenings in *Helianthus* led Newcomb (1973) to state that the thickenings were hardly filiform or thread-like and, indeed, appear to be principally thickenings of the synergid common wall. This wall lacks the increases in surface area usually associated with cell wall ingrowths but may still serve the typical functions of the FA and therefore should be regarded as such. The essentially universal occurrence of the FA and its restriction to synergids has suggested numerous functional attributes, including specializations for pollen tube reception, nutrition of the egg apparatus, and chemotropic attraction of the pollen tube. Usually, any transfer cell walls present in the synergid are restricted to the FA; however, in *Nicotiana* an unusual transfer cell wall develops attached to the outside of the synergid cell wall in the hook region penetrating into the central cell (Mogensen and Suthar, 1979).

Developmentally, the FA appears to represent an area of cell wall accretion that is deposited on previous layers of FA. Usually, the base of the FA is restricted to that area in contact with the embryo sac wall (Folsom and Cass, 1989). In *Torenia* (Van der Pluijm, 1964) and in other plants having an FA with a poorly developed basal region, the FA develops through a progressive thickening of the micropylar synergid common wall. Although it appears that the FA is generally secreted by dictyosomes (Sumner and Van Caeseele, 1989), the mature FA may contain numerous forms of inclusions. The completion of the FA is one of the last features of the synergid to develop, according to the most detailed structural studies of late megagametogenesis (Cass *et al.,* 1985, 1986; Folsom and Cass, 1990). If this wall is formed in response to solute fluxes, as in typical transfer cells (Gunning and Pate, 1969), the transport of solutes in the synergid is acquired late in maturation.

The FA is frequently observed as a biphasic structure in TEM preparations of *Aquilegia, Beta, Capsella, Epidendrum, Gossypium, Ornithogalum, Petunia, Zea,* and numerous other flowering plants, but the details of their organization differ in each species. In all of the plants listed except *Aquilegia,* the outer phase is electron-lucent and the inner phase is

electron-dense and more organized; however, in *Aquilegia* this pattern is reversed (Vijayaraghavan *et al.*, 1972). The description of the FA in cotton represents one of the most detailed available, reporting the presence of included fibrils, tubules, osmiophilic materials, differing phases of consistency, and even apparent plasmodesmata within the FA (Jensen, 1965a). In *Capsella*, which is typical of a number of plants, the inner phase of the FA consists of a tightly stacked mass of microfibrils and the peripheral layer contains fewer loosely organized microfibrils in a translucent matrix (Schulz and Jensen, 1968). Using light microscopy combined with the periodic acid-Schiff's (PAS) reaction for insoluble polysaccharides, *Scilla* is reported to have a triphasic structure consisting of a thin outer layer, less dense middle layer, and intensely staining inner layer (Bhandari and Sachdeva, 1983). In *Nicotiana* the FA is composed of a single phase with a heterogenous construction, containing lipids, aligned fibrils, and a generally electron-lucent matrix (Mogensen and Suthar, 1979).

In contrast, *Torenia* has an essentially homogenous FA, as do a number of Asteraceae (Godineau, 1969; Newcomb, 1973; Kuroiwa, 1989). Previous descriptions of the FA observed in chemically fixed tissues, however, may be significantly altered during preparation. Using freeze-substituted material, the biphasic structure of the transfer cell walls in the sporophyte–gametophyte transition in mosses was much better preserved than with comparable chemically fixed transfer cell walls (Browning and Gunning, 1977); these showed signs of considerable swelling and suggested that the electron-lucent layer should not exist in well-fixed materials. The most accurate descriptions of FA organization will likely require such methods.

The histochemistry of the FA differs among the angiosperms studied, but all FAs have a PAS-positive component. The principal component is believed to be hemicellulose in *Paspalum* (Chao, 1971) and *Plumbago* (Russell, unpublished results). Numerous reports indicate that the FA contains acidic and insoluble polysaccharides (*Zea*, Diboll and Larson, 1966; *Brassica*, Sumner and Van Caeseele, 1989), carbohydrates and pectin (cotton, Jensen, 1965a), and Calcofluor white-positive β-linked glucans or polyglucans (*Brassica*, Sumner and Van Caeseele, 1989; *Glycine*, Folsom and Cass, 1990; *Plumbago*, Huang *et al.*, 1990). In *Brassica* (Sumner and Van Caeseele (1989), Calcofluor white positivity could be eliminated by incubation in cellulase, suggesting that cellulose is a principal component in the FA of this plant; the additional presence of α-1,4-polyuronides in the FA is indicated by staining with alcian blue. In *Petunia* FA material subjected to 15 minutes in 100°C acetic acid and hydrogen peroxide did not eliminate the fibrils and revealed that their distribution was random in surface replicas (Van Went, 1970a). Rifot (1972) and Fougère-Rifot (1975) reported that cellulose is not present in the FA of *Aquilegia*. Protein is a

frequent component of the FA and has been reported in *Agave, Aquilegia, Eschscholtzia, Gossypium, Ornithogalum, Paspalum, Proboscidea,* but not in *Capsella, Hordeum, Nicotiana,* and *Vanda.* Nucleic acids have been reported in the FA of *Ornithogalum* (Tilton, 1981); however, it is unclear whether these are localized in the FA or possibly in the fingers of cytoplasm that are intwined in the intricate FA of this plant. Lipid droplets have been reported in the FA of *Nicotiana* (Mogensen and Suthar, 1979) and *Triticale* (Hause and Schröder, 1986), and osmiophilic droplets are evident in numerous other plants. One of the most unexpected findings is that of intense aniline blue-induced fluorescence, indicative of callose, in the FA and micropylar cell wall of the young synergid of *Torenia* (Tiwari, 1982) and in the FA of *Gasteria* (Willemse and Franssen-Verheijen, 1988). The presence of callose is usually associated with incompatibility in pollen (Knox, 1984) and is indicative of zygotic abortion when present in the ovules of *Rhododendron* (Williams *et al.,* 1982). In *Torenia* callose may be an adaptation to the FA being exerted from the integuments and may represent a chemical barrier to the entry of materials from the ovarian chamber (Tiwari, 1982).

Most synergids lack a cell wall adjacent to the egg and central cell and thereby expose their plasma membranes directly to the female cells that ultimately fuse with the male gametes (Jensen, 1974; Kapil and Bhatnagar, 1981). Undoubtedly, this is an adaptation that facilitates fertilization once the male gametes are deposited in the synergid. Although there are exceptions to this rule, the common theme is that if a wall is present, it is depauperate in cellulose fibrils, or these are interrupted by spaces large enough to permit the passage of sperm. In the synergid of *Epidendrum,* a fenestrated chalazal wall is reported bordering the egg and central cells (Cocucci and Jensen, 1969a,b). In *Capsella* the synergid cell wall is thin and irregular prior to fertilization but continuous; however, PAS staining of the synergid reveals that the periodate-sensitive phase of this wall forms a honeycomb-like configuration (Schulz and Jensen, 1968). This cell wall apparently does not form a barrier to the passage of sperm cells, as the force of pollen tube (PT) discharge alone is believed to be enough force to displace the wall (Schulz and Jensen, 1968).

Plasmodesmatal connections within synergid cell walls are variable in number and position in the flowering plants examined to date. Plasmodesmata are reported in all female germ unit cell walls in *Agave* (Tilton and Mogensen, 1979), *Beta* (Bruun, 1987), *Brassica* (Sumner and Van Caeseele, 1989), *Capsella* (Schulz and Jensen, 1968), *Glycine* (Folsom and Peterson, 1984; Dute *et al.,* 1989), *Helianthus* (Yan *et al.,* 1991), *Spinacia* (Wilms, 1981b), and *Zea* (Diboll and Larson, 1966), with the exception of the external embryo sac wall. In *Glycine* many plasmodesmata are reported between the two synergids and the egg cell and each synergid,

but fewer between the synergids and central cell (*Glycine*, Folsom and Peterson, 1984; Dute *et al.*, 1989). In *Epidendrum* plasmodesmata are reported between the two synergids, but not with either the egg or central cell (Cocucci and Jensen, 1969a). Frequent plasmodesmatal connections are reported between the synergids and the egg cell in *Nicotiana* (Mogensen and Suthar, 1979), whereas in *Proboscidea* (Mogensen, 1978b) plasmodesmata are not found between the egg and synergids and are found only occasionally between the synergids and the central cell. In *Arabidopsis* (Mansfield *et al.*, 1991) plasmodesmata are not observed between the synergid and central cells, although synergid-to-synergid plasmodesmata are frequently observed in *Arabidopsis*, *Gossypium* (Jensen, 1965a), and in *Helianthus*, prior to the thickening of the synergid common wall (Newcomb, 1973). Only rarely are plasmodesmata observed in the synergids of *Petunia* (Van Went, 1970a), and none were observed in the synergids of *Crepis* (Kuroiwa, 1989), *Oryza* (Dong and Yang, 1989), and *Quercus* (Mogensen, 1972).

b. Cytoplasmic Features   The cytoplasm of the synergid is strongly polarized, with the nucleus usually located toward the center of the cell, and most of the vacuoles toward the chalazal end (Fig. 5). The nucleus tends to be smaller than the nuclei of the egg and central cells but is often larger than nucellar nuclei, is irregular in profile, and possesses a nucleolus. Synergids are typically rich in mitochondria, but they may be distributed differently. In some species, such as *Arabidopsis*, *Capsella*, *Gossypium*, *Triticale*, and *Zea*, there is a strongly polarized distribution of mitochondria with many located near the FA, whereas in *Petunia*, *Proboscidea*, *Spinacia*, and *Quercus* mitochondria are uniformly distributed. In all angiosperms studied to date, the abundance and well-developed nature of mitochondrial cristae suggest that the mature synergids are metabolically active in support of either a secretory or absorptive function for synergids.

In *Glycine*, *Gossypium*, *Ornithogalum*, and other plants containing elaborate or intricate FA margins, fingers of synergid cytoplasm may penetrate into the FA; although some of the synergid cytoplasm remains in continuity with the rest of the cytoplasm, further cell wall thickening in the FA may actually result in the incorporation of cytoplasmic structures as inclusions. The occurrence of synergid haustoria has been reported in the Asteraceae in *Calendula*, *Cortedaria*, *Cotula*, *Mutsia*, and *Ursinia*, and in *Quinchamalium* in the Santalaceae; in *Cortedaria* the micropylar end encroaches into the nucellus and contains numerous transfer cell walls (Philipson, 1981). The persistent synergid haustorium remains functional after fertilization and displays even more elaborate cell wall ingrowths after fertilization; it therefore appears to play a role in the nutrition of the embryo sac and the embryo.

   Plastids, ER, and dictyosomes all possess characteristic distributions as well. Plastids are abundant and contain starch in *Gossypium, Helianthus, Nicotiana, Scilla, Spinacia,* and *Zea,* less abundant with small or diminishing starch grains in *Arabidopsis, Quercus,* and *Spinacia,* and are without starch in *Aquilegia, Capsella, Epidendrum, Hordeum, Linum, Ornithogalum, Petunia,* and *Populus;* in *Agave* and *Stipa* plastids were numerous but rarely contained starch. Plastids are more abundant in the micropylar end in *Aquilegia, Beta, Crepis, Gossypium,* and *Ornithogalum* but occur more frequently in the chalazal end in *Agave* and *Glycine.* The occurrence of micropylar plastids often coincides with a micropylar nuclear position and, therefore, may be more appropriately characterized as perinuclear. Long segments of ER are commonly oriented parallel with the long axis and are particularly abundant in the synergids of *Aquilegia* (Rifot, 1972; Fougère-Rifot, 1975, 1989), *Brassica* (Sumner and Van Caeseele, 1989), *Conium* (Dumas, 1978), and *Plantago* (Vannereau, 1978), and are found in association with the FA. Groups of extensive stacked arrays of parallel ER are reported in *Aquilegia, Arabidopsis, Brassica, Capsella,* and other plants. This distribution is particularly evident in synergids stained with OsFeCN (Sumner and Van Caeseele, 1989). In *Capsella* numerous connections between the ER and plasma membrane are observed near the FA (Schulz and Jensen, 1968), further supporting a transfer cell function. Concentric lamellae of RER are frequently seen in the synergid cytoplasm in *Glycine* (Folsom and Peterson, 1984). Dictyosomes are frequently elaborate, containing abundant stacked cisternae and are associated with numerous vesicles (Kapil and Bhatnagar, 1981; Willemse and Van Went, 1984). Lipid droplets or spherosomes are reported in *Calendula* (Plisko, 1971), *Gossypium, Proboscidea,* and *Zea.* Small lipid bodies appear to be associated with the FA in *Zea* (Diboll and Larson, 1966) and *Paspalum* (Yu and Chao, 1979).

   The distribution of vacuoles in the synergid is typically chalazal. In some species there is usually just a single large vacuole (*Capsella* and *Quercus*), but in a number of grasses (*Hordeum, Paspalum, Stipa, Triticale, Triticum,* and *Zea*) and in *Brassica, Helianthus,* and *Petunia* vacuoles are small and, in some grasses, exceedingly numerous. Mixed large and small vacuoles are present in the synergids of *Arabidopsis, Epidendrum,* and *Nicotiana,* among other plants. Flocculent material has been observed in the synergid vacuole of a number of species (*Glycine,* Dute *et al.,* 1989; *Gossypium,* Jensen, 1965a; *Populus,* Russell *et al.,* 1990). In *Gossypium* the material is reported to contain protein and a substance (presumably calcium) that remains after microincineration (Jensen, 1965a), and after freeze-substitution a water-soluble polysaccharide is detected (Fisher and Jensen, 1969; Section IV,B,3).

   During the formation of the FA the presence of active dictyosomes associated with numerous vesicles and dilated ER has been described in

numerous plants (Kapil and Bhatnagar, 1981; Vijayaraghavan and Bhat, 1983). Dictyosomes are usually abundant around the FA and in the micropylar region, as in *Capsella, Glycine,* and *Ornithogalum* but appear random in *Aquilegia* and *Petunia.* ER was found to be a particularly dynamic structure during synergid development in *Aquilegia* and becomes progressively vesiculated during degeneration (Fougère-Rifot, 1989). Ribosomes are essentially always abundant and are frequently seen free or attached to ER. Numerous RER cisternae in continuity with the plasma membrane of the FA accentuate the polarity of the synergid. The release of PA-TCH-SP–positive vesicles from the maturing face of dictyosomes adjacent to the FA and the fusion of PA-TCH-SP–positive dictyosome-like vesicles at its expanding edge suggests that dictyosomes may be involved in the formation of the FA in *Brassica* (Sumner and Van Caeseele, 1989). The association of synergid ER, dictyosomes, and PA-TCH-SP–positive vesicles could also have a nutrient absorption function (Sumner and Van Caeseele, 1989). Although the proposed function of nutrient absorption is difficult to demonstrate, ER-dictyosome-vesicle associations are frequently observed during FA formation in *Helianthus* and *Gasteria* and are not as frequently observed at maturity. Whether the synergids function in a secretory or absorptive capacity may vary in different species. The presence of starch in the synergid suggests that starch may serve as a dynamic nutritional source in some angiosperm synergids, but this is by no means universal because many synergids lack starch. The pattern of cellular contacts and plasmodesmata, the presence or absence of appressed nucellar cells, the distribution of a cuticle around the embryo sac, and the presence of an integumentary tapetum or other specialized absorptive structures will all modify potential reliance on the synergid for embryo sac nutrition. Possibly, a universal function of the synergid, however, is the secretion of materials to attract the pollen tube and to signal embryo sac receptivity through the degeneration of one or both of the two synergids.

c. Histochemistry and Cytochemistry of Synergid   The results of PAS reactions in the synergid cytoplasm and in the FA (Section III,B,1,a) have been examined in a number of plants. A large number of histochemical studies show that the synergid cytoplasm is also rich in protein and RNA. The concentration of cytoplasmic RNA and protein in *Gossypium* (Jensen, 1965a) and *Vanda* (Alvarez and Sagawa, 1965) is higher in the proximity of the FA. The activity of a number of enzymes in mature synergids was reported in *Zephyranthes* and *Lagenaria* (Malik and Vermani, 1975). Intense cytochrome oxidase, succinate dehydrogenase, and phosphatase activity was also localized in synergids in the previous study. Membrane-associated ATPases failed to be localized in the FA of *Saintpaulia* and

instead line the plasma membrane of the integumentary tapetum in this plant. These findings indicate that integumentary cells nearest to the embryo sac appear to actively load metabolites and that the embryo sac cells are more passive than surrounding sporophytic cells in terms of active transport function, presumably passively secreting a pollen tube-attracting substance (Mogensen, 1981a).

Based on the structural and cytochemical information, the synergid cells may be considered to be transfer cells (Gunning and Pate, 1969). It seems probable that the FA is involved in transport into synergids and that through the connecting symplast of the egg apparatus and central cell, these transfer cells supply the egg, synergids, and central cell with a portion of the nutrition needed for growth and development (Folsom and Cass, 1986, 1990). The callose present at the micropylar end of embryo sac of *Torenia fournieri* acts as a barrier to the flow of metabolites between the exposed portion of embryo sac and its surroundings (Tiwari, 1982) and may explain the variable patterns of uranin uptake observed in living ovules of this plant (Mogensen, 1981b). In *Antirrhinum majus* (Yang, 1989) the wall of the embryo sac is acetolysis- and enzyme-resistant and is resistant to strong oxidizers, which suggests that the wall sac functions as a barrier to prevent damage of the young embryo and endosperm from hydrolytic enzymes secreted by the integumentary tapetum while presumably not acting as a barrier for nutrient flow through the chalazal and micropylar apertures into the embryo sac. A nutritional function is also proposed in the synergid of *Cortaderia* (Philipson, 1977, 1981), in which the tip of the synergids elongates and grows along and out of the micropyle. A similar absorption function has been attributed to the synergids of *Brassica, Capsella, Stipa, Spinacia,* and *Glycine.* Alternately, the primary function of the synergid of *Petunia* (Van Went, 1970a) was considered to be the secretion of a chemotropic attractant of pollen tubes. A secretory function has also been attributed to the synergids of *Agave, Brassica, Helianthus, Ornithogalum,* and *Spinacia.* Enzyme production and secretory activity of the synergid have been postulated in *Paspalum,* where it may synthesize at least some of a PAS-soluble material believed to attract pollen tubes (Chao, 1971, 1977). Secretory activity of the synergid is usually considered to be related to the chemotropically directed growth of the pollen tube within the ovary (Willemse and Van Went, 1984). A thick deposit of cuticle over the micropylar wall of the embryo sac in *Jasione montana* (Berger and Erdelská, 1973) suggests that the synergids are unlikely to be involved in the transport of nutrients in this plant. Another important role attributed to the synergid is the entrance and discharge of pollen tube into the embryo sac and the conveyance of the two sperm cells to fertilize with egg cell and central cell, respectively. The degenerated cytoplasm of synergid may arrest pollen tube growth, build up pressure

inside the tube, and ultimately force the discharge of pollen tube contents. Presumably, pollen tube penetration may be influenced by the area of least resistance to the tube tip; therefore, the entry of the pollen tube into the more degenerate of the two synergids seems likely (Russell *et al.*, 1990). In cross-sectional view, it is possible that closer examination will show that the tube grows along the edge of FA but does not penetrate the fibril-rich matrix prior to its termination within the degenerated synergid.

2. Egg Cell

The egg cell is usually somewhat similar in shape to the synergid but is attached to the micropylar megagametophyte wall slightly chalazal to the attachment point of the synergids. Frequently, the egg also contains a hook region or apical pocket of central cell extending micropylar to the attachment between the egg and embryo sac wall (Fig. 7). As discussed previously (Section III,B), the egg cell shares a common wall surface with the synergids on two sides and with the central cell toward the chalazal end. The egg cell is typically surrounded by a cell wall in the micropylar one-half to two-thirds of the cell, and the remaining chalazal portion has large areas of exposed plasma membrane next to the synergids and central cell, as in *Agave, Beta, Brassica, Calendula, Cichorium, Crepis, Gossypium, Helianthus, Linum, Nicotiana, Oenothera, Oryza, Petunia, Picris, Proboscidea, Quercus, Spinacia, Stellaria, Stipa, Torenia, Triticale,* and *Triticum* (Kapil and Bhatnagar, 1981; Tilton, 1981; Willemse and Van Went, 1984; You and Jensen, 1985; Hause and Schröder, 1986; Bruun, 1987; Dong and Yang, 1989; Sumner and Van Caeseele, 1989). In some plants all of the cell walls of the egg apparatus were stained positively by PA-TCH-SP. Staining with PAS and alcian blue suggests the presence of substances with vicinal hydroxyl groups and acidic polysaccharides (Sumner and Van Caeseele, 1989). Exceptions, in which the egg apparatus is entirely enclosed by a PAS-positive wall, have been reported in *Capsella* (Schulz and Jensen, 1968), *Epidendrum* (Cocucci and Jensen, 1969b), *Papaver* (Olson and Cass, 1981), *Ornithogalum* (Tilton, 1981), and *Scilla* (Bhandari and Sachdeva, 1983). In some plants characteristic dense bodies are present at irregular intervals within the cell wall of the egg and central cell as in *Arabidopsis, Beta, Capsella,* and *Glycine;* these typically appear rectangular in profile in the former plants and elliptical in *Brassica*. That these are specifically localized between the egg and central cell is intriguing, but no specific functional significance is known for these regions during the fertilization process.

In most species the polarity of the egg is inverted compared with that of the synergid. The nucleus is frequently located at the chalazal end or the center of the egg cell. Two different patterns seem to predominate:

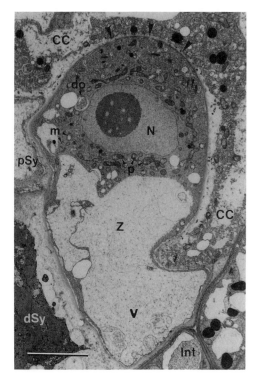

FIG. 7   Newly fertilized egg (Z) showing the typical features of the egg of cottonwood, *Populus deltoides*. Plastids (p) and mitochondria (m) are frequently found at the chalazal end of the egg near the nucleus (N), opposite the highly vacuolate (V) micropylar end. The apex of the egg essentially lacks a cell wall at the time of fertilization (arrowheads), forming a region adjacent to the central cell where the sperm cells may enter. An invagination of central cell (CC) forms an egg hook at the edge of the megagametophyte on the right side. do, degenerating organelle; dSy, degenerated synergid; pSy, persistent synergid. Bar, 5 μm. Reproduced with permission from Russell *et al.* (1990).

(1) In the case of a micropylar or central vacuolated region, the egg nucleus is located chalazally (Fig. 7), as, for example, in *Agave, Arabidopsis, Beta, Brassica, Capsella, Gossypium, Helianthus, Linum, Monotropa, Nicotiana, Ornithogalum, Papaver, Populus, Probosicidea, Quercus, Spinacia, Torenia* (Kapil and Bhatnagar, 1981), and others visualized at the light microscopic level (Maheshwari, 1950; Davis, 1966); the vacuolated region may consist of a single central vacuole or a larger number of smaller vacuoles toward the micropylar pole. (2) In the case that the nucleus occupies the central area, typically it is surrounded by numerous, small vacuoles located peripherally, and sometimes with a chalazal or micropylar size gradient, as in the grasses studied to date (*Hordeum, Oryza,*

Stipa, Triticale, Triticum, and *Zea*), and some other angiosperms (e.g., *Epidendrum* and *Paspalum*).

The egg nucleus is characteristically large and has a single large nucleolus, with some species displaying conspicuous nucleolar vacuoles, namely, *Helianthus* (Newcomb, 1973), *Hordeum* (Cass and Jensen, 1970), *Jasione* (Berger and Erdelská, 1973), *Stellaria* (Pritchard, 1964), and *Stipa* (Maze and Lin, 1975). The nucleolus of *Capsella* is organized with an interior of loosely packed amorphous material interwoven with fibrous matter with an exterior phase consisting of dense, tightly packed granules of approximately the same size as ribosomes (15–20 nm) (Schulz and Jensen, 1968). A biphasic nucleolus was also reported in *Petunia* (Van Went, 1970b) and *Epidendrum* (Cocucci and Jensen, 1969a). The dense fibrillar phase is now believed to be the site of RNA molecules in the process of transcription, whereas the granular component is believed to contain maturing ribosomal precursor particles (Alberts *et al.,* 1989). Based on the size of the two phases, this interpretation would indicate that the nucleolus is active in these plants, although the reduced granular phase in *Capsella* suggests that few ribosomal aggregates are actively maturing in the unfertilized egg (Schulz and Jensen, 1968), or they are being rapidly dispersed. The large size of the nucleoli in the egg cell may indicate that rDNA amplification as occurs in animal oocytes (Alberts *et al.,* 1989) may also be present in plants. A high density of cytoplasmic ribosomes has been reported in all flowering plant eggs examined to date (Kapil and Bhatnagar, 1981). Ribosomes may be either free or attached, but the scarcity of polysomes presumably indicates that ribosomes are rarely involved in active protein synthesis in the mature unfertilized egg.

The organization and ultrastructure of the egg cytoplasm vary considerably among the different species examined. Although most are depauperate of dictyosomes, there is variability in the distribution and frequency of plastids, mitochondria, ER, and lipid bodies. In *Brassica* the dictyosomes in the egg consisted of 5–7 flattened cisternae that do not appear active in vesicle production (Sumner and Van Caeseele, 1989); it seems likely that this is the case in most mature unfertilized egg cells. The plastids in the egg cell are usually perinuclear, variable in size and shape, and usually contain rudimentary lamellae. Starch grains have been reported in the plastids of most plants observed to date, including *Beta, Capsella, Gossypium, Hordeum, Linum, Nicotiana, Oryza, Persea, Plumbago, Populus, Quercus, Scilla, Spinacia, Stellaria, Stipa, Triticale, Triticum,* and *Zea;* starch may be variable in *Petunia* and is absent in *Agave* and *Ornithogalum.* In *Glycine max* starch grains are reported in egg plastids of cv. 'Gnome (Folsom and Cass, 1989), but are absent in egg plastids of cv. 'Bragg (Folsom and Peterson, 1984). If starch in the egg is a dynamic storage material, however, both the amount and presence of starch may

vary on a diurnal cycle or with other changes in light or culture conditions. In *Impatiens glandulifera* there are reportedly 150–480 plastids in the egg (Richter-Landmann, 1959), whereas in *Plumbago zeylanica,* in which the egg is considerably larger, there is an average of 730 plastids (Russell, 1987). *Oenothera erythrosepala,* however, which displays frequent biparental inheritance, contains 25–32 plastids (Meyer and Stubbe, 1974). Egg mitochondria tend to be spheroidal to roundly ellipsoidal, perinuclear in distribution, and abundant in all the species studied, except *Petunia* and *Agave* (Tilton, 1981). In *Spinacia* the number of mitochondria is reported to increase significantly before fertilization (Wilms, 1981a,b,c). The number of mitochondria in the egg of *Impatiens glandulifera* is 1000–2500, whereas *Plumbago zeylanica* contains approximately 39,900 ± 3100 mitochondria. Frequently, the mitochondria are reported to have few cristae. ER is typically relatively scarce in *Arabidopsis, Brassica, Capsella, Gossypium, Helianthus, Linum, Nicotiana, Oryza, Petunia, Triticum,* and *Zea.* In *Beta* and *Glycine,* however, ER is relatively abundant and in *Epidendrum* ER is commonly appressed to the plasma membrane. Lipid bodies are infrequent but present in *Beta, Capsella, Glycine, Linum, Nicotiana, Triticum,* and *Zea,* and are reported to be absent at maturity in *Oryza.*

Based on ultrastructural characteristics, the typical unfertilized egg cell appears to contain the potential for a highly synthetic cellular physiology but appears to be quiescent in comparison with the synergid. The infrequency of dictyosomes and dictyosome vesicles clearly suggests that they are not involved in significant amounts of either uptake or secretion in the mature egg prior to fertilization. The egg cells of *Capsella, Gossypium, Nicotiana, Petunia, Stellaria, Stipa,* and *Zea* are proposed to be relatively inactive cells because of their highly vacuolated cytoplasm, mitochondria with a few and short cristae, reduced numbers of dictyosomes and plastids, poorly developed ER, and the scarcity of polysomes. The eggs of *Beta, Glycine,* and *Helianthus* appear to be unusually active, containing larger amounts of ER, more dictyosomes, and greater ultrastructural evidence of activity. In *Nicotiana alata* cytoplasmic movement is conspicuous in isolated synergids and central cells, but isolated egg cells display significantly less movement (Huang *et al.,* 1992), thus confirming reports of quiescence in the egg cell.

The organization of the embryo sac of *Plumbago* (Cass, 1972, Cass and Karas, 1974, Russell, 1987, Huang *et al.,* 1990) and *Plumbagella micrantha* (Russell and Cass, 1988) appears to represent a remarkably reduced form, in which synergids are lacking and the egg alone occupies the micropylar end of the embryo sac, combining the functions of the synergid with the egg. The occurrence of synergid-lacking embryo sacs in angiosperms appears to be restricted to the *Plumbago* and *Plumbagella* types of devel-

opment (principally in the Plumbaginaceae). The egg of these plants represents the only example of the occurrence of an FA in any cell other than a synergid. The FA of *Plumbago* displays a biphasic fibrillar organization that is similar to the FA in synergids (Section III,B,1,a). The external phase is electron-lucent and amorphous and the internal phase contains abundant aligned fibrils (Cass and Karas, 1974) that are Calcofluor white-positive, indicative of β-1,4-polyglucan linkages (Huang *et al.*, 1990). These fibers are presumably cellulosic and appear to be oriented during development through the action of longitudinally aligned cortical microtubules (Cass and Karas, 1974). At the chalazal pole the cell wall is continuous and the constitution of the wall is amorphous. Interestingly, the egg–central cell boundary contains similar dense bodies to those reported in other flowering plants. These structures are reportedly associated with microtubules that appear to maintain the distance between cells at a regular distance, as has been described at the junction of endodermal cells (Cass and Karas, 1974). These dense bodies may stabilize the egg–central cell boundary during fertilization, holding both female cells in proximity with the deposited sperm cells during recognition (Russell, 1983). Observations using intermediate voltage electron microscopy (IVEM) of whole-mounted isolated embryo sacs indicate that the cell is considerably more electron-dense than the other cells (Russell *et al.*, 1989); this reflects in part the slightly thicker cytoplasmic layer of the egg and the greater density of its cytoplasmic contents. The egg of *Plumbago* (Cass and Karas, 1974; Russell, 1987) contains numerous mitochondria, plastids, RER lamellae, and dictyosomes and, more similar to the synergid, its ultrastructural appearance reflects a cell with an active physiological role.

3. Central Cell

The central cell is the largest cell within the embryo sac, occupying more than 75% of the volume of the gametophyte in some species (Russell, 1987). This cell shares common walls with the micropylar egg apparatus at one end of the embryo sac and the chalazal antipodal cells at the other. Typically, the central cell is highly vacuolated at maturity, containing a prominent central vacuole and a peripheral cytoplasm that is connected with the perinuclear region by transvacuolar strands. Small vacuoles are reported in the exceptionally reduced central cells of the orchid *Epidendrum* (Cocucci and Jensen, 1969a) and *Monotropa* (Olson, 1991), which produce among the smallest seeds in angiosperms, and in *Brassica,* which contains unusually dense food stores (Sumner and Van Caeseele, 1990) (Fig. 8). The organization of the transvacuolar strands observed using IVEM indicates that these emerge in an arborescent pattern from the perinuclear region and that individual strands differ in form, ranging from

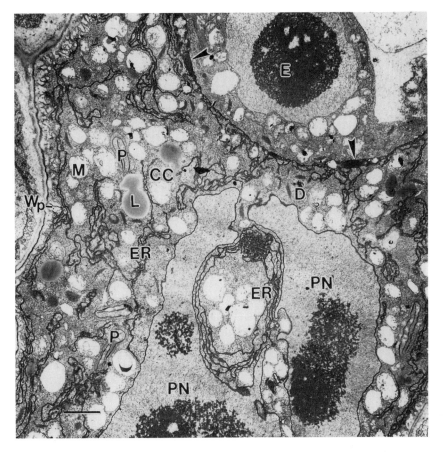

FIG. 8 Organization of the central cell (CC) and adjacent egg (E) in canola-rapeseed, *Brassica campestris,* during apposition of the two polar nuclei (PN), which do not complete fusion until fertilization. Typical organelles include mitochondria (M), plastids (P), lipids (L), and dictyosomes (D). Endoplasmic reticulum (ER) is prominent in the cytoplasm and in association with the cell wall, polar nuclei, and egg cell. Between the wall-lacking area of the egg and central cell are electron-dense bodies characteristic of this region (arrowheads). Wall projections (Wp), suggesting a transfer cell-like function, are present at the edge of the embryo sac. This material was prepared using the iron ferricyanide-osmium tetroxide method, which, interestingly, enhances membranes and ER in the central cell but not the egg. Bar, 2 μm. Reproduced with permission from Sumner and Van Caeseele (1990).

slender tubes to large and fenestrated sheets containing numerous organelles (Russell *et al.,* 1989). Unlike the other cells of the embryo sac and female germ unit, this cell is almost always multinucleate in flowering plants; the main exception is the *Oenothera* type of embryo sac development, in which only the micropylar quartet of mitotic products functions and the

central cell is uninucleate from inception (Maheshwari, 1950). The nuclei typically originate from the poles of the cell and they therefore have been called polar nuclei since the earliest accurate descriptions of megagameto-genesis. The number and disposition of these nuclei depend on the species and the pattern of embryo sac development displayed. Monosporic and bisporic embryo sacs, normally contain only two such polar nuclei, and these may or may not fuse prior to fertilization to form a diploid nucleus. In tetrasporic development, however, the number of nuclei ranges from two, as in the *Adoxa, Drusa, Fritillaria,* and *Plumbagella* types, to as many as 15 haploid nuclei in *Peperomia hispidula* (Johnson, 1914). These nuclei frequently fuse before fertilization, forming a diploid nucleus in typical *Adoxa* and *Drusa* types, a tetraploid nucleus in typical *Fritillaria, Penaea, Plumbago,* and *Plumbagella* types, an octaploid nucleus in the classical *Peperomia* type, and up to a 15-ploid nucleus in *Peperomia hispidula* (Maheshwari, 1950).

In most angiosperms the polar nuclei approach each other during matu-ration and eventually begin to fuse prior to fertilization. In *Gossypium* the first contact between the nuclei is through ER membranes that are continuous with the outer layer of the nuclear envelope. As the nuclei converge, the outer membrane of the nuclear envelope fuses next forming a membrane bridge, followed by the fusion of the inner envelope membrane. Ultimately, slender connections of nucleoplasm are formed (nuclear brid-ges), which gradually enlarge and coalesce until the nuclei are finally fused (Jensen, 1964). Frequently, this process traps small segments of cytoplasm, which are expelled later during the division of the primary endosperm nucleus (Jensen, 1974). This mechanism has also been de-scribed in numerous other plants (Kapil and Bhatnagar, 1981, 1983; Tilton, 1981).

The timing of polar nucleus fusion may be before or after fertilization in species with monosporic development. The partial fusion of the polar nuclei is indicated by nuclear bridges located between the nuclei but is not completed until fertilization in the following plants: *Brassica* (Sumner and Van Caeseele, 1990) (Fig. 8), *Glycine* (Folsom and Peterson, 1984; Dute *et al.,* 1989), *Gossypium* (Jensen, 1965b), *Hordeum* (Cass and Jensen, 1970; Mogensen, 1982, 1988), *Ornithogalum* (Tilton and Lersten, 1981), *Oryza* (Dong and Yang, 1989), *Spinacia* (Wilms, 1981b), *Stipa* (Maze and Lin, 1975), *Triticale* (Hause and Schröder, 1986), *Triticum* (You and Jensen, 1985), and *Zea* (Diboll and Larson, 1966). *Spinacia* is unusual in displaying slender extensions of the nuclear envelope (nucleoplasm runners) that extend into the cytoplasm and appear to entwine the polar nuclei, possibly aiding in the reception of the sperm nucleus (Wilms, 1981a).

The fusion of the polar nuclei is completed prior to fertilization in *Agave* (Tilton and Mogensen, 1979), *Arabidopsis* (Mansfield *et al.,* 1991), *Beta*

(Bruun, 1987), *Capsella* (Schulz and Jensen, 1973), *Helianthus* (Newcomb, 1973; Yan *et al.*, 1990, 1991), *Linum* (Vazart and Vazart, 1966), *Monotropa* (Olson, 1991), and *Rhododendron* (Palser *et al.*, 1989); the fused product is termed a secondary nucleus. Nuclear fusion is slow at first in these monosporic plants and is apparently initiated early in the maturation of central cell (Bruun, 1987), but the fusion process accelerates as the egg approaches maturity (Schulz and Jensen, 1973). In *Petunia* the fusion of polar nuclei is apparently intermediate because these nuclei were observed to be in various stages of fusion at the time of fertilization (Van Went, 1970b). *Scilla,* an example of bisporic development, also displays complete fusion prior to fertilization (Bhandari and Sachdeva, 1983). If the embryological literature reviewed by Davis (1966) is accurate, most, if not all, bisporic angiosperms are characterized by the formation of a secondary nucleus; however, the distinction between partially and completely fused polar nuclei is uncertain in material prepared using paraffin embedding and observed with light microscopy.

The nucleoli of the central cell are generally large and dense, sometimes containing prominent nucleolar vacuoles. Typically, the nucleolus is similar to the egg cell in general organization (Kapil and Bhatnagar, 1983), containing a large, dense fibrillar region and a significant granular region. A micronucleolus is often associated with the nucleolus in *Glycine* (Folsom and Peterson, 1984). A broad granular region is evident in the nucleolus of *Brassica,* suggesting the accumulation of a greater abundance of preribosomal particles (Alberts *et al.*, 1989) than present in the egg (Sumner and Van Caeseele, 1990).

The central cell wall projections frequently occur at the edge of the embryo sac similar in structure to those of classical transfer cells (Gunning and Pate, 1969). Frequently, specific areas of the embryo sac seem to display more extensive wall ingrowths than others. In *Linum* numerous, deep ingrowths are located at the micropylar end of the embryo sac and particularly thick projections are observed in the synergid hook (Secor and Russell, 1988); in contrast, the chalazal pole has an unelaborated wall except near the antipodals (Vazart, 1968). Extensive cell wall projections are also reported in the micropylar and chalazal embryo sac walls in *Euphorbia* (Gori, 1977), *Helianthus* (Newcomb and Steeves, 1971; Newcomb, 1973), and *Zea* (Diboll and Larson, 1966). Ingrowths appear to be restricted to the micropylar cell wall in *Aquilegia* (Fougère-Rifot, 1978), *Brassica* (Sumner and Van Caeseele, 1990), *Cortaderia* (Philipson, 1977), *Crepis* (Godineau, 1969), *Eschscholtzia* (Negi, 1974), *Glycine* (Folsom and Peterson, 1984), *Haemanthus* (Newcomb, 1978), *Hibiscus* (Ashley, 1975), *Lobelia* (Torosian, 1972), *Oryza* (Dong and Yang, 1989), *Paspalum* (Yu and Chao, 1979), *Scilla* (Bhandari and Sachdeva, 1983), *Stellaria* (Newcomb and Fowke, 1973), *Triticum* (Morisson *et al.*, 1978; You and Jensen, 1985), and *Triticale* (Hause and Schröder, 1986). Cell wall projections,

however, were not observed before fertilization in *Arabidopsis* (Mansfield *et al.*, 1991), *Capsella* (Schulz and Jensen, 1973), and *Ornithogalum* (Tilton and Lersten, 1981), although they are well developed and abundant after fertilization. The close proximity of dictyosomes to the developing wall projections suggests that dictyosomes and dictyosome vesicles directly participate in their formation in *Helianthus* (Newcomb and Steeves, 1971) and *Brassica* (Sumner and Van Caeseele, 1990) (Fig. 8). These projections are reported to be PA-TCH-SP–positive in *Brassica* and *Glycine* (Folsom and Peterson, 1984) and display a strong PAS-positive reaction in *Scilla* (Bhandari and Sachdeva, 1983). Because the plasma membrane of the central cell follows the contours of wall projections, thus greatly expanding its internal surface area, a transfer cell function seems likely. One unusual feature that has been noted is that in *Jasione* the entire embryo sac is apparently surrounded by a thick cuticle (Berger and Erdelská, 1973). Normally, cuticle is present on the outside of the ovule, frequently located between the integument and the nucellus during development; however, maturation often disrupts this layer, which then remains as an interrupted cuticle in this region (Sumner and Van Caeseele, 1988) that would not represent a barrier to the movement of solutes into the embryo sac. Plasmodesmatal connections between the central cell and antipodals are reported in *Arabidopsis* (Mansfield *et al.*, 1991), *Beta* (Bruun, 1987), *Capsella* (Schulz and Jensen, 1973), *Helianthus* (Newcomb, 1973), and *Zea* (Diboll and Larson, 1966).

The central cell cytoplasm typically contains large numbers of dictyosomes, plastids, mitochondria, lipid droplets, ribosomes, and RER, as in *Aquilegia*, *Brassica*, *Capsella*, *Crepis*, *Epidendrum*, *Glycine*, *Gossypium*, *Helianthus*, *Plumbago*, *Scilla*, *Stipa*, and *Triticum* (Kapil and Bhatnagar, 1981; Bhandari and Sachdeva, 1983; Folsom and Peterson, 1984; You and Jensen, 1985; Sumner and Van Caeseele, 1990). Dictyosomes are scarce and inactive in *Hordeum* and *Epidendrum* but abundant and well developed in *Brassica*, *Capsella*, *Glycine*, *Gossypium*, *Helianthus*, and *Zea*. Plastids in the central cell are typically large proplastids containing starch, as in *Arabidopsis*, *Gossypium*, *Nicotiana*, and *Spinacia*. There is abundant starch in the central cell of *Nicotiana rustica* before fertilization (Sehgal and Gifford, 1979) and large multigrain starch packets in the central cell of soybean (Folsom and Peterson, 1984). In *Helianthus* and *Linum*, however, the plastids do not contain starch, and in *Capsella* (Schulz and Jensen, 1973) and *Brassica* (Sumner and Van Caeseele, 1990) (Fig. 8) the plastids contain 3–5 lamellate grana and appear to be starch-containing chloroplasts. Mitochondria are numerous and well developed. In *Plumbago*, the central cell contains an average of 1840 ± 740 plastids and $178,700 \pm 17,700$ mitochondria (Russell, 1987). ER is commonly present and abundant. In *Brassica* (Sumner and Van Caeseele, 1990) OsFeCN

was used to accentuate ER, revealing numerous ornate arrays of ER in association with the nuclear envelope (Fig. 8). ER lamellae are also located throughout the cytoplasm and in association with cell wall ingrowths and the egg apparatus. As evidenced by its disposition and abundance in the cytoplasm, ER appears to be actively involved in the metabolism of the cell. Frequently, lipid bodies are present in addition to starch as nutrient reserves in the central cell. Microbodies found associated with lipid bodies, as in *Brassica* (Sumner and Van Caeseele, 1990) and *Capsella* (Schulz and Jensen, 1973), presumably represent glyoxysomes, containing the enzymes of the glyoxylate cycle required for lipolysis (Huang *et al.,* 1983).

The transport of nutrients into the embryo sac is likely facilitated by morphological structures penetrating into the central cell, including the hypostase (as in *Agave*) and postament (as in *Quercus*) (Tilton, 1981). Using the apoplastic dye uranin (disodium fluorescein), Mogensen (1981b) demonstrated that mineral nutrients present in the vacular bundle are transmitted through the integuments and surrounding tissue of the ovule in *Nicotiana* and *Polygonum*. In the anatropous ovule of *Nicotiana tabacum,* the pathway of nutrients bifurcates, first producing a micropylar gradient, soon followed by a chalazal gradient. In the orthotropous ovules of *Polygonum capitatum,* a gradient is generated first at the chalazal end, proceeding to the micropyle. Although the interior of the embryo sac could not be directly visualized, likely the cell wall projections present in the central cell would be a major site of uptake. The presence of extensive localized wall ingrowths, dictyosomes with many vesicles, numerous plastids and mitochondria, increased amounts of ER, and abundant ribosomes indicate that the central cell is a metabolically active cell that appears to be well prepared to take up available nutrients from the ovule before fertilization and afterwards to initiate endosperm development and assume the nutritional requirements of the embryo.

IV. Isolation and Characterization of Isolated Megagametophytes

A. Isolation of Embryo Sacs and Their Component Cells

Considering the crucial role of the female germ unit in double fertilization, the isolation of a physiologically intact angiosperm female gametophyte or gamete has been an elusive goal that has only been attained in the past few years (Russell, 1990; Theunis *et al.,* 1991). In the 1970s the embryo sac was partly isolated, allowing the process of fertilization and the first postfertilization divisions to be observed microcinematographically in the

embryo sacs of *Jasione montana* and *Galanthus nivalis* (Erdelská, 1974, 1983). Isolation of embryo sacs was first reported in 1975 by Russian groups (Enaleeva and Dushaeva, 1975; Tyrnov *et al.*, 1975), who introduced the technique of enzyme maceration for isolating the embryo sac of *Nicotiana* and electron microscopic observation; however, they reported no details of their technique and provided no illustrations. By the mid-1980s, a number of researchers had independently developed parallel methods resulting in published reports on 18 species (Zhou and Yang, 1985, 1986; Hu *et al.*, 1985; Mól, 1986; Huang and Russell, 1989; Wagner, *et al.*, 1989a,b; Van Went and Kwee, 1990). Currently, the isolation of fixed and viable embryo sacs and their component cells has been reported in 27 species (Table I; Fig. 9). The ultrastructure of isolated embryo sacs has been reported in three species at the TEM level, *Nicotiana* (Sidorova, 1985), *Plumbago* (Russell *et al.*, 1989), and *Zea* (Wagner *et al.*, 1988). The isolation of protoplasts of embryo sac cells was first reported in *Nicotiana tabacum* (Hu *et al.*, 1985), followed by *Atropa belladonna* (Li and Hu, 1986), *Plumbago* (Huang and Russell, 1989) (Fig. 9A), *Petunia* (Van Went and Kwee, 1990), *Nicotiana alata* (Huang *et al.*, 1992) (Fig. 9B), and *Zea mays* (Kranz *et al.*, 1991a). In *Nicotiana* these protoplasts are currently being used to develop a model for the study of synergid degeneration (Fig. 9C; Huang and Russell, 1992b). Significant progress has been made on the characterization of the female germ unit, which includes the first successful *in vitro* fertilization attempts (Kranz *et al.*, 1990; Kranz *et al.*, this volume).

1. Technique and Protocols

Various different methods for isolating embryo sacs and their component cells have been developed (Table II). These involve three general techniques: (1) micromanipulation (Allington, 1985), (2) enzymatic maceration (Zhou and Yang, 1985, 1986; Hu *et al.*, 1985; Mól, 1986; Wagner *et al.*, 1989a,b; Van Went and Kwee, 1990), and (3) enzyme maceration followed by brief micromanipulation (Huang and Russell, 1989). Although the first technique can avoid the effects of enzymes or other chemicals, it is time-consuming and delicate handling requires more skill and instruments (Allington, 1985); this method also often damages the cells and is therefore of limited utility. The enzymatic maceration technique provides an easier approach to isolate the female gametophyte (Zhou and Yang, 1985). The main principle of this technique is to use enzymes to gently digest the cell walls and middle lamellae of cells surrounding the embryo sac. The adherent tissues are then removed from the surface of the embryo sac by mechanical agitation. The important parameters for enzymatic maceration include the combination of enzymes, selection of pH, selection of osmotic medium, duration of enzyme incubation time, temperature, and force

of mechanical agitation. These vary in different species and at different developmental stages (Table II). It is important to search for the least traumatic combination of enzymes to separate the tissues from the embryo sac. Among the enzymes that have been used, cellulase and pectinase are considered to be the most universally effective (Zhou and Yang, 1985; Huang and Russell, 1989). Workers have also reported the utility of using hemicellulase, driselase, pectolyase, β-glucuronidase, snailase, and xylosidase, among others. In *Nicotiana tabacum* (Hu *et al.*, 1985) and *Petunia* (Van Went and Kwee, 1990), the embryo sac was released using only driselase. The viability and optimum recovery of unfixed materials seem to depend on minimizing exposure to macerating enzymes and maintaining the pH within the physiological limits of the tissue (Huang and Russell, 1989; Wagner *et al.*, 1989b). Low pH (pH 5–6) is optimum for the activity of the enzymes but may impair the physiology of the cells (Van Went and Kwee, 1990). The use of some micromanipulation to reduce the exposure to enzyme maceration solutions seems to prolong viability (Huang and Russell, 1989; Huang *et al.*, 1992). Protoplasts can be released by increasing the duration of enzymatic digestion to separate the cells from the embryo sac wall. The release of egg cells, synergids, central cells, and antipodals requires 8 hours of incubation in *Nicotiana tabacum* (Hu *et al.*, 1985), more than 2 hours in *Petunia* (Van Went and Kwee, 1990), but in *Nicotiana alata* only 0.5–1 hour is required with some micromanipulation (Huang *et al.*, 1992). The osmotic medium used for the isolation of the embryo sac and its component cells has varied among the different species studied and has included principally mannitol, sorbitol, and sucrose (Table II). The life expectancy of isolated embryo sacs and their component cells seems strongly improved by supplementing the medium with other nutrients and using an appropriate plant cell culture medium (Mól, 1986; Huang and Russell, 1989; Van Went and Kwee, 1990; Kranz *et al.*, 1991a,b). Ultimately, as more is learned about the conditions of embryo sac culture, it seems likely that separate modified media will be required to suit the different nutritional needs of each species. Also, different media may be required at each step of the procedure (i.e.,isolation, fusion with other cells, culture, and regeneration) to ensure greatest success.

2. Assessment of Viability and Quality in Isolated Embryo Sacs and Their Component Cells

The viability and quality of the isolated embryo sacs and their component cells have been assessed mainly for structural intactness, membrane integrity, enzyme activity, and the movement of organelles. Structural integrity has been evaluated using TEM and SEM to observe the condition of the cytoplasm and intactness of cellular membranes (Table I). The ultrastruc-

TABLE I

Summary of Observations from Isolated Angiosperm Embryo Sacs

Species	Method	Observations	Reference
Adenophora axilliflora	ENZ	Observed fertilization, zygote and endosperm	Wu and Zhou, 1988
Antirrhinum majus	ENZ, fixed	Preservation good at LM level	Zhou and Yang, 1984
	ENZ, fixed and living	FCR$^+$, good preservation	Zhou and Yang, 1984, 1985
	ENZ, living	FCR$^+$	Zhou, 1985
Arabidopsis thaliana	ENZ-SQ, fixed, antitubulin	Distribution of MTs in ES development	Webb and Gunning, 1990
Atropa belladonna	ENZ-SQ, living	Isolated gametes, attempted culture	Li and Hu, 1986
Belamcnada chinensis	ENZ-SQ, fixed	Observed megasporogenesis and megagametogenesis	Li and Hu, 1985
Brassica campestris (var. purpurea)	ENZ, fixed	ES in different stages at LM level	Zhou and Yang, 1982
Carica papaya	ENZ-SQ, fixed	Observed megasporogenesis and megagametogenesis	Qiu *et al.*, 1991
Galanthus nivalis	EES-MM, living	Partial isolation of ES, cytoplasmic streaming observed superb, fertilization recorded by microcinematography	Erdelská, 1983
Helianthus annuus	ENZ-MM, fixed, DAPI fluorescence	Nuclear events of fertilization seen	Zhou, 1985, 1987
Hevea brasiliensis	ENZ-SQ, fixed	Observed megasporogenesis and megagametogenesis	Qiu *et al.*, 1991
Hordeum vulgare	MM	Preservation marginal, viability unproven	Allington, 1985
Jasione montana	EES, living	Cytoplasmic streaming, minimal preparation	Erdelská, 1974
Lilium longiflorum	ENZ, living	LM observations of ES, selected stages	Wagner *et al.*, 1989a
Nicotiana tabacum	ENZ, fixed	LM observations of ES	Zhou and Yang, 1982
	ENZ-SQ, fixed and living	Isolated ES, apparatus and CC, FCR$^+$	Hu *et al.*, 1985
	ENZ-SQ, fixed, TEM	Results very good, cells intact at TEM level	Sidorova, 1985
	ENZ, fixed and living	LM observations of ES, FCR$^+$, H33258$^+$	Zhou, 1985; Zhou and Yang, 1985

272

Species	Method	Observations	Reference
	ENZ-MM, fixed and living	Isolated ES, egg apparatus, CC, FCR⁺, cytoskeleton in fertilization and synergid degeneration	Huang and Russell, 1991
Nicotiana alata	ENZ-MM, living	Isolated ES, egg apparatus, CC, FCR⁺, vigorous cytoplasmic streaming	Huang et al., 1992
Oenothera odorata	ENZ-SQ, fixed	Observed megasporogenesis and megagametogenesis	Li and Hu, 1985
Paulownia sp.	ENZ, fixed	Observed megasporogenesis and megagametogenesis	Yang and Zhou, 1984
Petunia hybrida	ENZ-MM, living	Isolated ES, egg apparatus, CC, FCR⁺, viability 80 hr 4°C in Brewbaker-Kwack medium	Van Went and Kwee, 1990
Platycodon grandiflorus	ENZ-SQ, fixed	Observed megasporogenesis and megagametogenesis	Li and Hu, 1985
Plumbago zeylanica	ENZ-MM, fixed and living	Isolated ES, egg, CC and lateral cells, FCR⁺ to 24 hr unmodified medium	Huang and Russell, 1989
	ENZ-MM, fixed, TEM, whole ES	Observed w/STEM, BEI; stereopairs of ES and CC	Russell et al., 1989
	ENZ-MM, fixed, tubulin	Distribution of MTs, lipids, and wall fibrils in mature ES	Huang et al., 1990
	ENZ-MM, fixed and living	Distribution of MTs and actin during fertilization	Huang et al., unpublished
	ENZ-MM, fixed	Behavior and polarity of nuclear and plastid DNA during megasporogenesis and megagametogenesis	Huang and Russell, 1992a
Secale cereale	MM	Preservation marginal, viability unproven	Allington, 1985
Sesamum indicum	ENZ, fixed	Observed megasporogenesis and megagametogenesis	Yang and Zhou, 1984
Torenia fournieri	EES, living	Cytoplasmic streaming, minimal preparation	Erdelská, 1974
	ENZ-MM, living	Culture using feeder cells, alive 8 days without special medium	Mól, 1986
	EES, living	Cytoplasmic streaming, tried to enter CW w/microinjection needle	Keijzer et al., 1988
Triticosecale	MM	Preservation marginal, viability unproven	Allington, 1985
Triticum aestivum	MM	Preservation marginal, viability unproven	Allington, 1985
Vanilla fragrans	ENZ-SQ, fixed	Observed megasporogenesis and megagametogenesis	Li and Hu, 1985
Vicia faba	ENZ, fixed	LM observations of ES	Zhou and Yang, 1982

(continued)

TABLE I (*continued*)

Species	Method	Observations	Reference
Zea mays	ENZ-MM, fixed, TEM	ES appears intact, preservation shows some extraction	Wagner *et al.*, 1988
	ENZ-MM, fixed and unfixed	FCR$^+$ in vacuoles only	Wagner *et al.*, 1989b
	ENZ-MM, living	FCR$^+$, fusion of isolated gametes; products divide to form microcallus, isolated gametes cultured up to 22 days	Kranz *et al.*, 1991a,b

BEI, backscattered electron imaging; CC, central cell; CW, cell wall; DAPI, 4′,6-diamidino-2-phenylindole; EES, emergent embryo sac (no treatment required); EES-MM emergent embryo sac with integuments removed; ENZ, enzyme-treated embryo sac; ENZ-MM, enzyme micromanipulation; ENZ-SQ, enzyme followed by squash; ES, embryo sac; FCR$^+$, accumulates fluorescein in fluorochromatic reaction; H33258$^+$, fluorescent in Hoechst 33258 stain; LM, light microscopy; MM, micromanipulation; MSB, buffer (0.1 M PIPES, pH 6.8, 0.05 M KCl, 2 mM MgCl$_2$, 10 mM EGTA, 1 mM EDTA, 10% DMSO); MT, microtubule; STEM, scanning TEM; TEM, transmission electron microscopy.

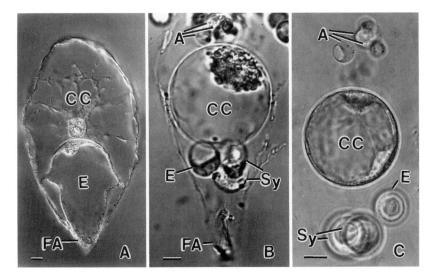

FIG. 9 Isolated embryo sacs and gametophytic cells as released by enzymatic digestion.
(A) Isolated embryo sac of the synergid-lacking angiosperm *Plumbago zeylanica* from
aldehyde-fixed material. (B) Living gametophytic cells within the gametophyte wall of *Nicotiana alata,* showing the antipodals (A), central cell (CC), egg (E), and synergids (Sy). The
filiform apparatus (FA) remains attached to the embryo sac walls as the enzyme treatment
causes the cells to become spherical protoplasts. (C) Extensive emzymatic digestion in the
embryo sac of *Nicotiana tabacum* results in disappearance of the surrounding wall and
release of protoplasts. Bars, 10 μm. (A) Reproduced with permission from Huang and Russell
(1989). (C) Reproduced with permission from Huang *et al.* (1992).

ture of isolated and intact embryo sacs of *Nicotiana tabacum* revealed
no significant changes in internal structure during the isolation process
(Sidorova, 1985). Similarly, isolated embryo sacs of *Plumbago* observed
as whole-mounted specimens using IVEM revealed few changes from the
conventional organization of the embryo sac (Russell *et al.,* 1989) (Fig.
10) and significantly extended information about the organization of transvacuolar strands in the central cell (Section III,B,3).

Enzyme activity and membrane integrity are readily examined by the
so-called fluorochromatic reaction (FCR). In this reaction fluorescein diacetate is readily absorbed by living cells but only becomes evident if the
fluorescent product of fluorescein diacetate cleavage accumulates within
the cell. This test provides an assay of both esterase activity and membrane
intactness. Viability of isolated embryo sacs and their component cells
has been evaluated in many species using this technique (Table I). Another
approach that should be more sensitive to cellular vigor, as opposed to
viability, is the assessment of cellular activity, cytoplasmic streaming, and

TABLE II

Summary of Methods for Embryo Sac Isolation in Angiosperms

Species	Isolation medium[a]	Enzyme combination[b]	Technique[c]	Time and temperature	Viability–observations[d]	Yield[e]	Reference
Adenophora axilliflora	FPA	3% P/2% C	Sh	2.5–3.5 hr 28–30°C	Fixed	NA	Wu and Zhou, 1988
Antirrhinum majus	FPA	2% P/1–1.5% C	Sh	2–3 hr 28–30°C	Fixed	NA	Zhou and Yang, 1984
	2–8% Sucr, 0.2% PDS, pH 5–5.5	2% P/1–1.5% C	Sh	2–3 hr 28–30°C	FCR[+]	NA	Zhou and Yang, 1984, 1985; Zhou, 1985
Arabidopsis thaliana	4% PFA/PM5E buffer	2% Cl/2% Mc, pH 6.8	Sq	1 h RT	MT fixation	NA	Webb and Gunning, 1990
Atropa belladonna	0.6 M Man	1% Dri	Sq	2.5–4 hr 28°C	FCR[+]	NA	Li and Hu, 1986
Belamcanda chinensis	FPA, FAA, Carnoy	2% Dri	Sh	3–6 hr 28°C	Fixed	NA	Li and Hu, 1985
Brassica campestris	FPA	2.5% P/2.5–3% C	Sh	5 hr 28–30°C	Fixed	NA	Zhou and Yang, 1982
Carica papaya	FAA or FPA	3% P/3% C or 3–4% Sn	Sh	4–6 hr 28–30°C	Fixed	NA	Qiu *et al.*, 1991
Galanthus nivalis	15% saccharose/1% agar	NA	Mm	NA	CS[+]	NA	Erdelská, 1983
Helianthus annuus	7–10% Sucr, pH 5–5.5	2–3% P/2–3% C/ 1–2% Sn/0.4–1% PLy	Sh	Several hours 28–30°C	FCR[+]	NA	Zhou, 1985
Hevea brasiliensis	10% Sucr, 0.02% KH$_2$PO$_4$	3% P/2% C/2% Sn/ 1% PLy	Sh	4–5 hr 28°C	H33258[+] (F)	NA	Zhou, 1987
	FAA or FPA	3% P/3% C or 3–4% Sn	Sh	4–6hr 28–30°C	Fixed	NA	Qiu *et al.*, 1991
Hordeum vulgare	GHPT buffer, pH 8.4	NA	Mm	NA	Unproven	NA	Allington, 1985
Jasione montana	5% Sucr/1% aga	NA	Mm	NA	CS[+]	NA	Erdelská, 1974

276

Species	Fixative/Buffer	Enzyme	Type	Time/Temp	Test	Yield	Reference
Lilium longiflorum	3.8% Man 0.05% CaCl$_2$ pH 5.2	1.5% P/0.6% C/0.5% H/4% PLy	Sh	1–1.5 hr 30°C	FCR+	5–8%	Wagner et al., 1989a
Lolium perenne	9% sucr, pH 5.0	C/P	Sq	NA	NA	NA	Van der Maas and Zaal, 1990
Nicotiana tabacum	Fixed, no details						Enaleeva and Dushaeva, 1975 Tyrnov et al., 1975
	Fixed, no details FPA	2.5% P/2.5–3% C	Sh	5 hr 28–30°C	Fixed	NA	Zhou and Yang, 1982
	0.2 M Man 10% Sucr	2% P/1% Dri/0.3% Xy	Sh	NA	Fixed, TEM	NA	Sidorova, 1985
	10% Sucr pH 5–5.5	1–2% Sn/0.4–1% PLy	Sh	Several hours 28–30°C	FCR+	NA	Zhou, 1985
	0.65 M Man	2–3% P/2–3% C	Sh	3–3.5 hr (ES) 4–8 hr (cells)	FCR+	NA	Hu et al., 1985
	0.65 M Sor	0.5% C/0.5% P	sMm	0.5–1 hr RT	FCR+, CS+	30% (ES), 35% (E), 30% (CC)	Huang et al., 1992
	4% PFA/PMEG buffer, pH 6.8	2% P/2% C	sMm	1 hr 37°C	CKT (F)	NA	Huang and Russell, 1991
Nicotiana alata	0.65 M Sor	0.5% C/0.5% P	sMm	0.5–1 hr RT	FCR+, CS+	35% (ES) 40% (E) 35% (CC)	Huang et al., 1992
Oenothera odorata	FPA, FAA, Carnoy	2% Dri	Sh	3–6 hr 28°C	Fixed	NA	Li and Hu, 1985
Paulownia sp.	FPA	3% P/1.5% C	Sh	5 hr 28–30°C	Fixed	NA	Yang and Zhou, 1984
Petunia hybrida	0.1% MES/8% Man to 10% BKM, pH 6.5	3% Dri	Sh	2 hr 30°C	FCR+	20–25%	Van Went and Kwee, 1990
Platycodon grandiflorus	FPA, FAA, Carnoy	2% Dri	Sh	3–6 hr 28°C	Fixed	NA	Li and Hu, 1985
Plumbago zeylanica	FPA	3% C/H/P, 1% β-Gluc	Sh/sMm	2 hr 37°C	Fixed	60%	Huang and Russell, 1989

(continued)

277

TABLE II (continued)

Species	Isolation medium[a]	Enzyme combination[b]	Technique[c]	Time and temperature	Viability–observations[d]	Yield[e]	Reference
	0.6 M sorbitol, 4%	2% C/H/P/0.5% PLy	Sh/sMm	2 hr RT	FCR+ (egg 33 hr) (CC 29 hr)	10% (E)	Huang and Russell, 1989
	glucose, 0.05 M Cacl$_2$	1% β-Gluc pH 5.8				6% (CC), 4% (LC)	
	2.5% GA/PFA, 0.1 M Pipes	3% C/3% P/3% H/1% β-Gluc	sMm	2 hr 37°C	Fixed, IVEM	40%	Russell et al., 1989
	MSB buffer, pH 6.8	2% P/2% C/1% H	sMm	0.5–1 hr 37°C	Fixed, MT	NA	Huang et al., 1990
	PMEG buffer, pH 6.8	2% C/2% P	sMm	0.5–1 hr 37°C	Fixed, CKT	NA	Huang et al., unpublished
	2.5% GA/PFA, 0.1 M Pipes	2% C/2% H/2% P/1% β-Gluc	sMm	1 hr 37°C	Fixed	NA	Huang and Russell, 1992a
Secale cereale	GHPT buffer, pH 8.4	NA	Mm	NA	Unproven	NA	Allington, 1985
Sesamum indicum	FPA	3% P/1.5% C	Sh	5 hr 28–30°C	Fixed	NA	Yang and Zhou, 1984
Torenia fournieri	5% Sucr/1% agar	NA	Mm	NA	CS+	NA	Erdelská, 1974
	0.7 M Man or 0.57 M Man/20% sea water	2% C/1% H/0.05% PLy	NA	5 hr RT	FCR+ (5 wk cult)	NA	Mól, 1986
	CC-medium	0.5% Macerozyme	Mm	NA	AF	NA	Keijzer et al., 1988
Triticosecale	GHPT buffer, pH 8.4	NA	Mm	NA	NA	NA	Allington, 1985

Species	Fixative	Enzyme mixture	Method	Time/Temp	Treatment	Yield	Reference
Triticum aestivum	GHPT buffer. pH 8.4	NA	Mm	NA	NA	NA	Allington, 1985
Vanilla fragrans	FPA, FAA, Carnoy	2% Dri	Sh	3–6 hr 28°C	Fixed	NA	Li and Hu, 1985
Vicia faba	FPA	2.5% P/2.5–3% C	Sh	5 hr 28–30°C	Fixed	NA	Zhou and Yang, 1982
Zea Mays	4% GA/2% OsO$_4$ 0.1 M PO$_4$	Ch/P/C/H/PLy	Sh	1–3 hr	Fixed, TEM	NA	Wagner et al., 1988
	9% Sucr, pH 4.5–5	2% Ch/3% P/2% Cy/1% PLy	Sh	1–3 hr 28°C	FCR$^+$ (vac)	30–50%	Wagner et al., 1989b
	570 mM/kg H$_2$O Man pH 5.0	0.75% P/0.25 PLy	Sh	23.5–24.5°C,	IVF	50% (E)	Kranz et al., 1991a,b
		0.5% H/0.5C					

[a] BKM, Brewbaker and Kwack's medium with 10% mannitol; PDS, potassium dextran sulphate; PM5E, 50 mM Pipes, 2 mM MgSO$_4$, 5 mM EGTA; GHPT buffer, 100 mM glycine, 1% hexyleneglycol, 1% propan-2-ol 0.1% Triton X-100; PMEG, 50 mM Pipes, 5 mM MgSO$_4$, 5 mM EGTA, 4% glycerol; CC-medium, 0.57 M mannitol, 20% sea water; FPA, formalin:propionic acid:alcohol (1:1:18); FAA, formalin:glacial acetic acid:alcohol (1:1:18); Carnoy, Carnoy's absolute alcohol:glacial acetic acid (3:1); GA, glutaraldehyde; Man, mannitol; MES, 2(N-morpholina) ethanesulfonic acid; MSB, microtubule-stablizing buffer; PFA, paraformaldehyde; PO$_4$, phosphate buffer; Sucr, sucrose; Sor, sorbitol.

[b] C, cellulase; Cl, cellulysin; H, hemicellulase; Dri, driselase; β-Gluc, β-glucuronidase; Mc, macerase; P, pectinase; PLy, pectolyase; Sn, snailase; Xy, xylosidase; Ch, cytohelicase; Cy, caylase rT-cellulase.

[c] Mm, micromanipulation; sMm, some micromanipulation; Sh, shaking or agitation; Sq, squash; Cells, gametophytic cells; ES, embryo sac; RT, room temperature.

[d] CC, central cell; CS$^+$, cytoplasmic streaming reported; Cult, culture; E, egg; FCR$^+$, cytoplasm positive in fluorochromatic reaction; FCR$^+$ (vac), only vacuole was FCR$^+$; IVEM, intermediate voltage electron microscopy; TEM, transmission electron microscopy; no technique given indicates light microscopy; MT, microtubule; CKT, cytoskeleton; AF, artificial fertilization; IVF, *in vitro* fertilization; F, fertilization.

[e] CC, central cell; E, egg; NA, not available; unmarked indicates yield of embryo sacs.

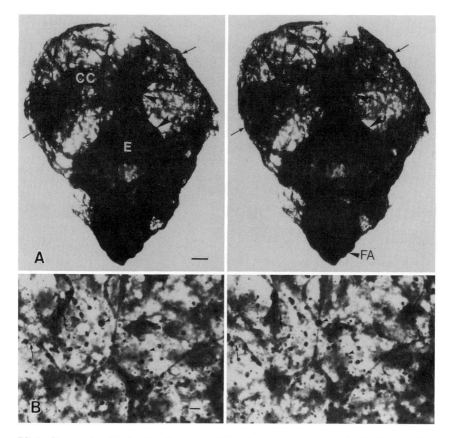

FIG. 10   Stereopairs of isolated embryo sacs of *Plumbago zeylanica* viewed using intermediate
voltage electron microscopy. (A) Egg (E) with filiform apparatus (FA) and central cell (CC)
are evident in this synergid-lacking embryo sac. Arrowhead indicates circumferential band
of cytoplasm in the central cell. Bar, 10 μm. (B) Higher magnification view of transvacuolar
strands (arrows) and organelles (arrowheads) in central cell cytoplasm. Bar, 1 μm. Reproduced with permission from Russell, *et al.* (1989).

organelle movement in the isolated cells (Huang *et al.*, 1992). In *Nicotiana
alata* the organelles showed vigorous movement within 2–4 hours after
embryo sac isolation, decreasing greatly in activity after 8 hours in the
unsupplemented medium. In *Petunia* isolated embryo sacs can be stored
as long as 80 hours in culture medium at 4°C (Van Went and Kwee, 1990).
In *Torenia* the embryo sacs survived for 2 weeks only when the culture
medium was supplemented with tissue culture nutients (Mól, 1986). Using
a more sophisticated breeder cell system in *Zea mays,* cells were still
viable after 22 days in culture (Kranz *et al.*, 1991a,b).

B. Characterization of Isolated Embryo Sacs and Their
   Component Cells

1. Cytological Characterization

At the light microscopic level, isolated embryo sacs have been used to
study megasporogenesis and megagametogenesis (Li and Hu, 1985; Yang
and Zhou, 1984; Webb and Gunning, 1990), double fertilization (Zhou,
1987; Huang *et al.*, 1990), and early embryogenesis (Huang *et al.*, 1990;
Webb and Gunning, 1991). The advantages of using isolated embryo sacs
are evident in studies of the cytoskeleton (Section V), the distribution of
cytoplasmic and nuclear DNA, the organization of the cell wall, and
deployment of the cytoplasm within the embryo sacs, which are difficult
to reconstruct from sectioned material.

Conventional TEM has been used to examine the organization of the
isolated embryo sac in *Nicotiana* (Sidorova, 1985), *Plumbago* (Huang and
Russell, 1989; Russell *et al.*, 1989), and *Zea* (Wagner *et al.*, 1988). IVEM
of the whole embryo sac (Russell *et al.*, 1989) and SEM of sectioned
embryo sacs (Huang *et al.*, 1990) provide ultrastructural data that complement and extend observations of embryo sacs in intact ovules (Russell, 1983). The advantages of IVEM are evident in portraying the three-
dimensional organization of the complete embryo sac, the fine structure
of the FA, the far denser cytoplasmic matrix of the egg (Fig. 10A), the
fibrillar bundles of the egg, and the multiple planes of organelles throughout
the embryo sac (Russell *et al.*, 1989) (Fig. 10B). The organization of the
FA is also far more easily evident in the embryo sac, as shown in both
Plumbago (Fig. 9A) and *Nicotiana* (Figs. 9B,C). The coordinated examination of the FA of the egg in *Plumbago* using Calcofluor white and SEM
revealed that the prominent strands evident in the cytoplasm fluoresce
using Calcofluor white and therefore appear to contain β-1,4-glucan fibrils
(Huang *et al.*, 1990). Studies of isolated embryo sac organization in *Zea*
mays indicate that, although the enzyme maceration technique does not
appear to alter the cellular and nuclear organization of isolated embryo
sac, the nature of physical associations between the constitutive cells of
the embryo sac are modified (Wagner *et al.*, 1988). This reflects a limitation
of the isolation technique that must be carefully examined and may influence the ability of isolated gametes to fuse (Kranz *et al.*, this volume).

2. Cytochemical Characterization

DNA probes, such as DAPI and Hoechst 33258, may be readily used to
label the nuclear and plastid DNA of either living or fixed isolated embryo
sacs and their component cells using concentrations as small as 1 μg/ml

(Theunis *et al.*, 1991). Hoechst 33258 was used to trace fertilization events of nuclear DNA in living embryo sacs of *Helianthus* (Zhou, 1987), confirming previous reports on the passage of the nuclei. An examination of cytoplasmic DNA during embryo sac formation of *Plumbago zeylanica* reveals that plastid nucleoids frequently associate with the most micropylar nucleus and may determine its ability to differentiate into an egg (Huang and Russell, 1992a). A reexamination of the distribution of cytoplasmic and nuclear DNA during megasporogenesis in a normal *Polygonum*-type embryo sac could be interesting as a method for defining the polarity of cytoplasmic DNA involved in megaspore abortion.

One component that appears to be missing in the isolated embryo sac is the presence of strong Calcofluor white staining in the embryo sac walls; this may be attributed to the use of cellulase in the enzyme maceration technique: a slight staining of the embryo sac cell wall remains in *Zea* (Wagner *et al.*, 1989b), reflecting the presence of polysaccharides; however, no staining is evident in *Antirrhinum* (Zhou and Yang, 1985). In *Plumbago* (Huang *et al.*, 1990) and *Nicotiana* (Huang and Russell, 1992b), however, the FA fluoresces strongly, indicating that the β-1,4-glucan component of these ingrowths is protected from enzymatic digestion during isolation.

Lipid bodies are readily stained using the fluorescent dye 3,4-benzpyrene-caffeine and are mainly distributed in the micropylar half of egg cell of *Plumbago* before fertilization and in the apical call of the embryo after fertilization (Huang *et al.*, 1990). Sudan black B, however, failed to stain the embryo sac of *Zea* (Wagner *et al.*, 1989b). In *Antirrhinum majus* (Zhou and Yang 1985) the embryo sac appears to be surrounded by a surprisingly resistant wall sac, which displays intense staining with Auramine O. Although there is no evidence for such a resistant cell wall in any other embryo sacs, the wall of this embryo sac is resistant to enzymes, strong oxidants, and even acetolysis, indicating that the wall may contain cutin or sporopollenin (Yang, 1989).

3. Elemental Characterization

Calcium and other elements were investigated using energy-dispersive x-ray microanalysis in rapid frozen wheat and pearl millet ovaries (Chaubal and Reger, 1990, 1992), revealing that the synergid cells contain a relatively high concentration of calcium and low levels of phosphorous, potassium, and sulfur, whereas the egg cell and central cell have relatively low levels of calcium and higher levels of phosphorous, potassium, and sulfur (Fig. 11). Among the different compartments within the ovule, the micropylar end of the embryo sac consistently displayed the highest calcium concentration compared with the nucellus, ovary wall, central cell, and antipodal

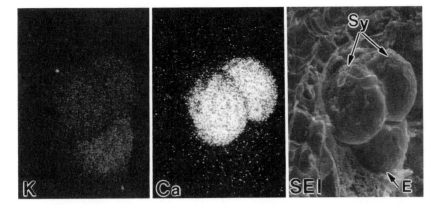

FIG. 11   Energy-dispersive x-ray analysis maps of potassium (K) and calcium (Ca) distribution in the rapid frozen, freeze-substituted egg (E) and synergids (Sy) of wheat, *Triticum aestivum,* and a coordinated secondary electron image (SEI) of these cells viewed using SEM. The synergids contain accumulations of calcium in their vacuoles, whereas the egg contains large quantities of potassium. Reproduced with permission from Chaubal and Reger (1990).

cells. According to Chaubal and Reger (1990, 1992), these elements are mainly distributed in a cytosolic or loosely bound exchangeable state. Work on the synergids of *Nicotiana* using chlorotetracycline fluorescence, which has an affinity for membrane-bound calcium, reveals intense staining in one synergid after degeneration (Huang and Russell, 1992b), suggesting that accumulations of bound calcium may be involved in synergid degeneration. Antimonate localizations also reveal dense precipitates of calcium in the synergids of *Pennisetum* (Chaubal and Reger, 1992). Antimonate precipitates occur in the degenerated organelles of the synergid and may be involved in the cytoplasmic breakdown of the synergid. Calcium is known to attract pollen tubes (Mascarenhas and Machlis, 1962) and in cotton may be a major component of the ash formed by microincineration in the vacuole of the synergid (Jensen, 1965a). Correct orientation of the pollen tube may be directed by a calcium gradient in the vicinity of the synergids and may induce the arrest and rupture of the pollen tube to release the sperm near the egg.

4. Biochemical and Immunological Characterization

So far, no reports about the biochemical characteristics of the embryo sac have been published because of the significant difficulties involved in isolating a large enough sample (Theunis *et al.,* 1991). However, a sensitive technique has been developed for detecting minute amounts of glycoprotein in the embryo sac of *Plumbago zeylanica,* in which more than 30

bands are detectable between M_r 29,000–205,000 using as few as 250 embryo sacs. For a single one-dimensional polyacrylamide electrophoretic gel, the minimum required number of embryo sacs for *Plumbago* is 150 for polypeptides; many species have smaller embryo sacs and would require more (Huang and Russell, unpublished data).

The immunological characterization of the embryo sac has been restricted to cytoskeletal labeling (Section V) and not to the unique immunological attributes that the surfaces of the egg and central cell may possess. Once again, the effort required to obtain an adequate sample for immunological characterization has been too great to elicit any studies. As an example of the nature of the problem, the entire embryo sac of *Plumbago* contains an average of 70 ng protein (determined by micro-Lowry assay, Huang and Geltz, unpublished data); isolation of the component cells multiplies the difficulties because the yields are decreased by this procedure and the demands of micromanipulation and sample preparation increase (Huang and Russell, 1989).

V. Cytoskeletal Organization in the Megagametophyte

The organization of the microtubular cytoskeleton during megasporogenesis has been reported in *Arabidopsis thaliana* (Webb and Gunning, 1990) and *Chamaenerion angustifolium* (Bednara *et al.*, 1988), and for microtubules and actin in *Gasteria verrucosa* (Willemse and Van Lammeren, 1988; Bednara *et al.*, 1988, 1990). Each of these belongs to monosporic patterns of development (*Polygonum* type in *Arabidopsis* and *Gasteria;* *Oenothera* type in *Chamaenerion*). Microtubules and actin tend to be randomly oriented and co-distributed in short fibers during early prophase I, elongate during late prophase, and become independently distributed during metaphase and anaphase, reaching their greatest length during division (Willemse and Van Lammeren, 1988; Bednara *et al.*, 1988, 1990). Meiotic division in these species, although involving a traditional spindle, is not preceded by the formation of a preprophase band of microtubules during megasporogenesis (Bednara *et al.*, 1988; Willemse and Van Lammeren, 1988; Webb and Gunning, 1990). Preprophase bands in normal higher plant cell division typically determine the location of the nucleus prior to mitosis and predict the position of cell wall formation during cytokinesis (Gunning and Hardham, 1982). The absence of a preprophase band may be related to the irregular pattern of cell wall formation between the meiotic products and the callosic nature of cell walls in the nonfunctional megaspores.

During later megasporogenesis microtubules frequently radiate from the nucleus, possibly aiding in the development of the micropylar–chalazal polarity of cytoplasmic organelles (Willemse and Van Lammeren, 1988; Webb and Gunning, 1990). Less frequent cortical microtubules are found randomly oriented at the edge of the cell, suggesting that the expansion of the functional megaspore during later development is not controlled by the megaspore itself but possibly by the surrounding cells of the nucellus and integuments (Willemse and Van Lammeren, 1988; Webb and Gunning, 1990), which maintain a conspicuous parallel system of cortical microtubules (Willemse and Van Lammeren, 1988). In the absence of strongly aligned cortical microtubules within the functional megaspore and young embryo sac, the pattern of developmentally programmed cell death in surrounding cells could be a crucial factor in the expansion of the embryo sac (Russell, 1979; Schulz and Jensen, 1981, 1986). Microtubules in abortive megaspores display an intense, fixed, and undynamic distribution, which, according to Bednara et al. (1988), may provide a signal for degeneration in Gasteria. Microtubules in the abortive megaspores are eventually lost as the cells degenerate (Webb and Gunning, 1990).

The microtubular cytoskeleton also appears to perform specific roles during megagametogenesis in Gasteria (Willemse and Van Lammeren, 1988) and Arabidopsis (M. C. Webb and B. E. S. Gunning, personal communication), and in the mature embryo sacs of these plants, and Plumbago (Huang et al., 1990) and Nicotiana (Huang and Russell, 1991). Microtubules are involved in spindle formation during the mitotic divisions of megasporogenesis (Rutishauser, 1969) and during cell wall formation of the cenocytic embryo sac (Cass et al., 1985, 1986). Dense random microtubules are present during expansion of the young synergid; these become longitudinally oriented in bundles during the formation of the FA and may persist at maturity (Willemse and Van Lammeren, 1988; Webb and Gunning, 1991; Huang and Russell, 1991) (Fig. 12). In Plumbago, in which the egg performs synergid-like functions, microtubules are organized in a dense cap-like structure next to the FA, some fibers extending longitudinally to the nucleus (Huang et al., 1990). Microtubules are randomly distributed in the central cell (Fig. 12A; M. C. Webb and B. E. S. Gunning, personal communication) or in cytoplasmic strands (Huang et al., 1990) but are rare to nonexistent in antipodal cells (Willemse and Van Lammeren, 1988; Huang and Russell, 1991). Microtubules and actin appear to play an important role in pollen tube guidance and sperm transport during fertilization (Russell, this volume).

One particularly interesting finding concerning the cytoskeleton is that preprophase bands of microtubules are absent throughout female reproductive function, including megasporogenesis, megagametogenesis, and cellularization of the embryo sac and endosperm but resume during the first

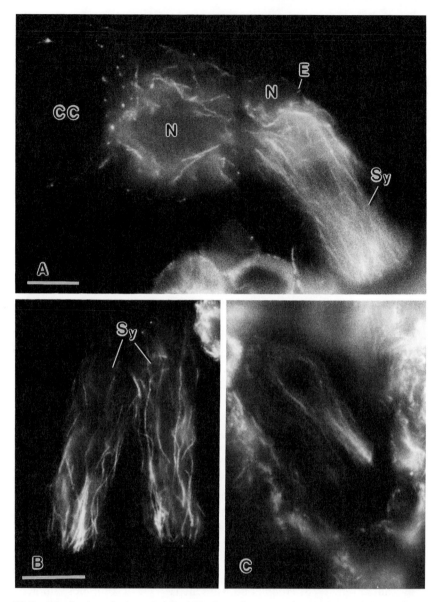

FIG. 12   Distribution of microtubules within the embryo sac of *Arabidopsis thaliana* viewed using immunofluorescence of antitubulin. (A) Abundant microtubules are detected in longitudinal orientation in the synergid (Sy). Microtubules are concentrated around the egg nucleus (N), particularly abutting the synergids. They are also found throughout the rest of the egg cell (E) and central cell (CC). (B and C) In synergids longitudinally oriented microtubules run the length of the cell in isolated (B) and cryostat sectioned material (C). Bar, 10 μm. Unpublished fluorescence micrographs kindly provided by M. C. Webb and B. E. S. Gunning.

division of zygote and presumably persist throughout later development (Webb and Gunning, 1991). The suspension of preprophase band formation has also been reported during similar stages of male meiosis, microspore division, and pollen grain formation (Van Lammeren *et al.*, 1985), and in the regeneration of protoplasts and the division of endosperm cells (Gunning and Hardham, 1982).

VI. Conclusion and Prospects

The isolation of female gametophytes and gametes provides a new approach for understanding embryological development and fertilization, allowing detailed studies to be conducted from a structural, biochemical, and molecular perspective. The intact three-dimensional organization of the embryo sac provides the opportunity to visualize cell-to-cell interactions directly rather than reconstructing the embryo sac, either physically or mentally, after sectioning. These techniques also allow more of the dynamism of megagametophyte behavior to be examined, including the distribution of cytoskeletal elements and organelles during development and fertilization events, *in vivo* and *in vitro* (Kranz *et al.*, this volume, *in vitro* fertilization *ex ovulo*). Isolated embryo sacs provide excellent material suitable for both chemical fixation and physical fixation using rapid freezing, freeze-substitution, and immunocytochemistry, and for studies of the relationships of subcellular components *in vivo* using video-enhanced contrast microscopy or confocal laser scanning microscopy (Fredrikson *et al.*, 1988; Fredrikson, 1990).

Once the technical problems of collecting large numbers of isolated embryo sacs are resolved and sufficiently sensitive techniques are developed, biochemical and molecular approaches can be applied to identify critical transition points during development, the expression of specific polypeptides, and the production of unique glycoproteins in the embryo sac and female gamete. These studies will undoubtedly contribute to a better understanding of gamete recognition, fusion, and activation. Further questions will need to address the problems of the regulation of megasporogenesis, megagametogenesis, fertilization, initiation of embryogenesis, and the induction of apomictic development.

Culturing isolated embryo sacs and their component cells may provide unique opportunities to manipulate haploid plants through biotechnology, using a genetically definable source. If techniques to prolong the viability of these cells are resolved, fundamental questions about the biology of the embryo sac may be examined experimentally to elucidate, for example, the nutritional relationship between sporophyte and gametophyte cells and

numerous other questions. Although the artificial fusion of embryo sac component cells has been achieved *in vitro* and microcalli have been obtained, further questions remain concerning whether gametes can fuse *in vitro* without electroporation and how recognition molecules may function in flowering plant gametes. Once these and the problems of obtaining optimum culture conditions for regeneration are resolved, gametic cells could be an attractive cell source for transformation. If these concepts are put into practice in the future, a useful system for recombinant progeny and examining fertilization can be realized.

Acknowledgments

The authors thank Drs. N. N. Bhandari, D. D. Cass, R. Chaubal, R. Dute, B. J. Reger, M. J. Sumner, and M. Webb for providing drawings and photographs used in this review. Portions of this research were supported by operating grants to Scott D. Russell from NSF (PCM 8208466 and DCB-8409151) and USDA (86-CRCR-1-1978, 88-37261-3761, and 91-37304-6471), and through equipment support by the University Research Council.

References

Alberts, B., Bray, D., Lewis, J., Raff, M., Roberts, K., and Watson, J. D. (1989). "Molecular Biology of the Cell," 2nd ed. Garland, New York.
Allington, P. M. (1985). *In* "Experimental Manipulation of Ovule Tissues" (G. P. Chapman, S. H. Mantell, and R. W. Daniels, eds.), pp. 39–51. Longman, London.
Alvarez, M. R., and Sagawa, Y. (1965). *Caryologia* **18,** 241–249.
Ashley, T. (1975). *Planta* **108,** 303–317.
Bannikova, V. P., Plyushch, T. A., and Gvilava, M. N. (1987). *Phytomorphology* **37,** 291–298.
Bednara, J. (1977). *Acta Soc. Bot. Pol.* **46,** 603–616.
Bednara, J., Van Lammeren, A. A. M., and Willemse, M. T. M. (1988). *Sex. Plant Reprod.* **1,** 164–172.
Bednara, J., Willemse, M. T. M., and Van Lammeren, A. A. M. (1990). *Acta Bot. Neerl.* **39,** 43–48.
Benavente, R. S., Skorupska, H., Palmer, R. G., and Shoemaker, R. C. (1989). *Am. J. Bot.* **76,** 1759–1768.
Berger, C., and Erdelská, O. (1973). *Caryologia* **25**(Suppl.), 109–120.
Bhandari, N. N., and Chitralekha, P. (1989). *Can. J. Bot.* **67,** 1325–1330.
Bhandari, N. N., and Sachdeva, A. (1983). *Protoplasma* **116,** 170–178.
Brown, R. C., and Lemmon, B. E. (1991). *Protoplasma* **163,** 9–18.
Browning, A. J., and Gunning, B. E. S. (1977). *Protoplasma* **93,** 7–26.
Bruun, L. (1987). *Nord. J. Bot.* **7,** 543–551.
Cass, D. D. (1972). *Am. J. Bot.* **59,** 279–283.
Cass, D. D., and Jensen, W. A. (1970). *Am. J. Bot.* **57,** 62–70.
Cass, D. D., and Karas, I. (1974). *Protoplasma* **81,** 49–62.
Cass, D. D., Peteya, D. J., and Robertson, B. L. (1985). *Can. J. Bot.* **63,** 2164–2171.

Cass, D. D., Peteya, D. J., and Robertson, B. L. (1986). *Can. J. Bot.* **64**, 2327–2336.
Chao, C. Y. (1971). *Am. J. Bot.* **58**, 649–654.
Chao, C. Y. (1977). *Am. J. Bot.* **64**, 922–930.
Chaubal, R., and Reger, B. J. (1990). *Sex. Plant Reprod.* **3**, 98–102.
Chaubal, R., and Reger, B. J. (1992). *Sex. Plant Reprod.* **5**, 34–46.
Cocucci, A. E., and Jensen, W. A. (1969a). *Kurtziana* **5**, 23–38.
Cocucci, A. E., and Jensen, W. A. (1969b). *Am. J. Bot.* **56**, 629–640.
Davis, G. L. (1966). "Systematic Embryology of the Angiosperms." Wiley, New York.
De Boer-de-Jeu, M. J. (1978). *Meded. Landbouwhogesch. Wageningen* **78**, 1–128.
Diboll, A. G., and Larson, D. A. (1966). *Am. J. Bot.* **53**, 391–402.
Dickinson, H. G. (1986). *In* "The Chondriome" (S. H. Mantell, G. P. Chapman, and P. F. S. Street, eds.), pp. 37–60. Longman, London.
Dickinson, H. G., and Heslop-Harrison, J. (1977). *Philos. Trans. R. Soc. Lond. [Biol.]* **277**, 327–342.
Dickinson, H. G., and Potter, U. (1978). *J. Cell Sci.* **29**, 147–169.
Dong, J., and Yang, H. Y. (1989). *Acta Bot. Sin.* **31**, 81–88.
Dumas, A. (1978). *Bull. Soc. Bot. Fr. Actual. Bot.* **125**, 193–199
Dumas, C., Knox, R. B., McConchie, C. A., and Russell, S. D. (1984). *What's New Plant Physiol.* **15**, 17–20.
Dute, R. R., Peterson, C. M., and Rushing, A. E. (1989). *Ann. Bot.* **64**, 123–135.
Ehdaie, M., and Russell, S. D. (1984). *Phytomorphology* **34**, 221–225.
Enaleeva, N. K., and Dushaeva, N. A. (1975). *In* "Apomixis and Cytoembryology of Plant" (in Russian), Vol. 3, pp. 171–175. Seratov University Press, Seratov, USSR.
Erdelská, O. (1974). *In* "Fertilization in Higher Plants" (H. F. Linskens, ed.), pp. 191–195. North Holland, Amsterdam.
Erdelská, O. (1983). *In* "Fertilization and Embryogenesis in Ovulated Plants" (O. Erdelská, ed.), pp. 49–54. VEDA, Bratislava, Czechoslovakia.
Eyme, J. (1965). *Botaniste* **48**, 99–155.
Fisher, D. B., and Jensen, W. A. (1969). *Planta* **84**, 122–133.
Folsom, M. W., and Cass, D. D. (1986). *Can. J. Bot.* **64**, 965–972.
Folsom, M. W., and Cass, D. D. (1989). *Can. J. Bot.* **67**, 2841–2849.
Folsom, M. W., and Cass, D. D. (1990). *Can. J. Bot.* **68**, 135–2147.
Folsom, M. W., and Peterson, C. M. (1984). *Bot. Gaz.* **145**, 1–10.
Fougère-Rifot, M. (1975). *C. R. Seances Acad. Sci. [III]* **280**, 2445–2447.
Fougère-Rifot, M. (1978). *Bull. Soc. Bot. Fr. Actual. Bot.* **125**, 207–213.
Fougère-Rifot, M. (1987). *Bull. Soc. Bot. Fr. Actual. Bot.* **134**, 113–160.
Fougère-Rifot, M. (1989). *Ann. Sci. Nat. Bot.* **10**, 49–62.
Fredrikson, M. (1990). *Am. J. Bot.* **77**, 123–127.
Fredrikson, M., Carlsson, K., and Franksson, O. (1988). *Nord. J. Bot.* **8**, 369–374.
Fukui, H., Nishimoto, K., and Nakamura, M. (1989). *J. Jpn. Soc. Hort. Sci.* **57**, 615–619.
George, R. A., George, G. P., and Herr, J. M., Jr. (1988). *Am. J. Bot.* **75**, 353–368.
Gifford, E. S., Jr., and Foster, A. S. (1988). "Comparative Morphology of Vascular Plants" 3rd ed. Freeman, New York.
Godineau, J. C. (1969). *Rev. Cytol. Biol. Vég.* **32**, 209–226.
Gori, P. (1977). *Isr. J. Bot.* **26**, 202–208.
Gunning, B. E. S., and Hardham, A. R. (1982). *Annu. Rev. Plant Physiol.* **33**, 651–698.
Gunning, B. E. S., and Pate, J. S. (1969). *Protoplasma* **68**, 107–133.
Habermann, A. (1906). *Beihefte Bot. Zentralbl.* **20**, 300–317.
Haig, D. (1986). *J. Theor. Biol.* **123**, 471–480.
Haig, D. (1990). *Bot. Rev.* **56**, 236–274.
Haig, D., and Westoby, M. (1986). In "Pollination '86" (E. G. Williams, R. B. Knox, and D. Irvine, eds.), pp. 211–214. School of Botany, University of Melbourne, Melbourne.

Haskell, D. A., and Postlethwait, S. N. (1971). *Am. J. Bot.* **58**, 595–603.

Hause, G., and Schröder, M. B. (1986). *Biol. Zentralbl.* **105**, 511–517.

Hepher, A., Sherman, A., Gates, P., and Boulter, D. (1985). In: "Experimental Manipulation of Ovule Tissues" (G. P. Chapman, S. H. Mantell, and R. W. Daniels, eds.), pp. 52–63. Longman, New York.

Herr, J. M. (1972). *Adv. Plant Morphol.* **1972**, 92–101.

Hu, S. Y., Li, L. G., and Zhou. C. (1985). *Acta Bot. Sin.* **27**, 337–344.

Huang, A. H., Trelease, H. C., and Moore, T. S. (1983). "Plant Peroxisomes," pp. 100–102. Academic Press, San Diego.

Huang, B. Q., and Russell, S. D. (1989). *Plant Physiol.* **90**, 9–12.

Huang, B. Q., and Russell, S. D. (1991). *Am. J. Bot.* **78**(Suppl.), 26 (Abstr.).

Huang, B. Q., and Russell, S. D. (1992a). *Sex. Plant Reprod.* **5**, in press.

Huang, B. Q., and Russell, S. D. (1992b). *Sex. Plant Reprod.* **5**, 151–155.

Huang, B. Q, Russell, S. D., Strout, G. W., and Mao, L. J. (1990). *Am. J. Bot.* **77**, 1401–1410.

Huang, B. Q., Pierson, E., Russell, S. D., Tiezzi, A., and Cresti, M. (1992). *Sex. Plant Reprod.* **5**, 156–162.

Jalouzot, M. F. (1978). *Bull. Soc. Bot. Fr. Actual. Bot.* **125**, 167–170.

Jensen, W. A. (1964). *J. Cell Biol.* **23**, 669–672.

Jensen, W. A. (1965a). *Am. J. Bot.* **52**, 238–256.

Jensen, W. A. (1965b). *Am. J. Bot.* **52**, 781–797.

Jensen, W. A. (1974). *In* "Dynamic Aspects of Plant Ultrastructure" (A. W. Robards, ed.), pp. 481–503. McGraw-Hill, New York.

Jensen, W. A., and Fisher, D. B. (1968). *Planta* **78**, 158–183.

Johnson, D. S. (1914). *Am. J. Bot.* **1**, 323–339.

Kapil, R. N., and Bhatnagar, A. K. (1981). *Int. Rev. Cytol.* **70**, 291–341.

Kapil, R. N., and Bhatnagar, A. K. (1983). *Phytomorphology* **23**, 157–167.

Keijzer, C. J., Reiders, M. C., and Lefrink-ten Klooster H. B. (1988). *In* "Sexual Reproduction in Higher Plants" (M. Cresti, P. Gori, and E. Pacini, eds.), pp. 119–124. Springer-Verlag, Berlin.

Kennell, J. C., and Horner, H. T. (1985). *Am. J. Bot.* **72**, 1553–1564.

Knox, R. B. (1984). *In* "Embryology of Angiosperms" (B. M. Johri, ed.), pp. 197–271. Springer-Verlag, Berlin.

Kranz, E., Bautor, J., and Lörz, H. (1990). *In* "Progress in Plant Cellular and Molecular Biology" (A. H. J. Nijamp, L. H. W. van der Plas, and J. van Aartrijk, eds.), pp. 252–257. Kluwer, Dordrecht.

Kranz, E., Bautor, J., and Lörz, H. (1991a). *Sex. Plant Reprod.* **4**, 12–16.

Kranz, E., Bautor, J., and Lörz, H. (1991b). *Sex. Plant Reprod.* **4**, 17–21.

Kuroiwa, H. (1989). *Bot. Mag. Tokyo* **102**, 9–24

Langlet, O. (1927). *Svensk. Bot. Tidskrift* **21**, 478–485.

Li, L. G., and Hu, S. Y. (1985). *Acta Bot. Sin.* **27**, 561–568.

Li, L. G., and Hu, S. Y. (1986). *Acta Biol. Exp. Sin.* **19**, 256–259.

Lintilhac, P. M. (1974a). *Am. J. Bot.* **61**, 135–140.

Lintilhac, P. M. (1974b). *Am. J. Bot.* **61**, 230–237.

Maheshwari, P. (1950). "An Introduction to the Embryology of Angiosperms." McGraw-Hill, New York.

Malik, C. P., and Vermani, S. (1975). *Acta Histochem.* **53**, 244–280.

Mansfield, S. G., Briarty, L. G., and Erni, S. (1991). *Can. J. Bot.* **69**, 447–460.

Mascarenhas, J. P., and Machlis, L. (1962). *Nature* **196**, 292–293.

Maze, J., and Lin, S. C. (1975). *Can. J. Bot.* **53**, 2958–2977.

Medina, F. J., Risueño, M. C., and Rodríguez-Garcia, M. I. (1981). *Planta* **151**, 215–225.

Meyer, B., and Stubbe, W. F. (1974). *Ber. Dtsch. Bot. Ges.* **87**, 29–38.

Mogensen, H. L. (1972). *Am. J. Bot.* **59**, 931–941.
Mogensen, H. L. (1978a). *Am. J. Bot.* **65**, 953–964.
Mogensen, H. L. (1978b). *Phytomorphology* **28**, 114–122.
Mogensen, H. L. (1981a). *Am. J. Bot.* **68**, 183–194.
Mogensen, H. L. (1981b). *Am. J. Bot.* **68**, 195–199.
Mogensen, H. L. (1982). *Carlsberg Res. Commun.* **47**, 313–354.
Mogensen, H. L. (1988). *Proc. Natl. Acad. Sci. U.S.A.* **85**, 2594–2597.
Mogensen, H. L., and Suthar, H. K. (1979). *Bot. Gaz.* **140**, 168–179.
Mól, R. (1986). *Plant Cell Rep.* **3**, 202–206.
Morrison, I. N., O'Brien, T. P., and Kuo, J. (1978). *Planta* **140**, 19–30.
Negi, D. (1974). Ph.D. Dissertation, University of California, Berkeley.
Newcomb, W. (1973). *Can. J. Bot.* **51**, 863–878.
Newcomb, W. (1978). *Can. J. Bot.* **56**, 483–501.
Newcomb, W., and Fowke, L. C. (1973). *Bot. Gaz.* **134**, 236–241.
Newcomb, W., and Steeves, T. A. (1971). *Bot. Gaz.* **132**, 367–371.
Olson, A. R. (1991). *Am. J. Bot.* **78**, 99–107.
Olson, A. R., and Cass, D. D. (1981). *Am. J. Bot.* **68**, 1333–1341.
Palser, B. F., Philipson, W. R., and Philipson, M. N. (1989). *Bot. J. Linn. Soc.* **101**, 363–393.
Philipson, M. N. (1977). *N. Z. J. Bot.* **15**, 777–778.
Philipson, M. N. (1981). *Acta Bot. Soc. Pol.* **50**, 151–152.
Plisko, M. A. (1971). *Bot. Zh.* **56**, 582–598.
Porsch, O. (1907). "Versuch einer phylogenetischen Erklärung des Embryosackes und der doppelten Befruchtung der Angiospermen." Jena.
Pritchard, H. N. (1964). *Am. J. Bot.* **51**, 371–378.
Qiu, D. B., Lu, F., and Xie, S. W. (1991). *Acta Bot. Sin.* **33**, 350–355.
Rembert, D. H., Jr. (1971). *Phytomorphology* **21**, 1–9.
Richter-Landmann, W. (1959). Planta **53**, 162–177.
Rifot, M. (1972). *Ann. Univ. A.R.E.R.S. (Reims)* **9**, 66–72.
Rodkiewicz, B. (1970). *Planta* **93**, 39–47.
Rodkiewicz, B. (1978). *Postepy Biologii Komorki* **5**, 135–154.
Rodkiewicz, B., and Mikulska, E. (1963). *Flora (Jena)* **154**, 383–387.
Rodkiewicz, B., and Mikulska, E. (1965a). *Flora (Jena)* **155**, 341–346.
Rodkiewicz, B., and Mikulska, E. (1965b). *Planta* **67**, 297–304.
Russell, S. D. (1979). *Can. J. Bot.* **57**, 1093–1110.
Russell, S. D. (1983). *Am. J. Bot.* **70**, 416–434.
Russell, S. D. (1987). *Theor. Appl. Genet.* **74**, 693–699.
Russell, S. D. (1990). *In* "Mechanism of Fertilization" (B. Dale, ed.), pp. 1–15. Springer-Verlag, Berlin.
Russell, S. D. (1991). *Annu. Rev. Plant Physiol. Plant Mol. Biol.* **42**, 189–204.
Russell, S. D., and Cass. D. D. (1988). *Am. J. Bot.* **75**, 778–781.
Russell, S. D., Huang, B. Q., and Strout, G. W. (1989). *In* "Some Aspects and Actual Orientations in Plant Embryology" (J. Pare and M. Bugnicourt, eds.), pp. 109–119. Université de Picardie, Picardie, France.
Russell, S. D., Rougier, M., and Dumas, C. (1990). *Protoplasma* **155**, 153–165.
Rutishauser, A. (1969). "Embryologie und Fortpflanzungsbiologie der Angiospermen." Springer-Verlag, Vienna.
Schacht, H. (1857). *Jahrbücher Wissenschaftliche Botanik* **1**, 193–232.
Schulz, P., and Jensen, W. A. (1973). *J. Cell Sci.* **12**, 741–763.
Schulz, P., and Jensen, W. A. (1981). *Protoplasma* **107**, 27–45.
Schulz, P., and Jensen, W. A. (1986). *Can. J. Bot.* **64**, 875–884.
Schulz, S. R., and Jensen, W. A. (1968). *Am. J. Bot.* **55**, 541–552.

Secor, D. L., and Russell, S. D. (1988). *Am. J. Bot.* **75**, 114–122.

Sehgal, C. B., and Gifford, E. M., Jr. (1979). *Bot. Gaz.* **140**, 180–188.

Sidorova, N. (1985). *Dokl. Akad. Nauk. Ukr. Ssr. Ser. B.* **12**, 63–66.

Strasburger, E. (1878). "Über Befruchtung und Zellteilung." Herman Dabis, Jena.

Sumner, M. J., and Van Caeseele, L. V. (1988). *Can. J. Bot.* **66**, 2459–2469.

Sumner, M. J., and Van Caeseele, L. V. (1989). *Can. J. Bot.* **67**, 177–190.

Sumner, M. J., and Van Caeseele, L. V. (1990). *Can. J. Bot.* **68**, 2553–2563.

Theunis, C. H., Pierson, E. S., and Cresti, M. (1991). *Sex. Plant Reprod.* **4**, 145–154.

Tilton, V. R. (1981). *New Phytol.* **88**, 505–531.

Tilton, V. R., and Lersten, L. R. (1981). *New Phytol.* **88**, 477–504.

Tilton, V. R., and Mogensen, H. L. (1979). *Phytomorphology* **29**, 338–350.

Tiwari, S. C. (1982). *Protoplasma* **110**, 1–4.

Tobe, H. (1989). *Bot. Mag. Tokyo* **102**, 351–367.

Torosian, C. (1972). Ph.D. Dissertation, University of California, Berkeley.

Tyrnov, V. S. , Enaleeva, N. K., and Knokhlov, S. S. (1975). *In* "Theses of Reports, XII International Botanical Congress," p. 266. Nauka, Leningrad.

Van der Maas, H. M., and Zall, M. A. C. M. (1990). *Acta Bot. Neerl.* **39**, 339.

Van der Pluijm, J. E. (1964). *In* "Pollen Physiology and Fertilization" (H. F. Linskens, ed.), pp. 8–16. North Holland, Amsterdam.

Van Lammeren, A. A. M., Keijer, C. J., Willemse, M. T. M., and Kieft, H. (1985). *Planta* **165**, 1–11.

Van Went, J. L. (1970a). *Acta Bot. Neerl.* **19**, 121–132.

Van Went, J. L. (1970b). *Acta Bot. Neerl.* **19**, 313–322.

Van Went, J. L., and Kwee, H. S. (1990). *Sex. Plant Reprod.* **3**, 257–262.

Van Went, J. L., and Willemse, M. T. M. (1984). *In* "Embryology of Angiosperms" (B. M. Johri, ed.), pp. 273–317. Springer-Verlag, Berlin.

Vannereau, A. (1978). *Bull. Soc. Bot. Fr. Actual. Bot.* **125**, 201–205.

Vazart, B., and Vazart, J. (1966). *Rev. Cytol. Biol. Vég.* **24**, 251–266.

Vazart, J. (1968). *C. R. Seances Acad. Sci. [III]* **266**, 211–219.

Vijayaraghavan, M. R., and Bhat, U. (1983). *Proc. Ind. Nat. Sci. Acad. Ser. B* **46**, 674–680.

Vijayaraghavan, M. R., Jensen, W. A., and Ashton, M. E. (1972). *Phytomorphology* **22**, 144–159.

Wagner, V. T., Song, Y, Matthys-Rochon, E., and Dumas, C. (1988). *In* "Sexual Reproduction in Higher Plants" (M. Cresti, P. Gori, and E. Pacini, eds.), pp. 125–130. Springer-Verlag, Berlin.

Wagner, V. T., Kardolus, J. P., and Van Went, J. L. (1989a). *Sex. Plant Reprod.* **2**, 219–224.

Wagner, V. T., Song, Y, Matthys-Rochon, E., and Dumas, C. (1989b). *Plant Sci.* **59**, 127–132.

Wagner, V. T., Dumas, C., and Mogensen, H. L. (1990). *Theor. Appl. Genet.* **79**, 72–76.

Webb, M. C., and Gunning, B. E. S. (1990). *Sex. Plant Reprod.* **3**, 244–256.

Webb, M. C., and Gunning, B. E. S. (1991). *Planta* **184**, 187–195.

Willemse, M. T. M. (1981). *Phytomorphology* **31**, 124–134.

Willemse, M. T. M., and Bednara, J. (1979). *Phytomorphology* **29**, 156–165.

Willemse, M. T. M., and Franssen-Verheijen, M. A. W. (1978). *Bull. Soc. Bot. Fr. Actual. Bot.* **125**, 187–191.

Willemse, M. T. M., and Franssen-Verheijen, M. A. W. (1988). *In* "Sexual Reproduction in Higher Plants" (M. Cresti, P. Gori, and E. Pacini, eds.), pp. 357–362. Springer-Verlag, New York.

Willemse, M. T. M., and Van Lammeren, A. A. M. (1988). *Sex. Plant Reprod.* **1**, 74–82.

Willemse, M. T. M., and Van Went, J. L. (1984). *In* "Embryology of Angiosperms" (B. M. Johri, ed.), pp. 159–196. Springer-Verlag, Berlin.

Williams, E. G., Knox, R. B., and Rouse, J. L. (1982). *J. Cell Sci.* **53**, 255–277.

Wilms, H. J. (1981a). *Acta Bot. Neerl.* **30,** 75–99.

Wilms, H. J. (1981b). *Acta Soc. Bot. Pol.* **50,** 165–168.

Wilms, H. J. (1981c). *Acta Bot. Neerl.* **30,** 101–122.

Woodcock, C. L. F., and Bell, P. R. (1968). *J. Ultrastruct. Res.* **22,** 546–563.

Wu, Y., and Zhou, C. (1988). *Acta Bot. Sin.* **30,** 210–212.

Yan, H., Yang, H. Y., and Jensen, W. A. (1990). *Acta Bot. Sin.* **32,** 165–171.

Yan, H., Yang, H. Y., and Jensen, W. A. (1991). *Can. J. Bot.* **69,** 191–202.

Yang, H. Y. (1989). *Acta Bot. Sin.* **31,** 817–823.

Yang, H. Y., and Zhou, C. (1984). *Acta Bot. Sin.* **26,** 342–346.

You, R. L., and Jensen, W. A. (1985). *Can. J. Bot.* **63,** 163–178.

Yu, S. H., and Chao, C. Y. (1979). *Caryologia* **32,** 147–160.

Zhou, C. (1985). *Acta Bot. Sin.* **27,** 258–262.

Zhou, C. (1987). *Plant Sci.* **52,** 147–152.

Zhou, C., and Yang, H. Y. (1982). *Acta Bot. Sin.* **24,** 403–407.

Zhou, C., and Yang, H. Y. (1984). *Acta Biol. Exp. Sin.* **17,** 141–147.

Zhou, C., and Yang, H. Y. (1985). *Planta* **165,** 225–231

Zhou, C., and Yang, H. Y. (1986). *In* "Haploids of Higher Plants *In Vitro*" (H. Hu and H. Y. Yang, eds.), pp. 192–203. Chinese Academic, Beijing.

Part III
Progamic Phase and Fertilization

A Dynamic Role for the Stylar Matrix in Pollen Tube Extension

Luraynne C. Sanders[1] and Elizabeth M. Lord

Department of Botany and Plant Sciences, University of California, Riverside, Riverside, California 92521

I. Introduction

Sexual reproduction in flowering plants occurs by the processes of pollination and fertilization. Pollination involves transfer of the pollen grain (the male gametophyte) to the stigma, its adhesion, hydration, and production of a pollen tube. The pollen tube extends through the gynoecium (stigma, style, and ovary), carrying the sperm cells that are delivered ultimately to the embryo sac (the female gametophyte) within the ovule. The extension of the pollen tube follows a defined secretory path within the gynoecium called the transmitting tract or stylar matrix. Pollen tube extension is typically restricted to this specialized extracellular matrix (Knox, 1984).

The role of the gynecium in pollen tube growth has been examined almost solely in self-incompatible species. Self-incompatibility is the inability of self-pollen to effect fertilization and is under genetic control in plants (Lewis, 1979; Heslop-Harrison, 1983). The genes involved are called *S* genes and their products are thought to be responsible for stopping the extension of pollen tubes by a biochemical recognition event (Lewis, 1979). This type of recognition event in plants has often been compared with the immune system in animals because each is capable of distinguishing between self and nonself (Mascarenhas, 1978; Nasrallah, 1989). In this model it is proposed that a component of the stylar matrix recognizes something in the pollen grain or tube that causes repression of pollen germination or tube extension. This role is purportedly met by complementary *S* gene products in the pollen and style. The *S* gene products have been identified and localized in the secretory matrix of the style and stigma, but as of yet the pollen *S* gene product has not been identified (Nasrallah

[1] Present address: Department of Immunology, Scripps Clinic and Research Foundation, 10666 North Torrey Pines Road, La Jolla, CA 92037.

297

et al., 1985; Anderson *et al.*, 1986, 1989; Cornish *et al.*, 1987; Kandasamy *et al.*, 1989; Dickinson *et al.*, this volume). In *Nicotiana alata* (tobacco) the stylar *S* gene product has been shown to be a ribonuclease (McClure *et al.*, 1989). The suggestion has been made that *S* gene products may also have a role associated with compatible pollination (Sanders and Lord, 1989; Nasrallah, 1989). Because self-incompatibility is likely a derived condition in plants (Nasrallah, 1989), a better knowledge of the events occurring during pollination in compatible systems, which predominate in the angiosperms, may shed light on the mechanisms of self-incompatibility. There is no doubt that the gynoecium plays a role in screening pollen in self-incompatible systems, but data are accumulating that it plays this role as well in self-compatible systems. Genetic data suggest that gynoecial factors may determine the pattern of pollen tube behavior, such as order of fertilization and rates of pollen tube growth (Ottaviano *et al.*, 1975; Stephenson and Bertin, 1983), and there is *in vitro* evidence that the gynoecium can screen compatible pollen (Malti and Shivanna, 1985).

Although the exogenous growth requirements were verified for pollen tubes many years ago and *in vitro* pollen tube growth is now commonly achieved, the tubes typically do not grow as fast, or as long, in culture as they do *in vivo* and often exhibit erratic changes in direction or wandering (Shivanna *et al.*, 1979; Heslop-Harrison and Heslop-Harrison, 1986). The present model for pollen tube extension is based on *in vitro* observations, and although it clearly accounts for and describes the capability of pollen tubes to extend in culture (Steer and Steer, 1989), it falls short of explaining the variety of pollen tube behaviors *in vivo*. The focus of this model is on the growth of the pollen tube wall at the tip and not on the movement of the cells inside the tube. In a review on the role of actin in tip growth, Steer (1990) postulated that the pollen tube tip "retains many of the attributes of primitive amoeboid motion." As we proposed in our model (Sanders and Lord, 1989), the vegetative cell and two sperm cells do not grow as such; rather they move through the stylar matrix to the ovules leaving a trail of cell wall behind. Therefore, the original question remains unanswered: What governs the *in vivo* rate and directionality of the pollen tube? Why are we unable to mimic the gynoecial environment in culture?

The secretions of the stigma and style are thought to supply nutrients to the pollen tube; hence they are viewed as culture medium (Labarca and Loewus, 1973; Knox, 1984). The unstated assumption here is that the interaction of the gynoecium with the pollen tube is biologically passive. The notable exception to this is the stylar S glycoprotein involved in the self-incompatibility reaction. However, in animal systems it has long been recognized that the extracellular matrix (ECM), a conventional term used to describe secretory matrices, performs active biological roles. For example, the ECM is known to facilitate cell migration in certain animal systems

(Dufour *et al.*, 1988; Hay, 1981a; Hynes, 1981). Could the stylar matrix facilitate pollen tube extension in some similar manner? During animal embryogenesis cells migrate from one area to another to carry out their biological functions. Similarly, the pollen tube functions to move the vegetative cell and the two sperm cells through the gynoecium to the ovules where these cells ultimately carry out their biological function. The protoplasm of the vegetative cell and sperm cells are sequestered at the tip of the extending tube, and there is no appreciable increase in cytoplasmic volume (Shivanna *et al.*, 1979). Consequently, the pollen tube could be considered analogous to a group of migrating cells, leaving a cell wall trail. In this chapter we present data to support this view of pollination and provide a model for how the pollen tube may be actively guided to the ovules by the secretions of the style. We first consider the principles of cell–ECM interactions in animal systems because they are better understood and there may be underlying similarities in the fundamental mechanisms involved in both.

II. Extracellular Matrices

A. The Extracellular Matrix in Animal Systems: Structure and Function

Cells of all organisms produce compounds that are secreted into the environment to form an ECM. The ECM in animals has long been recognized as having a twofold function: one in support and anchorage and another more dynamic role in growth and development (Grinnell, 1978; Hay, 1981b; Edelman, 1988; Adair and Mecham, 1990). The ECM in animal systems has been shown to play an active role in developmental processes, such as cellular polarity, differentiation, cell division, cell death, and cell migration (Hay, 1981b). The ECM of animal cells is composed of a variety of molecules, including various collagens, glycoproteins, and proteoglycans. These ECMs have a fibrillar nature when viewed with a transmission electron microscope (TEM). Molecules that are thought to confer these dynamic capabilities to the ECM belong to a family of secretory adhesive glycoproteins called substrate adhesion molecules (SAMs) (Edelman, 1988). Several SAMs have been described and localized in animal tissues; these include fibronectin, von Willebrand factor, fibrinogen, and vitronectin (Hayman *et al.*, 1985; Edelman, 1988). Such glycoproteins interact with receptors in the plasma membrane and mediate cell adhesion via a common Arg-Gly-Asp (RGD) sequence (Pierschbacher and Ruoslahti, 1984; Cheresh, 1987; Cheresh and Spiro, 1987). These RGD-directed receptors belong

to a superfamily of cell surface receptors called integrins (Hynes, 1987; Ruoslahti and Pierschbacher, 1987). The integrins are believed to interact with the cytoskeleton via actin-associated proteins, such as vinculin and talin (Hynes, 1987; Burridge *et al.*, 1988), forming a link between the inside and outside compartments of the cell.

Some of these SAMs are multifunctional glycoproteins that also occur as plasma proteins, namely, fibronectin and vitronectin. Both glycoproteins are involved in processes, such as wound healing, hemostasis, and opsonization (Furcht, 1983; Preissner, 1989). Fibronectin and vitronectin are SAMs that have been implicated in facilitating cell spreading and/or cell migration (Grinnell, 1978; Hynes, 1981; Barnes *et al.*, 1983; Preissner, 1989). Fibronectin, probably the most extensively characterized SAM, is believed to be the component in the ECM responsible for facilitating cell migration during embryogenesis (Hynes, 1981; Dufour *et al.*, 1988). Fibronectin is broadly distributed in most ECMs and is ubiquitous in the embryo. Inert latex particles placed within the neural crest region of a chicken embryo translocate in the ECM in a pattern that mimics neural crest cell migration (Bronner-Fraser, 1982); however, if these particles are coated with fibronectin, translocation does not occur (Bronner-Fraser, 1985). *In vitro* experiments showed that cells and latex particles preferentially migrate on substrates containing fibronectin (Newman *et al.*, 1985). Vitronectin, also known as serum spreading factor, S protein, and epibolin (Stenn, 1981; Tomasini and Mosher, 1986; Preissner, 1989), is thought to be biologically more active *in vitro* with regard to cell spreading, cell attachment, and growth than fibronectin (Hayman *et al.*, 1985; Underwood and Bennett, 1989). Vitronectin is also broadly distributed and has been localized in the ECM of human embryonic lung, fetal membranes, striated skeletal muscle, skin, and kidney (Hayman *et al.*, 1983). The *in vivo* function of vitronectin within the ECM is not well understood, although it has been suggested that it may be involved in cell migration (Preissner, 1989). It has been demonstrated that vitronectin is involved in promoting extension of retinal neurons during chick embryo development (Neugebauer *et al.*, 1991).

B. Extracellular Matrices in Plants

1. The Cell Wall

The ECM that encompasses each plant cell is typically referred to as the cell wall (Lamport and Catt, 1981). The structural components of the wall, or ECM, in plant cells are composed of a variety of polysaccharides and glycoproteins (Ericson and Elbein, 1980; Varner and Hood, 1988; Roberts, 1990). Historically, the cell wall has been viewed as inert support material

(Darvill *et al.*, 1980), consequently conferring a passive biological role to it. This view has been changing and a more dynamic role has been proposed for the plant cell wall, one involving active participation in developmental processes (Roberts, 1989, 1990; Adair and Mecham, 1990). The functions of the cell wall with regard to cell shape, morphogenesis, and defense have become active areas of research (Roberts, 1989).

It has been suggested that an interaction between the cell wall and plasma membrane is important in plant developmental processes (Green, 1986; Roberts, 1989; Schindler *et al.*, 1989). The portion of the ECM immediately outside the plasma membrane, defined as the cell surface, is known to contain a specific family of arabinogalactan proteins (AGPs) (Knox *et al.*, 1989; Pennell *et al.*, 1989). One group of these AGPs, which are identified by a monoclonal antibody specific to a particular sugar moiety, was determined to be a plasma membrane-associated protein and was found in all tissues examined in several species of flowering plants (Pennell *et al.*, 1989). Specific functions have not yet been attributed to AGPs; however, it has been proposed that this particular family of AGPs, localized to the cell surface, may function in attaching the plasma membrane to the cell wall polysaccharides (Pennell *et al.*, 1989). These AGPs may also be important in determining sexual development in plants because these glycoproteins are absent in the germ tissue of the anther and ovule and reappear in the embryo at approximately the heart stage (Pennell and Roberts, 1990). A group of glycoproteins that are secreted into the culture medium from *Daucus carota* (carrot) somatic embryos was isolated and shown to be important in inducing somatic embryogenesis. Absence, or aberrant forms, of these secretory proteins impedes *Daucus carota* somatic embryogenesis (De Vries *et al.*, 1988; Lo Schiavo *et al.*, 1990). ECM molecules are also thought to be important in developmental processes in algae. In the green alga *Chlamydomonas* cell wall proteins are known to be critical in recognition events during mating (Adair, 1988). In the brown alga *Fucus* the cell wall is necessary before axis fixation of the embryo can occur (Kropf *et al.*, 1988). Consequently, the function of the ECM in plant development may rely on molecular components similar to those of animal systems.

To establish the presence in plants of SAMs known to be involved in cell movement in animals, we used antibody probes to human vitronectin and verified cross-reactivity with a 55-kDa protein in four species of flowering plants (Sanders *et al.*, 1991). The presence of such molecules suggests an underlying mechanistic similarity between plant and animal ECMs.

2. The Stylar Matrix: A Specialized Type of Extracellular Matrix

a. Structure and Chemical Composition   There is extensive literature on the structure and chemical composition of the stylar ECM that supports

pollen tube growth (Knox, 1984). This ECM is produced on the surface of the cells along which the pollen tube travels. In some cases the tubes may travel in the wall itself (Jensen and Fisher, 1969; Rendle and Murray, 1988) or in a portion of the disintegrating outer wall, as well as in the secretions (Sanders *et al.*, 1990). In many cases, in which the ultrastructure of the transmitting tract has been examined, the boundary between the secreting matrix and the wall of the transmitting tract cells is not well defined. Often the secretions themselves have a fibrillar structure (Fig. 1) (Bell and Hicks, 1976; Hill and Lord, 1987; Sanders *et al.*, 1990), providing a surface for pollen tube interaction. Further structural studies are warranted to define the nature of this specialized matrix in styles.

The secretions produced by transmitting tract cells are frequently described as mucilage or as a fluid composed of various ingredients. Analyses

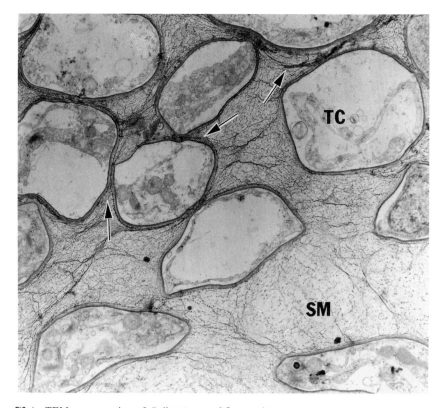

FIG. 1   TEM cross section of *Collomia grandiflora* style. Arrows are on regions where the wall is unraveling. ×6400. SM, stylar matrix; TC, transmitting tract cell. Reproduced with permission from Sanders *et al.*, (1990); copyright 1990 by Springer-Verlag.

of these secretions have demonstrated the presence of a variety of compounds that vary depending on the species examined. In general, these include sugars, amino acids, peptides, phenolic compounds, fatty acids, lipids, and glycolipids (Knox, 1984; Heslop-Harrison, 1983). The high molecular weight compounds are polysaccharides (mucilages, pectic compounds, and proteoglycans) and complex proteins (glycoproteins and lipoproteins) (Heslop-Harrison, 1983). Arabinogalactans appear to be widespread in such secretions and are thought to play a role in adhesion on the stigma (Clarke *et al.*, 1979), but no functional role has previously been assigned to them in the style (Clarke, 1981; Sedgley *et al.*, 1985).

b. Role in Pollination   Pollen tube extension has long been observed to follow the secretions of the transmitting tract, although the fact that they are normally restricted only to this tissue is less appreciated. The pollen tubes never cross free space *en route,* and in tracking the secretions from the stigma to the micropyle their route conforms precisely to the transmitting tract without penetrating secreting cells (Hill and Lord, 1987; Lord and Kohorn, 1986). The path may vary from that in a solid style, which is intercellular, to that of an epidermis in a hollow style. No role has been assigned to the ECM of the transmitting tract cells other than that of a nutrient medium and hence pollen tube guide. The analogy often made is that of a nutrient trail guiding a fungal hypha *in vitro.* This suggests a passive role for the gynoecium and one that could be mimicked easily *in vitro* by a defined medium.

The failure to achieve an adequate *in vitro* system was discussed earlier. The general assumption is that the ingredients of the medium are incomplete, and other, as yet undiscovered, components must be added to achieve a proper *in vitro* system. Chemotropic signals were once proposed to explain the apparent guidance of the tube to the ovules, and a Ca^{2+} gradient was shown to have this effect on pollen tubes *in vitro* (Mascarenhas, 1973; Rosen, 1964; Shivanna *et al.*, 1979). No such gradients were seen *in vivo,* however, and pollen tubes were shown to track secretions in either direction on a stylar segment *in vivo* (Mulcahy and Mulcahy, 1987). The generally accepted model now is one of guidance by the restricted location of the nutritive secretions in the transmitting tract (Heslop-Harrison and Heslop-Harrison, 1986).

One avenue of research that contradicts this conclusion is that of prezygotic gametophytic competition (Mulcahy, 1975; Ottaviano *et al.*, 1975; Stephenson and Bertin, 1983; Hill and Lord, 1986). Data accumulating in this field are suggestive of a role for the gynoecium (female choice) in determining pollen tube behavior in the style. Because of these experiments involving compatible pollinations, we have a better appreciation for the normal variability in pollen tube growth rates in fertilization. It is now

established that variation in these rates *in vivo* is a function of both maternal and paternal genotype. That the secretions of the gynoecium may act as a selective force on the pollen tubes suggests a recognition event in self-compatible crosses that plant cell biologists have not explored.

Other observations from compatible crosses further suggest an active role for the gynoecium in pollen tube growth. One such example is the common observation that many more pollen tubes germinate on the stigma than penetrate the ovary (Sedgley, 1976; Lord and Kohorn, 1986; Beck and Lord, 1988; Rendle and Murray, 1988). This bottleneck in the upper style is striking in legumes where the stigma is broad but the stylar matrix is narrow and can support only a limited number of pollen tubes (Lord and Kohorn, 1986). In species with only one ovule, penetration of one pollen tube into the ovary is frequently seen even though many pollen grains germinate on the stigma (Sedgley, 1976; Rendle and Murray, 1988; Beck and Lord, 1988). In some species primary pollinations have been observed to severely limit pollen tube growth in secondary pollinations (Eenink, 1982; Marshall and Ellstrand, 1985; Epperson and Clegg, 1987). These observations together suggested a more complex interaction between the pollen tube and the ECM of the gynoecium than has been proposed in the past. To examine this proposition more closely, we applied inert latex particles to the stylar ECM in three plant species to determine whether these secretions function in a manner similar to those of the ECM in chick embryos during neural crest cell migration (Sanders and Lord, 1989). We also examined the occurrence and distribution of a variety of classical SAMs known to be involved in cell movement, in the stylar secretions of *Vicia faba* (Sanders *et al.*, 1991).

III. Evidence for an Active Biological Role of the Stylar Matrix

A. Latex Bead Translocation

Given the parallels between the ECM of animals and plants, and the way in which pollen tube growth can also be viewed as a special case of cell movement, we have proposed that the stylar ECM actively facilitates pollen tube extension. To test this hypothesis, we initially placed 6-μm red-dyed latex beads (because a 6-μm bead approximates the size of a pollen tube tip) onto the stigmatic end, or the cut stigmatic end, of gynoecia from three species of flowering plants (Figs. 2A and B) (Sanders and Lord, 1989). The three species used were *Hemerocallis flava* (daylily), *Raphanus raphanistrum* (wild radish), and *Vicia faba* (broad bean). In all

three cases bead translocation did occur. In *Hemerocallis flava* the red beads could be seen through the translucent tissue of the style, moving in a front composed of hundreds of beads. Measurements of this movement were made and rates calculated (Table I). In *Hemerocallis flava* pollen tube extension also occurs in a front containing hundreds of tubes. The normal site of pollen tube extension in daylily is along the transmitting tract cells of the secretory epidermis, beneath the cuticle of the hollow style. The latex beads traveled on the same path (Figs. 2C and D). In *Raphanus raphanistrum,* which has a solid style, the secretory matrix is intercellular and pollen tube extension is restricted to this area (Hill and Lord, 1987), as was bead translocation. In *Vicia faba* the beads traveled only on the localized transmitting tract that normally supports pollen tubes in the hollow portion of the style and ovary (Figs. 2E and F). In all three species bead translocation mimicked pollen tube extension in three critical ways: (1) **kinetics,** in all cases the rate of bead translocation was not significantly different from pollen tube extension rates *in vivo;* (2) **location,** in all cases bead translocation occurred in the same area as pollen tube extension on the transmitting tract; and (3) **numbers,** the number of beads translocated in each case was similar to the number of pollen tubes the style typically supported.

B. The Stylar Matrix of *Vicia faba*

Vicia faba has an intermediate type of style; the upper part is solid and the lower part is hollow (Fig. 2B). In the hollow part of the style and in the ovary, the few transmitting tract cells line one side of the gynoecium, and it is to this area that pollen tube extension is restricted (Fig. 2E,5D).

Pollen tubes may exhibit polarity and are often divided into four zones (Cresti *et al.,* 1977; Steer and Steer, 1989), each of which can be distinguished at the ultrastructural level. The most distal zone, referred to as the apical zone, has densely staining cytoplasm and contains numerous vesicles (Fig. 3), which migrate to the tip and secrete cell wall material. The subapical zone is composed of numerous Golgi bodies, mitochondria, endoplasmic reticulum, and vesicles. The nuclear zone contains the vegetative nucleus, generative cell (or two sperm), and few other organelles. And, proximally, the vacuolar zone occurs, which is also the area of callose plug formation. These zones are all easily distinguishable at the ultrastructural level and have been further subdivided by other researchers (Steer and Steer, 1989). Because of this polarity, interaction between certain zones of the pollen tube and the stylar matrix can be resolved. Our preliminary data in *Vicia faba* suggest that a tight association occurs between the pollen tube and the stylar matrix away from the tube apex in

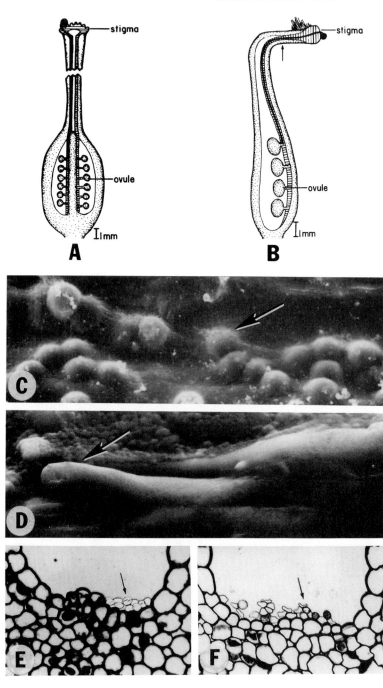

the area of the vacuolar zone (Fig. 3). At the extreme pollen tube tip in the apical zone, the interaction between the tube and stylar matrix is loose in appearance (Fig. 3).

C. Occurrence of the Substrate Adhesion Molecule Vitronectin in Plants

Based on the bead data and other similarities observed between a pollen tube and migrating cells, we hypothesized that the stylar matrix actively facilitates pollen tube extension via a biochemical recognition–substrate adhesion system similar to that proposed in animal cell migration during embryogenesis by Dufour *et al.* (1988). This would imply the presence of SAMs and their receptors in plants. To address this question we probed plant tissue with existing commercially available animal SAM probes. We started with probes to fibronectin because of its implication in cell migration and because it is the most researched SAM; results were inconclusive (Sanders *et al.*, 1991). Schindler *et al.* (1989), using a vitronectin receptor polyclonal antibody, detected proteins similar to the β-subunit of the human vitronectin receptor (an integrin) in plasma membranes of cultured soybean cells.

To establish whether there are vitronectin-like proteins in plants, we used immunoblots containing protein extractions from four plant species: *Lilium longiflorum* (Easter lily), *Vicia faba* (broad bean), *Glycine max* (soybean), and *Lycopersicon esculentum* (tomato), and two organs: leaf and root (Sanders *et al.*, 1991). When incubated with human vitronectin antiserum, these blots showed a strongly resolved single 55-kDa protein band (Fig, 4A). These protein bands were not detected if the blots were incubated with nonimmune serum (Fig. 4B). When purified IgG from human vitronectin antiserum is preincubated in a 10-molar excess of human vitronectin, the immunological cross-reactivity of the 55-kDa plant

---

FIG. 2 Diagrams, SEMs, and plastic-embedded sections of beads and pollen tubes in the transmitting tract. (A, B) Diagrams of longitudinal sections, showing the path pollen tubes travel in compatible pollinations. The transmitting tract is represented by cross-hatching. (A) *Hemerocallis flava*. (B) *Vicia faba*. Arrow indicates where gynoecium was cut for bead application. (C, D) SEMs of *Hemerocallis flava*. (C) Surface of transmitting tract with beads (arrow) beneath the cuticle that overlies the stylar matrix. ×1150. (D) Pollen tubes (arrow) on the transmitting tract. ×1150. (E and F) Plastic-embedded cross sections of *Vicia faba* showing beads and pollen tubes on the transmitting tract. (E) Beads (arrow). ×210. (F) Pollen tubes (arrow). ×210. Reproduced with permission from Sanders and Lord (1989); copyright 1989 by the AAAS.

TABLE I

Rate of Bead Movement versus Pollen Tube Extension (μm/min ± SD)[a]

	Beads	Pollen tubes
Hemerocallis flava	59.5 ± 17.6 (n = 7)	57.2 ± 23.9 (n = 5)
Raphanus raphanistrum	20.7 ± 10.8 (n = 15)	16.1 ± 7.3 (n = 20)
Vicia faba	10.8 ± 3.2 (n = 6)	15.4 ± 4.8 (n = 12)

[a] Rates of bead movement versus pollen tube extension in the style show no significant differences at $p < .02$, Mann-Whitney U Test. From Sanders and Lord (1989), with permission; copyright 1989 by the AAAS.

protein was drastically reduced. Immunoblots containing plant protein extracts, when incubated in a monospecific antibody purified from human vitronectin, also showed cross-reactive bands at 55 kDa. Alternatively, immunoblots incubated in a monospecific antibody purified from the 55-kDa protein of *Lilium longiflorum* and *Vicia faba* roots recognized the 55-kDa protein in leaf extracts as well as the 65- and 75-kDa human vitronectin proteins (Sanders *et al.*, 1991). These data demonstrated that the 55-kDa proteins from these four plant species are immunologically

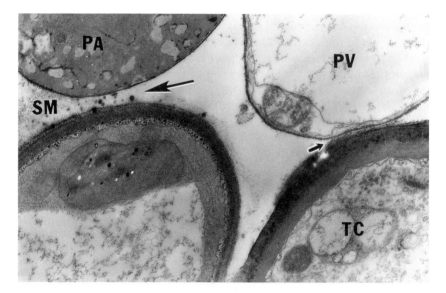

FIG. 3   TEM cross section of *Vicia faba* gynoecium. ×11,800. Large arrow indicates area where pollen tube–stylar matrix interactions appear loose. Small arrow indicates area where pollen tube–stylar matrix interactions appear tight. SM, stylar matrix; TC, transmitting tract cell; PA, apical zone of a pollen tube; PV, vacuolar zone of a pollen tube.

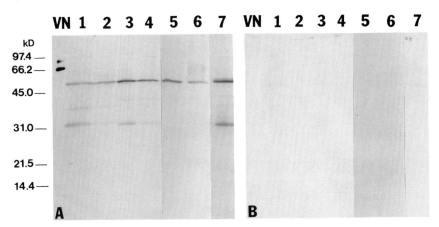

FIG. 4 Detection of a 55-kDa plant protein by human vitronectin antiserum. Lanes 1–7 are protein extractions from plant tissue: lane 1, *Lilium longiflorum* leaf; lane 2, *Lilium longiflorum* root; lane 3, *Vicia faba* leaf; lane 4, *Vicia faba* root; lane 5, *Glycine max* leaf; lane 6, *Glycine max* root; lane 7, *Lycopersicon esculentum* leaf. The lane labeled VN contains 100 ng purified human vitronectin. (A) Protein blot incubated with human vitronectin antiserum. (B) Protein blot incubated with nonimmume serum. Reproduced with permission from Sanders *et al.* (1991); copyright 1991 by the ASPP.

related to human vitronectin. Although there is a difference in the molecular weight between the human vitronectin (65–75 kDa) and plant vitronectin (55 kDa), the molecular weights of animal vitronectins vary from 56 to 80 kDa (Kitagaki-Ogawa *et al.*, 1990). This 55-kDa plant protein is at the lower end of this range.

Frozen sections of *Lilium longiflorum* leaf and *Vicia faba* gynoecium were used to localize the vitronectin-like protein by immunofluorescence (Fig. 5). *Lilium longiflorum* leaf sections incubated in human vitronectin antiserum revealed an ubiquitous pattern of fluorescence in the tissue. The fluorescence appeared in patches in the innermost part of the wall around each cell, which has been referred to as the cell surface (Figs. 5A–C). Cross sections of *Vicia faba* gynoecial tissue incubated in human vitronectin antiserum exhibited a similar distribution of fluorescence in all cells, except for the transmitting tract cells in which fluorescence was more intense and continuous around the cells (Figs. 5D–F).

To determine whether vitronectin-like sequences could be detected in plants, plant genomic DNA blots were hybridized with a ^{32}P-labeled human vitronectin cDNA probe. Several DNA fragments in *Vicia faba*, *Glycine max*, and *Lycopersicon esculentum* were visualized with the human vitronectin probe (Fig. 6). We were unable to detect any vitronectin-like sequences in *Lilium longiflorum* using this probe; however, the genome

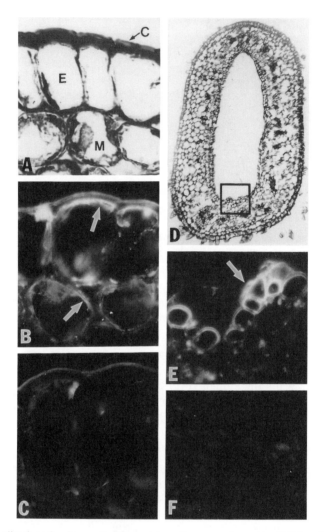

FIG. 5   Localization of a vitronectin-like protein on frozen sections. (A–C) *Lilium longiflorum* leaf cross sections, (D–F) *Vicia faba* gynoecium cross section. (A) Section stained with toluidine blue O. ×400. (B) Section incubated with human vitronectin antiserum. ×400. (C) Section incubated with nonimmune serum. ×400. (D) Section of gynoecium stained with toluidine blue O. Boxed area indicates transmitting tract. ×50. (E) Section of transmitting tract incubated with human vitronectin antiserum. ×495. (F) Section of transmitting tract incubated with nonimmune serum. ×495. E, epidermis; M, mesophyll; C, cuticle; white arrows indicate areas of immunoreactivity. Reproduced with permission from Sanders *et al.* (1991); copyright 1991 by the ASPP.

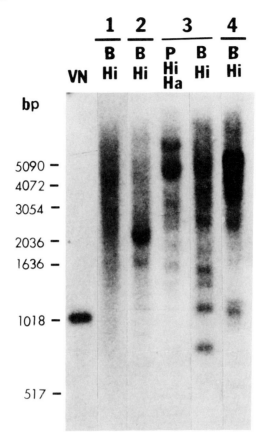

FIG. 6   Detection of vitronectin-like sequences using DNA hybridization blot analysis. Lanes 1–4 represent genomic DNA from *Lilium longiflorum, Vicia faba, Glycine max,* and *Lycopersicon esculentum,* respectively. All lanes were hybridized with a human ^{32}P-labeled vitronectin cDNA probe. Two picograms of human vitronectin cDNA served as a positive control (VN). The position of the 1kb DNA ladder (BRL) markers is indicated. B, *Bam*HI; Ha, *Hae*III; Hi, *Hind*III; P, *Pvu*II. Reproduced with permission from Sanders *et al.* (1991); copyright 1991 by the ASPP.

of lily is 10–50 times larger than the other three species (Bennett and Smith, 1976), which may explain why no strong hybridization signals were observed. The immunoblots (Fig. 4A) and the immunofluorescence (Fig. 5B) indicated that a vitronectin-like product also exists in *Lilium longiflorum.* Together, the genomic DNA blot, immunoblots, and immunofluorescence studies provide strong evidence that a molecule similar to human vitronectin exists in plants. Sequencing of the vitronectin-like genes and gene products in plants is necessary to determine the degree of

similarity between plant and human vitronectins. Two recent studies have confirmed these results: one in plants (Zhu *et al.*, 1991) and one in algae (Quatrano *et al.*, 1991). In both cases antiserum to human vitronectin was used and showed cross-reactivity to plant proteins extracted from cultured tobacco cells (Zhu *et al.*, 1991) and from *Fucus* embryos (Quatrano *et al.*, 1991).

D. A Model for Pollen Tube Extension Via the Stylar Matrix

The fact that a SAM-like molecule, which has been implicated in facilitating cell spreading and/or migration in animal systems occurs in plants, lends credence to the proposal that pollen tube extension is a special case of cell migration in plants (Sanders and Lord, 1989). Immunolocalization with the human vitronectin antiserum showed a strong reaction on the transmitting tract cell surfaces that function to support pollen tubes (Fig. 5E). Based on these results, as well as the extensive literature on pollination, we have proposed a model for compatible pollen tube extension in the gynoecium (Fig. 7). Our proposal is based on a model reported by Dufour *et al.* (1988) to explain the role of SAMs in both stationary cell adherence as well as the transitory adherence that occurs during cell motility in animals. According to this model, in the stationary cell the

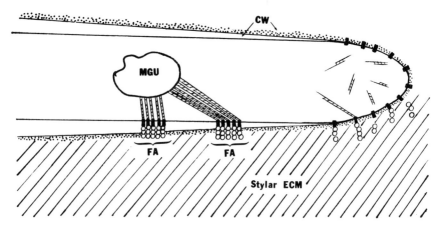

FIG. 7   A model for the role of SAMs in cell movement during pollination. The pollen tube tip is diagrammed as it progresses through the stylar ECM. The cell wall of the pollen tube (CW) is thin at the tip (stippling) and thicker and more complex back from the tip. The male germ unit (MGU) is composed of the tube cell nucleus and the generative cell (or two sperm). ■, integrin; ⌐⌐⌐⌐ , microfilament; ○, SAM (i.e., vitronectin); FA, focal adhesion.

binding affinity of the SAM for its receptor is strong and the microfilaments of the cell bind to the receptor indirectly, forming a connection with the cytoskeleton and creating focal adhesions (Burridge *et al.*, 1988). In the motile cell the affinity between the SAM and its receptor is weak and only transitory attachments occur. Inside the cell microfilaments form a loose arrangement.

We have adopted this model for pollen tube extension because both conditions probably occur at once in the pollen tube: a zone of motility in the extreme tip, which is extending forward and, back from the tip, a stationary adhering zone (Fig. 7). Literature on the cytoskeleton of the pollen tube shows a loose arrangement of microfilaments at the growing tip (Lancelle *et al.*, 1987; Tiwari and Polito, 1988). Further back the microfilaments of the cytoskeleton are more organized. Only when pollen tubes have been grown *in vivo* are focal adhesions observed (Pierson *et al.*, 1986). Focal adhesions in animal cells are links between the cytoskeleton and the ECM (Bissell *et al.*, 1982; Burridge *et al.*, 1988). A connection between the microfilaments surrounding the nucleus of the pollen tube and its associated generative cell (or two sperm cells) (Tiwari and Polito, 1988) with the cortical cytoskeleton of the tube cell has been described in pollen tubes (Heslop-Harrison and Heslop-Harrison, 1989).

The pollen tube cell produces its own ECM, or wall, which occurs in three layers back from the tip: an inner callosic layer overlaid by cellulose, which is covered by pectins (Heslop-Harrison, 1987; Steer and Steer, 1989). The wall at the expanding tip of the pollen tube is thin and primarily pectic in nature. We propose that at the pollen tube tip integrin-like molecules in the plasma membrane are exposed to the stylar ECM due to the loosely arranged pectic wall. Here, the pollen tube resembles a migratory animal cell (Dufour *et al.*, 1988). Further back from the tip the focal adhesions form, perhaps induced by a change in the composition of the pollen tube wall (Fig. 7).

The actual mechanism for cell movement is unknown, but at least two major hypotheses exist: one that advocates a motive force inherent in the SAM (Newman *et al.*, 1985; Wolff and Lai, 1989) and the other that suggests that contractile microfilaments at the focal adhesions provide the motive force (Abercrombie, 1980). A modification of this latter theory is that actin at the tip can act along with osmotic pressure to generate the force for advancement (Small, 1989). In our model for compatible pollen tube extension, we propose that substrate interactions constitute a recognition event occurring between the pollen tube and the stylar ECM, which actively facilitates the extension of the pollen tube over associated cells in the gynoecium. Any of the aforementioned hypotheses for a motive force could explain this movement.

IV. Conclusions

It has been proposed that the ECM of plants consists of at least three separate networks: a protein network, a pectin network, and a cellulosic network (Roberts, 1989). This separation classically describes the plant cell wall based on the main biochemical groups that comprise it. However, this is not descriptive with regard to the functional aspects of the plant ECM. In light of recent research, it is reasonable to suggest that plant ECMs have at least two functional networks: a biologically passive network that functions in support and a biologically active network that gives the ECM its dynamic character, allowing it to function in primary developmental processes (i.e., cell division) and cellular communication. In the highly specialized ECM of the style, one active capability may be that of pollen tube extension. A model like the one presented here could explain the increase in rate and length, as well as the directionality, observed in pollen tubes grown *in vivo* versus those grown *in vitro*. In view of this model, the term haptotaxis may best describe the process involved in the directionality of pollen tube extension. Chemotropism, often used to explain this phenomenon, is defined as growth toward a chemical substance, generally assumed to be a soluble factor; haptotaxis is defined as directional cell migration along a gradient of a substratum bound factor (Basara *et al.*, 1985).

Detection of the vitronectin-like protein in all tissues examined indicates that it may have a basic function in plant cells, other than facilitating pollen tube extension. Perhaps it is this cell factor that links the cytoskeleton to the cell wall. Green's (1986) work on the biophysics of organogenesis in plants suggests a connection between the cytoskeleton and cellulose microfibrils in the wall. The implication of SAMs as morphoregulators involves them in a wide range of animal developmental processes other than cell migration (Edelman, 1988). In animal cells SAMs function as links between the cytoskeleton and the ECM via trans-membrane integrins (Burridge *et al.*, 1988; Dufour *et al.*, 1988). We propose that a similar model may be operating in plants and that plants contain a class of SAMs, which act as morphoregulators. These molecules, classified as plant SAMs, may exist in the ECM or wall of the plant cell adhering to their specific receptor or integrin in the membrane, which, in turn, interacts with the cytoskeleton (Schindler *et al.*, 1989). Thus, gene expression during primary developmental processes may be affected by mechanochemical means, in a manner analogous to that proposed in animal systems (Bissell *et al.*, 1982; Edelman, 1988; Ingber and Folkman, 1989). The importance of the plant cell wall in the primary developmental process of cell division has been well recognized (Meyer and Abel, 1975). In light

of this, it seems almost inappropriate to refer to secretory matrices as extracellular, which implies that the matrix is not part of the cell. By way of their active and dynamic function, these matrices seem to be an integral part of the cell, forming an identifiable compartment outside the plasma membrane. This exterior compartment may provide the plant cell with a dynamic scaffolding, which functions in development, communication, and, in a specialized case, pollen tube cell movement.

References

Abercrombie, M. (1980). *Proc. R. Soc. Lond. [Biol.]* **207**, 129–147.

Adair, S. W. (1988). *In* "Self-Assembling Architecture" (J. E. Varner, ed.), pp. 25–41. Liss, New York.

Adair, W. S., and Mecham, R. P. (1990). "Organization and Assembly of Plant and Animal Extracellular Matrix: Biology of Extracellular Matrix," A Series. (W. S. Adair and R. P. Mecham, eds.). Academic Press, San Diego.

Anderson, M. A., Cornish, E. C., Maa, S. L., Williams, E. G., Hoggart, R., Atkinson, A., Bonig, I., Grego, B., Simpson, R., Roche, P. J., Haley, J. D., Penshow, J. D., Niall, H. D., Tregear, G. E., Coghlan, J. P., Crawford, R. J., and Clarke, A. E. (1986). *Nature* **321**, 38–44.

Anderson, M. A., McFadden, G. I., Bernatzky, R., Atkinson, A., Orpin, T., Dedman, H., Tregear, G., Fernley, R., and Clarke, A. E. (1989). *Plant Cell* **1**, 483–491.

Barnes, D. W., Silnutzer, J., See, C., and Shaffer, M. (1983). *Proc. Natl. Acad. Sci. U.S.A.* **80**, 1362–1366.

Basara, M. L., McCarthy, J. B., Barnes, D. W., and Furcht, L. T. (1985). *Cancer Res.* **45**, 2487–2494.

Beck, N. G., and Lord, E. M. (1988). *Am. J. Bot.* **75**, 1913–1922.

Bell, J., and Hicks, G. (1976). *Planta* **131**, 187–200.

Bennett, M. D., and Smith, J. B. (1976). *Philos. Trans. R. Soc. Lond. [Biol.]* **274**, 227.

Bissell, M. J., Hall, H. G., and Parry, G. (1982). *J. Theor. Biol.* **99**, 31–68.

Bronner-Fraser, M. (1982). *Dev. Biol.* **91**, 50–63.

Bronner-Fraser, M. (1985). *Dev. Biol.* **108**, 131–145.

Burridge, K., Fath, K., Kelly, T., Nuckolls, G., and Turner, C. (1988). *Annu. Rev. Cell Biol.* **4**, 487–525.

Cheresh, D. A. (1987). *Proc. Natl. Acad. Sci. U.S.A.* **84**, 6471–6475.

Cheresh, D. A., and Spiro, R. C. (1987). *J. Biol. Chem.* **262**, 17701–17711.

Clarke, A. E. (1981). *In* "Plant Carbohydrates. II, Extracellular Carbohydrates" (W. Tanner and F. A. Loewus, eds.), pp. 577–582. Springer-Verlag, New York.

Clarke, A. E., Gleeson, P., Harrison, S., and Knox, R. B. (1979). *Proc. Natl. Acad. Sci. U.S.A.* **76**, 3358–3362.

Cornish, E. C., Pettitt, J. M., Bonig, I., and Clarke, A. E. (1987). *Nature* **326**, 99–102.

Cresti, M., Pacini, E., Ciampolini, F., and Sarfatti, G. (1977). *Planta* **136**, 239–247.

Darvill, A., McNeil, M., Albersheim, P., and Delmer, D. P. (1980). *In* "The Biochemistry of Plants," Vol. 1 (N. E. Tolbert, ed.), pp. 91–162. Academic Press, San Diego.

De Vries, S., Booij, H., Janssens, R., Vogels, R., Saris, L., Lo Schiavo, F., Terzi, M., and Van Kammen, A. (1988). *Gene Dev.* **2**, 462–476.

Dufour, S., Duband, J. L., Kornblihtt, A. R., and Thiery, J. P. (1988). *Trends in Genetics* **4**, 198–203.

Edelman, G. M. (1988). "Topobiology: An Introduction to Molecular Embryology. Basic Books, New York.

Eenink, A. H. (1982). *Euphytica* **31,** 773–786.

Epperson, B. K., and Clegg, M. T. (1987). *Heredity* **58,** 5–14.

Ericson, M. C., and Elbein, A. D. (1980). *In* "Biochemistry of Plants," pp. 589–616. Academic Press, San Diego.

Furcht, L. T. (1983). *In* "Modern Cell Biology" (B. H. Satir, ed.), Vol. 1, pp. 53–117. Liss, New York.

Green, P. B. (1986). *In* "Society for Experimental Biology Symposium 40" (D. H. Jennings and A. J. Trewavas, eds.), pp. 211–232. Company of Biologists, Cambridge, England.

Grinnell, F. (1978). *Int. Rev. Cytol.* **53,** 65–144.

Hay, E. D. (1981a). *In* "Cell Biology of Extracellular Matrix" (E. D. Hay, ed.), pp. 379–409. Plenum Press, New York.

Hay, E. D. (1981b). "Cell Biology of Extracellular Matrix." Plenum Press, New York.

Hayman, E. G., Pierschbacher, M. D., and Ruoslahti, E. (1983). *Proc. Natl. Acad. Sci. U.S.A.* **80,** 4003–4007.

Hayman, E. G., Pierschbacher, M. D., Suzuki, S., and Ruoslahti, E. (1985). *Exp. Cell Res.* **160,** 245–258.

Heslop-Harrison, J. (1983). *Proc. R. Soc. Lond. [Biol.]* **218,** 371–395.

Heslop-Harrison, J. (1987). *Int. Rev. Cytol.* **107,** 1–78.

Heslop-Harrison, J., and Heslop-Harrison, Y. (1986). *In* "Biology of Reproduction and Cell Motility in Plants and Animals" (M. Cresti and R. Dallai, eds.), pp. 169–174. University of Siena, Siena.

Heslop-Harrison, J., and Heslop-Harrison, Y. (1989). *J. Cell Sci.* **93,** 299–308.

Hill, J. P., and Lord, E. M. (1986). *Evolution* **40,** 1328–1333.

Hill, J. P., and Lord, E. M. (1987). *Am. J. Bot.* **74,** 988–997.

Hynes, R. O. (1981). *In* "Cell Biology of Extracellular Matrix" (E. D. Hay, ed.), pp. 295–334. Plenum Press, New York.

Hynes, R. O. (1987). *Cell* **48,** 549–554.

Ingber, D. E., and Folkman, J. (1989). *Cell* **58,** 803–805.

Jensen, W. A., and Fisher, D. A. (1969). *Planta* **84,** 97–121.

Kandasamy, M. K., Paolillo, D. J., Faraday, C. D., Nasrallah, J. B., and Nasrallah, M. E. (1989). *Dev. Biol.* **134,** 462–472.

Kitagaki-Ogawa, H., Yatohgo, T., Izumi, M., Hayashi, M., Kashiwagi, H., Matsumoto, I., and Seno, N. (1990). *Biochim. Biophys. Acta* **1033,** 49–56.

Knox, R. B. (1984). *Encyclopedia Plant Physiol.* **17,** 508–608.

Knox, J. P., Day, S., and Roberts, K. (1989). *Development* **106,** 47–56.

Kropf, D. L., Kloarge, B., and Quatrano, R. S. (1988). *Science* **239,** 187–190.

Labarca, C., and Loewus, F. (1973). *Plant Physiol.* **52,** 87–92.

Lamport, D. T. A., and Catt, J. W. (1981). *In* "Plant Carbohydrates II, Extracellular Carbohydrates" (W. Tanner and F. A. Loewus, eds.), pp. 134–165. Springer-Verlag, New York.

Lancelle, S. A., Cresti, M., and Hepler, P. K. (1987). *Protoplasma* **140,** 141–150.

Lewis, D. (1979). *Stud. Biol.* **110,** 1–59.

Lord, E. M., and Kohorn, L. U. (1986). *Am. J. Bot.* **73,** 70–78.

Lo Schiavo, R., Giuliano, G., De Vries, S. C., Genga, A., Bollini, R., Pitto, L., Cozzani, R., Nuti-Ronchi, V., and Terzi, M. (1990). *Mol. Gen. Genet.* **223,** 385–393.

Malti, and Shivanna, K. R. (1985). *Theor. Appl. Genet.* **70,** 684–686.

Marshall, D., and Ellstrand, N. (1985). *Am. Nature* **126,** 596–605.

Mascarenhas, J. P. (1973). *In* "Behavior of Microorganisms" (A. Perez-Miravete, ed.), pp 62–69. Plenum Press, London.

Mascarenhas, J. P. (1978). *In* "Taxis and Behavior: Receptors and Recognition" (G. L. Hazelbauer, ed.), Series B, pp. 171–203. Chapman and Hall, London.

McClure, B. A., Haring, V., Ebert, P. R., Anderson, M. A., Simpson, R. J., Sakiyama, F., and Clarke, A. E. (1989). *Nature* **342,** 955–957.

Meyer, Y., and Abel, W. D. (1975). *Planta* **123,** 33–40.

Mulcahy, D. L. (1975). *In* "Gamete Competition in Plants and Animals" (D. L. Mulcahy, ed.), pp. 1–4. North Holland, Amsterdam.

Mulcahy, G. B., and Mulcahy, D. L. (1987). *Am. J. Bot.* **74,** 1458–1459.

Nasrallah, J. (1989). *In* "Plant Reproduction: From Floral Induction to Pollination" (E. Lord and G. Bernier, eds.), pp. 156–164. American Society of Plant Physiologists, Rockville, Maryland.

Nasrallah, J. B., Kao, T. H., Goldberg, M. L., and Nasrallah, M. E. (1985). *Nature* **318,** 263–267.

Neugebauer, K. M., Emmett, C. J., Venstrom, K. A., and Reichardt, L. F. (1991). *Neuron* **6,** 345–358.

Newman, S. A., Frenz, D. A., Tomasek, J. J., and Rabuzzi, D. D. (1985). *Science* **228,** 885–888.

Ottaviano, E., Sari-Gorla, M., and Mulcahy, D. L. (1975). *In* "Gamete Competition in Plants and Animals" (D. L. Mulcahy, ed.), pp. 125–134. North Holland, Amsterdam.

Pennell, R. I., Knox, J. P., Scofield, G. H., Selvendran, R. R., and Roberts, K. (1989). *J. Cell Biol.* **108,** 1967–1977.

Pennell, R. I., and Roberts, K. (1990). *Nature* **344,** 547–549.

Pierschbacher, M. D., and Ruoslahti, E. (1984). *Nature* **309,** 30–33.

Pierson, E. S., Derksen, J., and Traas, J. A. (1986). *Eur. J. Cell Biol.* **41,** 14–18.

Preissner, K. T. (1989). *Blut* **59,** 419–431.

Quatrano, R. S., Brian, L., Aldridge, J., and Schultz, T. (1991). *In* "Molecular and Cellular Bases of Pattern Formation" (K. Roberts, E. Coen, C. Dean, J. Jones, K. Charter, R. Flanell, A. Wilkins, and N. Holder, eds.), pp. 11–16. Company of Biologists, Cambridge.

Rendle, H., and Murray, B. G. (1988). *N. Z. J. Bot.* **26,** 467–471.

Roberts, K. (1989). *Curr. Opin. Cell Biol.* **1,** 1020–1027.

Roberts, K. (1990). *Curr. Opin. Cell Biol.* **2,** 920–928.

Rosen, W. G. (1964). *In* "Pollen Physiology and Fertilization" (H. F. Linskens, ed.), pp. 159–166. North Holland, Amsterdam.

Ruoslahti, E., and Pierschbacher, M. D. (1987). *Science* **238,** 491–497.

Sanders, L. C., and Lord, E. M. (1989). *Science* **243,** 1606–1608.

Sanders, L. C., Eckard, K. J., and Lord, E. M. (1990). *Protoplasma* **159,** 26–34.

Sanders, L. C., Wang, C. S., Walling, L. L., and Lord, E. M. (1991). *Plant Cell* **3,** 629–635.

Schindler, M., Meiners, S., and Cheresh, D. A. (1989). *J. Cell Biol.* **108,** 1955–1965.

Sedgley, M. (1976). *New Phytol.* **77,** 149–152.

Sedgley, M., Blesing, M. A., Bonig, I., Anderson, M. A., and Clarke, A. E. (1985). *Micron Microsc. Acta* **16,** 247–254.

Shivanna, K. R., Johri, B. M., and Sastri, D. C. (1979). "Development and Physiology of Angiosperm Pollen." Today and Tomorrow's, Delhi.

Small, J. V. (1989). *Curr. Opin. Cell Biol.* **1,** 75–79.

Steer, M. W. (1990). *In* "Tip Growth in Plant and Fungal Cells" (I. B. Heath, ed.), pp. 119–145. Academic Press, San Diego.

Steer, M. W., and Steer, J. M. (1989). *New Phytol.* **111,** 323–358.

Stenn, K. S. (1981). *Proc. Natl. Acad. Sci. U.S.A.* **78,** 6907–6911.

Stephenson, A. G., and Bertin, R. L. (1983). *In* "Pollination Biology" (L. Real, ed.), pp. 109–149. Academic Press, San Diego.

Tiwari, S. C., and Polito, V. S. (1988). *Protoplasma* **174,** 100–112.

Tomasini, B. R., and Mosher, D. F. (1986). *Blood* **68,** 737–742.
Underwood, P. A., and Bennett, F. A. (1989). *J. Cell Sci.* **93,** 641–649.
Varner, J. E., and Hood, E. E. (1988). *In* "Self-Assembling Architecture" (J. E. Varner, ed.), pp. 97–103. Liss, New York.
Wolff, C., and Lai, C. S. (1989). *Arch. Biochem. Biophys.* **268,** 536–545.
Zhu, J. K., Shi, J., Singh, U., and Carpita, N. C. (1991). *Plant Physiol.* **53**(Suppl.), 10.

Double Fertilization in Nonflowering Seed Plants and Its Relevance to the Origin of Flowering Plants

William E. Friedman

Department of Botany, University of Georgia, Athens, Georgia 30602

I. Introduction

Double fertilization in angiosperms, in which one sperm fertilizes an egg nucleus while a second sperm fuses with the two polar nuclei of the female gametophyte (embryo sac), was first reported independently by Navaschwin (1898) and Guignard (1899). Within 2 years this seemingly unusual phenomenon had been shown to occur in 19 additional species of flowering plants (Sargant, 1900). Since that time, the occurrence of double fertilization, followed by the subsequent formation of a (usually) triploid endosperm tissue, has been shown to be ubiquitous among angiosperms. During the greater part of the Twentieth Century, double fertilization and endosperm have been considered to be unique and defining characteristics of flowering plants (Stebbins, 1976; Cronquist, 1988), associated with the evolutionary success of angiosperms (Stebbins, 1974; Tiffney, 1981). However, the potential for multiple fertilization events within a single female gametophyte or ovule is not limited to flowering plants and may occur among various groups of nonflowering seed plants.

Among extant nonflowering seed plants (Coniferales, Cycadales, *Ginkgo biloba,* and Gnetales), multiple fertilization events within a single ovule can result from two distinct processes. The first involves the fertilization of separate eggs within two or more archegonia of a single ovule and results in the production of multiple embryos, a phenomenon referred to as simple polyembryony. A second manner in which multiple fertilization events can occur within a single ovule in nonflowering seed plants involves a phenomenon referred to as intra-archegonial double fertilization. During this process separate sperm fertilize an egg nucleus and a second female nucleus within the same archegonium. This review will focus exclusively

319

on what is known of the process of intra-archegonial double fertilization in nonflowering seed plants.

In the years following the initial discovery of double fertilization in flowering plants, several reports of intra-archegonial double fertilization events among nonflowering seed plants appeared. These double fertilization events typically were characterized by the presence of a female nucleus (in addition to the egg nucleus) within the egg cell cytoplasm with which a second sperm nucleus was believed to fuse. The second fertilization product was reported either to degenerate, produce free nuclei, or give rise to additional embryos.

Intra-archegonial double fertilization phenomena in nonflowering seed plants (or the potential for such events) have been reported for members of the Coniferales, Cycadales, *Ginkgo biloba,* and Gnetales. Four years after the initial documentation of a process of double fertilization in angiosperms, Land (1902) suggested that a fertilization event, in addition to the fusion of a sperm and egg nucleus, could occur in *Thuja occidentalis,* a conifer. Evidence of intra-archegonial double fertilization in conifers has also been proposed for *Abies balsamea* (Hutchison, 1915) and *Pseudotsuga taxifolia* (Allen, 1943). Among certain cycads and *Ginkgo biloba,* the ventral canal nucleus (which normally degenerates prior to the time of fertilization) may undergo development similar to the egg nucleus within the confines of the egg cell (Sedgewick, 1924; Bryan and Evans, 1957). Under these circumstances it has been suggested that the ventral canal nucleus can serve as a potential second egg nucleus to be fertilized by a second sperm nucleus.

Perhaps, the most intriguing and persistent suggestions of a process of intra-archegonial double fertilization among nonflowering seed plants have been advanced for members of the genus *Ephedra,* within the Gnetales. A number of reports of possible double fusion events in the genus *Ephedra* appeared between 1907 and 1978 (Land, 1907; Herzfeld, 1922; Maheshwari, 1935; Khan, 1940, 1943; Mulay, 1941; Narang, 1955; Moussel, 1978). Most of these workers provided evidence that was suggestive of an occasional second fertilization event during sexual reproduction in several species of *Ephedra.* Recently, however, intra-archegonial double fertilization has been shown to be a regular feature of the reproductive process in *Ephedra nevadensis* (Friedman, 1990a,b) and *Ephedra trifurca* (Friedman 1991, 1992). Although both other genera of the Gnetales, *Gnetum* and *Welwitschia,* are highly derived and lack archegonia, it has been postulated that double fertilization events might occur in members of these two taxa (Pearson, 1909; Vasil, 1959).

The establishment of a process of intra-archegonial double fertilization in nonflowering seed plants (assuming that a single fertilization event per archegonium involving an egg nucleus and sperm nucleus is plesiomorphic)

is dependent upon a number of specific developmental preconditions: (1) presence of a second female nucleus at the site of normal fertilization—the ventral canal nucleus (which is degenerate in many groups of nonflowering seed plants) may serve this role by being persistent through the time of normal fertilization; (2) acquisition of egg-like features by the second female nucleus that allow it to function as a female gamete and attract and fuse with a sperm nucleus; and (3) consistent entry of a second functional sperm nucleus into the archegonial cavity to participate in a second fertilization event. As will be seen, these three preconditions for a second fertilization event may be met in a number of nonflowering seed plants, although in several of the reported instances, double fertilization would appear to be teratological.

The goal of this review is to examine critically the evidence for the occurrence of double fertilization events in nonflowering seed plants and to analyze the cell-biological events that are essential for the establishment of such a process. In addition, given the many historical questions that remain as to how double fertilization and endosperm might have evolved in the earliest angiosperms or their immediate ancestors, double fertilization events in nonflowering seed plants will be analyzed with respect to their significance to the resolution of the origin of flowering plants.

II. Double Fertilization in Conifers

A. Basic Aspects of Conifer Reproduction

Development of the female gametophyte in conifers is remarkably uniform. The female gametophyte is monosporic and initially undergoes a free nuclear pattern of development. After producing numerous nuclei, the resulting cenocytic female gametophyte initiates a process of cellularization to produce a relatively large (compared with the embryo sac of angiosperms) multicellular organism (Maheshwari and Singh, 1967). A variable number of archegonia (one to as many as several hundred, although typically between 2 and 10) are initiated by the cellular female gametophyte (Chamberlain, 1935; Willson and Burley, 1983). Each archegonium contains a large central cell and a central cell nucleus that will ultimately divide to produce a ventral canal nucleus and egg nucleus prior to fertilization.

Following mitosis of the central cell nucleus, the ventral canal nucleus typically remains at the extreme micropylar end of the former central cell cytoplasm. The egg nucleus migrates in a chalazal direction and occupies a central region of the egg cell. In conifers where mitosis of the central cell

nucleus is followed by a process of cytokinesis (e.g., *Pinaceae;* Chamberlain, 1935), cleavage of the central cell results in the formation of a small ventral canal cell at the apex of the former central cell and a large egg cell that occupies most of the archegonial cavity. Among most groups of conifers, mitosis of the central cell nucleus is not followed by cytokinesis, and a binucleate egg cell, containing a ventral canal nucleus and egg nucleus, is formed (Chamberlain, 1935; Maheshwari and Sanwal, 1963).

In many conifers the ventral canal nucleus or cell degenerates between the time of its formation and the time of fertilization (Herzfeld, 1922; Chamberlain, 1935; Maheshwari and Sanwal, 1963). As a consequence, only one viable female nucleus, the egg nucleus, is present within the confines of the archegonium at the time of arrival of sperm. Nevertheless, in a number of conifers, the ventral canal nucleus or cell regularly persists up through the time of fertilization (Herzfeld, 1922). The persistence of the ventral canal nucleus until the arrival of sperm within the archegonium is a prerequisite for the occurrence of intra-archegonial double fertilization.

Following pollination in conifers, the male gametophyte germinates from a pollen grain and develops within the sporophytic tissues of the ovule (and sometimes also the female strobilus). The pollen tube functions as a siphonogamous conduit to transfer nonmotile sperm cells (typically two) to the egg. In some conifers the male gametophyte produces a single binucleate sperm cell that is delivered to the archegonium (Singh, 1978). The sperm cells or nuclei may be dimorphic in certain groups of conifers, with only one of the two sperm of a pollen tube being truly functional (Doyle, 1957; Willson and Burley, 1983). Those conifers in which only one functional sperm cell or sperm nucleus is regularly brought to the egg are likely to be incapable of intra-archegonial double fertilization.

In conifers sexual fusion of a sperm nucleus and an egg nucleus produces a zygote nucleus. The resulting zygote nucleus (in all conifers except *Sequoia sempervirens*) enters into a series of free nuclear divisions. The number of free nuclei produced varies among species and may be as few as four or as many as 64 (Singh, 1978). Free nuclear development of the embryo in conifers is followed by a process of cellularization at the base (chalazal end) of the former egg cell to produce a single multicellular proembryo. Elongation of the suspensor pushes the embryo proper in a chalazal direction into the nutritive tissues of the female gametophyte where further development occurs.

B. Reports of Intra-Archegonial Double Fertilization
    in Conifers

Regular or teratological intra-archegonial double fertilization events have been reported in a small number of conifers. Most of these studies, how-

ever, were performed in the early part of the Twentieth Century and predate modern histological techniques as well as any photographic representation of the described events. As a result, these reports are often difficult to interpret and evaluate. Nevertheless, in at least two cases (*Thuja occidentalis* and *Abies balsamea*), the evidence strongly suggests that a consistent pattern of intra-archegonial double fertilization may occur.

The first evidence of a process of intra-archegonial double fertilization in a conifer was reported 4 years after the initial discovery of double fertilization in flowering plants. In a study of reproduction in *Thuja occidentalis (Cupressaceae),* Land (1902) suggested that in addition to the fusion of an egg nucleus and sperm nucleus, the ventral canal nucleus was,

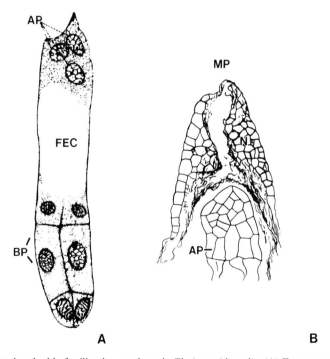

A   B

FIG. 1   Putative double fertilization products in *Thuja occidentalis*. (A) Former egg cell with cellular basal proembryo (BP) at the 12 cell stage of development (six cells are visible) and a free nuclear apical proembryo (AP) (three of the four free nuclei are visible). The basal proembryo is derived from the zygote nucleus, whereas the apical proembryo is hypothesized to have developed from a second fusion product of the ventral canal nucleus and a second sperm nucleus. FEC, fertilized egg cell. (B) Later stage of development of an apical proembryo (AP). The proembryo is cellular and is beginning to grow into the nucellus toward the micropyle of the ovule. The normal basal proembryo (from the fertilized egg nucleus) is not shown in this figure. MP, micropylar region. Reproduced from Land (1902).

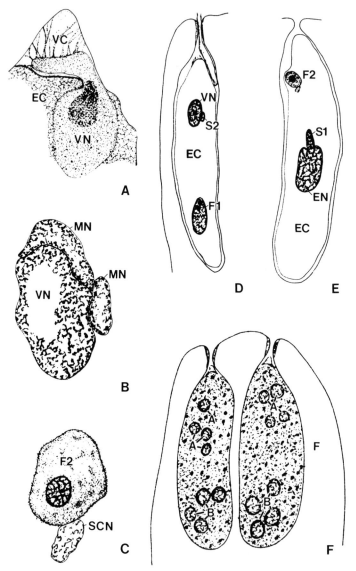

FIG. 2   Double fertilization in *Abies balsamea*. (A) Escape of the ventral canal nucleus (VN) from the confines of the ventral canal cell (VC) into the egg cell (EC). (B) Ventral canal nucleus with two male nuclei (MN) adjacent to it. One of the male nuclei is a sperm nucleus and the other is a sterile cell nucleus. ×675. (C) Second fertilization product (F2) resulting from the fusion of the ventral canal nucleus and a sperm nucleus. The sterile cell nucleus (SCN) is still adjacent to the fusion product. ×725. (D) Double fertilization within a single egg cell. The first fertilization event (F1) between the egg nucleus and the first sperm nucleus is taking place at the base of the egg cell. The second fertilization event (VN and S2) is occurring approximately one-fourth of the distance from the apex of the egg cell and is drawn

on occasion, fertilized by a second sperm nucleus within an individual archegonium. Land presented figures (camera-lucida drawings) of supernumerary embryos within individual archegonia in support of his hypothesis that separate fertilization events had occurred within individual archegonia (Fig. 1).

In one archegonium examined by Land (1902), a cellular proembryo (consisting of 12 cells) was situated at the base of the former egg cell, while a group of four nuclei, which appeared to be at an earlier stage of free nuclear proembryogenesis, were located at the apex (micropylar end) of the former egg cell (Fig. 1). An additional archegonium was described (without figures) by Land (1902) that contained a supernumerary proembryo comprising eight nuclei at the apex of the former egg cell; this proembryo was in addition to the normal zygotically derived proembryo at the base (chalazal end) of the former egg cell. As further evidence of a process of intra-archegonial double fertilization, Land (1902) described occasional seeds of *Thuja occidentalis* in which a cellular embryo was found growing from the apex of the former egg cell in a micropylar direction into the nucellus (Fig. 1). This supernumerary embryo was in addition to the zygotically derived embryo that develops at the base of the former egg cell and grows in a chalazal direction into the central region of the female gametophyte. Land (1902) noted that micropylar embryos that grow into the nucellus (instead of the female gametophyte) are less vigorous than those produced at the base of the archegonium. No figures, however, showed evidence of a sperm and ventral canal nucleus in the process of fusing.

The most convincing evidence of a process of double fertilization in a conifer was presented in a study of reproduction in *Abies balsamea* (Pinaceae) by Hutchinson (1915). Hutchinson indicated that after mitosis and cytokinesis of the central cell to produce a ventral canal cell and egg cell within the archegonium, the ventral canal nucleus regularly "breaks through" the cell walls that separate the ventral canal cell and egg cell (Fig. 2). The ventral canal nucleus then migrates in the same direction as did the egg nucleus at the conclusion of mitosis of the

---

at a higher magnification in B. ×125. (E) Later stage of intra-archegonial double fertilization. The second fertilization event (between ventral canal nucleus and second sperm nucleus) has been completed (and is seen at higher magnification in C), while the first fertilization event (egg nucleus and first sperm nucleus, S1) is still taking place. (F) Two archegonia showing early stages of proembryogenesis. The four free nuclei at the base of each former egg cell (B) are derived from the zygote nucleus and will produce the normal embryo. The four free nuclei in the apical portion of each former egg cell (A) are derived from the second fertilization event between the ventral canal nucleus and the second sperm nucleus. F, female gametophyte. Reproduced from Hutchinson (1915).

central cell nucleus. Following entry of the ventral canal nucleus into the cytoplasm of the egg cell, this nucleus was reported to increase in size and acquire a "structure" similar to that of the egg nucleus (Hutchinson, 1915).

In addition to the normal fusion of a sperm and egg nucleus in *Abies balsamea,* Hutchinson (1915) observed a sperm nucleus in the process of fusing with the ventral canal nucleus on several occasions. Figures (camera-lucida drawings) were presented of the close approximation of a second sperm nucleus and ventral canal nucleus (Fig. 2). Examination of these figures indicates that the fusion of a second sperm nucleus and ventral canal nucleus occurs approximately one-fourth the distance from the apex of the egg cell, whereas the fusion of the first sperm nucleus and egg nucleus takes place in a more basal location within the egg cell (Fig. 2).

The resulting second fertilization product apparently undergoes two successive waves of mitosis to produce a total of four free nuclei that are similar in appearance to the four free nuclei derived from the zygotic nucleus during early proembryogenesis (Fig. 2). The four free nuclei derived from the second fertilization event were situated within the micropylar end of the former egg cell, whereas the four nuclei derived from the true zygote nucleus were located in the chalazal end of the former egg cell (Hutchinson, 1915). No descriptions of later stages of embryo development were reported at that time. Although the ultimate fate of the second fertilization product in *Abies balsamea* was not studied, Hutchinson (1915) concluded that the ventral canal nucleus in this species can function as an additional egg nucleus and is the evolutionary homolog of the true egg nucleus (similar ideas were advanced by Chamberlain, 1899, 1935). However, a more recent study of reproduction in *Abies amabilis* (Owens and Molder, 1977) did not uncover evidence of a second fertilization event between the ventral canal nucleus and a second sperm nucleus.

In *Pseudotsuga taxifolia* (Pinaceae) Allen (1943) reported that the ventral canal cell typically degenerates prior to the time of fertilization. However, on rare occasions the ventral canal nucleus may escape from the confines of its cell walls and undergo considerable enlargement within the cavity of the egg cell. Allen (1943) suggested it was possible in certain instances, in which supernumerary nuclei were found near the apex of a fertilized egg, that the ventral canal nucleus might have been fertilized by a second sperm nucleus. No figures of a fusion process involving the ventral canal nucleus and a sperm nucleus were presented.

In addition to the descriptions of possible intra-archegonial double fertilization events in *Abies, Thuja,* and *Pseudotsuga,* Chamberlain (1899)

reported that in *Pinus laricio* (Pinaceae) the ventral canal nucleus, which typically degenerates prior to fertilization, may in exceptional cases pass through a series of specific developmental stages that resemble those of the egg nucleus. In some instances the ventral canal nucleus escapes from the ventral canal cell and enters into the cytoplasm of the egg cell. Chamberlain (1899) hypothesized that when the ventral canal nucleus undergoes a development parallel to that of the egg nucleus (instead of degenerating), fertilization of the ventral canal nucleus by a sperm nucleus might occur, although he did not observe such events. In *Cedrus atlantica* (Pinaceae) (Smith, 1923), instances have been reported in which the walls between the ventral canal cell and egg cell break down and the two nuclei come to lie near each other in the archegonial cavity. It is not known whether intra-archegonial double fertilization events occur in *Cedrus*.

III. Potential Double Fertilization in *Ginkgo biloba* and Cycads

A. Basic Aspects of Reproduction in *Ginkgo biloba* and Cycads

In *Ginkgo biloba* and cycads development of the female gametophyte is similar to that of conifers. The female gametophyte is monosporic and undergoes a period of free nuclear development followed by a process of cellularization (Chamberlain, 1935). Between two and four archegonia are initiated per female gametophyte in *Ginkgo biloba* (Herzfeld, 1927; Lee, 1955; Favre-Duchartre, 1956). In the female gametophyte of cycads, from one archegonium (*Cycas circinalis,* Rao, 1961) to more than 100 archegonia (*Microcycas calocoma,* Downie, 1928) may be formed (Willson and Burley, 1983). In *Ginkgo* a true ventral canal cell is formed but typically degenerates prior to the time of fertilization (Lee, 1955; Favre-Duchartre, 1956). Among cycads a binucleate egg cell results from the mitotic division of the central cell nucleus (Chamberlain, 1935). With only a few known exceptions, the ventral canal nucleus in cycads degenerates prior to the time of fertilization.

In both *Ginkgo biloba* and cycads, a pollen tube is produced that is strictly vegetative and functions to absorb nutrients from the host tissues of the sporophyte (Friedman 1987; Choi and Friedman, 1991). In these seed plants motile sperm (typically two) are produced and released by the male gametophyte to effect fertilization (zooidogamy). In *Ginkgo biloba* and some cycads more than one sperm may enter an archegonium; however, only one sperm has been reported to penetrate deeply into the egg

cell and fertilize the egg nucleus (Chamberlain, 1935; Lee, 1955; Favre-Duchartre, 1956).

B. Potential for Intra-Archegonial Double Fertilization in *Ginkgo biloba* and Cycads

There have been no reports of actual intra-archegonial double fertilization events in *Ginkgo biloba* or cycads. Most likely, this is a consequence of the fact that in both of these groups of seed plants, the ventral canal nucleus (or cell) typically degenerates prior to the time of fertilization (Chamberlain, 1935). However, several workers have reported anomalous egg-like development of the ventral canal nucleus in cycads (*Dioon edule,* Chamberlain, 1906; *Ceratozamia mexicana,* Chamberlain, 1912; *Encephalartos* sp., Sedgewick, 1924; *Zamia umbrosa,* Bryan and, Evans, 1957) and, in one instance, *Ginkgo biloba* (Ikeno, 1901).

In *Encephalartos* Sedgewick (1924) reported that in most cases the ventral canal nucleus enlarges (instead of degenerates) and develops in a fashion similar to the egg nucleus. Bryan and Evans (1957) documented that the ventral canal nucleus regularly persists in *Zamia umbrosa.* These workers showed that the ventral canal nucleus often migrates in a chalazal direction to occupy a more central position within the egg cell, in proximity to the egg nucleus (Fig. 3). In many cases (Bryan and Evans, 1957) the ventral canal nucleus acquires cytological features similar to the egg nucleus, and it was hypothesized that the ventral canal nucleus might function as an egg nucleus. In both *Encephalartos* and *Zamia,* in which egg-like development of the ventral canal nucleus occurs with some regularity, the ventral canal nucleus was displaced from its normal apical position within the egg cell cytoplasm to a more central position closer to the egg nucleus.

In *Ginkgo biloba* Ikeno (1901) reported an instance in which the egg cell contained two large nuclei and hypothesized that the second nucleus was the ventral canal nucleus (Fig. 3). Ikeno suggested that this phenomenon in *Ginkgo* was fundamentally similar to the earlier reported (Chamberlain, 1899) egg nucleus-like development of the ventral canal nucleus in *Pinus laricio.*

IV. Double Fertilization in *Ephedra*

A. Basic Features of Reproduction in *Ephedra*

During the last century, there have been many studies of reproduction in the genus *Ephedra* (Land, 1904, 1907; Berridge and Sanday, 1907; Ber-

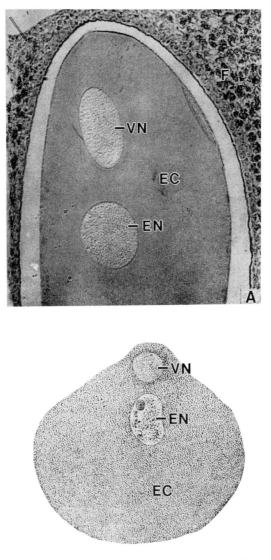

FIG. 3 Egg nucleus-like development of the ventral canal nucleus in zooidogamous seed plants. VN, ventral canal nucleus; EC, egg cell; EN, egg nucleus; F, female gametophyte. (A) *Zamia umbrosa*. ×64. Reproduced with permission from Bryan and Evans (1957). (B) *Ginkgo biloba*. ×140. Reproduced from Ikeno (1901).

ridge, 1909; Herzfeld, 1922; Maheshwari, 1935; Mulay, 1941; Khan, 1940, 1943; Narang, 1955; Favre-Duchartre, 1959; Singh and Maheshwari, 1962; Lehmann-Baerts, 1967; Moussel, 1977, 1978, 1980). Development of the female gametophyte in *Ephedra* is typical of that which occurs in most nonflowering seed plants (Maheshwari and Singh, 1967). The female gametophyte is monosporic and initially undergoes a period of free nuclear development. This first phase of growth is followed by a period of centripetal wall formation (alveolarization), giving rise to a multicellular female gametophyte similar to those of conifers, cycads, and *Ginkgo*. Following cellularization archegonia are initiated at the micropylar end of the gametophyte. Each archegonial initial divides to produce a primary neck cell (which will give rise to a multitiered neck) and a central cell. Archegonia remain at the central cell stage until further development is stimulated by pollination of the ovule (Moussel, 1977; Friedman, 1990b).

At the time of pollination, the ovule of *Ephedra* contains a female gametophyte with one to six archegonia, each of which is typically surrounded by a multilayered jacket (Maheshwari and Sanwal, 1963). Jacket cells are binucleate or multinucleate (Land, 1907; Maheshwari, 1935; Khan, 1943; Lehmann-Baerts, 1967; Moussel, 1977) and are densely cytoplasmic (in contrast to the adjacent cells of the female gametophyte). Nuclei of the jacket cells have been shown to migrate into the central cell or egg cell cytoplasm (Fig. 4) before or shortly after fertilization through pores or degenerate areas of the common cell walls between the central cell or egg cell and the surrounding jacket cells (Land, 1907; Berridge and Sanday, 1907; Khan, 1943; Narang, 1955; Moussel, 1977; Friedman, 1990b). Thus, female nuclei, in addition to the egg nucleus and ventral canal nucleus, may be present within an archegonium at the time of fertilization. Land (1907) suggested that migratory jacket cell nuclei might participate in supernumerary fertilization events (Section IV,B). However, this possibility has been eliminated on the basis of microspectrophotometric analysis of DNA content of the nuclei involved. Nuclei of the egg and sperm are strictly 2C in *Ephedra nevadensis* at the time of gamete fusion, whereas those of archegonial jacket cells are in excess of 2C and easily discriminated (Friedman, 1990b).

Elapsed time between pollination and fertilization in *Ephedra* is extremely brief (10 hours in *Ephedra trifurca* [Land, 1907] and 16 hours in *Ephedra distachya* [Moussel, 1977]). During this time a number of coordinated events within the ovule appear to be set in motion. Within hours of pollen reception (*Ephedra distachya*, Moussel, 1977), the central cell nucleus, which is located at the micropylar end of the central cell, divides to produce a ventral canal nucleus and egg nucleus. The ventral canal nucleus remains *in situ* at the extreme micropylar end of the egg cytoplasm (Lehmann-Baerts, 1967; Moussel, 1977; Friedman, 1990a,b,

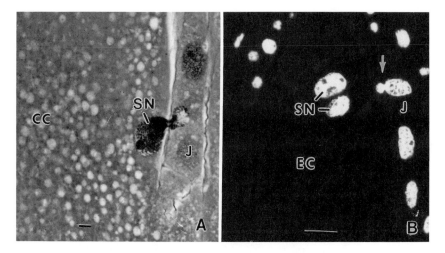

FIG. 4   Migration of nuclei from jacket cells (J) of the archegonium into the central cell (CC) or egg cell (EC) in *Ephedra nevadensis*. (A) Migratory nucleus moving from a jacket cell through two pores into a central cell (CC). Bar, 10 μm. (B) Fluorescence view (stained with DAPI) of migration event from a jacket cell into an egg cell (shown at arrow), as well as two supernumerary nuclei (SN) within the egg cell that are originally from a jacket cell. Microspectrophotometric data indicate that migratory nuclei do not participate in the second fertilization event in *Ephedra nevadensis*. Bar, 50 μm. Reproduced with permission from Friedman (1990b).

1991) and persists through the time of fertilization (Berridge, 1909; Narang, 1955; Lehmann-Baerts, 1967; Moussel, 1977; Friedman 1990a,b, 1991, 1992), although some workers have reported that it may be in a state of degeneration (Maheshwari, 1935; Khan, 1943). Studies of *Ephedra nevadensis* and *Ephedra trifurca* indicated that the ventral canal nucleus is present and clearly visible at the apex of the egg cell cytoplasm just prior to and at the time of entry of sperm into the egg cell (Fig. 5) (Friedman, 1990a,b, 1991, 1992).

In *Ephedra* the egg nucleus migrates in a chalazal direction toward the center of the archegonium where it enters and becomes immersed in an unusual column of cytoplasm (Fig. 5) (Land, 1907; Khan, 1943; Lehmann-Baerts, 1967; Moussel, 1977; Friedman, 1990a,b, 1991) that is rich in both mitochondria and plastids (Moussel, 1977). In contrast, the micropylar region of the egg cytoplasm is vacuolate and largely devoid of most cytoplasmic organelles (Moussel, 1977).

Following germination of the pollen grain and the production of a pollen tube, the spermatogenous cell divides mitotically to yield a binucleate sperm cell (Moussel, 1977; Friedman 1990b, 1992). Both sperm nuclei enter the egg cell cytoplasm (Land, 1907; Berridge and Sanday, 1907;

FIG. 5   Mature egg cell (EC) just prior to fertilization in *Ephedra nevadensis*. The egg nucleus (EN) is located within a cytoplasmically dense zone (CDZ) and the ventral canal nucleus (VN) is located at the extreme apex of the egg cell. The general tissues of the female gametophyte (F) surround the archegonium. J, jacket cells. Bar, 100 μm. Reproduced with permission from Friedman (1990b).

Berridge, 1909; Herzfeld, 1922; Khan, 1943; Friedman, 1990a,b, 1991, 1992). All workers have found that fertilization of the egg nucleus by one of the two sperm nuclei ultimately yields a zygote nucleus that is situated at the base (chalazal end) of the former egg cell.

Proembryo development (Section IV,C) appears to follow a pattern that is typical of most nonflowering seed plants: initial free nuclear development is succeeded by a process of cellularization (Friedman, 1990b, 1992). However, unlike conifers, cycads, and *Ginkgo,* in which cellularization of the free nuclei within the confines of the former egg cell yields a single multicellular embryo, each free nucleus in *Ephedra* produces a spherical cell wall and becomes a separate single-celled proembryo (Land, 1907;

Khan, 1943; Lehmann-Baerts, 1967; Moussel, 1977). Typically eight free nuclei are found within a fertilized archegonium and, hence, eight proembryos may be initiated from a fertilized egg cell. As will be discussed (Section IV,D), the origin of each of the eight proembryos within an archegonium was, until recently, incorrectly interpreted. Ultimately, only one embryo will persist within a mature seed.

B. Early Reports of Intra-Archegonial Double Fertilization in *Ephedra*

The first suggestion of a possible participation of the second sperm nucleus in a second fertilization event in *Ephedra* was made by Land (1907). Land believed that in *Ephedra trifurca,* formation of an ephemeral tissue after fertilization within the archegonial cavity might be the joint result of chromatin of the second sperm nucleus and jacket cell nuclei that had entered the egg cytoplasm. Porsch (1907), interpreting Land's plates, which show the second sperm nucleus directly adjacent to the ventral canal nucleus (Fig. 6), suggested that the ephemeral tissue reported by Land (1907) might have resulted from a fusion of the ventral canal nucleus and the second sperm nucleus. Although the ephemeral tissue within the archegonium has since been shown to be a noncellular product resulting from the cleavage and breakdown of egg cell cytoplasm (Lehmann-Baerts, 1967; Moussel, 1977; Friedman, 1990b), Land's figure showing the second sperm nucleus adjacent to a persistent ventral canal nucleus (Fig. 6) is nevertheless suggestive of the possibility that intra-archegonial double fertilization may occur upon occasion in *Ephedra trifurca.*

In 1922 Herzfeld reported that the second sperm nucleus often fuses with the ventral canal nucleus in *Ephedra campylopoda.* Herzfeld presented a drawing of an egg cell in which a sperm nucleus is in the process of fusing with the egg nucleus, and a second sperm nucleus is coupled with the ventral canal nucleus (Fig. 6). Based on a single histological preparation from *Ephedra campylopoda,* in which fusion of the egg nucleus with a sperm nucleus was occurring, Narang (1955) believed that the large size of the ventral canal nucleus suggested that a second fertilization event between a sperm nucleus and the ventral canal nucleus had taken place.

In *Ephedra foliata* several workers have hypothesized that intra-archegonial double fertilization can occur on occasion. Maheshwari (1935) suggested that in exceptional cases the ventral canal nucleus may behave like an egg nucleus if displaced from its normal apical position within the egg cytoplasm to a more central position (no figures accompanied this publication). Based upon evidence from a preparation showing a second sperm nucleus next to the ventral canal nucleus, Khan (1940, 1943) pro-

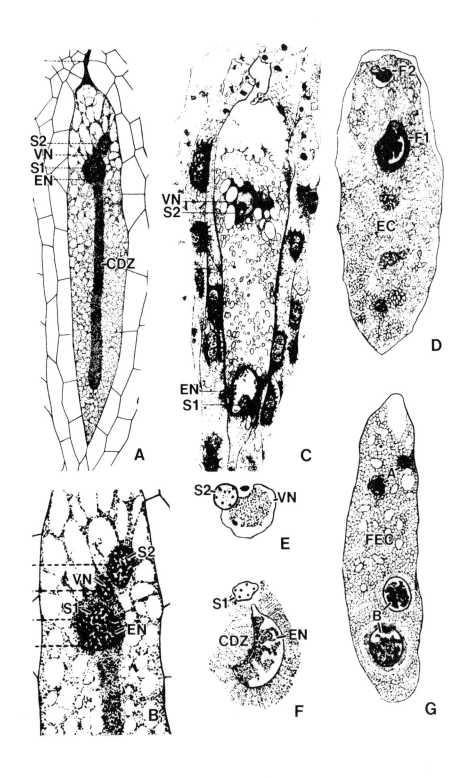

posed that double fertilization occasionally takes place in *Ephedra foliata* (Fig. 6). Khan (1940, 1943) believed that supernumerary nuclei, occasionally found in the apical portion of the egg cytoplasm (Fig. 6), were derivatives of such a second fertilization event and suggested that these nuclei might form additional proembryos, although evidence to this effect was not presented. Mulay (1941) reported that fusion of the ventral canal nucleus with the second sperm occurs in *Ephedra foliata,* although no figures were published.

Moussel (1978) reported a case in *Ephedra distachya* in which a supernumerary nucleus (in addition to the normal zygote nucleus) was observed with an ultrastructure similar to that of a zygote nucleus (Fig. 7). This supernumerary nucleus was displaced from its normal (apical) position within the archegonium and was situated more deeply within the egg cytoplasm, near the cytoplasmically rich zone that usually contains only the egg nucleus. A ventral canal nucleus was not found elsewhere within this particular archegonium. Moussel (1978) suggested that the supernumerary nucleus was a diploid nucleus formed by the fertilization of the ventral canal nucleus by a sperm nucleus. He believed that the second fertilization product (when formed) eventually gave rise to supernumerary proembryo nuclei. Moussel (1978) noted, however, that only microspectrophotometric analysis (of DNA content) could resolve whether the second nucleus was diploid and had resulted from a second fertilization event.

FIG. 6   Early published drawings of putative double fertilization events in *Ephedra*. (A) Egg cell of *Ephedra trifurca* with the first sperm nucleus (S1) in contact with the egg nucleus (EN) and the second sperm nucleus (S2) next to the ventral canal nucleus (VN), which has migrated from its extreme apical position to a location deeper within the egg cell. COZ, cytoplasmically dense zone. ×165. (B) Higher magnification of gamete nuclei in egg cell in A. ×370. (C) Putative intra-archegonial double fertilization in *Ephedra campylopoda*. The first sperm nucleus is next to the egg nucleus, and both nuclei are situated at the base of the egg cell. The ventral canal nucleus has been displaced from its apical position and is in the process of fusing with a sperm nucleus. (D) Putative intra-archegonial double fertilization in *Ephedra foliata*. The first fertilization event has been initiated between the egg nucleus and a sperm nucleus. The second sperm nucleus and the ventral canal nucleus have established contact near the apex of the egg cell. EC, egg cell; F1, first fertilization product; F2, second fertilization product. ×220. (E) Higher magnification view of second fertilization event between the second sperm nucleus and the ventral canal nucleus shown in D. ×575. (F) Higher magnification view of first fertilization event shown in D. ×555. (G) Putative free nuclear products of double fertilization in *Ephedra foliata*. The two apical nuclei (A) are derived from the second fertilization event. The two basal nuclei (B) are derived from the zygote nucleus. FEC, fertilized egg cell ×220. (A and B) Reproduced from Land (1907). (C) Reproduced from Herzfeld (1922). (D–G) Reproduced with permission from Khan (1940).

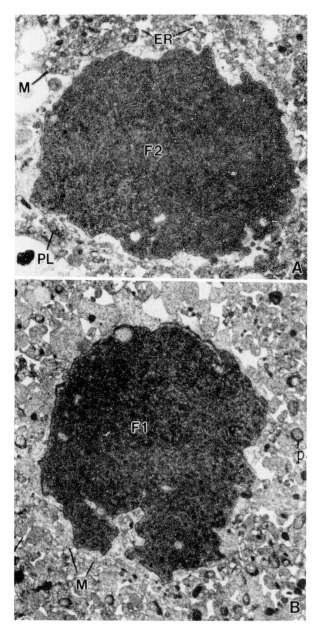

FIG. 7 Putative double fertilization products in *Ephedra distachya*. (A) Second fertilization product (F2) believed to have resulted from a fusion of the ventral canal nucleus and a second sperm nucleus. The ultrastructure of this nucleus is similar to that of the zygote nucleus. ×2400. (B) Zygote nucleus derived from the fusion of an egg nucleus and a sperm nucleus. ×2400. F1, first fertilization product; ER, endoplasmic reticulum; M, mitochondria PL, plastids. Reproduced with permission from Moussel (1978).

C. Recent Studies of Fertilization in *Ephedra:* Confirmation of a Regular Process of Intra-Archegonial Double Fertilization

Circumstantial evidence of intra-archegonial double fertilization in a number of species of *Ephedra* has existed for more than three-quarters of a century. Only recently, however, has a concerted effort been made to document the existence, regularity, and cell-biological features of such a process. Two species of *Ephedra* (*Ephedra nevadensis* and *Ephedra trifurca*) have been intensively studied with modern histological and cytological approaches (Friedman, 1990a,b, 1991, 1992) and ultrastructural techniques (Friedman, unpublished data).

In both *Ephedra nevadensis* and *Ephedra trifurca*, the ventral canal nucleus is persistent through the time of entry of sperm into the egg cell. Thus, the egg cell in *Ephedra nevadensis* and *Ephedra trifurca* is binucleate, containing a ventral canal nucleus at the apex and an egg nucleus situated within a cytoplasmically dense region that contains significant amounts of organellar DNA (Fig. 5). During the fertilization process two sperm nuclei from a single binucleate sperm cell enter the egg cell (Friedman, 1990b, 1992).

In *Ephedra nevadensis* and *Ephedra trifurca,* the first sperm nucleus immediately migrates to the egg nucleus. As contact is made between the two gamete nuclei, the egg nucleus becomes invaginated (Figs. 8–10).The paired first sperm nucleus and egg nucleus next initiate a pattern of migration to the base (chalazal end) of the former egg cell (Friedman, 1990a,b, 1991, 1992). During this time both the male nucleus and female nucleus maintain their integrity (Figs. 8 and 9). After the sperm nucleus and egg nucleus reach the base of the former egg cell, fusion takes place to yield a single nucleus, the zygote nucleus (Fig. 11) (Friedman, 1990a,b, 1991, 1992).

In *Ephedra nevadensis* and *Ephedra trifurca,* shortly after initial contact is made between the first sperm nucleus and egg nucleus, the ventral canal nucleus initiates a chalazal migration pattern in tandem with the second sperm nucleus (Friedman, 1990a,b, 1991, 1992). Contact is established between these nuclei (Figs. 8–10) and the paired nuclei come to rest in a zone of the egg cell approximately one-fourth to one-third of the distance from the apex. Similar to the egg nucleus, the ventral canal nucleus invaginates as a result of its contact with the second sperm nucleus (Figs. 8 and 9), and fusion between these two nuclei is delayed, with each nucleus maintaining its separate zone of nucleoplasm. Eventually, either slightly before or after the fusion of the egg nucleus and first sperm nucleus, the ventral canal nucleus and second sperm nucleus fuse to form a single nucleus (Fig. 11) (Friedman, 1990b, 1991).

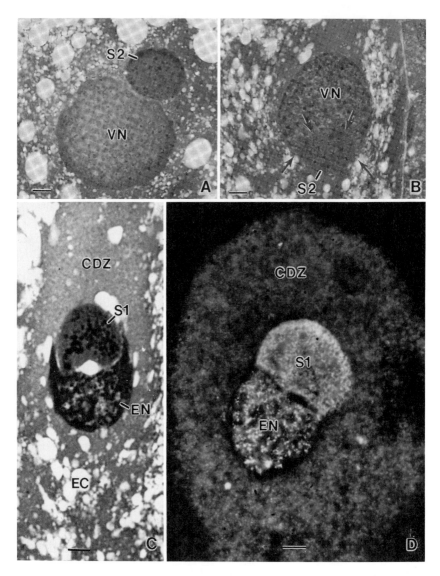

FIG. 8 Double fertilization in *Ephedra nevadensis*. (A) Initial contact between the ventral canal nucleus (VN) and the second sperm nucleus (S2). The ventral canal nucleus has invaginated in response to the approach of the sperm nucleus. Bar, 10 μm. (B) Later stage of fusion between ventral canal nucleus and second sperm nucleus. Arrows indicate area of contact between the two nuclei. Bar, 10 μm. (C) Invagination of egg nucleus (EN) in response to contact with first sperm nucleus (S1). EC, egg cell. Bar, 10 μm. (D) Fluorescence view (stained with DAPI) of first fertilization event showing cytoplasmically dense zone (CDZ) that surrounds the first sperm nucleus and egg nucleus and contains significant quantities of organellar DNA. Bar, 10 μm. (A and B) Reproduced with permission from Friedman (1990a), copyright 1990 by the AAAS. (C and D) Reproduced with permission from Friedman (1990b).

Intra-archegonial double fertilization in *Ephedra trifurca* was also examined with microspectrophotometric techniques (Friedman, 1991). Data from this analysis indicate that during the period of time the sperm nucleus and egg nucleus are in contact but have not fully fused, each nucleus passes (separately) through the S phase of the cell cycle and attains the 2C quantity of DNA (each sperm nucleus contains the 1C quantity of DNA while within the pollen tube). Complete fusion of the sperm nucleus and egg nucleus does not occur until both nuclei have doubled their respective DNA contents (Friedman, 1991). The result is a delay in the actual fusion of the egg nucleus and sperm nucleus from the time of their initial contact (in the upper portion of the egg cell), until both nuclei have completed DNA synthesis and each contains the 2C amount of DNA. At this point the paired nuclei are situated at the base of the former egg cell, and the product of the first fertilization event in *Ephedra trifurca* is a diploid nucleus that contains a 4C quantity of DNA (Friedman, 1991).

Microspectrophotometric data indicate that the ventral canal nucleus and second sperm nucleus each undergoes DNA synthesis (from 1C content to 2C content) prior to their complete fusion. Thus, the delay between initial contact of the second sperm nucleus and the ventral canal nucleus, and their ultimate fusion, is integrated with the cell cycle and is similar to the behavior of the egg nucleus and the first sperm nucleus. Measurements of the fully fused second fertilization product in *Ephedra trifurca* indicate that it is diploid and contains a 4C quantity of DNA (Friedman, 1991).

The quantitative fluorescence data, in conjunction with basic light and fluorescence images of the fertilization process in *Ephedra trifurca* and *Ephedra nevadensis,* indicate that each pair of nuclei involved in each of the two separate fertilization events within a single egg cell behaves similarly at the cell-biological level from three basic perspectives: (1) the ventral canal nucleus and the egg nucleus both invaginate when contact is established with a respective sperm nucleus; (2) complete fusion of male and female nuclei is delayed until DNA synthesis is completed by the egg nucleus and the first sperm nucleus, and the ventral canal nucleus and the second sperm nucleus; and (3) the product of each fertilization event is a diploid nucleus with a 4C content of DNA that is prepared to enter into mitosis (Figs. 11 and 12).

D. Fate of the Second Fertilization Product in *Ephedra*

The fate of the second fertilization product has been studied in *Ephedra trifurca* (Friedman, 1992). At the end of the intra-archegonial double fertilization process in *Ephedra trifurca,* two diploid nuclei reside within the former egg cell (Figs. 11, 13, and 14). The first fertilization product (derived

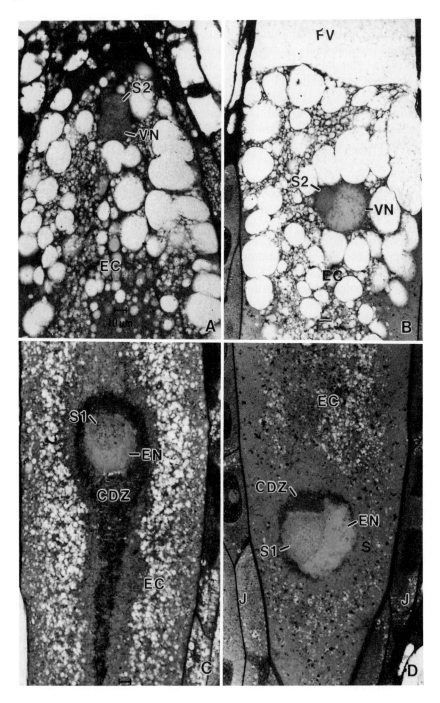

from a fusion of the egg nucleus and a sperm nucleus) is situated at the base of the former egg cell. The second fertilization product (derived from a fusion of the ventral canal nucleus and a second sperm nucleus) is located approximately one-fourth to one-third of the distance from the apex of the former egg cell (Figs. 11, 13, and 14) (Friedman, 1991, 1992). Each of the two fertilization products enters into mitosis (Fig. 13) to produce two daughter nuclei (and a total of four free nuclei within the former egg cell) (Fig. 14). A second set of synchronous mitotic divisions is initiated, and the formation of eight free nuclei results (Fig. 14). Four of the eight free nuclei are derived from the first fertilization product (the true zygote nucleus) and are situated near the base of the former egg cell. The other four free nuclei, derived from the second fertilization product, are found within the upper (chalazal) half of the former egg cell (Fig. 15) (Friedman, 1992).

The formation of eight (there may be more or fewer; Khan, 1943) free nuclei within the cytoplasm of the former egg cell after fertilization in *Ephedra* has been reported for all species within this genus that have been investigated. However, earlier workers (Land, 1907; Khan, 1943; Narang, 1955; Lehmann-Baerts, 1967; Moussel, 1977), unaware of the fact that regular intra-archegonial double fertilization occurs in *Ephedra,* inferred that all of the proembryo free nuclei were derived from a single zygote nucleus (Fig. 16). Thus, the zygote nucleus was believed to divide to produce two nuclei, from which four and then eight nuclei were formed through subsequent waves of mitotic activity.

All studies of reproduction in *Ephedra* indicate that following the 8-nucleate stage of postfertilization development, phragmoplasts form around each proembryo nucleus (Land, 1907; Lehmann-Baerts, 1967; Moussel, 1977). Cell walls are produced that separate each of the eight nuclei from the remaining cytoplasm of the former egg cell. As a consequence, eight proembryos are formed per fertilized archegonium. Four of the proembryos are derived from the true zygote nucleus and four from the second fertilization product (Friedman, 1992). All of the proembryos

---

FIG. 9   Double fertilization in *Ephedra trifurca*. (A) Initial contact between the second sperm nucleus (S2) and the ventral canal nucleus (VN). The ventral canal nucleus has invaginated. EC, egg cell. Bar, 10 μm. (B) Later stage of second fertilization event. The paired ventral canal nucleus and second sperm nucleus have migrated from the apex of the egg cell to a position deeper within the egg cell. FV, fertilization vacuole. Bar, 10 μm. (C) Early stage of first fertilization event. The first sperm nucleus (S1) has plunged into the egg nucleus (EN) and both nuclei are surrounded by the cytoplasmically dense zone (CDZ). Bar, 10 μm. (D) Late stage of the first fertilization event. The first sperm nucleus and egg nucleus have migrated to the extreme base of the egg cell. J, jacket cells. Bar, 10 μm. Reproduced with permission from Friedman (1991).

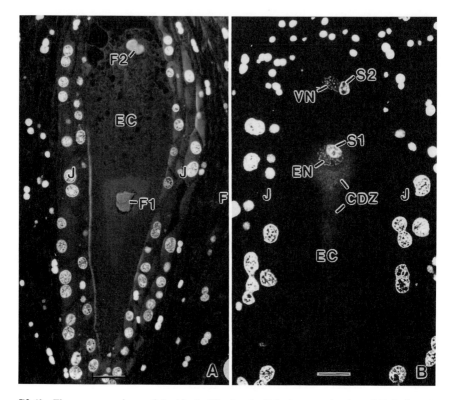

FIG. 10 Fluorescence views of double fertilization in *Ephedra nevadensis* and *Ephedra tri-furca*. (A) *Ephedra nevadensis*. F1, first fertilization event; F2, second fertilization event; EC, egg cell; F, female gametophyte; J, jacket cells. Bar, 100 μm. (B) *Ephedra trifurca*. EN, egg nucleus; S1, first sperm nucleus; S2, second sperm nucleus; VN, ventral canal nucleus; CDZ, cytoplasmically dense zone. Bar, 50 μm. (A) Reproduced with permission from Fried-man (1990b). (B) Reproduced with permission from Friedman (1991).

are genetically identical because the ventral canal nucleus and egg nucleus are mitotic derivatives of the central cell nucleus, and the two sperm nuclei (from a binucleate sperm cell) are also sister nuclei. Ultimately, only one embryo will persist and develop to maturity in a single seed.

V. Potential Double Fertilization in *Gnetum* and *Welwitschia*

Gnetum and *Welwitschia* comprise the remaining extant taxa within the monophyletic group, Gnetales. Both groups are considered to be highly

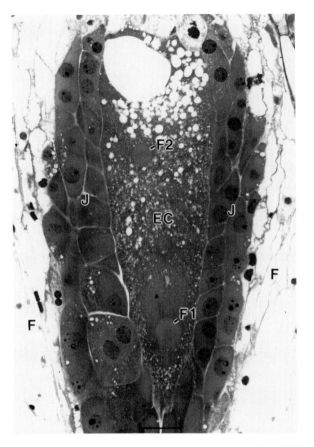

FIG. 11   End result of double fertilization in *Ephedra nevadensis*. The first fertilization product (F1) is located at the base (chalazal end) of the egg cell (EC) and the second fertilization product (F2) (from the fusion of the ventral canal nucleus and second sperm nucleus) is situated in the micropylar half of the egg cell. Bar, 50 μm. J, jacket cells; F, female gametophyte tissue. Reproduced with permission from Friedman (1990a), copyright 1990 by the AAAS.

derived with respect to the structure of the female gametophyte and the overall fertilization process (Doyle and Donoghue, 1986). In both *Gnetum* and *Welwitschia,* archegonial structure has been lost from the female gametophyte, and unusual modified structures are involved in the fertilization process.

In *Gnetum* the female gametophyte is tetrasporic (Maheshwari and Vasil, 1961) and initially undergoes a free nuclear period of development. Cellularization of the coenocytic female gametophyte is only partial, and hence is divergent from the normal pattern expressed in most nonflowering

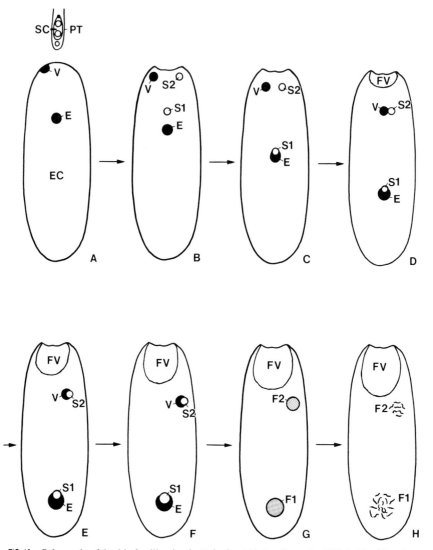

FIG. 12 Schematic of double fertilization in *Ephedra*. (A) A pollen tube (PT) (with a binucleate sperm cell, SC) approaches a binucleate egg cell (EC). E, egg nucleus; V, ventral canal nucleus. (B) Two sperm nuclei enter the egg cell. (C) The first sperm nucleus (S1) makes contact with the egg nucleus, and the ventral canal nucleus dislodges from its apical position. (D–F) Migration of each pair of male and female nuclei. During this time each nucleus maintains its separate identity and undergoes DNA synthesis. (G) Fusion of the first sperm nucleus and egg nucleus, and the second sperm nucleus and ventral canal nucleus to yield two diploid nuclei, each with 4C content of DNA. (H) Mitosis of the first (F1) and second fertilization products (F2). FV, fertilization vacuole. Reproduced with permission from Friedman (1991).

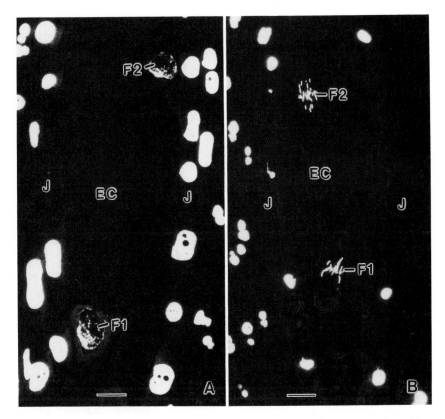

FIG. 13 End result of double fertilization in *Ephedra trifurca*. (A) Fluorescence view of first (F1) and second fertilization products (F2), each of which is diploid and contains a 4C quantity of DNA. EC, egg cell; J, jacket cells. Bar, 50 μm. (B) Mitosis of first and second fertilization products. Four nuclei will be formed: two nuclei will be located at the base of the former egg cell (derived from the zygote nucleus) and two nuclei will be situated in the micropylar half of the former egg cell (derived from the second fertilization product). Bar, 50 μm. Reproduced with permission from Friedman (1991).

seed plants. Ultimately, a female gametophyte is formed with a cellular chalazal end (each cell may be multinucleate, Maheshwari and Vasil, 1961) and a micropylar end that contains groups of uninucleate cells and free nuclei (Vasil, 1959; Waterkeyn, 1954; Maheshwari and Vasil, 1961). After a pollen tube has penetrated the apex of the female gametophyte, a cell (or nucleus) adjacent to the male gametophyte functions as a female gamete (Waterkeyn, 1954; Vasil, 1959) and fuses with a sperm nucleus. In rare instances two female cells (or nuclei) may fuse with the two sperm from a single pollen tube (Vasil, 1959). Unfortunately, most aspects of the fertilization process in *Gnetum* are poorly understood.

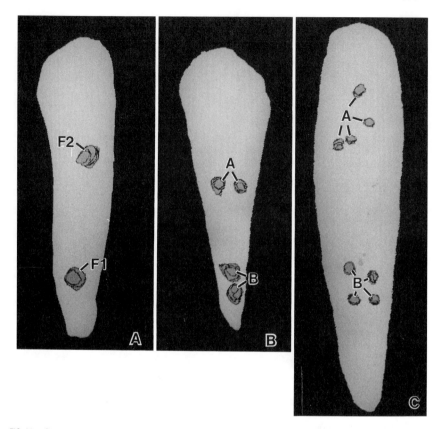

FIG. 14   Computer-reconstructed images (based on serial sections through archegonia) of free nuclear development of the two fertilization products in *Ephedra trifurca*. (A) First (F1) and second fertilization products (F2) at the end of the double fertilization process. (B) Four-nucleate stage of development with two apical nuclei (A) from the second fertilization product and two basal nuclei (B) from the zygote nucleus. (C) Eight-nucleate stage of development. The four apical nuclei are derived from the second fertilization event and the four basal nuclei are derived from the first fertilization event (zygote nucleus). Each of the eight nuclei typically will develop a cell wall and form a separate single celled proembryo. Fig. 14 similar to Fig. 2 from Friedman (1992).

In *Welwitschia* the female gametophyte is also tetrasporic in origin and initially undergoes a period of free nuclear development (Martens, 1971). Cellularization of the female gametophyte results in the formation of many multinucleate cytoplasmic masses, each delimited by a cell wall. Coeno-cytic cells at the apex (micropylar end) of the female gametophyte produce prothallial tubes; tube-like extensions that grow up through the nucellus and eventually meet and fuse with the tip of a downward growing pollen

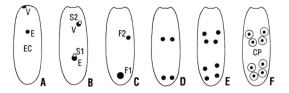

FIG. 15   Postfertilization development in *Ephedra trifurca*. (A) Binucleate egg cell (EC). E, egg nucleus; V, ventral canal nucleus. (B,C) Double fertilization of the egg nucleus and the ventral canal nucleus to produce two diploid fertilization products (F1 and F2). S1, first sperm nucleus; S2, second sperm nucleus. (D) Four-nucleate stage of postfertilization development after mitosis of each of the two fertilization products. (E) Eight-nucleate stage of postfertilization development following mitosis of 4-nucleate stage. (F) Cellularization of individual proembryo nuclei to produce eight cellular proembryos (CP). The four apical proembryos are derived from the second fertilization event and the four basal proembryos are from the zygote. All of the proembryos are genetically identical. Reproduced with permission from Friedman (1992), copyright by the AAAS.

tube (Martens, 1971). Martens (1971) believed that all of the female nuclei (there may be several) within a prothallial tube may function as sexual nuclei. Double fertilization events between the gametes of a prothallial tube and a pollen tube have not been observed, but Pearson (1909) suggested that such events were theoretically possible.

Clearly, more information and modern study of sexual reproduction in *Gnetum* and *Welwitschia* are needed to resolve whether double fertilization events do indeed occur and how they might relate to intra-archegonial double fertilization in *Ephedra*. If double fertilization is a primitive (symplesiomorphic) condition within the *Gnetales* (and not an autapomorphy

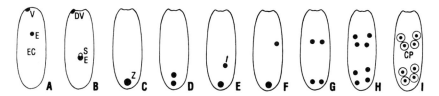

FIG. 16   Schematic of earlier hypotheses of proembryo development in *Ephedra*. (A) Binucleate egg cell (EC). (B,C) Single fertilization of egg nucleus by sperm nucleus to form zygote nucleus (Z). The ventral canal nucleus ultimately degenerates (DV). (D–F) Mitosis of zygote nucleus to produce two nuclei at base of former egg cell. Evidence of this single mitotic division was not found in *Ephedra trifurca* (Friedman, 1992). One of the nuclei was hypothesized to migrate to the micropylar half of the former egg cell to produce late 2-nucleate stage of proembryogenesis. (G) Four-nucleate stage of postfertilization development after mitosis of the two nuclei derived from the zygote nucleus. (H) Eight-nucleate stage of postfertilization development following mitosis of 4-nucleate stage. (I) Cellularization of individual proembryo nuclei to produce eight cellular proembryos (CP). It was believed that all of the proembryos were derived from a single fertilization event. Reproduced with permission from Friedman (1992), copyright 1992 by the AAAS.

of *Ephedra*), it is possible that such events will be found upon closer examination of the reproductive process in members of the genus *Gnetum* and in *Welwitschia mirabilis*.

VI. Cell Biological Features of Double Fertilization in Nonflowering Seed Plants

In most instances the ventral canal nucleus in nonflowering seed plants either degenerates prior to the time of fertilization or does not participate in a second fertilization event. The fact that the ventral canal nucleus is the sister nucleus of the egg nucleus (and hence both nuclei are genetically identical) suggests that the local cytoplasmic environments of the egg nucleus and the ventral canal nucleus are important determinants in the ultimate fate of each nucleus. Thus, competency to fuse, in those cases in which the ventral canal nucleus participates in a second fertilization event, may be an acquired feature that is correlated with a novel pattern of cell-biological behavior by the ventral canal nucleus within the egg cell cytoplasm.

The process of intra-archegonial double fertilization among nonflowering seed plants has been most closely investigated in the genus *Ephedra*. An examination of each figure presented by workers as evidence of double fertilization in *Ephedra* (Land, 1907; Herzfeld, 1922; Khan, 1940; Friedman, 1990a,b, 1991, 1992) reveals that in every case reported and figured, the ventral canal nucleus was displaced from its normal apical position within the egg cytoplasm and was situated more deeply within the archegonial cavity, close to or within a cytoplasmically dense zone of the egg cell. Ultrastructural studies by Moussel (1977) indicated that the egg cell cytoplasm in *Ephedra distachya* is extremely heterogeneous and that the normal cytoplasmic environments for the ventral canal nucleus and egg nucleus (prior to any migration of the ventral canal nucleus) differ significantly: the ventral canal nucleus is situated in a region of the egg cell cytoplasm that is vacuolate and devoid of most heritable organelles, whereas the egg nucleus is found in a cytoplasmically rich central zone that contains numerous plastids and mitochondria.

Moussel (1977) suggested that the different cytoplasmic environments of the ventral canal nucleus and the egg nucleus may be important determinants in the developmental potential and fate of each nucleus and that if displaced from the apical vacuolate region of the egg cell cytoplasm into a cytoplasmically rich zone, the ventral canal nucleus may acquire the ability to fuse with a male gamete. This hypothesis is consistent with earlier observations by Maheshwari (1935) concerning the egg nucleus-like development of the ventral canal nucleus in *Ephedra foliata*.

More broadly, egg nucleus-like development of the ventral canal nucleus has been reported in cycads (Chamberlain, 1906, 1912; Sedgewick, 1924; Bryan and Evans, 1957), *Ginkgo biloba* (Ikeno, 1901), and conifers (Chamberlain, 1899; Hutchinson, 1915). In each case in which the ventral canal nucleus acquires cytological features similar to those of the egg nucleus, the ventral canal nucleus had been displaced from its normal apical position within the egg cell cytoplasm to a more central region, closer to the egg nucleus. Thus, in all nonflowering seed plants (with archegonial structure), the local cytoplasmic environment may determine the developmental capacity of each of the daughter nuclei of the central cell nucleus (the egg nucleus and the ventral canal nucleus) to fuse with a sperm.

Microspectrophotometric data from intra-archegonial double fertilization in *Ephedra trifurca* (Friedman, 1991) suggested that fusion of the egg nucleus and sperm nucleus is tightly correlated with the cell cycle and that the fusion of the ventral canal nucleus and second sperm nucleus mimics the behavior of the first fertilization event with respect to the synthesis of DNA. Thus, the ability of a female nucleus to acquire fecundible characteristics may be integrated with specific cytoplasmic signals that are found only in certain portions of the egg cell. Given the large size of egg cells in nonflowering seed plants (hundreds to thousands of micrometers in length), it is likely that the cytoplasm of the egg cell is extremely heterogeneous and plays a vital role in the developmental process of sexual fusion.

Although the female gametophyte in angiosperms lacks archegonia and is much reduced, ultrastructural studies (Willemse and Van Went, 1984) have shown that within the largely vacuolate central cell of the embryo sac, the cytoplasm surrounding the polar nuclei (one of which is the sister nucleus to the egg), or secondary nucleus, contains substantial quantities of organelles. Thus, similar to the situation in nonflowering seed plants, the cytoplasmic environment in angiosperms may also be an important determinant of the ability of a specific nucleus within a female gametophyte to behave in a fecundible fashion. Ultimately, all of these observations concerning the egg-like potential of the sister nucleus of the egg nucleus in nonflowering seed plants and angiosperms may provide considerable insight, when examined more closely, into the fundamental parameters required for a particular nucleus to participate in a process of fertilization (Friedman, 1990b).

VII. Evolutionary Considerations

A. Comparisons among Nonflowering Seed Plants

The potentially wide phylogenetic distribution of intra-archegonial double fertilization events in nonflowering seed plants suggests that the prerequi-

sites for the evolution of a fertilization of the sister nucleus of the egg nucleus may have existed since the origin of seed plants. Persistence of the ventral canal nucleus through the time of fertilization may occur regularly, or occasionally, in conifers, cycads, *Ginkgo biloba,* and *Ephedra.* Entry of two functional sperm into a single archegonium has also been documented in certain conifers, cycads, *Ginkgo biloba,* and *Ephedra.* However, strong evidence for the existence of a regular pattern of intra-archegonial double fertilization in nonflowering seed plants would appear to be limited to members of the genus *Ephedra* (Friedman, 1990a,b, 1991, 1992).

Although limited by the primitive nature of histological techniques during the early part of the Twentieth Century, the evidence suggests that intra-archegonial double fertilization can occur in certain conifers. However, many of the cell-biological details of this process remain unclear, and modern confirmation of intra-archegonial double fusion events in conifers (where it has previously been reported) is highly desirable. In cycads and *Ginkgo biloba* evidence of an actual fusion of a sperm nucleus and ventral canal nucleus has not been presented. Nevertheless, reports of cytological development of the ventral canal nucleus similar to that of the egg nucleus suggest that intra-archegonial double fertilization events might occur in zooidogamous seed plants.

The totality of the evidence suggests that double fertilization events in conifers, *Ginkgo biloba,* and cycads are, for the most part, largely isolated and perhaps anomalous. However, intra-archegonial double fertilization events are a regular feature of the reproductive process in at least two (and probably more) species of *Ephedra* (Friedman, 1990a,b, 1991, 1992). In *Ephedra* the ventral canal nucleus is regularly persistent through the time of fertilization, two sperm nuclei always enter an individual egg cell, and the ventral canal nucleus mimics (at the cell-biological level) the behavior of the egg nucleus when in the vicinity of the second sperm nucleus (Friedman, 1990b, 1991, 1992). In contrast to angiosperms, the separate fertilization of the egg nucleus and its sister nucleus in *Ephedra* remained an evolutionary dead end, producing supernumerary embryos, but no elaborate tissue similar to endosperm.

B. Evolutionary Relationships of Double Fertilization in *Ephedra* and Angiosperms

The presence of regular (not teratological, as recently misinterpreted by LoConte and Stevenson, 1991) double fertilization events in *Ephedra* has profound evolutionary implications with respect to the origin of double fertilization and endosperm in angiosperms. Recent phylogenetic studies

have provided strong evidence that the Gnetales (*Ephedra, Gnetum,* and *Welwitschia*) are monophyletic (Crane, 1985; Doyle and Donoghue, 1986). These studies indicated that the Gnetales are the most closely related extant group of seed plants to the angiosperms and that *Ephedra* is basal within the Gnetales (Crane, 1985; Doyle and Donoghue, 1986) (Fig. 17).

Double fertilization events involving separate fusions of sperm with an egg nucleus and its sister nucleus are a regular feature of reproduction in

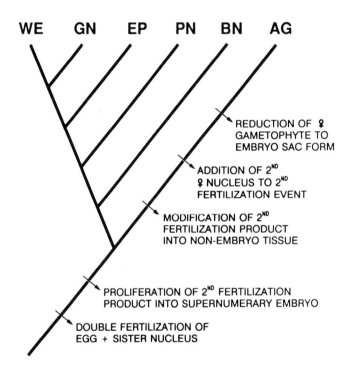

FIG. 17   Cladogram of the most parsimonious interpretation of the evolution of double fertiliza-tion and endosperm in angiosperms and their closest relatives. Phylogenetic relationships are based on work by Doyle and Donoghue (1986). Double fertilization of the egg nucleus and its sister nucleus is a synapomorphy of angiosperms (AG) and their sister group (*Ephedra,* EP; *Gnetum,* GN; *Welwitschia,* WE; character states of fossil groups *Pentoxylon,* PN, and *Bennettitales,* BN, are not known). Free nuclear to cellular proliferation of the second fertilization product into a supernumerary embryo also predates the origin and divergence of the angiosperms and their sister group. Autapomorphies of angiosperms are a modification of development of the second fertilization product into a nonembryo tissue (endosperm); the addition of a second female nucleus to the second fertilization event to produce a triploid fusion product; and the reduction of the female gametophyte to the simple flowering plant embryo sac form. The order in which these three evolutionary advances that characterize the flowering plants took place is not known. Reproduced with permission from Friedman (1992), copyright 1992 by the AAAS.

Ephedra and angiosperms. In *Ephedra* the ventral canal nucleus is the sister nucleus of the egg nucleus. In angiosperms with a primitive pattern of embryo sac development (*Polygonum* type) (Stebbins, 1974; Willson and Burley, 1983), one of the two polar nuclei with which the second sperm nucleus fuses is the sister nucleus of the egg nucleus (Brink and Cooper, 1947). In angiosperms, however, the second fertilization event involves not only the sister nucleus of the egg nucleus but an additional female nucleus, thereby yielding a triploid fusion product.

From a comparative perspective, the second fertilization product in *Ephedra* exhibits a developmental pattern that is fundamentally similar to endosperm in primitive groups of flowering plants (free nuclear endosperm) (Friedman, 1992). In *Ephedra* the second fertilization product yields four free nuclei that subsequently undergo a process of cellularization to produce proembryos. Among flowering plants with a primitive free nuclear pattern of endosperm development (Stebbins, 1974), the primary endosperm nucleus (formed from the fusion of a sperm nucleus with two polar nuclei) initiates a process of free nuclear proliferation (similar to *Ephedra*). This free nuclear period of development is usually followed by a process of cellularization of the endosperm tissue (Vijayaraghavan and Prabhakar, 1984). Although the extent of free nuclear development of endosperm in angiosperms is far more extensive (often hundreds of nuclei that later become cellular) than is the case with the derivatives of the second fertilization product in *Ephedra* (four nuclei that become cellular), the underlying pattern of free nuclear to cellular development is fundamentally similar (Friedman, 1992).

In view of the critical phylogenetic position of *Ephedra* and the recent confirmation of double fertilization in this genus, double fertilization and a subsequent free nuclear to cellular proliferation of a second fertilization product can no longer be assumed to be autopomorphies of angiosperms (Friedman, 1991, 1992). Friedman (1990a, 1992) and others (Meeuse, 1986; Donoghue, 1989) have proposed that a regular pattern of intra-archegonial double fertilization may have been established in a common ancestor of angiosperms and *Ephedra*.

If double fertilization and free nuclear to cellular development of the second fertilization product are evolutionarily homologous (synapomorphous) in *Ephedra* and angiosperms, the endosperm, as found in flowering plants, is likely to have evolved through an intermediate stage in which diploid embryos initially resulted from the second fertilization event (Friedman, 1992). This stage of evolutionary advancement would have been established in the common ancestor of *Ephedra* and flowering plants (Fig. 17).

The discovery that the second fertilization product in *Ephedra trifurca* regularly produces embryos (Friedman, 1992) provides support for a hypothesis concerning the evolutionary origin and history of endosperm

first proposed by Sargant in 1900, 2 years after the discovery of double fertilization in flowering plants. Sargant (1900) suggested that following the initial establishment of a process of double fertilization in the ancestors of flowering plants, the second fertilization product originally yielded a supernumerary embryo (which underwent free nuclear to cellular development, as is typical of most nonflowering seed plants). Sargant hypothesized that endosperm subsequently evolved through a modification of supernumerary embryo development into an aberrant nonembryo (endosperm) tissue. She further suggested that this developmental transformation occurred as a consequence of the addition of a second female nucleus to the second fertilization event. Although it is currently not possible to resolve whether endosperm tissue evolved prior to or after the addition of a third nucleus to yield a triploid second fertilization product (Friedman, 1992), the basic tenets of Sargant's hypothesis are supported by a study of fertilization and embryogeny in *Ephedra trifurca* (Friedman, 1992; see also discussion by Thomas, 1907).

The addition of a second female nucleus to the second fertilization event could have occurred prior to or after the evolution of a true embryo-nourishing tissue (endosperm) (Friedman, 1992). Thus, endosperm may originally have been diploid if the addition of a third nucleus to the second fertilization event occurred after the transformation of embryo development into a nutritive tissue (Fig. 17). In either case, with the establishment of a novel diploid or triploid nutritive tissue, the reduction of the female gametophyte (which in all nonflowering seed plants is typically large and functions in the nourishment of the developing embryo) to the embryo sac form in flowering plants must have been predicated upon the prior evolution of a novel embryo-nourishing tissue (Friedman, 1992). From this perspective it is likely that evolutionary advances relating to the embryo sac and endosperm, as now evident in angiosperms, were based on a preangiosperm origin of a process of intra-archegonial double fertilization (Friedman, 1990a,b, 1992).

The existence of a regular process of intra-archegonial double fertilization in *Ephedra* with subsequent free nuclear to cellular development of supernumerary embryos from the second fertilization product suggests that the first angiosperms inherited the basic foundations of a process of double fertilization from a common ancestor with *Ephedra* (Friedman, 1990a, 1992). In addition, the free nuclear to cellular development of the second fertilization product (present in proembryos of *Ephedra* and endosperm of primitive flowering plants) is consistent with the hypothesis that these features of reproduction are synapomorphous in *Ephedra* and angiosperms. These findings fundamentally alter previously held concepts about the origin of angiosperms and those reproductive features that traditionally have been believed to be unique to flowering plants.

In light of the potential evolutionary homologies (with respect to sexual reproduction) between *Ephedra* and angiosperms, the only distinguishing features of angiosperms with respect to sexual reproduction that can definitively be claimed to be unique to flowering plants are as follows: the developmental transformation of the second fertilization product from an embryo to endosperm; the addition of a second female nucleus to the second fertilization event; and the reduction of the female gametophyte to the characteristic seven-celled, 8-nucleate angiosperm embryo sac form (Fig. 17) (Friedman, 1992).

Acknowledgments

The author thanks Pam Diggle for many suggestions for the improvement of the manuscript and Beth Yao for assistance with photographic work. The research by the author reviewed in this paper was supported by NSF grant BSR 8818035 and an NSF Presidential Young Investigator Award BSR 9158182).

References

Allen, G. S. (1943). *Am. J. Bot.* **30,** 655–661.
Berridge, E. M. (1909). *Ann. Bot.* **23,** 509–512.
Berridge, E. M., and Sanday, E. (1907). *New Phytol.* **6,** 128–134.
Brink, R. A., and Cooper, D. C. (1947). *Bot. Rev.* **13,** 423–541.
Bryan, G. S., and Evans, R. I. (1957). *Am. J. Bot.* **44,** 404–415.
Chamberlain, C. J. (1899). *Bot. Gaz.* **27,** 268–279.
Chamberlain, C. J. (1906). *Bot. Gaz.* **42,** 321–358.
Chamberlain, C. J. (1912). *Bot. Gaz.* **53,** 1–19.
Chamberlain, C. J. (1935). "Gymnosperms, Structure and Evolution." University of Chicago Press, Chicago.
Choi, J. S., and Friedman, W. E. (1991). *Am. J. Bot.* **78,** 544–560.
Crane, P. R. (1985). *Ann. Mo. Bot. Gard.* **72,** 716–793.
Cronquist, A. (1988). "The Evolution and Classification of Flowering Plants." New York Botanical Garden, New York.
Donoghue, M. J. (1989). *Evolution* **43,** 1137–1156.
Downie, D. G. (1928). *Bot. Gaz.* **85,** 437–450.
Doyle, J. (1957). *Adv. Sci.* **14,** 120–130.
Doyle, J. A., and Donoghue, M. J. (1986). *Bot. Rev.* **52,** 321–431.
Favre-Duchartre, M. (1956). *Rev. Cytol. Biol. Veg.* **17,** 1–218.
Favre-Duchartre, M. (1959). *C. R. Acad. Sci. Ser. D* **249,** 1551–1553.
Friedman, W. E. (1987). *Am. J. Bot.* **74,** 1797–1815.
Friedman, W. E. (1990a). *Science* **247,** 951–954.
Friedman, W. E. (1990b). *Am. J. Bot.* **77,** 1582–1598.
Friedman, W. E. (1991). *Protoplasma* **165,** 106–120.
Friedman, W. E. (1992). *Science* **255,** 336–339.
Guignard, L. (1899). *C. R. Acad. Sci. Paris* **128,** 864–871.

Herzfeld, S. (1922). *Denk. Akad. Wiss. Wien Math. Natur. Klasse* **98**, 243–268.

Herzfeld, S. (1927). *Jahrb. Wiss. Bot.* **66**, 814–862.

Hutchison, A. H. (1915). *Bot. Gaz.* **60**, 457–472.

Ikeno, S. (1901). *Ann. Sci. Nat. Bot., Ser. VIII* **13**, 305–318.

Khan, R. (1940). *Curr. Sci.* **9**, 323–324.

Khan, R. (1943). *Proc. Nat. Acad. Sci. India* **13**, 357–375.

Land, W. J. G. (1902). *Bot. Gaz.* **34**, 249–259.

Land, W. J. G. (1904). *Bot. Gaz.* **38**, 1–18.

Land, W. J. G. (1907). *Bot. Gaz.* **44**, 273–292.

Lee, C. L. (1955). *Bot. Gaz.* **117**, 79–100.

Lehmann-Baerts, M. (1967). *Cellule* **67**, 53–87.

LoConte, H., and Stevenson, D. W. (1991). *Brittonia* **42**, 197–211.

Maheshwari, P. (1935). *Proc. Indian Acad. Sci.* **1**, 586–601.

Maheshwari, P., and Sanwal, M. (1963). *Mem. Indian Bot. Soc.* **4**, 104–119.

Maheshwari, P., and Singh, H. (1967). *Biol. Rev.* **42**, 88–130.

Maheshwari, P., and Vasil, V. (1961). "Gnetum." Council of Science and Industrial Research, New Delhi.

Martens, P. (1971). "Les Gnetophytes." Borntraeger, Berlin.

Meeuse, A. D. J. (1986). *Phytomorphology* **36**, 17–21.

Moussel, B. (1977). *Rev. Cytol. Biol. Veg.* **40**, 73–123.

Moussel, B. (1978). *Phytomorphology* **28**, 336–345.

Moussel, B. (1980). *Rev. Cytol. Biol. Veg. Bot.* **3**, 167–193.

Mulay, B. N. (1941). *J. Univ. Bombay* **10**, 56–69.

Narang, N. (1955). *Proc. Indian Sci. Cong.* **3**, 224.

Navaschwin, S. C. (1898). *Bull. Acad. Sci. St. Petersbourg* **9**, 4.

Owens, J. N., and Molder, M. (1977). *Can. J. Bot.* **55**, 2653–2667.

Pearson, H. H. W. (1909). *Philos. Trans. R. Soc. Lond. [Biol.]* **200**, 331–402.

Porsch, O. (1907). "Versuch einer Phylogenetischen Erklarung des Embryosackes und der Doppelten Befruchtung der Angiospermen." Gustav Fischer, Jena.

Rao, L. N. (1961). *Indian Bot. Soc.* **40**, 601–619.

Sargant, E. (1900). *Ann. Bot.* **14**, 689–712.

Sedgewick, P. J. (1924). *Bot. Gaz.* **77**, 300–310.

Singh, H. (1978). "Embryology of Gymnosperms." Gebruder-Borntraeger, Berlin.

Singh, H., and Maheshwari, K. (1962). *Phytomorphology* **12**, 361–372.

Smith, R. W. (1923). *Bot. Gaz.* **75**, 203–208.

Stebbins, G. L. (1974). "Flowering Plants: Evolution above the Species Level." Harvard University Press, Cambridge, Massachusetts.

Stebbins, G. L. (1976). *In* "Origin and Early Evolution of Angiosperms" (C. B. Beck, ed.), pp. 300–311. Columbia University Press, New York.

Thomas, E. N. (1907). *Sci. Prog.* **1**, 420–426.

Tiffney, B. H. (1981). *In* "Paleobotany, Paleoecology, and Evolution" (K. J. Niklas, ed.), pp. 193–230. Praeger, New York.

Vasil, V. (1959). *Phytomorphology* **9**, 167–215.

Vijayaraghavan, M. R., and Prabhakar, D. (1984). *In* "Embryology of Angiosperms" (B. M. Johri, ed.), pp. 319–376. Springer-Verlag, New York.

Waterkeyn, L. (1954). *Cellule* **56**, 103–146.

Willemse, M. T. M., and Van Went, J. L. (1984). *In* "Embryology of Angiosperms" (B. M. Johri, ed.), pp. 159–196. Springer-Verlag, Berlin.

Willson, M. F., and Burley, N. (1983). "Mate Choice in Plants." Princeton University Press, Princeton, New Jersey.

Double Fertilization

Scott D. Russell

Department of Botany and Microbiology, University of Oklahoma, Norman, Oklahoma 73019

I. Introduction

The process of double fertilization is a unique biological process in which one sperm fuses with the egg to produce the embryo while the second fuses with the central cell to form the endosperm, a nutritive tissue needed for growth of the succeeding generation. This characteristic defines the angiosperms as a natural systematic group. Although there is accumulating evidence for similar nuclear events within the Gnetophytes (especially *Ephedra;* Fricdman, this volume), the fate of the second sperm nucleus does not result in a developmentally distinct nutritive tissue.

The topics of double fertilization emphasized in this review include principally the following: (1) the arrival of the male gametophyte (pollen tube) in the female gametophyte (loosely termed the embryo sac [ES]), (2) the release of the male gametes into the female gametophyte, (3) the entry or plasmatic fusion of the sperm into the female reproductive cells, the egg cell and central cell, (4) the migration of sperm nuclei into alignment with female nuclei, and (5) their subsequent nuclear fusion. Previous works (Maheshwari, 1948, 1950; Steffen, 1963; Linskens, 1969; Jensen, 1973, 1974; Kapil and Bhatnagar, 1975; Van Went and Willemse, 1984; Fougère-Rifot, 1987; Mogensen, 1990) have reviewed the role of pollen tube (PT) growth in fertilization; however, with the expansion of the literature, the present review will be restricted specifically to gametophytic relationships during the process of fertilization.

A. General Historical Sketch

Historically, Giovanni Amici has been credited with the first report of the growth of a PT (Amici, 1824) and with the recognition that what he termed a "germinal vesicle" (i.e., egg) was already present within

the ovule prior to the fertilizing effect of the PT (first reported in 1842 and published in 1847). These results were definitively corroborated by the work of Wilhelm Hofmeister (1849). The recognition that the vegetative and sexual organs of flowering plants represented two generations of plants, the sporophyte (spore-producing) and gametophyte (gamete-producing) generations, respectively, is also credited to Hofmeister. The ES and pollen grain represent modern derivatives of free-living gametophytes, now relying on sporophytic organs of the flower (stigma, style, ovary, ovule) to function sexually. A recognition that the nuclear fusion of male and female gametes was central to the process of angiosperm reproduction was proven in Eduard Strasburger's monograph of 1884. This work embodied a number of major concepts of fertilization, including the observation that nuclear fusion was central to the process, that cytoplasm was not critical to fertilization and the male and female nuclei were true nuclei. Although the fusion of the polar nuclei of the central cell to form a secondary nucleus was reported as early as 1887 (Le Monnier, 1887), the involvement of a male nucleus in the formation of the endosperm was not described until the independent reports of Sergius Nawaschin (1898) and Leon Guignard (1899) that established double fertilization in *Lilium martagon*.

Research on double fertilization in the last 100 years has divided the century into two parts: the first half century was an intensive exploratory period during which the variability and diversity of reproduction in flowering plants were characterized (Maheshwari, 1950; Davis, 1966); and the second half century has, in contrast, been devoted to the development of new methods, from genetics to electron microscopy and descriptive molecular biology with the goal of refining our knowledge of the events of angiosperm reproduction and developing new concepts of embryogenesis.

Evidence has emerged that contradicts a number of basic ideas of the first half of this century, including the occurrence of dimorphic sperm cells, preferential fertilization, and the transmission of male cytoplasmic organelles, sometimes in significant numbers. The concept of the male germ unit (Mogensen, this volume) and female germ unit (Huang and Russell, this volume) has contributed an improved understanding of double fertilization and contributed to the live isolation of both male (Chaboud and Perez, this volume) and female gametes (Huang and Russell, this volume) and to pioneering work on the combination of gametes *in vitro* to produce artificial zygotes (Kranz *et al.*, this volume). Similarly, the role of male cytoplasm in the fertilization process has received increasing scrutiny, although comprehensive mechanisms of action are not yet well described.

B. Overview of Fertilization Events

In 1981 Tilton published the following general sequence of fertilization events in flowering plants, based on then available evidence and speculation: (1) pollen lands on the stigma; (2) gynecium responds via recognition reactions, nutrient mobilization, and hormone synthesis; (3) hormones reach the synergids; (4) hormones activate a synthetic mechanism in at least one synergid; (5) synergids synthesize both autolytic and chemotropic compounds; (6) one synergid begins to degenerate via autolysis; (7) chemotropic and autolytic compounds are secreted into the micropyle via the filiform apparatus (FA); (8) PTs reach the ovary and are attracted into the micropyle by a chemotropic agent(s); (9) a PT comes into contact with the autolytic compound(s) and its cytoplasm begins to degenerate; (10) a PT penetrates into the cytoplasm of the degenerated synergid and discharges its contents; (12) sperms migrate to their respective male nuclei; and (13) syngamy and double fertilization are effected. Jensen *et al.* (1983) proposed "crosstalk" between the male and female gametophyte consisting of relatively long distance hormone and chemotropic signals, initiated by PT growth, inducing changes in the female gametophyte that attract the tube and condition its receptivity; these events guide the PT into one synergid and trigger tube discharge. Evidence regarding the involvement of hormones, autolytic and chemotropic activity during fertilization *in vivo,* however, is yet unavailable, and therefore must be regarded as possible models or hypotheses, which will require considerable effort to prove. Hormones and lytic enzymes similarly have not been proven. A single model will be unlikely to account for the diversity of observations that have been published here and elsewhere and stresses the importance of continued careful experimental studies.

II. Receipt of the Pollen Tube

In the final pathway of their growth, PTs typically diverge from their more or less straight course within the style as they reach the ovule and turn at a 90 degree angle to enter the ovule. This is widely regarded as evidence for some form of PT guidance (i.e., chemotaxis, mechanical control, electrotaxis, etc.) (Dumas and Russell, this volume). Most often the PT arrives at the ovule and enters through the micropyle (porogamy). In relatively few plants the PT enters the ovule through the chalazal end (chalazogamy) or through the side of the ovule (mesogamy). In all patterns, however, the

PT tip is reported to enter the ES at the micropylar end, likely through a synergid.

A. Role of the Synergid in Pollen Tube Receipt

Synergids are present in every family of angiosperms examined to date and appear to have largely similar cytoplasmic organization and ultrastructure, especially at the family level and below (Huang and Russell, this volume). Such similarities suggest that synergids in angiosperms have generally similar functions during the reproductive process, although these functions are subject to controversy. The presence of cell wall ingrowths at the edge of the synergid are in contact with the ES wall, forming the FA, which resembles a transfer cell wall. Similar to a transfer cell, concentrations of mitochondria tend to be associated with the FA, suggesting an active and possibly secretory role in the ES, either directed toward the rest of the ES (e.g., Jensen, 1965) or toward the micropyle (e.g., Linskens, 1969). Products secreted through the micropyle purportedly attract PTs. The presence of a periodic acid-Schiff's positive product was reported in the micropyle of *Paspalum* (Chao, 1971, 1977), a presumed protein in *Lilium* (Welk *et al.*, 1965), and large concentrations of calcium have been reported in synergid vacuoles in cotton and several grasses (Jensen, 1965; Chaubal and Reger, 1990, 1992). Also, the FA (or a similar cell wall modification) occurs even in angiosperms lacking synergids, where the FA occurs at the base of the egg, suggesting a central role for this structure in the attraction and/or acceptance of the PT into the ES.

Numerous patterns of PT entry have been documented using light microscopic methods (Maheshwari, 1950). Kapil and Bhatnagar (1975) listed six general methods of PT entry: (1) between two intact synergids; (2) between the egg and synergid; (3) between ES wall and one or both synergids; (4) penetrating the synergid; (5) destroying, at contact, the synergid it encounters; and (6) directly entering the cytoplasm of the degenerated synergid. The synergid displays a number of structural modifications that appear to promote the fertilization process, including the presence of the FA and the typical presence of interrupted cell walls between the egg and central cell, which adapt this region for male gamete deposition, membrane apposition, and fusion. At the electron microscopic level, it is universally reported that the PT enters and discharges its contents into the degenerated synergid. Therefore, the receptivity of the synergid may be reflected by a degenerated condition.

Usually, just one synergid degenerates, although the stimulus for synergid degeneration is unclear. In a number of flowering plants synergid degeneration becomes evident only after pollination, when PTs are grow-

ing in the style (Table I; *e.g., Gossypium,* Jensen and Fisher, 1968; *Hordeum,* Cass and Jensen, 1970; *Linum,* Vazart, 1969; Russell and Mao, 1990; *Quercus,* Mogensen, 1972). However, synergid degeneration before pollination is reported in *Brassica* (Van Went and Cresti, 1988; Sumner and Van Caeseele, 1989) and *Triticum* (You and Jensen, 1985). In *Beta* (Bruun, 1987) and *Oryza* (Dong and Yang, 1989), synergid degeneration has been reported to occur even before anthesis. In plants containing degenerated synergids prior to PT penetration, the early stages in synergid degeneration are characterized by a general increase in the electron density of the cytoplasm, changes in the organization of the nucleus, and redistribution of some organelles (Fig. 1A). Later, the major cytoplasmic organelles of the synergid break down and the cytoplasm becomes essentially electron-opaque (Fig. 1B). At this stage the plasma membrane of the synergid is no longer intact, and in its absence degenerated synergid cytoplasm may leak between the egg and central cell. In cotton acid phosphatase activity increases in the synergid during degeneration, beginning with the plastids and dictyosomes, then spherosomes, and finally is present throughout the cell (Hill, 1977). High concentrations of calcium are typically found within the vacuole of the intact synergid (Jensen, 1965; Chaubal and Reger, 1990, 1992). Upon degeneration the receptive synergid loses the ability to retain fluorescein diacetate fluorescence and becomes intensely labeled with chlorotetracyline, which binds with the membrane-bound calcium that is abundant throughout the synergid cytoplasm (Huang and Russell, 1992). Electron-dense bodies spaced at intervals between the egg and central cell are noted after synergid degeneration in soybean (as shown in Fig. 1B, arrowheads; Dute *et al.,* 1989) and in numerous other plants (Huang and Russell, this volume). Although these bodies are frequently observed in angiosperms, their nature is unknown. One suggestion is that these bodies may have a role in maintaining the close proximity of the egg and central cell membranes and, in this regard, have been compared with the Casparian strip regions of endodermis (Cass and Karas, 1974).

In some plants, however, the degeneration of a synergid does not appear to precede the arrival of the PT (e.g., *Gasteria,* Willemse and Franssen-Verheijen, 1988; *Ornithogalum,* Van Rensberg and Robbertse, 1988; *Petunia,* Van Went, 1970), and in angiosperms lacking synergids, there is no degeneration of ES cells during fertilization. Once the PT arrives within a synergid, however, it immediately triggers the same cataclysmic changes in the synergid observed in other angiosperms. In *Nicotiana* only subtle changes in the density of the synergid cytoplasm are evident at the ultrastructural level: one synergid becomes slightly more dense after pollination and before PT arrival (Mogensen and Suthar, 1979). Cellular degradation, however, appears to occur only once the PT reaches the ovule, suggesting

TABLE I

Observations of Mode of Pollen Tube Entry and Fertilization in 36 Selected Species (and Cultivars) in 17 Families

Plant–species	Salient details and observations	References
Monocotyledons		
Liliaceae		
Ornithogalum caudatum	Single synergid degenerates after PT penetration	Van Rensberg and Robbertse, 1988
Orchidaceae		
Epidendrum scutella	Single synergid degenerates after pollination	Cocucci and Jensen, 1969
Epilobium palustre	Single synergid degenerates after pollination	Bednara, 1977
Epipactis atrorubens	Single synergid degenerates	Savina and Zhukova, 1983
Poaceae		
Hordeum vulgare (barley)	One or both synergids degenerate after pollination; when PT discharged into persistent synergid, gamete transfer does not occur (freeze-substitution LM study)	Cass and Jensen, 1970; Cass, 1981
cv. Atsel		
cv. Bonus	One or both synergids degenerate after pollination; unfused sperm cells observed in dSy; male cytoplasm not transmitted to egg, remaining in a cytoplasmic body; dSy proximal to vasculature in 67% of ovules; one or both synergids may degenerate after pollination	Mogensen, 1982, 1988 Mogensen, 1984
cv. Bomi	Single synergid degenerates 0–20 hr after receptivity of stigma	Engell, 1988a,b
Oryza sativa	Single synergid degenerates before anthesis	Dong and Yang, 1989
Stipa elmeri	Single synergid degenerates after PT penetration	Maze and Lin, 1975
Triticale	Single synergid degenerates; cytoplasm of dSy and PT penetrates between egg and central cell	Hause and Schröder, 1987
Triticum aestivum	Both synergids degenerate regardless of pollination	You and Jensen, 1985
Zea mays (maize)	Single synergid degenerates	Diboll, 1968
Dicotyledons		
Asteraceae		
Crepis tectorum	Single synergid degenerates	Godineau, 1969
Crepis capillaris	Single synergid degenerates after pollination	Kuroiwa, 1989
Gasteria verrucosa	Single synergid degenerates; callose in FA at receptivity	Willemse and Franssen-Verheijen, 1988
Helianthus annuus (sunflower)	Single synergid degenerates	Newcomb, 1973

362

Brassicaceae		
Arabidopsis thaliana	Single synergid degenerates	Mansfield and Briarty, 1991
Brassica campestris (rape)	Both synergids degenerate regardless of pollination; dSy and PT cytoplasm coat egg and central cell	Van Went and Cresti, 1988
Capsella bursa-pastoris	Single synergid degenerates; pSy degenerates soon after	Schulz and Jensen, 1968
Chenopodiaceae		
Beta vulgaris	Single synergid degenerates before anthesis	Brunn, 1987; Brunn and Olesen, 1988
Spinacia oleracea (spinach)	Single synergid degenerates after pollination; sperm fuse prior to transmission into egg and central cell	Wilms, 1981
Cucurbitaceae		
Cucumis sativus (cucumber)	Single synergid degenerates after pollination	Van Went et al., 1985
Fabaceae		
Glycine max (soybean)	Single synergid degenerates after pollination details of egg cell determination given: micropylar nucleus forms synergids, other forms egg–central cells	Dute et al., 1989; Folsom and Cass, 1986, 1990, 1992
Fagaceae		
Quercus gambelii	Single synergid degenerates after pollination; only one ovule of six in the ovary has a dSy and is fertilized	Mogensen, 1972
Linaceae		
Linum usitatissimum (flax)	Single synergid degenerates after pollination; dSy cytoplasm seen outside synergid between egg and CC; left dSy occurs in 60% of ES examined; 53% of dSy are proximal to vasculature; left synergid has smaller FA ($p<.01$), suggesting handedness in ES organization	Vazart, 1969; Russell and Mao, 1990
Malvaceae		
Gossypium hirsutum (cotton)	Single synergid degenerates after pollination with some exceptions (not viable); sperm cytoplasm remains in dSy synergid degeneration is dependent on chemical fixation; freeze-substituted synergids did not appear degenerate after PT discharge; pollen cytoplasm is seen outside of synergids, coating chalazal end of egg cell	Jensen and Fisher, 1968; Fisher and Jensen, 1969

(continued)

TABLE I (Continued)

Plant–species	Salient details and observations	References
Martineaceae *Proboscidea louisiana*	Single synergid degenerates after pollination; when two PTs enter the micropyle, one is accepted and the other apparently inhibited by the intact synergid	Mogensen, 1978
Plumbaginaceae *Plumbago zeylanica*	[Lacks synergids:] PT enters between egg and CC dimorphic sperm cells, with preferential fertilization male:female mitochondrial ratio stays equal during fertilization because of cytoplasmic heterospermy cytoskeletal (MT) guidance of PT pathway; MT & F-actin appear to guide gametic nuclei during fertilization	Russell, 1982, 1983 Russell, 1984, 1985 Russell, 1987 Huang et al., 1990 Huang, 1991
Plumbagella micrantha	[Lacks synergids:] PT enters between egg and CC	Russell and Cass, 1988
Ranunculaceae *Aquilegia formosa*	Single synergid degenerates	Vijayaraghavan et al., 1972
Scrophulariaceae *Jasione montana* *Torenia fournieri*	Single synergid degenerates Single synergid degenerates	Erdelská and Klasova, 1978, Van der Pluijm, 1964
Salicaceae *Populus deltoides*	Single synergid degenerates after pollination strong cytoplasmic diminution in male gametes, male cytoplasm excluded during fertilization	Russell et al., 1990 Rougier et al., 1991
Solanaceae *Lycopersicon esculentum* (tomato)	Single synergid degenerates after PT penetration	Kadej and Kadej, 1981, 1983, 1985
Petunia hybrida (petunia) *Nicotiana tabacum* (tobacco)	Single synergid degenerates after PT penetration Single synergid begins degeneration after pollination (in TEM), completely degenerates after PT penetration; dSy completed when PT is near ovule ($\sim$ 100 μm); two dSys rarely occur, without affecting function; strong chlorotetracycline labeling in dSy, indicating membrane-bound Ca in dSy; cytoskeletal involvement in PT guidance probable	Van Went, 1970 Mogensen and Suthar, 1979 Huang et al., 1992 Huang and Russell, 1992

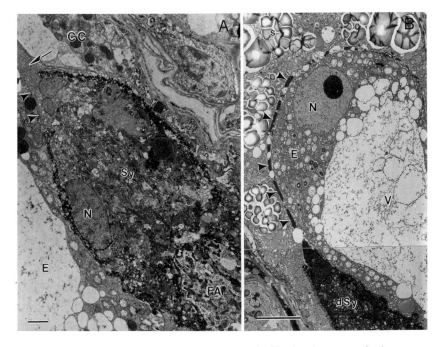

FIG. 1 Synergid degeneration in a pollinated but unfertilized embryo sac of *Glycine max* (soybean). (A) An early stage in synergid (Sy) degeneration. Electron density of the cytoplasm and nucleus (N) increases. Osmiophilic material in the synergid is mostly distributed to the periphery of the cell. An arrow indicates the site at which the synergid cell membrane has ruptured, exposing the cytoplasm to the wall-lacking egg cell plasma membrane (arrowheads). CC, central cell; E, egg cell; FA, filiform apparatus. Bar, 2.5 μm. (B) Degeneration of the synergid cytoplasm (dSy) becomes general in the intermediate stage of degeneration. Synergid cytoplasm infiltrates the chalazal part of the egg cell wall. Arrowheads indicate electron-dense patches of homogenous material between the egg (E) and central cell (CC) that are typically found near the site of future sperm cell fusion. Starch aggregates (s) are conspicuous in the central cell. N, nucleus; V, vacuole. Bar, 5 μm. Reproduced with permission from Dute *et al.* (1989).

that the stimulus for synergid degeneration in this plant acts at a short range (Huang and Russell, 1992). Interestingly, synergid degeneration in cotton appears to be fixation-dependent; synergid degeneration was only observed in chemically fixed ovules (Jensen and Fisher, 1968) but not in freeze-substituted material (Fisher and Jensen, 1969). It seems possible that chemical fixation in cotton allows physiological changes in the receptive synergid to be visible even when the organization of the cells is only slightly altered. More than one synergid has been reported to degenerate in *Aquilegia, Brassica, Hordeum, Linum, Nicotiana,* and *Triticum* (Table I); this, however, does not appear to impair fertilization.

The degenerated synergid is clearly the preferred PT entry point into the female gametophyte. Although chemotropic activity in this synergid may initially be responsible for the attraction of a PT into the ovule, as the tube approaches the ES, the lowered resistance of the degenerated synergid to penetration may become the most important factor in PT entry. Because the other female gametophyte cells are intact, a degenerated or weakened synergid may represent an attractive pathway for the PT compared with the more turgid persistent synergid, egg, or central cell. Once in contact with the receptive synergid, the exact site of PT penetration is usually described as being the FA, which covers much of the outer wall of the synergid attached to the wall of the ES, but the exact location of PT penetration is difficult to observe. In many cases the PT appears to penetrate through the FA; however, aligned fibrillar inclusions are frequently observed in the FA, suggesting that rather than growing directly into the center of this structure, the FA may instead redirect or deflect PTs to the site of discharge. The exact pathway may be evident only in a study combining the use of longitudinal and cross-sectional views; for example, the PT appears to penetrate the FA in longitudinal sections of the ES in *Populus* but clearly skirts the FA in cross section (Russell *et al.,* 1990). In *Torenia,* the PT may travel through the middle lamella between the FAs of the two synergids (Van der Pluijm, 1964). Microtubules (MTs) are associated with the FA in both *Plumbago* (Huang *et al.,* 1990) and *Nicotiana* (Huang, 1991), suggesting that MTs may serve as a dynamic component in synergid development. In *Nicotiana* MTs disappear when the synergid degenerates, as do the major organelles, vacuoles, nucleus, and plasma membrane; however, in *Plumbago* these appear to remain in the egg and guide the path of the PT (Huang, 1991).

The cessation of PT extension and the release of the sperm cells into the ES are triggered by the arrival of the tube within the receptive synergid. As it enters the synergid, undoubtedly there are numerous modifications in the immediate environment of the PT that might trigger rupture, including changes in osmotic potential, surrounding turgor pressure, mineral concentrations (e.g., calcium, Reiss *et al.,* 1983), oxygen tension (Stanley and Linskens, 1967), available nutrients (Knox, 1984), the ionic environment, and other factors of which we are not aware (e.g., secretion of a PT-dissolving factor by the synergid, as suggested by Haberlandt, 1927 and Van der Pluijm, 1964; or PT-lytic factors, such as those acting during incompatible rejection, De Nettancourt, 1977). Given the inherent fragility of an osmotically driven, self-electrophoretic system like that proposed for PT extension (Weisenseel *et al.,* 1975), any of these factors separately or in combination could be enough to trigger the discharge of the PT and the release of the sperm cells.

In most angiosperms, the discharge of the PT occurs through a terminal aperture (Fig. 2, arrows; Kapil and Bhatnagar, 1975). Apparently, PT

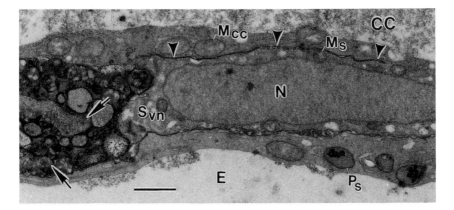

FIG. 2 Sperm cell soon after gametic fusion showing the sperm cytoplasm (S_{vn}) and nucleus (N) in the central cell cytoplasm of *Plumbago zeylanica,* a synergid-lacking angiosperm. The pollen tube aperture is indicated by unlabeled arrows. After fusion organelles originating in the sperm cytoplasm (M_s, mitochondrion; P_s, plastid) are observed within the central cell (CC) and egg (E), respectively. Unlabeled arrowheads indicate the partially fused sperm–central cell plasma membranes; center arrowhead illustrates the progressive vesiculation of fused sperm–central cell membranes. M_{cc}, mitochondrion of central cell origin. Bar, 1 μm. Reproduced with permission from Russell (1983).

discharge occurs when insufficient cell wall material polymerizes at the tip, and the osmotic pressure of the tube exceeds the ability of the wall to contain it. At that time the PT ruptures at the weakest point, which is usually near or at the tip; likely this follows the same pattern as osmotic bursting *in vitro*. Ultrastructural observations show that the tip is usually thinnest at the point of rupture and that incorporation of cell wall vesicles is irregular throughout the last several micrometers of growth. Fibrillar elements in the final 10 μm of PT growth typically reflect a less organized mode of fibril arrangement than is present in the proximal tube, indicating an apparent deterioration in normal cell wall deposition and incorporation (Russell, 1982). In some plants the tip forms an aperture immediately upon penetrating the FA and entering the synergid cytoplasm (e.g., *Nicotiana, Quercus*), but in other plants the tube may grow 10 μm or longer. The most elegant mechanism of PT rupture is that described in *Gossypium*, in which the PT is first arrested in development and forms a thick cap at the tip, while the side wall thins and ruptures. This forcefully expels the contents of the tube through a subterminal aperture, releasing the male gametes in a fountain of PT cytoplasm (Jensen and Fisher, 1968; Fisher and Jensen, 1969). In the case of cotton, which has particularly large synergids, it has been suggested that degradative compounds in the synergid modify the PT discharge by coagulating peripheral cytoplasm, thus

propelling the gametes to the chalazal end of the synergid (Fisher and Jensen, 1969).

The PT is subsequently sealed off by renewed cell wall formation within the synergid. In typical flowering plants the tube is sealed or visibly "pinched off" near the FA and not necessarily near the aperture. If the PT is compressed by surrounding cells at the base of the ES, the amount of cell wall deposition required to seal the tube is relatively small. In the subterminal aperture of *Gossypium*, a large and well-formed plug consisting of callose forms directly over the aperture of the PT, effectively capping it off from the rest of the ES (Jensen and Fisher, 1968). The opposite extreme is *Epidendrum*, in which a plug seems to be absent from the PT (Cocucci and Jensen, 1969). The formation of a PT plug may reflect, at least in part, a wound-healing response, which is not specifically related to the events of fertilization. The basal closure of the ES, however, may be required before further development of the zygote and endosperm occurs. As increased uptake of nutrients after fertilization occurs, the turgor of these cells typically increases and presumably exerts added pressure to all of the surrounding cells. Those cells located at the base of the gametophyte may be particularly susceptible to rupture, given their degenerated condition.

B. Selection of the Degenerated Synergid

The degeneration of synergids appears to be initiated in response to a yet unspecified signal released by the interaction of PTs and stylar cells with the ES (Jensen *et al.*, 1983). The exact nature of this stimulus is not yet known; however, in cotton, ovules grown in culture can be stimulated to mimic normal synergid degeneration when gibberellic acid and indole acetic acid are added to the culture medium (Jensen *et al.*, 1977). In cultured ovules, as in normal ovules grown *in vivo*, only one synergid responds to this signal. However, whether these phytohormones act directly in stimulating synergid degeneration *in vivo* is not clear, and it is not known whether they may serve as a primary or secondary signal in synergid degeneration or whether it possibly produces a false trigger for another, as yet undescribed signal or messenger. Synergid degeneration in cotton is triggered over a relatively long distance (half of the style length); for other flowering plants with only short distance responses, other mechanisms may need to be postulated. Ovules removed from the gynecium and pollinated *in vitro* also display synergid degeneration after PT penetration (Olson and Cass, 1981) but require the placenta to be present before fertilization.

Why one synergid degenerates instead of the other is one of numerous questions that remain to be solved. Does the ES (gametophyte) control the pattern of synergid degeneration or is the physiological environment of the ovule (sporophyte) more important? Is one specific synergid more likely to degenerate than the other? If a gametophytic predisposition exists between the two synergids, can it be overcome by local environmental conditions? Is there cross-talk between the two synergids (which are usually joined by plasmodesmata) to assure that typically only one of the synergids responds to the pollination stimulus?

Four studies provide statistical evidence for a sporophytic role in synergid degeneration. The degenerated synergid is usually the one closer to the vascular trace of the ovule in *Helianthus* (81%, Yan *et al.,* 1991) and *Hordeum* (67%, Mogensen, 1984; Engell, 1988a). This response in barley appeared to be variable in Mogensen's late season pollinations (43%) with a majority of the ovules (68%) not producing a degenerated synergid and remaining unfertilized. In contrast, sporophytic control is not evident in *Linum* (flax, Russell and Mao, 1990), where a nearly random response was noted with respect to ovular landmarks: 52% of the degenerated synergids were proximal and 53% faced toward the septum (not statistically significant). Differences in the proximity of the synergid to the vascular strand could predispose one synergid to degenerate preferentially if the vascular tissue: (1) provides nutrients required for the autolysis of the synergid or (2) carries a signal chemical that indicates the presence of PTs in the style. Mogensen (1984) concluded that before degeneration the synergids may differ either morphologically, physiologically, or both. If preferential synergid degeneration is triggered by a long-distance signal, as hypothesized by Jensen *et al.* (1977, 1983), it likely arrives through the vasculature.

One study provides evidence for gametophytic control of synergid degeneration, which is a possibility that can only be evaluated when the two synergids can be differentiated from one another (i.e., left versus right synergids as viewed from one direction). In *Linum* the left synergid (as viewed from the chalaza) preferentially degenerates in 60% of the ovules, a statistically significant difference ($p < .05$, Russell and Mao, 1990). The synergids did not differ in surface area and size, nor did they differ in their relative area of contact with the egg, ruling out the prospect of control by the egg. However, in *Linum* the left and right synergids differ significantly in surface area and volume of the FA ($p < .01$); the left synergid has 87.5% of the area of the right synergid and 81.5% of its volume. Although these are not large differences, their effect on the physiology of the synergid may be sufficient to predispose the left synergid to degenerate. No evidence for synergid handedness was found in cotton (Jensen, unpublished data) and *Nicotiana* (Huang and Russell, 1992). Therefore, the mechanism by which synergids are selected for degeneration is unclear

and apparently differs in various plants. One aspect of control that logically resides within the gametophyte, however, is the characteristic of single synergid degeneration, which occurs in most flowering plants as an integral part of the gametophytic program.

C. Pollen Tube Arrival in Synergid-Lacking Angiosperms

Synergids are found in all patterns of ES formation, with the exception of the *Plumbago* and *Plumbagella* types, in which synergids are invariably absent (Maheshwari, 1950). These are largely restricted to the subfamily Plumbaginoideae (genera *Ceratostigma, Dyerophytum, Plumbago,* and *Plumbagella*) of the Plumbaginaceae. Synergid functions are consequently transferred to the egg cell and are reflected in the occurrence of an FA at the micropylar end of this cell (Cass, 1972; Cass and Karas, 1974). When the PT arrives at the ES, no cellular degeneration is involved. Instead, the PT penetrates through the base of the ES at the FA and continues to grow between the egg and central cell until it reaches a region of strong curvature at the summit of the egg, near the nucleus where the PT discharges (Russell, 1982). Evidence from immunocytochemical localizations for both tubulin and F-actin indicates that the egg is a highly zonated structure (Huang *et al.,* 1990) and that MTs present near the FA may deflect and guide the direction of PT growth in the ES (Huang, 1991). Just below the summit of the egg, an aperture forms at the tip of the PT and pollen cytoplasm carrying the two gametes and the vegetative nucleus is released between the egg and central cell (Fig. 2). Any cell wall material remaining between these female cells is conveyed from the surrounding region by the force of PT discharge (Russell, 1982, 1983).

III. Delivery of the Male Gametes in the Embryo Sac and Their Condition Prior to Fusion

Unfused sperm cells have been observed within the ES in *Hordeum* (Cass and Jensen, 1970; Mogensen, 1982, 1988), *Lycopersicon* (Kadej and Kadej, 1985), *Plumbago* (Russell, 1983), *Populus* (Russell *et al.,* 1990), and *Spinacia* (Wilms, 1981) and appear to share the following common characteristics: (1) the male gametes are deposited in a male germ unit (gametes closely associated with the vegetative nucleus), which is severed during PT discharge; (2) the additional plasma membrane surrounding both of the sperm cells (i.e., the inner pollen plasma membrane) is removed during PT discharge, directly exposing the sperm plasma membrane to the envi-

ronment; (3) progressive loss of cell wall material during development, until there is no cell wall present at the time of gamete delivery; (4) male gametes are true cells containing a cytoplasm with usually a diminished number of organelles; vacuolization of sperm cells may increase during their passage in the PT (Russell *et al.*, 1990; Rougier *et al.*, 1991); (5) male gametes usually contact both the egg and central cell simultaneously prior to fusion; and (6) male gametes remain unfused and in contact relatively briefly (between 30 seconds and several minutes), as surmised by the relatively infrequent number of observations. A model for this developmental stage is presented in Figure 3A.

The observation that the sperm plasma membrane is typically exposed just prior to fusion to both the egg and central cells suggests that recogni-

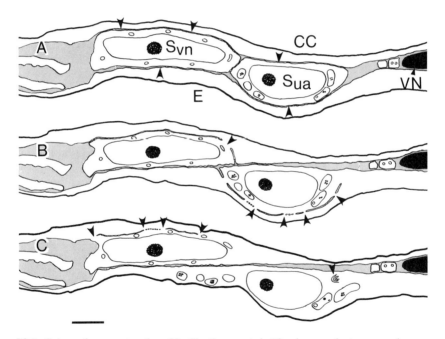

FIG. 3   Schematic reconstruction of fertilization events in *Plumbago zeylanica*, an angiosperm that lacks synergids. (A) Unfused sperm cells (S_{vn}, sperm associated with vegetative nucleus; S_{ua}, sperm unassociated with vegetative nucleus) lose their outer pollen plasma membrane during discharge from the pollen tube and become directly appressed to the egg (E) and central cell (CC), having plasma membrane contacts with both cells (arrowheads). (B) Gametic fusion is initiated at one site, as illustrated by the unlabeled arrowhead in the central cell. This is followed by several fusion events at other sites in the membranes (unlabeled arrowheads in egg). (C) Subsequently, the fusion membranes of the central cell begin to break down (arrowheads in central cell), whereas in the egg these membranes are quickly reduced to a residual whorl (arrowhead in egg), which soon entirely disintegrate. VN, vegetative nucleus. Bar, ~3 μm. Reproduced with permission from Russell (1983).

tion of gametes may occur in all of the flowering plants described to date. One universal characteristic of fertilization is that *in vivo* sperm cells only seem to express their fusigenic potential with female target cells. To this extent gamete recognition is directed and successful. In plants with dimorphic sperm cells (Russell, 1991), gametic recognition can be combined with preferential fertilization, in which the sperm cell that will fuse with the egg cell may be distinct from the other cell. In the latter case it is unclear whether the two sperm cells are interchangeable or whether there has been a differentiation of function that may allow only one cell to be truly designated as a sperm cell.

IV. Gametic Fusion in Plants and Animals

Because fertilization in flowering plants is hidden within the ovule, the mechanism of gametic fusion is a matter of considerable speculation. Each of the participating reproductive cells is known to lack a cell wall and possesses just one delimiting membrane; so rather simple models may be useful in understanding the mechanism of membrane fusion in flowering plants. Gametic fusion in angiosperms may therefore resemble gamete fusion in other organisms or exocytosis of sufficiently large vesicles. Membrane fusion mechanisms have been described in numerous reviews (White, 1990). Yet the exact mechanism of membrane fusion in biological systems is still somewhat unclear. Capturing the initial stages of fusion has proven extremely difficult because the unstable intermediates that initiate fusion are difficult to capture even with ultra-rapid freezing devices known to freeze material in 1 ms (Ornberg and Reese, 1981).

Theoretical models for membrane fusion biophysics suggest that sexually based cellular fusions are based on proteins integral to the membrane that, upon recognizing complementary macromolecules on an opposite membrane, stimulate fusion. These molecules are usually glycosylated proteins, termed viral proteins, and are typically hydrophobic, tending to draw the two membranes together closely enough that the hydrophilic outer boundary is no longer a barrier to fusion. An alternative to this involves an initial destabilization of the hydrophilic–hydrophobic organization of phospholipids by divalent cations, such as calcium. The exact model by which fusion occurs in gametic systems has not been entirely resolved in any biological system (Yanagimachi, 1988).

A. Models of Fertilization and Membrane Fusion

1. Gametic Fusion in Animal Systems

a. Fertilization in Animals In animal fertilization the stoichiometry of membrane fusion is varied. In some animals the sperm cells directly fuse

with the egg plasma membrane, but in others fusion occurs first with the vitelline envelope. Essentially all animals and fucoid algae modify the vitelline envelope to form a fertilization membrane that blocks the fusion of supernumerary sperm cells. This appears to have no counterpart in angiosperms. Sperm–egg interactions appear to involve several glycoproteins: bindin serves to bind the cells to the egg surface in a number of species; in mammals zona pellucida proteins may both control recognition and bind the sperm to the egg (e.g., ZP3 in mouse) (Yanagimachi, 1988). These molecules do not, by themselves, control fusion but may confer specificity. In their absence species-specificity of sperm cells may be lost, but their ability to fuse is not. Fertilization in animal systems has been reported to be inhibited by the addition of antibodies specific to the recognition sites, glycosylation inhibitors, glycoprotein synthesis inhibitors, disulfide-reducing compounds (dithiothreitol), digestion with proteinase, absence of calcium, and the presence of overly numerous charged molecules (e.g., erythrosin B). Interestingly, the area of the sperm plasma membrane free of intramembrane particles is believed to form the fusion surface (Yanagimachi, 1988).

In numerous animals and in fucoid algae, the egg membrane is normally negatively charged prior to fusion with the sperm cell, but immediately after fusion the membrane undergoes a reversal known as depolarization. This alteration of charge immediately blocks the fusion of supernumerary sperm cells and constitutes a "fast block" to polyspermy (Jaffe and Cross, 1986). The evidence that changes in membrane polarity constitute a block to polyspermy is that: (1) if depolarization is prevented, supernumerary sperm cells fuse readily, and (2) if the unfertilized egg is electrically simulated to the same polarity as a fertilized egg, no sperm cells fuse. During less than 2 minutes after initial fusion, the numerous cortical granules fuse essentially simultaneously with the vitelline membrane, causing it to rise from the surface of the egg and form the conspicuous fertilization membrane, which is impervious to sperm cells and serves as a permanent physical block to polyspermy.

b. Mechanisms of Membrane Fusion   The fusion of membranes (or apposed lipid bilayers) is an energetically unfavorable reaction that requires overcoming a significant hydrophobic barrier between membranes. In animals this may require specific pH conditions, the involvement of divalent cations, specific energy molecules, such as guanosine triphosphate (GTP), and the presence of specific membrane-bound proteins with fusigenic sites. Membrane-targeting mechanisms inside (involved in exocytosis) and outside (endocytosis) of the cell are likely to be distinct: one requires the factors to be present on the cytoplasmic face and the other on the ectoplasmic face. In the case of cellular fusion, the surface may be large enough that surface curvature effects are minimal, but in vesicle

fusion curvature and surface tension may combine to make fusion reactions even less energetically favorable (Rand and Parsegian, 1986).

For membranes to approach each other closer than the normal 2–3 nm (Fig. 4B), the water associated with the hydrophilic outer layer of the membranes must be removed; this repulsive force increases exponentially to a distance of 0.2–0.3 nm. An energy requirement of 10–100 ergs/cm^2 is postulated for fusion. Given this high energy requirement, it seems likely that focal point fusions, as observed using freeze–fracture (Ornberg and Reese, 1981), are initiated at specific sites and propagate laterally to achieve cell fusion. The role of divalent cations may be to dehydrate polar groups and achieve close enough contact between the membranes to allow other molecules present to induce inverted micelles (Fig. 4A-1), which are believed to initiate fusion. Focused destabilization of membrane compo-

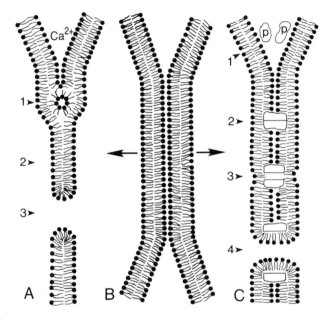

FIG. 4 Diagrammatic representation of two generalized models for membrane events during cell fusion: formation of inverted micelles by divalent cations (Ca^{2+}) and action of viral-like fusion proteins (A and C, respectively). (A) Apposed membranes become destabilized by the neutralization of hydrophilic groups with Ca^{2+} at the center of the micelle (1), forming a unilamellar fusion membrane (2). When the unilamellar membrane is destabilized, this results in a fusion pore (3). (B) Two apposed, but unfused membranes. (C) Partially assembled viral proteins (p) expose hydrophobic regions to the membrane (1) and become associated with the membrane as a functional protein unit (2). The functional unit then cleaves, with its outer surface becoming associated with the hydrophobic layer (3). As the distances continue to increase, the membrane flows around the protein forming a fusion pore (4).

nents may trigger fusion when the lipid composition of the contact surface is changed or when externally secreted lipases, pH changes, or specific membrane-bound glycoproteins are present on the ectoplasmic faces of one or both of the fusing membranes. This may cause the two membranes to form one lipid bilayer (Fig. 4A-2), which may then precipitate the breakdown of the membrane barrier in entirety (Fig. 4A-3). Between closely appressed lipid bilayers, transient structures have been observed using freeze–fracture techniques (Cullis and Hope, 1988) that may represent inverted micellar intermediates, such as in Figure 4A-1. According to freeze–fracture observations, the inverted micelle may represent a hexagonal phase component (H_{II}), while the rest of the membrane remains in the classical L_α bilayer configuration. The smallest pores observed during early fusion are as small as 5 nm, although most intermediates appear to be 20 nm (Almers, 1990).

The best defined model for fusigenic proteins is that of viral proteins, which are homogenous and abundant on the surface of a number of viruses (White, 1990). These have been widely studied because their surface is immeasurably easier to understand than typical cell membranes and for their obvious role in pathogenicity. In the simplest viral model the fusigenic protein is an oligomer (frequently a trimer) with all parts required for function. Based on examination of a number of unrelated viruses, fusigenic proteins in general appear to share transmembrane domains near their C terminus and have fusion peptides, containing high alanine and glycine residues, toward their N terminus. Secondary processing of the molecule is typically necessary for function. To be effective in fusion, the viral proteins must be membrane bound and the characteristic number of oligomers must be correctly associated. One model is that free monomers present outside the membrane (Fig. 4C-1) undergo self-assembly, thereby exposing a hydrophobic region to the outer lipid layer in such a manner that it becomes associated with both membranes (Fig. 4C-2). When assembled, the aggregated fusion molecule would conceal its internal fusigenic region, allowing the opposing membranes to associate more closely. When conditions appropriate for fusion are met, the oligomers change their conformation, exposing their hydrophobic site (Fig. 4C-3), causing fatty acids to migrate into the bridge forming a focal fusion point. Once a pore is formed (Fig. 4C-4), the surface tension of the membranes would likely provide ample energy to expand the fusion point laterally until the membranes are fully coalesced.

2. Membrane Fusion in Angiosperms

In angiosperms both the egg and the sperm cells are contained within the confines of their respective gametophytes, and therefore until spontaneous

in vitro fusion of isolated gametes is possible, these cells are unavailable to many of the approaches used for examining animal cells. The best information on the gametic fusion event is based on electron microscopy of the synergid-lacking *Plumbago zeylanica* (Russell, 1982, 1983, 1985). In this plant the sperm cells are deposited between the egg and central cell as naked cells, not surrounded by any cell wall material and lacking the inner PT plasma membrane, which surrounds these cells prior to discharge from the tube (Russell and Cass, 1981). The removal of the tube plasma membrane from the sperm cells has been compared with capacitation in the sense that the sperm cell surface is modified to allow the cell to fuse with the egg. In animals capacitation refers to the enzymatic modification of the sperm surface required for fusion and is distinct from the acrosome reaction, which has no apparent counterpart in angiosperms (Yanagimachi, 1988).

The egg and central cells border one another in all angiosperms, but in *Plumbago* and *Plumbagella* the PTs travel through the egg cell wall (i.e., between the egg and the central cell) without disrupting either cell membrane (Russell, 1982; Russell and Cass, 1988). (As discussed previously, in typical flowering plants the PT enters through a degenerated synergid to gain access to this area.) Sperm cells released into this area in *Plumbago zeylanica* contact both the egg and central cell (Fig. 3A), with an intervening distance of from 20 to 70 nm separating their respective membranes, varying randomly along the contact surface, except where degenerated pollen cytoplasm is trapped (Russell, 1983). Upon fusion the sperm nucleus is transmitted directly into the cytoplasm of the egg or central cell, and the plasma membrane remaining from the fusion event rapidly dissipates. Apparently, fusion is initiated at numerous sites simultaneously, propagates through the membrane, and then the fusion membranes rapidly vesiculate.

In *Plumbago zeylanica* this remnant membrane is observed to consist of an interrupted boundary within the central cell, as seen in Figure 2 (arrowheads). The sperm nucleus is initially surrounded with the sperm cytoplasm after entry into the central cell and is still at approximately the same site at which it was deposited, as indicated by the nearby PT aperture (arrows, Figs. 2 and 3B). The fusion membrane continues to disintegrate until only an evanescent membrane whorl is observed (Fig. 3C). Because *Plumbago* has a stringently controlled timetable for fertilization events, it was possible to hand pollinate and collect flowers at 5-minute intervals from 8.3 to 8.7 hours and collect all of these stages of fusion. Based on the infrequent occurrence of sperm fusion stages (three observations of 200 ovules sectioned), this phase apparently occurs rapidly. Likely, this entire phase requires less than 1 minute (Russell, 1983) and is similar in this regard to some animal systems.

In plants with uniparental maternal inheritance, strong evidence for the exclusion of male organelles at the egg boundary has been presented (Section V); however, their mechanism of gametic fusion is at present unknown. This is an area that deserves closer examination.

B. Gametic Fusion

Likely, the environment of the synergid and the region between the egg and the central cell are highly modified to promote the fusion of the male and female gametes. Important features of this environment include the absence of a cell wall at the surface of the egg and central cell and the access of the sperm cells to this region. Once the sperm cells are deposited into this region, they appear to fuse *in situ,* without subsequent migration. The environment of the synergid is strongly modified by the discharge of the PT. The presence of high concentrations of free calcium described in the synergid before fertilization (Chaubal and Reger, 1990, 1992) may also influence the immediate environment of fusion in the degenerated synergid. Pollen tube materials discharged during the process of fusion may further condition the environment for gametic fusion.

Typically, double fertilization involves the fusion of male gametes from the same PT. However, in some plants multiple PTs are known to arrive in the ES and the male nuclei in this case may arise from different PTs, thus participating in heterofertilization (Robertson, 1984; Mogensen, 1990).

In *Plumbago* available evidence suggests that fusion to form the zygote and endosperm is not in a fixed temporal sequence—embryo sacs in which the sperm fused first with the egg were as frequent as those in which the sperm fused first with the central cell. Fusions, once initiated at a single site, appear to occur independently at multiple locations on the surface of the membrane (arrowheads, Figs. 2 and 3B). This pattern of multiple, independent fusions is typical of membrane fusion events in other biological systems (Ornberg and Reese, 1981).

C. Preferential Fertilization

The association of one sperm cell and the vegetative nucleus imposes polarity on the male germ unit, which may extend to cytoplasmic differences in sperm cells (unequal contents of heritable organelles, cytoplasmic heterospermy) or there may be differences in nuclear content (nuclear heterospermy) (Russell, 1985). Cytoplasmic heterospermy is apparently not uncommon in flowering plants and occurs in numerous examples (Russell, 1991), but nuclear differences have been reported only in *Zea*

mays, where B chromosomes (when they occur) often do not separate during sperm cell formation resulting in aneuploid cells (Roman, 1948). The most common form of cytoplasmic heterospermy consists of iniquities in mitochondria between the two sperm cells, with the sperm cell associated with the vegetative nucleus (S_{vn}) usually containing more mitochondria. Apparently, the most extreme form examined to date is *Plumbago,* in which both plastid and mitochondrial contents differ between the two cells (Corriveau and Coleman, 1988; Mogensen, this volume). The S_{vn} usually contains no plastids and more than 200 mitochondria, whereas the other sperm cell (S_{ua}) contains an average of 24 plastids and less than 50 mitochondria (Russell, 1984).

In these well-described cases in which the sperm cells have traceable differences, preferential fertilization occurs; one sperm cell has a greater probability of fusing with the egg than the other. In *Plumbago* the plastid-rich S_{ua} fuses with the egg in 94% of the cases examined (Russell, 1985), selectively transmitting its plastids into the zygote (Figs. 2 and 3). Interestingly, however, the ratio between male and female mitochondria remains about the same between the egg and central cell, at about a 1 : 1000 ratio (Russell, 1987). The difference in mitochondrial content between the S_{ua} and S_{vn} nearly compensates for the differences present between the egg and central cell. The significance of this observation in relation to cytoplasmic heterospermy in general is still somewhat unclear. The possibility remains, however, that this ratio is important and may represent a maximal permissible dose of sperm mitochondria to prevent sperm transmission of mitochondrial DNA or a minimum needed for recombination of mitochondrial DNA to be possible (Russell, 1987).

Preferential fertilization also occurs in *Zea,* in which about 65% of the cases, the sperm cell containing an extra set of B chromosomes fuses with the egg (Roman, 1948; Carlson, 1969, 1986). Ironically, the mere presence of supernumerary B chromosomes seems to confer a selective advantage to the sperm cell containing them, but too many B chromosomes negate this effect, as does introducing a specific B chromosome (TB-9b) into the egg (Carlson, 1969). That maternal genotype can alter preferential fertilization suggests that discrimination between sperm cell is maternally controlled and active under relatively strict conditions. Maternal control of fertilization is also suggested in barley in a mutant line in which seeds are produced that contain normally fertilized endosperm but unfertilized, haploid embryos. In this case one sperm cell fuses with the central cell, and this is not altered by using different pollen sources (Mogensen, 1982). There is obviously considerable benefit in using such mutants for further experimentation to understand the mechanisms underlying double fertilization.

V. Influence of Fusion on the Pattern of Cytoplasmic Inheritance

Once gametic fusion is accomplished, the nucleus of the sperm accounts for half of the nuclear complement of the young sporophyte; however, the influence of heritable cytoplasmic organelles present within the sperm cell (mitochondria and plastids) is variable in angiosperms. In many cases the complement of organelles originating from the male versus female gametes determines the ratio of extranuclear DNA expression in the next generation. Although organelles may be present (albeit infrequently in some plants), the number of organelles delivered is an important factor in cytoplasmic inheritance. The comparison of maternal versus paternal organelles in the zygote and endosperm appears to be critical to the pattern of cytoplasmic inheritance.

The vast majority of angiosperms have uniparental maternal cytoplasmic inheritance for both plastids (Gillham, 1978) and mitochondria (Chapman, 1986). A fluorescence microscopic survey of angiosperm pollen in 235 species indicated that plastid DNA was absent or undetected in the sperm and generative cells of 82% of the species examined (Corriveau and Coleman, 1988). In the remaining 18% of the species, plastid DNA was detected in the male germ lineage using the DNA fluorochrome 4'6-diamidino-2-phenylindole (DAPI) and could potentially result in biparental inheritance. In rare plants the maternal plastid DNA is eliminated during embryogenesis and the plastid DNA inherited in the embryo is strictly paternal. Just two unequivocal examples of uniparental, paternal inheritance have been described in flowering plants: *Medicago* (alfalfa, Fairbanks *et al.*, 1988; Schumann and Hancock, 1989; Masoud *et al.*, 1990) and *Daucus* (Boblenz *et al.*, 1990).

A. Uniparental, Maternal Cytoplasmic Inheritance Mechanisms

In the majority of plants, cytoplasmic DNA fails to be inherited in the embryo, resulting in uniparental maternal cytoplasmic inheritance. This pattern is typically attributed to a reduction in the amount of DNA-containing organelles during the formation of the generative cell or the elimination of organellar DNA during development (Hagemann and Schröder, 1989). During generative cell development in *Solanum*, ultrastructural observations suggest that plastids are eliminated from the male germ lineage (Clauhs and Grun, 1977), but the content of mitochondria

fluctuates. A decreased content of detectable organellar DNA was observed during microsporogenesis in a number of plants examined using DAPI fluorescence (Miyamura *et al.*, 1987). The mechanism for the elimination of plastids is still unclear. Alterations in cytoplasmic DNA occur that also appear to limit the ability of the cells to be cultured after the microspore stage (Day and Ellis, 1984). In addition to possible changes in the DNA, the organelles, themselves, may be altered during development (Vaughn *et al.*, 1980). In barley reduction of the male cytoplasm occurs through "pinching-off" of cytoplasmic bodies during sperm cell maturation (Mogensen and Rusche, 1985). The generation of enucleated cytoplasmic bodies (ECBs) and vesicle-containing bodies (VCBs) that reduce the volume of the cytoplasm throughout pollen maturation and tube elongation have been observed in *Nicotiana* (Yu *et al.*, 1992) and in other plants (Russell and Yu, 1991). At the time of gametic fusion, the sperm cell is still cellular (Fig. 5A), but the male cytoplasm is later excluded from the

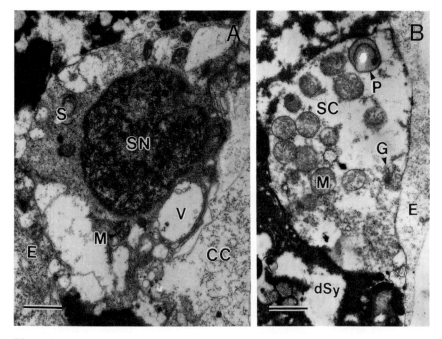

FIG. 5   Ultrastructural evidence for a barrier to the entry of sperm organelles during fertilization in *Hordeum vulgare* (barley). (A) Intact, unfused sperm cell (S) between the egg (E) and central cell (CC). The sperm cell contains mitochondria (M), a prominent nucleus (SN), and several vacuoles (V) of varying sizes. (B) After the transmission of the nucleus, the sperm cytoplasm (SC) and sperm organelles (G, Golgi body; M, mitochondrion; P, plastid) form a cytoplasmic body, which remains in the degenerated synergid (dSy) and is excluded from the egg (E). Bars, 1 μm. Reproduced with permission from Mogensen (1988).

egg during gametic fusion (Mogensen, 1988). The shed male organelles (Fig. 5B) remain outside of the egg in ECBs that closely match the number of plastids and mitochondria present in a typical barley sperm cell. In barley only one ECB was noted in each ES observed, suggesting that uniparental inheritance may be restricted to the zygote, while the endosperm receives both paternal and maternal cytoplasmic organelles (Mogensen, 1988); this pattern is represented schematically in Figure 6A. Thus, fertilization in barley indicates that the egg is cytoplasmically restrictive: male cytoplasmic organelles are shed during maturation of the male lineage (Mogensen and Rusche, 1985), and the remaining sperm cytoplasm is excluded from incorporation at the egg surface (Mogensen, 1988). The central cell of the same plant, however, appears to be cytoplasmically permissive, allowing the entry of sperm organelles into the central cell. In

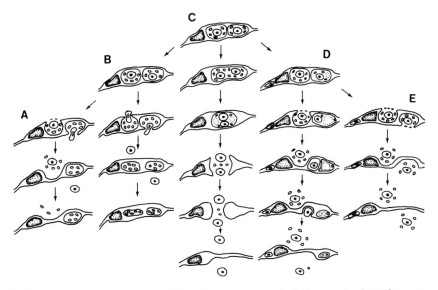

FIG. 6 Schematic representation of five alternative gametic fusion mechanisms for cytoplasmic transmission in selected flowering plant species. In all figures the sperm nuclei are originally surrounded by a cytoplasm containing organelles (top, center); dark object is vegetative nucleus, which is excluded from the egg (below, each drawing) and central cell (above). (A–D) Uniparental maternal transmission patterns. (A) Sperm cell cytoplasm is excluded from the egg but not central cell in barley (Mogensen, 1982, 1988). (B) Sperm cytoplasm of both egg and central cell is excluded in cotton (Jensen and Fisher, 1968). (C) Fusion of two sperm cells is followed by mutual double fusion and disintegration of cytoplasm in spinach (Wilms, 1981). (D) Vacuolization of the unfused sperm cell separates the cell from much of its surrounding cytoplasmic organelles in cottonwood (Russell *et al.*, 1990). (E) Biparental transmission pattern; in this composite model essentially all of the cytoplasm surrounding the sperm cells enters the egg and central cell (Van Went, 1970; Russell, 1983). Reproduced with permission from Russell *et al.* (1990).

Gossypium two ECBs of apparent sperm origin were observed in the degenerated synergid (one ECB at the surface of the egg and the other ECB at the surface of the central cell), suggesting that the cytoplasm of both sperm may be excluded (Jensen and Fisher, 1968; Fig. 6B). Wilms (1981) proposed that in *Spinacia* the two sperm cells fuse to form a binucleated cell that then simultaneously fuses with both the egg and central cells (Fig. 6C). In *Populus* the sperm cells develop large vacuoles during their descent in the PT; during their passage into the degenerated synergid, the cytoplasm separates from the sperm nucleus and apparently excludes an ECB (Fig. 6D; Russell *et al.*, 1990).

B. Paternal Cytoplasmic Inheritance Mechanisms

A simple alternative to maternal cytoplasmic inheritance is the transmission of the sperm cytoplasm, as reported in *Petunia* (Van Went, 1970) and *Plumbago* (Fig. 6E); this form of cytoplasmic permissiveness in the egg presumably is the general case for plants with biparental and paternal cytoplasmic inheritance. In the case of biparental inheritance, the representation of male and female cytoplasmic genomes depends on their ratio, how it is affected by replication, and on the random or nonrandom sorting-out of individual organelles in succeeding cell lineages. Because both male and female cytoplasmic organelles are present from the zygote, the tissues of the plants may display variegation and mosaicism.

Whereas the control of exclusively maternal cytoplasmic inheritance appears to reside in the behavior of the sperm cells, paternal cytoplasmic inheritance appears to reside largely, though not exclusively, in the egg. Ultrastructural studies revealed extremely high plastid content in the sperm cells of *Medicago*, but there were no obvious differences between plastid content in "strong" and "weak" males observed in either qualitative (Zhu *et al.*, 1990) or quantitative studies (Zhu *et al.*, 1991b). However, weak females that allow paternal plastid inheritance were characterized by fewer plastids (16.6 versus 25.5 plastids) and larger volumes per plastid (17.1 μm^3 versus 9.6 μm^3). Additionally, the eggs of these weak females were characterized by a largely micropylar distribution of plastids instead of the typical perinuclear distribution of plastids (Zhu *et al.*, 1991a). The micropylar distribution of plastids in the weak females may result in their transmission into the basal cell in the first division of the zygote, which forms the suspensor, a terminal cell lineage that does not contribute to the embryo proper. Interestingly, the weak females lacked the prominent micropylar vacuole that is present in both strong females of alfalfa (Zhu *et al.*, 1991a) and the majority of angiosperms (Huang and Russell, this volume).

A similar pattern of paternal inheritance is observed in crosses using *Daucus muricatus* as the female and *Daucus carota* and *Daucus aureus* as the male, but not in reciprocal crosses (Boblenz *et al., 1990*). A structural investigation of *Daucus muricatus* revealed that only a small number of plastids (8–12) were present in the egg and that some apparent plastids at the periphery of the egg appeared to be degenerating (Hause, 1991). The eggs of *Daucus muricatus* further displayed a relatively high degree of vacuolization compared with the eggs of *Daucus carota* and *Daucus aureus*. The exact means by which maternal plastids are excluded, however, may result from the sorting-out of the minority (maternal) plastid genome, differential plastid replication, or the degeneration of maternal plastids. In light of the preponderance of paternal inheritance of plastids in conifers (Owens and Morris, 1991), it is difficult to explain why angiosperms display such an overwhelming representation of maternal organellar DNA during fertilization. Because the pattern of plastid inheritance in angiosperms appears to be conserved at the genus, but not at the family level (Corriveau and Coleman, 1988), cytoplasmic inheritance appears to be a variable feature in flowering plants and one that is difficult to ascribe clear significance.

VI. Nuclear Fusion

Nuclear fusion (or karyogamy) in flowering plants is initiated by the migration of the sperm nuclei within the egg and central cell into alignment with their respective female target nucleus. The migration usually follows the shortest path in most plants; in *Gossypium,* however, the sperm nucleus may enter the egg and travel the circumference of the egg before alignment (Jensen and Fisher, 1967). Cytoskeletal elements, such as MTs and F-actin, are known to undergo significant changes in distribution during the process of PT delivery and gametic fusion in *Plumbago* and *Nicotiana* (Huang, 1991), and MTs in particular are likely to be involved in nuclear alignment. The process of karyogamy is initiated by the apposition and fusion of endoplasmic reticulum connected with the outer envelopes of the two nuclei. These layers coalesce at several contact points forming bridges between the outer layers of the two nuclear envelopes. Next, the inner layers of the nuclear envelopes come into contact and fuse, causing continuity of the nucleoplasm. Nuclear bridges continue to expand until the nuclei are completely fused. The membranes of both nuclei contribute to the new nuclear envelope. As a consequence of this method of nuclear fusion, frequently small apparent islands of cytoplasm are observed in sections of the fusing nuclei. Whether these are continuous or discontinu-

ous with surrounding cytoplasm is difficult to determine without serial sectioning, but according to Jensen (1964), these regions are never isolated within the nucleus in *Gossypium*.

In many regards the fusion of the polar nuclei prior to fertilization is similar to the events of sexual karyogamy; however, the fusion of polar nuclei is frequently arrested at small points of coalescence (particularly in monosporic plants) until the arrival of the sperm nucleus (Huang and Russell, this volume), when the nuclei participate in triple fusion to form the primary endosperm nucleus. Similar membrane events have been described in *Petunia* (Van Went, 1970) and *Capsella* (Schulz and Jensen, 1973). During the process of karyogamy in *Triticale,* small lipid droplets are present at some of the points of coalescence between the male and female nuclear envelopes, but it is unclear whether these may precipitate fusion or merely reflect lipid instabilities that may then be captured during chemical fixation (Hause and Schröder, 1987).

The nuclear fusion events described herein conform to the premitotic mode of karyogamy, in which the male and female nuclei merge prior to zygotic mitosis (Gerassimova-Navashina, 1960). In this form of karyogamy, there is no structural barrier to the co-mingling of male and female genetic material. This is representative of the majority of flowering plants. However, a significant minority of plants display either postmitotic or an intermediate form of nuclear fusion. In postmitotic fusion the male and female nuclei approach one another and remain in close proximity until mitotic division ensues. When the nuclear envelope breaks down, a common metaphase plate is formed, and during the course of mitosis the nuclear complement of the zygote is constituted and divided. Prior to fusion the smaller sperm nucleus swells until it and its nucleolus expand to the size of the egg nucleus and its nucleolus. Examples of this form of nuclear fusion are frequent in the Liliaceae. An intermediate form of nuclear fusion is that in which the male nucleus enters mitosis precociously. The sperm nucleus is tightly appressed to the egg nucleus and their membranes may fuse, but the chromatin remains separated until mitosis. As the egg enters mitosis, groupings of male and female chromosomes remain separated until late in metaphase, when they align on a common plane and divide according to normal patterns. This is reported in detail in *Impatiens* and appears to be restricted in occurrence (Gerassimova-Navashina, 1960).

An exciting area of research that has not been fully exploited is the observation of fertilization in living ovules. By choosing ovules with ESs exerted through the micropyle, it is possible to observe the cells of the egg apparatus directly (Erdelská, 1974) and examine the events of fertilization, nuclear migration, and nuclear fusion directly. Impressive films have been released showing early postfertilization events in *Galanthus nivalis* and

endosperm development in *Jasione montana* (Erdelská, 1983); these two films are distributed by the Institut für den Wissenschaftlichen Film (Göttingen, Germany). These microcinematographic works illustrate phases of gametic interaction that have rarely been observed. Because all of the other approaches are subject to the artifacts of fixation and preparation, this approach offers the possibility of obtaining unique insights into the fertilization process *in vivo*. By combining this technique with modern probes, additional progress can be made in understanding the subcellular behavior underlying the receipt of the PT, transmission of the sperm nuclei, and the initiation of nuclear fusion. Continued progress on this topic will require technical innovation, carefully selected systems, and the use of modern techniques to understand the physical and molecular bases of double fertilization.

Acknowledgments

The author thanks Dr. R. Dute and Prof. H. L. Mogensen for providing some of the photographs used in this review. Portions of this research were supported by operating grants from NSF (PCM 8208466 and DCB-8409151) and USDA (86-CRCR-1-1978, 88-37261-3761, and 91-37304-6471).

References

Almers, W. (1990). *Annu. Rev. Physiol.* **52,** 607–624.

Amici, G. (1824). *Ann. Sci. Nat. Bot.* **2,** 41–70, 211–248.

Amici, G. (1847). *Ann. Sci. Nat. Bot.* **7,** 193–205.

Bednara, J. (1977). *Acta Soc. Bot. Pol.* **46,** 603–615.

Boblenz, K., Nothnagel, T., and Metzlaff, M. (1990). *Mol. Gen. Genet.* **220,** 489–491.

Bruun, L. (1987). *Nord. J. Bot.* **7,** 543–551.

Bruun, L., and Olesen, P. (1988). *In* "Sexual Reproduction in Higher Plants" (M. Cresti, P. Gori, and E. Pacini, eds.), p. 462. Springer-Verlag, New York.

Carlson, W. R. (1969). *Genetics* **62,** 543–554.

Carlson, W. R. (1986). *Crit. Rev. Plant Sci.* **3,** 201–226.

Cass, D. D. (1972). *Am. J. Bot.* **59,** 279–283.

Cass, D. D. (1981). *Acta Bot. Soc. Pol.* **50,** 177–179.

Cass, D. D., and Jensen, W. A. (1970). *Am. J. Bot.* **57,** 62–70.

Cass, D. D., and Karas, I. (1974). *Protoplasma* **81,** 49–62.

Chao, C. Y. (1971). *Am. J. Bot.* **58,** 649–654.

Chao, C. Y. (1977). *Am. J. Bot.* **64,** 922–930.

Chapman, G. P. (1986). *In* "The Chondriome: Chloroplast and Mitochondrial Genomes" (S. H. Mantell, G. P. Chapman, and P. F. S. Street, eds.), pp. 61–68. Longman, London.

Chaubal, R., and Reger, B. J. (1990). *Sex. Plant Reprod.* **3,** 98–102.

Chaubal, R., and Reger, B. J. (1992). *Sex. Plant Reprod.* **5,** 34–46.

Clauhs, R. P., and Grun, P. (1977). *Am. J. Bot.* **64,** 377–383.

Cocucci, A. E., and Jensen, W. A. (1969). *Am. J. Bot.* **56,** 629–640.

Corriveau, J. L., and Coleman, A. W. (1988). *Am. J. Bot.* **75,** 1443–1458.
Cullis, P. R., and M. J. Hope. (1988). *In* "Molecular Mechanisms of Membrane Fusion" (S. Ohki, D. Doyle, T. D. Flanagan, S. W. Hui, and E. Mayhew, eds.), pp. 37–51. Plenum Press, New York.
Davis, G. L. (1966). "Systematic Embryology of the Angiosperms." Wiley, New York.
Day, A., and Ellis, T. H. N. (1984). *Cell* **39,** 359–368.
De Nettancourt, D. (1977). "Incompatibility in Angiosperms." Springer-Verlag, Berlin.
Diboll, A. G. (1968). *Am. J. Bot.* **55,** 787–806.
Dong, J., and Yang, H. Y. (1989). *Acta Bot. Sin.* **31,** 81–88.
Dute, R. R., Peterson, C. M., and Rushing, A. E. (1989). *Ann. Bot.* **64,** 123–135.
Engell, K. (1988a). *In* "Sexual Reproduction in Higher Plants" (M. Cresti, P. Gori, and E. Pacini, eds.), pp. 383–388. Springer-Verlag, New York.
Engell, K. (1988b). *In* "Sexual Reproduction in Higher Plants" (M. Cresti, P. Gori, and E. Pacini, eds.), p. 471. Springer-Verlag, New York.
Erdelská, O. (1974). *In* "Fertilization in Higher Plants" (H. F. Linskens, ed.), pp. 191–195. North Holland, Amsterdam.
Erdelská, O. (1983). *In* "Fertilization and Embryogenesis in Ovulated Plants" (O. Erdelská, ed.), pp. 49–54. VEDA, Bratislava, Czechoslovakia.
Erdelská, O., and Klasova, A. (1978). *Actual. Bot. Soc. Bot. Fr.* **25,** 249–252.
Fairbanks, D. J., Smith, S. E., and Brown, J. K. (1988). *Theor. Appl. Genet.* **76,** 619–622.
Fisher, D. B., and Jensen, W. A. (1969). *Planta* **84,** 122–133.
Folsom, M. W., and Cass, D. D. (1986). *Can. J. Bot.* **64,** 965–972.
Folsom, M. W., and Cass, D. D. (1990). *Can. J. Bot.* **68,** 2135–2147.
Folsom, M. W., and Cass, D. D. (1992). *Am. J. Bot.* **79,** in press.
Fougère-Rifot, M. (1987). *Bull. Soc. Bot. Fr. Actual. Bot.* **134,** 113–160.
Gerassimova-Navashina, H. (1960). *Nucleus* **3,** 111–120.
Gillham, N. W. (1978). "Organelle Heredity." Raven Press, New York.
Godineau, J. C. (1969). *Rev. Cytol. Biol. Vég.* **32,** 209–226.
Guignard, L. (1899). *Rev. Gén. Bot.* **11,** 129–135.
Haberlandt, G. (1927). *Sitzungsberichte der Preus. Akad. Wissenschaft. Physik-math. Klasse* (*Berlin*) **1927,** 33–47.
Hagemann, R., and Schröder, M. B. (1989). *Protoplasma* **152,** 57–64.
Hause, G. (1991). *Sex. Plant Reprod.* **4,** 288–292.
Hause, G., and Schröder, M. B. (1987). *Protoplasma* **139,** 100–104.
Hill, R. A. (1977). Ph.D. Dissertation, University of California, Berkeley.
Hofmeister, W. (1849). "Die Entstehung des Embryo der Phanerogamen." Leipzig.
Huang, B. Q. (1991). Ph.D. Dissertation, University of Oklahoma, Norman.
Huang, B. Q., and Russell, S. D. (1992). *Sex. Plant Reprod.* **5,** 151–155.
Huang, B. Q., Russell, S. D., Strout, G. W., and Mao, L. J. (1990). *Am. J. Bot.* **77,** 1401–1410.
Huang, B. Q., Pierson, E., Russell, S. D., Tiezzi, A., and Cresti, M. (1992). *Sex. Plant Reprod.* **5,** 156–162.
Jaffe, L. A., and Cross, N. L. (1986). *Annu. Rev. Physiol.* **48,** 191–200.
Jensen, W. A. (1964). *J. Cell Biol.* **23,** 669–672.
Jensen, W. A. (1965). *Am. J. Bot.* **52,** 238–256.
Jensen, W. A. (1973). *Bioscience* **23,** 21–27.
Jensen, W. A. (1974). *In* "Dynamic Aspects of Plant Ultrastructure" (A. W. Robards, ed.), pp. 481–503. McGraw-Hill, New York.
Jensen, W. A., and Fisher, D. B. (1967). *Phytomorphology* **17,** 261–269.
Jensen, W. A., and Fisher, D. B. (1968). *Planta* **78,** 158–183.
Jensen, W. A., Schulz, P., and Beasley, C. A. (1977). *Planta* **133,** 179–189.
Jensen, W. A., Ashton, M. E., and Beasley, C. A. (1983). *In* "Pollen: Biology and Implications for Plant Breeding" (D. L. Mulcahy and E. Ottaviano, eds.), pp. 67–72. Elsevier Biomedical Press, New York.

Kadej, A., and Kadej, F. (1983). *In* "Fertilization and Embryogenesis in Ovulated Plants" (O. Erdelská, ed.), pp. 249–252. VEDA, Bratislava, Czechoslovakia.

Kadej, A., and Kadej, F. (1985). *In* "Sexual Reproduction in Seed Plants, Ferns and Mosses" (M. T. M. Willemse and J. L. Van Went, eds.), p. 149. PUDOC, Wageningen, The Netherlands.

Kadej, F., and Kadej, A. (1981). *Acta Bot. Soc. Pol.* **50**, 139–142.

Kapil, R. N., and Bhatnagar, A. K. (1975). *Phytomorphology* **25**, 334–368.

Knox, R. B. (1984). *In* "Embryology of Angiosperms" (B. M. Johri, ed.), pp. 197–271. Springer-Verlag, Berlin.

Kuroiwa, H. (1989). *Bot. Mag. Tokyo* **102**, 9–24

Le Monnier, G. (1887). *J. Bot. (Paris)* **1**, 140–142.

Linskens, H. F. (1969). *In* "Fertilization: Comparative Morphology, Biochemistry and Immunology" (C. B. Metz and A. Monroy, eds.), pp. 189–253. Academic Press, San Diego.

Maheshwari, P. (1948). *Bot. Rev.* **14**, 1–56.

Maheshwari, P. (1950). "An Introduction to the Embryology of Angiosperms." McGraw-Hill, New York.

Mansfield, S. G., and Briarty, L. G. (1991). *Can. J. Bot.* **69**, 461–476.

Masoud, S. A., Johnson, L. B., and Sorensen, E. L. (1990). *Theor. Appl. Genet.* **79**, 49–55.

Maze, J., and Lin, S. C. (1975). *Can. J. Bot.* **53**, 2958–2977.

Miyamura, S., Kuroiwa, T., and Nagata, T. (1987). *Protoplasma* **141**, 149–159.

Mogensen, H. L. (1972). *Am. J. Bot.* **59**, 931–941.

Mogensen, H. L. (1978). *Am. J. Bot.* **65**, 953–964.

Mogensen, H. L. (1982). *Carlsberg Res. Commun.* **47**, 313–354.

Mogensen, H. L. (1984). *Am. J. Bot.* **71**, 1448–1451.

Mogensen, H. L. (1988). *Proc. Natl. Acad. Sci. U.S.A.* **85**, 2594–2597.

Mogensen, H. L. (1990). *In* "Reproductive Versatility in the Grasses" (G. P. Chapman, ed.), pp. 76–99. Cambridge University Press, Cambridge, England.

Mogensen, H. L., and Rusche, M. L. (1985). *Protoplasma* 128, 1–13.

Mogensen, H. L., and Suthar, H. K. (1979). *Bot. Gaz.* **140**, 168–179.

Nawaschin, S. G. (1898). *Bull. Acad. Imp. Sci. St. Petersburg* **9**, 377–382.

Newcomb, W. (1973). *Can. J. Bot.* **51**, 879–890.

Olson, A. R., and Cass, D. D. (1981). *Am. J. Bot.* **68**, 1333–1341.

Ornberg, R. L., and Reese, T. S. (1981). *J. Cell Biol.* **90**, 40–45.

Owens, J. N., and Morris, S. J. (1991). *Am. J. Bot.* **78**, 1515–1527.

Rand, R. P., and Parsegian, V. A. (1986). *Annu. Rev. Physiol.* **48**, 201–212.

Reiss, H. D., Herth, W., Schnepf, E., and Nobiling, R. (1983). *Protoplasma* **115**, 153–159.

Robertson, D. S. (1984). *J. Hered.* **75**, 457–462.

Roman, H. (1948). *Proc. Natl. Acad. Sci. U.S.A.* **34**, 36–42.

Rougier, M., Jnoud, N., Saïd, C., Russell, S., and Dumas, C. (1991). *Protoplasma* **162**, 140–150.

Russell, S. D. (1982). *Can. J. Bot.* **60**, 2219–2230.

Russell, S. D. (1983). *Am. J. Bot.* **70**, 416–434.

Russell, S. D. (1984). *Planta* **162**, 385–391.

Russell, S. D. (1985). *Proc. Natl. Acad. Sci. U.S.A.* **82**, 6129–6134.

Russell, S. D. (1987). *Theor. Appl. Genet.* **74**, 693–699.

Russell, S. D. (1991). *Annu. Rev. Plant Physiol. Plant Mol. Biol.* **42**, 189–204.

Russell, S. D., and Cass, D. D. (1981). *Protoplasma* **107**, 85–107.

Russell, S. D., and Cass, D. D. (1988). *Am. J. Bot.* **75**, 778–781.

Russell, S. D., and Mao, L. J. (1990). *Planta* **182**, 52-57.

Russell, S. D., and Yu, H. S. (1991). *Am. J. Bot.* **78** (Suppl. 6), 34.

Russell, S. D., Huang, B. Q., and Strout, G. W. (1989). *In* "Some Aspects and Actual Orientations in Plant Embryology" (J. Pare and M. Bugnicourt, eds.), pp. 109–119, Université de Picardie, Faculté des Sciences, Picardie.

Russell, S. D., Rougier, M., and Dumas, C. (1990). *Protoplasma* **155,** 153–165.

Savina, G. I., and Zhukova, G. Y. (1983). *In* "Fertilization and Embryogenesis in Ovulated Plants" (O. Erdelská, ed.), pp. 207–210. VEDA, Bratislava, Czechoslovakia.

Schulz, P., and Jensen, W. A. (1973). *J. Cell Sci.* **12,** 741–763.

Schulz, S. R., and Jensen, W. A. (1968). *Am. J. Bot.* **55,** 541–552.

Schumann, C. M., and Hancock, J. F. (1989). *Theor. Appl. Genet.* **78,** 863–866.

Stanley, R. G., and Linskens, H. F. (1967). *Science* **157,** 833–834.

Steffen, K. (1963). *In* "Recent Advances in the Embryology of Angiosperms" (P. Maheshwari, ed.), pp. 105–133. University of Delhi, Delhi.

Strasburger, E. (1878). "Über Befruchtung und Zellteilung." Herman Dabis, Jena.

Sumner, M. J., and Van Caeseele, L. V. (1989). *Can. J. Bot.* **67,** 177–190.

Tilton, V. R. (1981). *New Phytol.* **88,** 505–531.

Van der Pluijm, J. E. (1964). *In* "Pollen Physiology and Fertilization" (H. F. Linskens, ed.), pp. 8–16. North Holland, Amsterdam.

Van Rensburg, J. G. J., and Robbertse, P. J. (1988). *S. Afr. Tydskr. Plantk.* **54,** 196–202.

Van Went, J. L. (1970). *Acta Bot. Neerl.* **19,** 468–480.

Van Went, J. L., and Cresti, M. (1988). *Sex. Plant Reprod.* **1,** 208–216.

Van Went, J. L., and Willemse, M. T. M. (1984). *In* "Embryology of Angiosperms" (B. M. Johri, ed.), pp. 273–317. Springer-Verlag, Berlin.

Van Went, J. L., Theunis, C. H., and Nijs, A. P. M. den (1985). *In* "Sexual Reproduction in Seed Plants, Ferns and Mosses" (M. T. M. Willemse and J. L. Van Went, eds.), pp. 153–154. PUDOC, Wageningen, The Netherlands.

Vaughn, K. C., DeBonte, L. R., Wilson, K. G., and Schaffer, G. W. (1980). *Science* **208,** 196–198.

Vazart, J. (1969). *Rev. Cytol. Biol. Vég.* **32,** 227–240.

Vijayaraghavan, M. R., Jensen, W. A., and Ashton, M. E. (1972). *Phytomorphology* **22,** 144–159.

Weisenseel, M. H., Nuccitelli, R., and Jaffe, L. F. (1975). *J. Cell Biol.* **66,** 556–567.

Welk, M. S., Millington, W. F., and Rosen, W. G. (1965). *Am. J. Bot.* **52,** 774–781.

White, J. A. (1990). *Annu. Rev. Physiol.* **52,** 675–697.

Willemse, M. T. M., and Franssen-Verheijen, M. A. W. (1988). *In* "Sexual Reproduction in Higher Plants" (M. Cresti, P. Gori, and E. Pacini, eds.), pp. 357–362. Springer-Verlag, New York.

Wilms, H. J. (1981). *Acta Bot. Neerl.* **30,** 101–122.

Yanagimachi, R. (1988). *In* "Membrane Fusion in Fertilization, Cellular Transport, and Viral Infection" (N. Düzgünes and F. Bronner, eds.) *Curr. Top. Membrane Transplant.* **32,** 3–43.

Yan, H., Yang, H. Y., and Jensen, W. A. (1991). *Can. J. Bot.* **69,** 191–202.

You, R. L., and Jensen, W. A. (1985). *Can. J. Bot.* **63,** 163–178.

Yu, H. S., Hu, S. Y., and Russell, S. D. (1992). *Protoplasma* in press.

Zhu, T., Mogensen, H. L., and Smith, S. E. (1990). *Protoplasma* **158,** 66–72.

Zhu, T., Mogensen, H. L., and Smith, S. E. (1991a). *In* "Proceedings of the 49th Annual Meeting of the Electron Microscopy Society of America" (G. W. Bailey, ed.), pp. 210–211. San Francisco Press, San Francisco.

Zhu, T., Mogensen, H. L., and Smith, S. E. (1991b). *Theor. Appl. Genet.* **81,** 21–26.

Part IV
Manipulation in Pollination and Fertilization Mechanisms

In Vitro Pollination: A New Tool for Analyzing Environmental Stress

Isabelle Dupuis
CIBA-GEIGY Biotechnology Research
Research Triangle Park, North Carolina 27709

I. Introduction

Unlike animals, plants cannot migrate to avoid unfavorable fluctuations of the environment. Levitt (1980) defined a biological stress as any environmental factor capable of inducing a potentially injurious strain to a living organism. Several types of abiotic stress have been shown to affect plant growth and development: anaerobiosis, salt, heavy metals, water, and cold or heat stress; and plants have developed complex stress response systems to reduce the impact of these environmental changes (Matters and Scandalios, 1986; Nover, 1989; Chapin, 1991). Environmental stress represents the most limiting factor to agricultural productivity. For major crop plants the average losses in productivity due to stressful environment is estimated at 50–80% of their genetic potential (Boyer, 1982). In this view stress offers not only an interesting system to study molecular response in plant cells but also has important economic implications.

One of the most extensively studied and characterized stress responses is that produced by high temperatures. The definition of heat stress is complex because the effect of temperature on the plant is dependent on the interaction of other factors, including light, hydric state, soil, and air composition (Bourdu, 1984). Furthermore, heat stress has to be defined as a function of the intensity and the length of the stress. The threshold of temperature above which a plant is subjected to stress conditions depends on the normal growth temperature of the plant; 10–15°C above optimal growth temperature for a period of 15 minutes to a few hours is usually considered stressful (Schöffl et al., 1988; Brodl, 1990). Although plants are able to tolerate some variations in their optimal growth conditions, temperature becomes lethal to the plant above a certain limit.

The reproductive phase in the plant life cycle appears to be particularly sensitive to heat stress (Herrero and Johnson, 1980). Several factors interact to determine fertilization success; among these are female receptivity, pollen viability, and environmental influences (Sadras *et al.*, 1985; Struik *et al.*, 1986). Because of the number of factors involved, reproductive stress physiology studies are difficult to conduct in the field. In this view *in vitro* pollination appears to be a model technique to analyze the effects of environmental stress on different phases of plant reproduction under controlled culture conditions. In addition to the physiological consequences of heat stress on the reproduction success, high temperatures induce a response at the cellular level. Heat shock response is a highly conserved biological phenomenon that appears to occur universally among a wide range of organisms (Craig, 1985). In plants heat shock response (particularly the induction of heat shock proteins [HSPS]) has been extensively studied in vegetative tissues (Kimpel and Key, 1985; Schöffl *et al.*, 1988; Ho and Sachs, 1989). However, little is known about the response of male and female reproductive tissues to high temperatures. Such studies might provide a comprehensive view of the heat stress resistance phenomenon at the gametophyte level. The focus of this chapter will be on the effects of temperature, and particularly heat stress, on plant reproductive functions using maize as a model system.

II. Influence of Temperature Stress on the *in Vivo* Reproductive Process

Whole plant studies show that heat and/or drought stress (these two factors often being related in the field) can drastically reduce crop productivity (Moss and Downey, 1971; Herrero and Johnson, 1981; Schoper *et al.*, 1986, 1987a). Because of its economic importance, maize stress physiology has been extensively studied. Previous research reveals that floral physiology and reproduction are particularly sensitive to stress. Some of the critical stages in maize are tasseling and silking (Shaw, 1977; Hall *et al.*, 1981; Westgate and Boyer, 1986), pollination (Schoper *et al.*, 1986), and the ear filling stage (Shaw, 1977). Furthermore, high temperatures seem to have a more negative effect on the male organs than on the female organs (Schoper *et al.*, 1986). The effects of heat shock on pollen include a shortened duration of pollen shedding and reduced viability and amount of pollen shed (Struik *et al.*, 1986). Pollen sensitivity to heat stress is evident in experiments in which male and female partners are separately submitted to a 4-hour temperature stress (Fig. 1). The dramatic effects of high temperature on seed set extend not only to the isolated pollen but

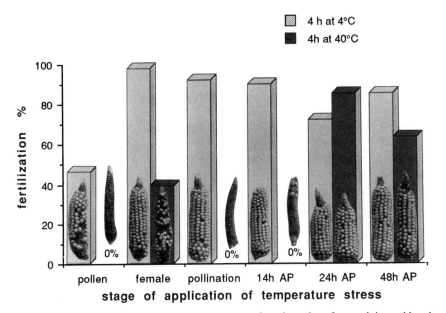

FIG. 1 Influence of temperature stress on *in vivo* seed set in maize after applying cold and heat stress at different stages of development. Isolated pollen was incubated 4 hours at 4°C or 40°C and then used for pollination. For pollination and postpollination stages 14, 24, and 48 hours after pollination (AP), temperature stress was applied locally on the ear using a water bath system maintained at the appropriate cold or hot temperatures. The success of fertilization is indicated in percent based on seed set. From Federation Nationale des Producteurs de Semences de Maïs et de Sorgho (FNPSMS)/Association generale des Producteurs de Maïs (AGPM), Pau, France.

also to the pollination and early postpollination period. By comparison, high temperature applied selectively to the female before pollination or exposure to cold temperature throughout the fertilization process has only a moderate effect on seed set (Fig. 1).

Heat stress selectively applied to isolated tassels and pollen confirms the negative effect of high temperature on pollen physiology. Heat stress applied to tassels prior to pollen shedding reduces pollen quality as estimated by seed set experiments (Schoper *et al.*, 1986) or by *in vitro* tests. Herrero and Johnson (1980) have shown that high temperature stress (27–38°C) during anthesis can nearly eliminate *in vitro* pollen germination. Furthermore, pollen tube elongation *in vitro* is reduced by 60% when pollen is germinated at 40°C (Binelli *et al.*, 1985; Frova *et al.*, 1986). In other species, such as tobacco (Shivanna and Cresti, 1989; Van Herpen *et al.*, 1989), lily (Van Herpen *et al.*, 1989), and pine (Frankis and Grayson, 1990), *in vitro* germination of pollen is also altered when temperature is simply elevated 10°C above the optimal growth temperature.

Genotypes differ in their response to temperature, and some genotypes appear more tolerant to heat stress than others (Herrero and Johnson, 1980; Schoper *et al.*, 1987b). It is interesting to note that inbred lines used in the 1970s germinated significantly better as a group than the lines widely used in the 1930s and 1950s (Herrero and Johnson, 1980). The selection process for agronomic production traits has also apparently selected for heat stress tolerance.

Pollen quality seems to be particularly affected by high temperature stress in the field. A more precise characterization of pollen heat sensitivity requires a controlled *in vitro* system.

III. *In Vitro* Pollination System

Because fertilization success is determined by a number of factors including pollen viability, female receptivity, and environmental influences that are impractical to test in the field, *in vitro* pollination is an attractive technique for studying the fertilization process. Both pollination and culture are conducted under controlled nutritional and environmental conditions using this approach.

A. *In Vitro* Pollination in Maize

In vitro pollination and subsequent recovery of seeds have been achieved in a wide range of species (Zenkteler, 1990). The focus here is on a maize model because *in vitro* culture techniques in this plant have been particularly well studied. *In vitro* pollination and successful fertilization in maize were first reported by Sladky and Havel (1976) and Gengenbach (1977a,b). The basic protocol for *in vitro* pollination and fertilization in maize is that published by Gengenbach (1977a,b). Unpollinated ears are removed from the plant, and segments of the cob are dissected under sterile conditions, and usually kept in a paired arrangement with two rows of ovaries. The segments (containing 4–30 ovaries per segment with their silks intact) are then placed in a small Petri dish containing a solid culture medium (modified from Murashige and Skoog, 1962). The silks are allowed to extend outside of the dish and are cut 4.5–9 cm in length. This procedure allows nonsterile pollen to be used without contaminating the culture. Pollen is then applied in saturating amounts to the distal part of the silks (Fig. 2A). This is then enclosed within a larger Petri dish for culture.

Successful *in vitro* fertilization has been obtained with as few as only one pollen grain per silk (Raman *et al.*, 1980; Hauptli and Williams, 1988;

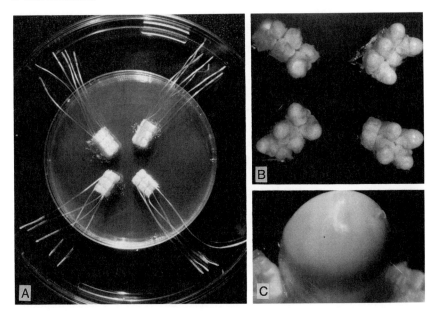

FIG. 2   *In vitro* fertilization chamber used to pollinate and culture maize kernels. (A) Each chamber contains four segments of a maize ear, containing six spikelets pollinated on the distal part of their trimmed silks. (B) Developing kernels are shown 20 days after *in vitro* pollination. (C) One normal kernel developed *in vitro*.

Kranz and Lörz, 1990). Usually, the fertilization rates are lower with a single pollen grain than following a mass pollination (Raman *et al.*, 1980; Hauptli and Williams, 1988); however, Kranz and Lörz (1990) reported similar fertilization rates between mass and single grain pollinations. After a 24- to 48-hour treatment at 26–28°C in the dark, silks are removed and cultures are incubated 35–45 days in the same conditions, which allows for kernel development (Figs. 2B, C). Kinetics and autoradiographic studies show that fertilization is accomplished 8 hours after *in vitro* pollination using 4.5-cm length silks (Dupuis and Dumas, 1989). The *in vitro* elongation rate of the pollen tubes in the silks in culture (6.5–9 mm/hour) is similar to those previously reported *in vivo* (Dupuis and Dumas, 1989).

The average frequency of fertilization, estimated by ovule swelling a few days after pollination, can be relatively high (more than 80%) (Gengenbach, 1977b). Significant differences were observed in fertilization frequency depending on the genotype of both the ovary and the pollen source (Gengenbach, 1977b; Bajaj, 1979; Higgins and Petolino, 1988). Although a wide range of abnormal, incomplete, and normal kernels are observed later in development (from small embryos with no endosperm to full

kernels) (Bajaj, 1979; Gengenbach, 1977a; Higgins and Petolino, 1988; Dupuis *et al.,* 1988; Dupuis and Dumas, 1989), normal, mature embryos can be routinely obtained in a percentage of the kernels using this technique. Explant size (number of spikelets per cob segment) and ovule to cob ratio (number of spikelets allowed to develop on a cob segment) appear to have a significant effect on the completion of kernel formation *in vitro* (Higgins and Petolino, 1988). Furthermore, kernel abortion can be associated with an inadequate sugar supply, reducing dry weight accumulation (Hanft and Jones, 1986a), suggesting that developing kernels compete for nutrient assimilation. As suggested by Higgins and Petolino (1988), it is subsequent kernel development and filling, rather than fertilization *per se,* which is the limiting step to the recovery of plants using this technique.

B. Control of Male and Female Quality

The quality of male and female cells and tissues used for culture are necessary factors to ensure optimum *in vitro* fertilization. Pollen quality can be defined as the ability of pollen to fertilize a receptive and compatible pistil (Kerhoas and Dumas, 1988) or as the competence of individuals of a pollen population to deliver sperm cells to the embryo sac (Heslop-Harrison *et al.,* 1984). Routine pollen tests, such as *in vitro* germination (Pfahler, 1967) and the fluorochromatic reaction (Heslop-Harrison and Heslop-Harrison, 1970), have been used routinely to assess pollen quality prior to *in vitro* pollination (Hauptli and Williams, 1988; Dupuis and Dumas, 1989).

Female receptivity can be defined as the period of time when, in presence of viable pollen, the silk can support pollen germination and pollen tube growth; and the embryo sac receive and incorporate the two sperm in double fertilization. Female maturity has a significant effect on *in vitro* fertilization success for a given genotype (Gengenbach, 1977a; Higgins and Petolino, 1988). In the intact ear the number of days after silk emergence is useful in estimating relative female age, the presence of a maturational gradient means that spikelets originating from the same ear might have different levels of receptivity. Analysis of female receptivity by individual spikelet allows for a more precise estimation of receptivity. The developmental stage can be estimated by the total silk length and condition of each spikelet after dissection of the ear. Spikelets with different silk lengths have been tested using *in vitro* pollination, and the resultant fertilization rate is correlated with the maturation of the spikelets (Dupuis and Dumas, 1990a). Three different sequences can be distinguished in the female development in relationship to the receptivity: (1) immature, partially receptive spikelets, (2) mature, fully receptive spikelets, and (3) senescent, nonre-

ceptive spikelets. For some genotypes the receptivity can appear before the emergence of the silks from the husks (Dupuis and Dumas, 1990a). Therefore, the pattern of female receptivity has to be defined for each genotype.

In conclusion, the *in vitro* pollination system supports fertilization and kernel development in a manner similar to the natural *in vivo* process. Using careful control of male and female quality and a standardized environment, this system allows for the selective manipulation of individual factors to observe their influence on the reproductive process.

IV. Influence of Temperature Stress at Different Phases of the *In Vitro* Reproductive Process

As emphasized in whole plant studies, fertilization is a critical stage in the plant life cycle and is particularly sensitive to temperature stress. *In vitro* pollination provides a convenient method to study the heat sensitivity of male and female tissues on different phases of the reproductive process.

A. Postpollination

To analyze the influence of temperature on *in vitro* fertilization success, spikelets pollinated *in vitro* were transferred for various lengths of time from the control temperature of 28°C to either a cold stress (4°C), or heat stress condition (32–40°C) (Dupuis and Dumas, 1990b). Subsequent scoring of fertilization rates revealed that cold treatment of up to 8 hours and exposure to moderately high temperature (under 36°C) did not affect fertilization. By contrast, fertilization was immediately and irreversibly altered at 40°C (Fig. 3) (Dupuis and Dumas, 1990b). The sensitivity threshold of fertilization in this system was situated between 36°C and 40°C. However, when heat stress was delayed by 4 hours and applied 4 hours after pollination at the normal temperature, exposure to 40°C stress for 4 hours did not appear to affect *in vitro* fertilization (Dupuis and Dumas, 1990b). This suggests that the first 4 hours following *in vitro* pollination is a critical period of heat shock sensitivity. Mitchell and Petolino (1988) found that the *in vitro* fertilization rate was also significantly reduced after a 48-hour treatment at 38°C.

To better understand the failure of fertilization under stress conditions, radiolabeled pollen was used to visualize the extent of pollen tube growth in the silks after pollination. Five hours after pollination at 28°C, pollen tubes have progressed in the silks and are close to the ovary. By contrast,

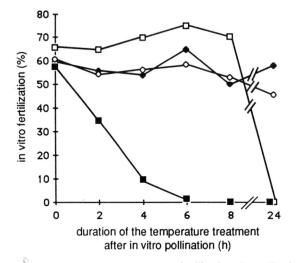

FIG. 3   Influence of temperature stress on *in vitro* fertilization. Immediately after *in vitro* pollination, spikelets were exposed to a temperature of 4°C (□), 32°C (♦), 36°C (◇), or 40°C (■) during 2–24 hours, and then returned to 28°C. Each point represents the average fertilization score of all spikelets submitted to the same temperature treatment. Reproduced with permission from Dupuis and Dumas (1990b).

no pollen tube growth was observed after incubation at 4°C or 40°C (Dupuis and Dumas, 1990b). Under cold shock conditions the inhibition of pollen tube growth is reversible when the system is returned to 28°C; however, if a 40°C heat shock is applied, pollen function is irreversibly altered.

B. Stress on Reproductive Tissues

To analyze more precisely the effect of temperature stress on the male tissues, different types of stress treatments were designed. A 38°C thermal stress was applied during the time between tassel harvest and pollen collection (Mitchell and Petolino, 1988); or the mature shed pollen was collected and incubated 4 hours at 4°C or 40°C (Dupuis and Dumas, 1990b). Short-term cold treatment does not reduce the fertilization ability of the pollen. Similarly, when whole tassels were exposed to high temperatures during pollen shedding, the fertilization rate was not greatly affected (Mitchell and Petolino, 1988). By contrast, the isolated pollen loses its fertilization ability after even brief heat stress treatment (Dupuis and Dumas, 1990b). This suggests that shed pollen is sensitive to high temperatures. Once liberated from the anther, pollen is directly exposed to the

high temperatures and no longer benefits from a possible protective effect of the tassel.

In the female tissues a similar temperature treatment was applied to the female spikelets before *in vitro* pollination to assess their sensitivity to stress. Female receptivity, however, was not affected by any of the same treatments that drastically affected the pollen (Dupuis and Dumas, 1990b). Therefore, the female reproductive organ appears to be more tolerant to thermal stress than the pollen.

C. Kernel Development

After successful fertilization the last component involved in seed set determination is the development and filling of the kernel; this process can also be altered by high temperature stress. Complete *in vitro* kernel formation rate is altered by previous heat stress during the different phases of the reproductive process (pollen shedding, silk receptivity, pollen–stigma interaction, and early kernel formation) (Mitchell and Petolino, 1988). Elevated temperatures during the first week of kernel development also inhibit complete kernel formation (Mitchell and Petolino, 1988). Temperature has a great effect on both the rate and the duration of grain filling (Jones *et al.*, 1981). Kernels induced to abort by high temperatures *in vitro* failed to enter a period of linear dry matter accumulation (Hanft and Jones, 1986a) and seem to have a reduced capacity to synthesize starch in the endosperm (Hanft and Jones, 1986b).

Studies using *in vitro* reproduction have indicated differences in sensitivity to heat stress among different genotypes (Mitchell and Petolino, 1988). *In vitro* pollination experiments were conducted using parental genotypes with variable heat stress sensitivity as expressed by their pollen germination rates (Petolino *et al.*, 1990). The progeny were obtained from heat stress tolerant and sensitive parents. After *in vitro* pollination under normal (28°C) and selecting conditions (38°C), the selected progeny showed superior agronomic performance and increased pollen heat tolerance in most cases than the nonselected progeny. Therefore, high temperature stress at the time of pollen function can reduce the heat sensitivity of pollen in the next generation, suggesting that selection for this trait has occurred.

D. Pollen Selection

Although pollen has a highly specialized structure and function, most of the genes expressed in pollen are also expressed in the vegetative parts of

a plant (Willing and Mascarenhas, 1984; Sari-Gorla *et al.*, 1986; Willing *et al.*, 1988; Mascarenhas, 1988, 1989). This overlap between sporophytic and gametophytic genomes offers the possibility to use pollen as a selection mechanism for tolerance and resistance to various environmental factors. Mulcahy (1979) proposed that selection during male gametophyte function could have a positive effect on the following sporophytic generation (Zamir, 1983; Bino and Stephenson, 1988). Positive experimental results for gametophytic selection have been obtained for environmental selective factors, such as tolerance of low temperature in *Lycopersicon* (Zamir, 1983) or high temperature in cotton (Rodriguez-Garay and Barrow, 1988). Environmental stress, such as temperature, is an interesting possible application for the use of pollen selection because of the deleterious effect of temperature on the agronomic performance of plants. *In vitro* pollination represents a good system to induce a specific selection pressure during pollen function under controlled conditions and might thus be used as a tool in selective breeding programs.

V. Heat Shock Response in Reproductive Tissues

When plants are exposed to high temperature, they react in a complex manner (Burke, 1990). One of the most spectacular results of exposure to high temperature is the synthesis of a new specific set of proteins, HSPs, corresponding to rapid expression of the heat shock genes. This heat shock response was first characterized in *Drosophila* (Tissieres *et al.*, 1974), but induction of HSPs has been identified in a wide range of prokaryotic and eukaryotic organisms (Craig, 1985), including plants (Kimpel and Key, 1985; Schöffl *et al.*, 1988; Ho and Sachs, 1989). HSPs in plants can be divided into several groups: high molecular mass (80–100 kDa) proteins and the HSP70 family, which are well conserved in all organisms (Schöffl *et al.*, 1988), and low molecular mass (15–20 kDa) proteins, which are particularly prominent in plants (Mansfield and Key, 1987). The major HSPs have been strongly conserved in structure throughout evolution, indicating that their function in the survival of the organism under stress might be vital (Schlesinger, 1986). Their presence appears to enhance the cells' ability to recover from stress (Schlesinger, 1988; Krishnan *et al.*, 1989), but there is no simple relationship between HSPs and high temperature tolerance (Laszlo, 1988; Vierling, 1990).

Whereas heat shock response has been extensively studied in vegetative tissues of plants, little is known about the heat shock response in reproductive tissues. Plant HSPs are not tissue-specific; a typical set of HSPs is synthesized by all vegetative tissues tested. In maize, for example, HSPs

are induced in seedlings, roots, leafs, and cultured cells (Baszczynski *et al.*, 1982; Cooper and Ho, 1983; Cooper *et al.*, 1984; Bonham-Smith *et al.*, 1988). Furthermore, the female reproductive tissues, silks, and ovaries in maize (Fig. 4) (Dupuis and Dumas, 1990b) and pistils in *Sorghum* (Frova *et al.*, 1991) respond in a similar way. In contrast, the male gametophyte has recently received some attention, and it appears that pollen gives a specific heat shock response.

Characterization of the male gametophytic response seems to vary during three general phases of pollen life: pollen development, maturity, and pollen germination. During early stages of pollen development in maize, synthesis of a set of HSPs is induced prior to microspore mitosis, whereas only a reduced number are expressed later, in binucleate and trinucleate stages (Frova *et al.*, 1989). Only two of these HSPs correspond to sporo-

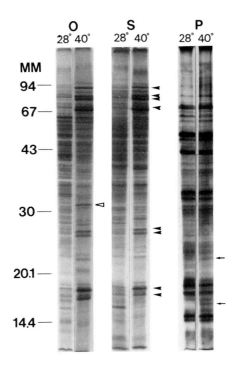

FIG. 4 Fluorography of the proteins synthesized in male and female reproductive organs at control (28°C) and heat shock (40°C) temperatures. The proteins were labeled using [^{35}S]methionine during 4 hours at the appropriate temperature. O, ovary; S, silk; P, pollen, MM, molecular mass. The standard heat shock proteins are indicated by larger arrows. An ovary-specific heat shock protein is indicated by an open arrow; additional proteins synthesized in pollen under heat shock condition are indicated with small arrows. Reproduced with permission from Dupuis and Dumas (1990b).

phytic HSPs, whereas others appear to be characteristic of the postmeiotic phase. Similarly, immature pollen grains of *Sorghum* synthesize HSPs under stress conditions (Frova *et al.*, 1991). In contrast, no HSPs were synthesized in the mature pollen of maize (Fig. 4) (Dupuis and Dumas, 1990b). In germinating pollen no major HSPs appear to be synthesized during heat shock, despite minor changes in gene expressional patterns in *Tradescantia* (Mascarenhas and Altschuler, 1983; Xiao and Mascarenhas, 1985), maize (Cooper et al., 1984; Frova *et al.*, 1989), *Lilium* (Schrauwen *et al.*, 1986; Van Herpen *et al.*, 1989), and *Petunia* (Schrauwen *et al.*, 1986). By contrast, some heat-induced polypeptide synthesis has been detected after heat shock in germinating pollen of tobacco (Van Herpen *et al.*, 1989), *Sorghum* (Frova *et al.*, 1991), and pine (Frankis and Grayson, 1990). Some of these polypeptides are classical HSPs, whereas others seem to be characteristic of the male gametophyte.

Overall, the male gametophyte heat shock response differs from the typical sporophytic response. Induction of HSPs in pollen appears to be regulated developmentally. Furthermore, when HSPs are expressed at the pollen level, two types of HSPs are observed: those common with the sporophyte and those characteristic of the male gametophyte.

The synthesis of HSPs is primarily regulated by transcriptional control, allowing rapid induction of mRNA and preferential translation during heat shock (Schöffl *et al.*, 1988). Mature pollen is relatively dormant in gene expression, whereas immature and germinating pollen are more active (Mascarenhas, 1989). This could explain the differences observed in HSP expression during different developmental stages and the altered ability of pollen to synthesize typical HSPs. This low level of HSP synthesis could correlate to the sensitivity of pollen to heat. However, in the case of *Tradescantia,* thermotolerance can be achieved by germinating pollen, independently of the presence of HSPs (Mascarenhas and Altschuler, 1983; Xiao and Mascarenhas, 1985). It will be necessary to extend studies to the mRNA level to better understand the heat shock response in pollen.

VI. Conclusions

The reproductive process (and particularly pollen function) has proven to be sensitive to high temperature stress as revealed by *in vivo* and *in vitro* studies. *In vitro* pollination–fertilization represents a powerful tool for manipulation of the reproductive process under standardized conditions. This technique has been used successfully in maize to analyze the effect of heat stress, as well as to apply high temperatures at the time of pollination as a selection pressure. The application of *in vitro* pollination for

genetic manipulation should be extended to other types of selection conditions as well as to other species.

Pollen is characterized by its sensitivity to heat stress and by the specificity of its heat shock response. Under heat stress pollen does not synthesize a typical set of HSPs, unlike vegetative or female reproductive tissues. Synthesis of pollen heat shock proteins seems to be developmentally regulated, and HSPs synthesized in pollen appear to be specific. This could reveal sporophyte–gametophyte differences. It is interesting to note that pollen thermosensitivity may correlate with the lack of synthesis of the classical HSPs because these proteins might have a role in protection of the cells against the effects of temperature. However, it is likely that thermotolerance is a complex phenomenon encompassing parameters in addition to the synthesis of HSPs. Furthermore, mechanisms for the protection of pollen against high temperature might be different from those active in sporophytic tissues. Further analysis will be necessary to understand the heat shock response in pollen and potentially to improve heat tolerance in important crop species.

Acknowledgments

The author thanks Drs. V. T. Wagner (University of North Carolina, Chapel Hill) for his critical reading of the manuscript, C. Dumas (RCAP, Lyon) for supporting the maize heat stress project, and G. M. Pace (Ciba-Geigy, Research Triangle Park) for allowing the time to write this chapter.

References

Bajaj, Y. P. S. (1979). *Indian. J. Exp. Biol.* **17,** 475–478.
Baszczynski, C. L., Walden, D. B., and Atkinson, B. G. (1982). *Can. J. Biochem.* **60,** 569–579.
Binelli, G., Vieira De Manincor, E., and Ottaviano, E. (1985). *Genet. Agr.* **39,** 269–281.
Bino, R. J., and Stephenson, A. G. (1988). *In* "Plant Sperm Cells as Tools for Biotechnology" (H. J. Wilms and C. J. Keijzer, eds.), pp. 125–135. Pudoc, Wageningen.
Bonham-Smith, P. C., Kapoor, M., and Bewley, J. D. (1988). *Can. J. Bot.* **66,** 1883–1890.
Bourdu, R. (1984). *In* "Physiologie du maïs," pp. 389–424. INRA, Paris.
Boyer, J. S. (1982). *Science* **218,** 443–448.
Brodl, M. R. (1990). *In* "Environmental Injury to Plants" (F. Katterman, ed.), pp. 113–135. Academic Press, San Diego.
Burke, J. J. (1990). *In* "Stress Responses in Plants: Adaptation and Acclimation Mechanisms" (R. G. Alscher and J. R. Cumming, eds.), Plant Biology, Vol. 12, pp. 295–309. Wiley–Liss, New York.
Chapin, F. S., III. (1991). *Bioscience* **41,** 29–36.
Cooper, P., and Ho, T. D. (1983). *Plant Physiol.* **71,** 215–222.
Cooper, P., Ho, T. D., and Hauptmann, R. M. (1984). *Plant Physiol.* **75,** 431–441.

Craig, E. A. (1985). *C.R.C. Crit. Rev. Biochem.* **18,** 239–280.

Dupuis, I., and Dumas, C. (1989). *Sex. Plant Reprod.* **2,** 265–269.

Dupuis, I., and Dumas, C. (1990a). *Plant Sci.* **70,** 11–19.

Dupuis, I., and Dumas, C. (1990b). *Plant Physiol.* **94,** 665–670.

Dupuis, I., Zhao, Z. X., and Dumas, C. (1988). *In* "Sexual Reproduction in Higher Plants" (M. Cresti, P. Gori, and E. Pacini, eds.), pp. 309–311. Springer-Verlag, New York.

Frankis, R. C., and Grayson, G. K. (1990). *Sex. Plant Reprod.* **3,** 195–199.

Frova, C., Binelli, G., and Ottaviano, E. (1986). *In* "Biotechnology of Pollen" (D. L. Mulcahy, G. Mulcahy, and E. Ottaviano, eds.), pp. 33–38. Springer-Verlag, New York.

Frova, C., Taramino, G., and Binelli, G. (1989). *Dev. Genet.* **10,** 324–332.

Frova, C., Taramino, G., and Ottaviano, E. (1991). *Plant Sci.* **73,** 35–44.

Gengenbach, B. G. (1977a). *Crop Sci.* **17,** 489–492.

Gengenbach, B. G. (1977b). *Planta* **134,** 91–93.

Hall, A. J., Lemcoff, J. H., and Trapani, N. (1981). *Maydica* **26,** 19–38.

Hanft, J. M., and Jones, R. J. (1986a). *Plant Physiol.* **81,** 503–510.

Hanft, J. M., and Jones, R. J. (1986b). *Plant Physiol.* **81,** 511–515.

Hauptli, H., and Williams, S. (1988). *Plant Sci.* **58,** 231–237.

Havel, L., and Novak, F. J. (1981). *Plant Cell Rep.* **1,** 26–28.

Herrero, M. P., and Johnson, R. R. (1980). *Crop Sci.* **20,** 796–800.

Herrero, M. P., and Johnson, R. R. (1981). *Crop Sci.* **21,** 105–110.

Heslop-Harrison, J., and Heslop-Harrison, Y. (1970). *Stain Tech.* **45,** 115–120.

Heslop-Harrison, J., Heslop-Harrison, Y., and Shivanna, K. R. (1984). *Theor. Appl. Genet.* **67,** 367–375.

Higgins, R. K., and Petolino, J. F. (1988). *Plant Cell Tissue Organ Cult.* **12,** 21–30.

Ho, T. H. D., and Sachs, M. M. (1989). *In* "Plants under Stress" (H. G. Jones, T. J. Flowers, and M. B. Jones, eds.), pp. 157–180. Cambridge University Press, Cambridge, Massachusetts.

Jones, R. J., Gengenbach, B. G., and Cardwell, V. B. (1981). *Crop Sci.* **21,** 761–766.

Kerhoas, C., and Dumas, C. (1988). *In* "Plant Sperm Cells as Tools for Biotechnology" (H. J. Wilms and C. J. Keijzer, eds.), pp. 97–104. Pudoc, Wageningen.

Kimpel, J. A., and Key, J. L. (1985). *Trends Biochem. Sci.* **10,** 353–357.

Krishnan, M., Nguyen, H. T., and Burke, J. J. (1989). *Plant Physiol.* **90,** 140–145.

Kranz, E., and Lörz, H. (1990). *Sex. Plant Reprod.* **3,** 160–169.

Laszlo, A. (1988). *Exp. Cell Res.* **178,** 401–414.

Levitt, J. (1980). *In* "Responses of Plants to Environmental Stresses, Vol. I, Chilling, Freezing, and High Temperature Stresses." Academic Press, San Diego.

Mansfield, M. A., and Key, J. L. (1987). *Plant Physiol.* **84,** 1007–1017.

Mascarenhas, J. P. (1988). *In* "Temporal and Spatial Regulation of Plant Genes" (D. P. S. Verma and R. B. Goldberg, eds.), pp. 97–115. Springer-Verlag, New York.

Mascarenhas, J. P. (1989). *Plant Cell* **1,** 657–664.

Mascarenhas, J. P., and Altschuler, M. (1983). *In* "Pollen: Biology and Implications for Plant Breeding" (D. L. Mulcahy and E. Ottaviano, eds.), pp. 3–8. Elsevier, New York.

Matters, G. L., and Scandalios, J. G. (1986). *Dev. Genet.* **7,** 167–175.

Mitchell, J. C., and Petolino, J. F. (1988). *J. Plant Physiol.* **133,** 625–628.

Moss, G. I., and Downey, L. A. (1971). *Crop Sci.* **11,** 368–372.

Mulcahy, D. L. (1979). *Science* **206,** 20–23.

Murashige, T., and Skoog, F. (1962). *Physiol. Plant.* **15,** 473–498.

Nover, L. (1989). *In* "Heat Shock and Other Stress Response Systems in Plants" (L. Nover, D. Neumann, and K. D. Scharf, eds.), pp. 87–104. Springer-Verlag, Berlin.

Petolino, J. F., Cowen, N. M., Thompson, S. A., and Mitchell, J. C. (1990). *J. Plant Physiol.* **136,** 219–224.

Pfahler, P. L. (1967). *Can. J. Bot.* **45,** 839–845.

Raman, F., Walden, D. B., and Greyson, R. I. (1980). *J. Hered.* **71,** 311–314.

Rodriguez-Garay, B., and Barrow, J. R. (1988). *Crop Sci.* **28,** 857–859.

Sadras, V. O., Hall, A. J., and Schlichter, T. M. (1985). *Maydica* **30,** 37–47.

Sari-Gorla, M., Frova, C., Binelli, G., and Ottaviano, E. (1986). *Theor. Appl. Genet.* **72,** 42–47.

Schlesinger, M. J. (1986). *J. Cell Biol.* **103,** 321–325.

Schlesinger, M. J. (1988). *Biochemistry* **1,** 161–164.

Schöffl, F., Baumann, G., and Raschke, E. (1988). *In* "Plant Gene Research: Temporal and Spatial Regulation of Plant Genes" (D. P. S. Verma and R. B. Goldberg, eds.), pp. 253–273. Springer-Verlag, Vienna.

Schoper, J. B., Lambert, R. J., and Vasilas, B. L. (1986). *Crop Sci.* **26,** 1029–1033.

Schoper, J. B., Lambert, R. J., Vasilas, B. L., and Westgate, M. E. (1987a). *Plant Physiol.* **83,** 121–125.

Schoper, J. B., Lambert, R. J., and Vasilas, B. L. (1987b). *Crop Sci.* **27,** 27–31.

Schrauwen, J. A. M., Reijnen, W. H., De Leeuw, H. C. G. M., and Van Herpen, M. M. A. (1986). *Acta Bot. Neerl.* **35,** 321–327.

Shaw, R. H. (1977). *In* "Corn and Corn Improvement" (G. F. Sprague, ed.), Agronomy, Vol. 18, pp. 591–623. ASA, Madison, Wisconsin.

Shivanna, K. R., and Cresti, M. (1989). *Sex. Plant Reprod.* **2,** 137–141.

Sladky, Z., and Havel, L. (1976). *Biol. Plant* **18,** 469–472.

Struik, P. C., Doorgeest, M., and Boonman, J. G. (1986). *Netherlands J. Agri. Sci.* **34,** 469–484.

Tissieres, A., Mitchell, H. K., and Tracy, U. M. (1974). *J. Mol. Biol.* **84,** 389–398.

Van Herpen, M. M. A., Reijnen, W. H., Schrauwen, J. A. M., De Groot, P. F. M., Jager, J. W. H., and Wullems, G. J. (1989). *J. Plant Physiol.* **134,** 345–351.

Vierling, E. (1990). *In* "Stress Responses in Plants: Adaptation and Acclimation Mechanisms" (R. G. Alscher and J. R. Cumming, eds.), Plant Biology, Vol. 12, pp. 357–375. Wiley–Liss, New York.

Westgate, M. E., and Boyer, J. S. (1986). *Crop. Sci.* **26,** 951–956.

Willing, R. P., and Mascarenhas, J. P. (1984). *Plant Physiol.* **75,** 865–868.

Willing, R. P., Bashe, D., and Mascarenhas, J. P. (1988). *Theor. Appl. Genet.* **75,** 751–753.

Xiao, C. M., and Mascarenhas, J. P. (1985). *Plant Physiol.* **78,** 887–890.

Zamir, D. (1983). *In* "Isozymes in Plant Genetics and Breeding" (S. D. Tanksley and T. J. Orton, eds.), pp. 313–330. Elsevier, Amsterdam.

Zenkteler, M. (1990). *C.R.C. Crit. Rev. Plant Sci.* **9,** 267–279.

In Vitro Fusion of Gametes and Production of Zygotes

Erhard Kranz,* Horst Lörz,* Catherine Digonnet,† and Jean-Emmanuel Fauret†
* Universität Hamburg, Institut für Allgemeine Botanik, Angewandte
Molekularbiologie der Pflanzen II, Hamburg, Germany
† Reconnaissance Cellulaire et Amélioration des Plantes, Université Claude Bernard-Lyon 1,
Villeurbanne, France

I. Introduction

Progress in our understanding of fertilization mechanisms in higher plants
is mainly due to the availability of comprehensive information based on
studies using light and electron microscopy. The ability to handle, manipulate, and culture the female gametophyte, the female and male gametes,
and nongametic cells of the embryo sac under defined conditions will be
another valuable source of information.

Using maize the isolation of both male gametes (Dupuis et al., 1987;
Matthys-Rochon et al., 1987; Cass and Fabi, 1988; Wagner et al., 1989a;
Yang and Zhou, 1989) and female gametes (Wagner et al., 1988, 1989b)
has been described previously, and the isolation of egg cells has been
improved (Kranz et al., 1990, 1991a).

Using these isolated and individually selected egg cells and sperm cells
of maize, in vitro fertilization (IVF) can be performed by electrical fusion
of pairs of gametes (Kranz et al., 1991a). Fusion products can also be
created by fusion of single synergids as well as central cells with single
sperm cells (Kranz et al., 1991b). Artificially produced zygotes grow to
multicellular structures with a high frequency of division. Fertilization
processes can now be studied without the influence of other tissues at the
single cell level. Single gametes and in vitro produced zygotes can be
manipulated by means of electrofusion, electroporation, and microinjection. Gametes and artificial zygotes might be promising target cells for
transformation and as a useful model for molecular developmental studies.
The transmission of alien cytoplasm through the fertilization process may
provide an experimental approach in studies of cytoplasmic inheritance.

Because of their haploid nature and high fusion potential, sperm cells are highly suitable for cell reconstruction by fusion with somatic protoplasts or cytoplasts.

In this chapter we describe the methods and discuss the prospects and limitations of the technique of electrofusion-mediated IVF with single plant gametes with the goal of stimulating interest and offering suggestions for further research in this field. We also briefly review intergametic recognition and gamete fusion in animals to provide a context for evaluating the potential of intergametic recognition in plants as well as for developing a model of IVF involving the mixing of male and female gametes of higher plants without facilitation by any chemical or physical device.

II. Recognition and Gamete Fusion in Animals and Plants

In animals intergametic recognition leading to membrane fusion has been studied by the means of IVF systems. In flowering plants limited data are available on the occurrence of an intergametic recognition. Cytological and genetic studies in *Plumbago zeylanica* (Russell, 1985) and *Zea mays* (Roman, 1948) suggested that gametes might possess cell surface determinants involved in the fertilization process. To study recognition and gamete fusion in plants, an *in vitro* model of intergametic fusion must be developed. Recognition and fusion studies require a fusion model that is close to the natural reproduction system occurring *in vivo*. This model can be proposed from successful experiments in animals, especially mammals.

A. *In Vitro* Fertilization (IVF) in Animals

In the sea urchin fertilization is external, therefore, IVF may be easily reproduced in sea water with numerous male and female gametes that are produced in the medium.

In mammals fertilization occurs within the oviductal lumen, a complex and undefined biological environment. Nevertheless, fertilization and embryo development are feasible *in vitro* (Menezo *et al.*, 1984). The IVF method contrasts in a number of important respects with the *in vivo* fertilization process:

1. The fertilization process *in vivo* is continuous and involves numerous interactions between the male spermatozoa and the female reproductive tract (Yanagimachi, 1988). The IVF process is discontinuous: gametes have to be isolated and introduced into a chemically defined medium

containing compounds similar to secretions of the female genital tract (Menezo, 1985).

2. *In vivo* mammalian spermatozoa are selected within the tract, and very few sperm cells are found at the fertilization site. *In vitro* one oocyte is added to numerous spermatozoa, which have not been screened. The degeneration of the supernumerary spermatozoa leads to the necessary modification of the medium composition (Menezo, 1985).

3. *In vivo* mammalian spermatozoa are coated by epididymal and accessory gland secretions during maturation in the epididymis and during ejaculation. This coating phenomenon inhibits or masks the ability of spermatozoa to fertilize the egg (Sidhu and Guraya, 1989); spermatozoa recover this ability *in vivo* during passage through the female tract. *In vitro* sequential treatments to alter the coating must be performed to fully capacitate sperm (Menezo, 1985). In 1963 Yanagimachi and Chang first reported that mammalian spermatozoa could be capacitated *in vitro*.

There is not a single medium that can support IVF in different species. The media commonly used today for *in vitro* capacitation and fertilization are modifications of Tyrode's and Krebs-Ringer's solution supplemented with proper energy sources (glucose, lactate, pyruvate) and albumin (Yanagimachi, 1988; Hammitt *et al.*, 1990). Serum albumin may play some physiological role in the alteration of the sperm membrane components (Oura and Toshimori, 1990).

B. Sperm Capacitation in Animals and Plants

In mammals sperm capacitation is considered to involve a series of incompletely understood molecular changes that prepare the spermatozoa for the acrosome reaction and thus make the spermatozoa capable of fertilizing the egg (Yanagimachi, 1988). These changes result in surface alteration, intramembrane alteration, and physiological changes of the sperm.

Capacitation involves the removal of the epididymal and seminal plasma proteins that coat the sperm surface and possibly also an alteration in the glycoproteins of the sperm surface. (Sidhu and Guraya, 1989; Phelps *et al.*, 1990).

The phospholipid composition of the sperm plasma membrane might also change during capacitation (Yanagimachi, 1988). A decrease in the cholesterol–phospholipid ratio is observed in the sperm plasma membrane (Hoshi *et al.*, 1990). Some plasma membrane proteins migrate from the head plasma membrane into the flagellar plasma membrane in capacitated boar sperm; actin might be involved in this translocation (Saxena *et al.*, 1986). Areas also form that are depleted of membrane proteins, have no

or few sterol and anionic lipids, and might become the site of membrane fusion during the acrosome reaction (Sidhu and Guraya, 1989).

Physiological changes accompanying sperm capacitation are expressed in the modification of sperm motility, respiration, and substrate utilization (Sidhu and Guraya, 1989).

Capacitation is a unique phenomenon in mammals and perhaps in relatively few nonmammalian animals. Capacitation may be a result of an evolutionary adaptation of mammalian spermatozoa to internal fertilization (Yanagimachi, 1988).

In plants, in which fertilization occurs deep inside the sporophyte tissues, we have no data on sperm capacitation. Does it occur between pollen tube germination and its discharge into one of the two synergids? In this chapter we describe the cytological and biochemical changes that occur in the pollen tube and the male gametes during pollen tube growth through the gynecial tissues. All of these changes are susceptible to modify the fertilizing ability of the male gametes.

1. Pollen Tube Changes

Proteins are synthesized during pollen germination and early tube growth, utilizing the ribosomes and mRNAs transcribed and accumulated during pollen maturation. Some of these proteins are concerned with pollen germination and tube growth (Mascarenhas, 1990, this volume). During further tube growth *de novo* transcription and translation of genes occur (Frova and Padoani, 1990). It has been demonstrated *in vitro* that calcium ions of the stigma cells or exudate are taken up by germinating pollen grains in *Primula officinalis* and *Ruscus aculeatus* (Bednarska, 1991). Another metabolic change is induced by the heterotrophic growth of the pollen tube in the style. The polysaccharides of the style exudate are used by the pollen tube cytoplasm during its passage through the female tissue (Labarca and Loewus, 1972, 1973).

2. Male Gamete Changes

In barley after pollen activation, considerable changes occur in sperm cell ultrastructure on the stigma surface: after pollination but before the exit of the sperms from the pollen grain, an intimate association is formed between the vegetative nucleus and the sperm. Other postpollination changes are the occurrence of a distinct cell wall around each sperm and the production of a cell extension by both sperm (Mogensen and Wagner, 1987).

A process of cytoplasmic and organelle elimination occurs during sperm maturation in barley. The number of mitochondria per sperm cell is re-

duced by 50%. Sperm cell surface area and volume are reduced by 30% and 51%, respectively (Mogensen and Rushe, 1985). In spinach an apparent decrease in organelle numbers and male cytoplasm occurs during sperm migration in the pollen tube (Wilms and Leferink-ten Klooster, 1983). In poplar sperm mitochondria appear to become less frequent (Russell *et al.*, 1990). Thus, the amount of sperm cytoplasm declines during pollen maturation and pollen tube growth, but also during gamete fusion in barley (Mogensen, 1988), cotton (Jensen, 1974), and spinach (Wilms, 1981) when sperm cytoplasm is excluded by selective fusion.

Vacuolation of the sperm occurs during pollen tube growth in poplar: the sperm cells appear progressively hypertrophied during their passage in the pollen tube (Russell *et al.*, 1990). Vacuolation of the sperm cell inside the degenerated synergid has also been observed in barley (Mogensen, 1982, 1988) and spinach (Wilms, 1981).

After pollen tube discharge sperm cells lose the inner plasma membrane of the vegetative cell, making them suitable for intergametic fusion. *In vitro* this membrane is lost as soon as the gametic cells are released into the medium from the pollen grain.

All of the postpollination changes previously described and their possible influence on fertilization potential may be important considerations in developing procedures for IVF using isolated male and female gametes without external chemical agents or electrical pulses.

C. *In Vitro* Fertilization (IVF) in Higher Plants by Mixing Male and Female Gametes without Chemical or Physical Devices

To develop this model, large numbers of good quality gametes must be isolated. *Zea mays* L. (maize) emerges as the species of choice for developing a model of IVF in higher plants because of the considerable amount of genetic and cytogenetic information (Sheridan, 1982) and because of the extensive cytological data available on this economically important cereal.

The *Zea mays* pollen grain is tricellular; therefore, the two sperm cells are already present in the vegetative cell. Dupuis *et al.* (1987), Cass and Fabi (1988), and Wagner *et al.* (1989a) have succeeded in isolating quantities of sperm cells from this pollen. The quality and the functional state of the isolated gametes have been assessed by cytological, biochemical, and physiological tests, including transmission electron microscopy and adenosine triphosphate measurements (Dupuis *et al.*, 1987; Wagner *et al.*, 1989a; Roeckel, unpublished results). Functional and cytological integrity of the sperm plasma membrane is a prerequisite for the study of cell

recognition. This integrity was suggested by preliminary measurements of transmembrane currents recorded by patch-clamp experiments in attached cells and by the study of the membrane permeability using the fluorochromatic test (Roeckel, unpublished results). In addition, the membranes of isolated female and male gametes have been proven to be functional by successful electrofusion experiments (Kranz *et al.*, 1990, 1991a,b).

The zygote obtained artificially by electrofusion divides in culture medium (Kranz *et al.*, 1990, 1991a,b).

One question concerning IVF in maize has to be resolved. Are the male gametes released from the shed pollen grain competent to fuse with the female gametes without chemical agents or electrical pulses? *In vivo*, after germination of pollen grains on the silk, the pollen tubes grow into the silk down to the embryo sac in the ovule. It is important to note that the silk may attain 30 cm in length. Sperm capacitation may occur in the silk or in the synergid just after the pollen tube discharges into it. Data obtained by way of *in vitro* pollination and fertilization in maize suggest that male gametes do not require extended maturation during pollen tube elongation to complete fertilization (Dupuis and Dumas, 1989; Kranz and Lörz, 1990; Dupuis, this volume). Indeed, fertilization remains successful *in vitro*, although the silk length required for pollination is six times shorter than that required *in vivo*. Nevertheless, if sperm maturation in the pollen tube is not required in this model of *in vitro* pollination–fertilization, sperm alteration is still possible when the male gametes are discharged into one of the two synergids. Thus, it cannot be ascertained that male gametes isolated from pollen grains are actually capable of fusing with the female gametes.

For the IVF model the male and female isolated gametes should both be incubated in a medium depleted of any fusogenic elements to study cell recognition and fusion. In the laboratory the strategy for the analysis of gametic recognition is based on the construction of a monoclonal antibody library directed against the cell surface of viable gametes. Experiments using fusion inhibition by monoclonal antibodies in this *in vitro* model would provide a functional assay that would allow the examination of specific cell surface determinants involved in gametic recognition or fusion (Chaboud *et al.*, 1991). Whereas many studies have been done on pollen–stigma recognition (Hairing *et al.*, 1990), to date no data have been published concerning intergametic recognition in flowering plants. The only information available on this subject deals with intergametic recognition in algae and animals. In this chapter, our purpose is to introduce strategies used in both algae and animals to study intergametic recognition through the use of an IVF model. Different tools, such as competition assays, effects of monoclonal antibodies and enzymatic digestion, will presumably allow the demonstration of the molecular basis and the specificity of gamete recognition as described next.

D. Intergametic Recognition during the Fertilization Process

In both animals and plants fertilization consists of many steps, including species-specific gamete recognition, mediated by complementary proteins and carbohydrates on opposing gamete surfaces (Macek and Shur, 1988). IVF has been used as a tool to study these gamete interactions, especially in the mouse and the sea urchin.

Mammalian sperm first interacts *in vitro* in a species-specific way with the egg's thick extracellular coat, the zona pellucida, composed of three different glycoproteins in mice, called ZP1, ZP2, and ZP3 (Bleil and Wassarman, 1980a, 1983). *In vitro* competition assays with each glycoprotein of the zona pellucida have been performed in mice. Only sperm exposed to purified egg ZP3 (molecular weight: 83,000) are prevented from binding to the zona pellucida of unfertilized eggs (Bleil and Wassarman, 1980b). In addition, O-linked oligosaccharides of mouse egg ZP3 inhibit IVF in a same manner (Florman and Wassarman, 1985). Thus, the O-linked oligosaccharides of ZP3 have been identified as sperm receptors of the zona pellucida. The sperm receptor on mouse eggs binds specifically to the heads of sperm cells (Bleil and Wassarman, 1986). So it appears that the plasma membrane surrounding the sperm head contains ZP3 binding sites. This sperm mediator of sperm–egg binding is potentially galactosyl transferase, an enzyme recognizing N-acetylglucosamine residues on ZP3 (Lopez *et al.*, 1985).

IVF has also been used to test the effects of monoclonal antibodies and enzymatic digestion on fertilizability of eggs. Monoclonal antibodies to sperm cell membrane components that inhibit fertilization have been produced in numerous animals, such as mice (Sailing and Waibel, 1985; Sailing, 1986) and the guinea pig (Primakoff *et al.*, 1987). They allow the localization and identification of components involved in gamete recognition. Other complementary experiments are needed. In brown marine algae an egg cell is fertilized by small biflagellate sperm cells in a highly species-specific manner. Predigestion of eggs with low concentrations of carbohydrases, α-mannosidase or α-fucosidase, inhibits fertilization (Bolwell *et al.*, 1979; Callow, 1985). Therefore, components containing mannose and fucose on the egg surface may be involved in egg–sperm recognition. Similar results have been obtained in *Chlamydomonas eugametos*, a unicellular green alga. This alga produces isomorphic flagellated gametes of two sexes: mt^+ and mt^- (Van Den Ende *et al.*, 1990). The interaction of gametes of opposite mating types during the fertilization process is mediated by extrinsic glycoproteins of the flagellar membrane called agglutinins. Treatment of agglutinins of mt^+ gametes with α-mannosidase or agglutinins of mt^- gametes with α-galactosidase inhibits IVF, suggesting that mannose and galactose residues are involved in gamete agglutination (Samson *et al.*, 1987).

All these *in vitro* techniques have yielded information regarding the nature of sperm–egg interactions. These techniques also provide information about the specificity of gamete interactions. In the sea urchin gamete recognition during the fertilization process involves species-specific interactions between a sperm glycoprotein (bindin) and highly sulfated proteoglycan-like molecules on the surface of the egg (Rossignol *et al.*, 1984; De Angelis and Glabe, 1987). A fragment of the egg component that specifically inhibits fertilization in competition assays has been obtained by proteolysis and has been purified. Specificity can be abolished by pronase digestion of the fragment, suggesting that the polypeptide backbone of the fragment mediates specific gamete recognition (Foltz and Lennarz, 1990).

The analysis of experimental studies of intergametic recognition and fusion in animals and lower plants allows us to propose the development of an IVF model as previously described. This model is not yet in practice, but IVF by fusion of selected pairs of egg and sperm cells by an electrical pulse is now possible (Kranz *et al.*, 1990, 1991a,b).

III. Micromanipulation of Single Gametes, Synergids, and Central Cells

So far, micromanipulation methods have been used to mechanically isolate embryo sacs and nongametic nuclei from the embryo sacs of cereals (Allington, 1985) and to inject sperm cells into the embryo sac of *Torenia fournieri* (Keijzer *et al.*, 1988). IVF is possible using the techniques of individual, single cell transfer combined with the electrofusion of selected pairs of gametic protoplasts (Kranz *et al.*, 1990, 1991a,b). Pairs of female and male gametes or single male gametes with single nongametic cells of the embryo sac have been fused under microscopic observation and individually transferred into small dishes for growth (Fig. 1). One-to-one protoplast fusion avoids the unwanted fusions that would be generated by mass fusion and facilitates the further analysis of these defined fusion products.

A. Isolation and Selection

Sperm cells of various species have been isolated (Chaboud and Perez, this volume). For individual fusions the maize sperm cells were selected individually after release from the ruptured pollen grains using osmotic shock (Russell, 1986; Matthys-Rochon *et al.*, 1987; Cass and Fabi, 1988;

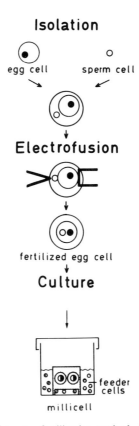

FIG. 1 Electrofusion-mediated *in vitro* fertilization method. After isolation single sperm and egg cells are transferred into the fusion droplet. Pairs of individually selected gametes are fused electrically after dielectrophoretical alignment on one of the electrodes. For culture fusion products are transferred into "millicell-CM" inserts surrounded by a feeder cell suspension. Reproduced with permission from Kranz *et al.* (1991a).

Yang and Zhou, 1989). They are then transferred immediately into the fusion medium.

The isolation of viable egg cells is more difficult. Isolation of viable eggs has been reported using embryo sacs of *Nicotiana* (Hu *et al.*, 1985), *Torenia* (Mól, 1986), *Zea* (Wagner *et al.*, 1988, 1989b), *Plumbago* (Huang and Russell, 1989), and *Petunia* (Van Went and Kwee, 1990). Egg cells of maize can be isolated by enzyme treatment using pectinase, pectolyase, hemicellulase, and cellulase, followed by mechanical isolation (Kranz *et al.*, 1990). Other enzymes used for isolating embryo sacs include driselase (Hu *et al.*, 1985; Van Went and Kwee, 1990), snailase (Hu *et al.*, 1985; Zhou, 1987; Zhou and Yang, 1985, 1986), macerozyme R10 (Mól, 1986),

β-glucuronidase (Huang and Russell, 1989), and cytohelicase (Wagner *et al.*, 1988, 1989b).

The exact time that ovular tissues are exposed to the enzyme incubation medium is important for retaining viability (Wagner *et al.*, 1989b; Huang and Russell, 1989; Kranz *et al.*, 1990). Once the optimal incubation time is determined, the yield of isolated egg cells appears to depend on the manual isolation step (Huang and Russell, 1989; Kranz *et al.*, 1990, 1991a).

Depending on the quality of the plant material, a yield of up to 50% viable egg cells could be obtained using maize. Routinely, five egg cells per 20 ovule pieces could be isolated (Kranz *et al.*, 1991a). Approximately 10% egg cells, 4% lateral cells, and 6% central cells could be isolated using unfixed ovules of *Plumbago zeylanica* (Huang and Russell, 1989).

An important step in the method of single gamete fusion and the culture of fusion products is the transfer of the single cells without damage. A suitable system is that of Koop and Schweiger (1985a) using a computer-controlled dispenser–diluter. The microcapillaries are connected by means of Teflon tubing filled with mineral oil.

B. Electrofusion

It is obvious that IVF using isolated gametes should be performed with selected pairs of gametes. Double fertilization performed *in vitro* requires the fusion of a single sperm cell with a single central cell, specifically, in addition to the artificial zygote. Therefore, to combine two particular cells with adequate control of fusion requires a method based on single-cell techniques. The individual cell fusion method developed for somatic cells (Koop *et al.*, 1983; Koop and Schweiger, 1985b; Schweiger *et al.*, 1987) has been adapted to gametic cells (Kranz *et al.*, 1990, 1991a,b). Fusion of defined pairs of gametes, sperm cells with synergids, and sperm cells with central cells is performed in microdroplets of fusion medium on a coverslip, covered by mineral oil to prevent evaporation (Fig. 1). Controlled electrical fusion is performed with a pair of electrodes that are fixed to an electrode support mounted under the condenser of the inverted microscope. The distance between the electrodes can be adjusted by moving the electrodes along the Z-axis with a computer-controlled stepping motor and a positioning system. The conditions generally used for the electrical fusion of protoplasts of somatic cells may be applied successfully to the gametic protoplasts of maize (Kranz *et al.*, 1990, 1991a) (Fig. 2).

Sperm cells have a high fusion potential. They can be electrically fused with the egg cell without using any cell wall-degrading enzyme treatment prior to fusion. This observation confirms the results of Dupuis *et al.*

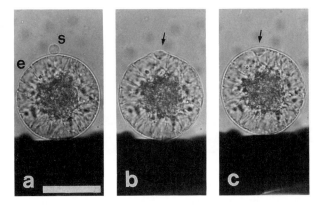

FIG. 2   Electrofusion of a single egg cell (e) with a single sperm cell (s) of *Zea mays*. After application of the DC pulse, the time interval from (a) to (c) was 3 seconds. Arrows indicate the fusion site of the sperm cell with the egg cell. Bar, 50 μm.

(1987), Matthys-Rochon *et al.* (1988), and Cass and Fabi (1988) that the sperm cells in the mature pollen grain of maize are protoplasts. However, when a sperm cell is brought into close contact with another sperm cell or with an egg cell, no spontaneous fusion could be observed. Fusion could only be achieved by an electrical pulse.

The method of IVF mediated by electrical fusion of single gametes can also be used for fusion of individual sperm cells with nongametic cells of the embryo sac. In maize the following partners were successfully fused with single sperm cells: (1) a synergid with adherent egg cell, (2) a synergid with adherent synergid and egg cell, (3) an egg cell with two adherent synergids, (4) an egg cell with adherent synergids and a central cell, and (5) a central cell (Kranz *et al.*, 1990, 1991b). This will facilitate studies of the function of the nongametic cells of the embryo sac at fertilization and postfertilization at the single cell level. It also offers the possibility to accomplish *in vitro* double fertilization by fusions using two sperm cells: one with the egg cell and one with the central cell because the egg cell, synergids, and central cell protoplasts can be isolated as a whole unit and used for fusion.

It is evident from genetic studies that cytoplasmic organelles are inherited maternally in the majority of plants. Electrofusion-mediated IVF might also be useful in studies of cytoplasmic inheritance. Maize sperm cells contain mitochondria (Mogensen *et al.*, 1990); so organelle transfer from the paternal partner is possible *in vitro* because the whole sperm cell is introduced into the egg cell by the electrical pulse. The exclusion of the entire male cytoplasm during *in vivo* gamete fusion is suggested by previ-

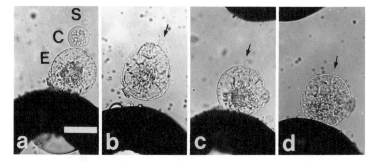

FIG. 3   Electrofusion of a selected egg (E), cytoplast (C), and sperm (S) of *Zea mays*. Time after application of the first DC pulse: (a) 0 seconds. (b) 4 seconds. Arrow indicates fusion site of the sperm cell with the cytoplast, which had already fused with the egg cell. (c) 6 seconds. Immediately before application of the second pulse. (d) 7 seconds. Fusion product of the egg cell, cytoplast, and sperm cell. Bar, 50 μm.

ous electron-microscopic studies on the process of syngamy in barley (Mogensen, 1988). Consequently, the result of *in vivo* fusion might be different from the *in vitro* result.

The method of single cell fusion is a valuable tool for the transmission of cytoplasmic organelles through the fertilization process. Alien cytoplasm has been transmitted using individually selected cytoplasts from maize mesophyll (Kranz *et al.*, 1991b) (Fig. 3). The possibility of transferring plastids and mitochondria into the fertilized egg cell will facilitate experimental studies on cytoplasmic inheritance.

Additionally, sperm cells are suitable for reconstituting cells with different ploidy levels and cytoplasms by fusing them individually with defined protoplasts and cytoplasts, respectively (Kranz *et al.*, 1991b).

IV. Culture of Egg Cells and Zygotes

Cell division and further development of isolated egg cells of higher plants in culture have not been reported so far. Cultivated egg cells of maize show protoplasmic streaming during 22 days of culture, but attempts to induce cell division have failed (Kranz *et al.*, 1990, 1991a). Cultured embryo sac protoplasts of *Torenia* were viable for 2 weeks, but no cell walls formed during culture in a liquid CC medium (Potrykus *et al.*, 1979), supplemented with 10% coconut water and 2 mg/l 2,4-D (Mól, 1986).

In maize when a sperm cell was fused with an egg cell, however, the fusion product started to divide within 2.5–3 days. Multicellular structures

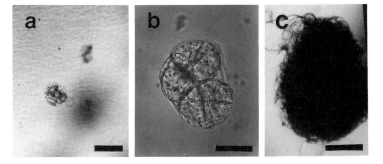

FIG. 4 Development of fusion products of an egg–sperm cell pair of *Zea mays* on the transparent, semipermeable membrane of a "millicell-CM" insert, surrounded by maize feeder cells. (a) First cell division, 2.5 days after fusion. Bar, 100 μm. (b) Multicellular structure 4 days after fusion. Bar, 50 μm. (c) Multicellular structure 18 days after fusion. Bar, 300 μm.

developed at a high frequency (mean frequency: 83%) (Fig. 4). The fusion of sperm cells and egg cells from different lines of maize also resulted in high frequencies of division and the formation of multicellular structures. So far two media have been used for the development of the zygote: (1) a modified MS medium (Murashige and Skoog, 1962), adjusted to 600 mosm/kg H_2O with glucose, and (2) a modified MS medium containing thiamine, L-lysine, L-glutamic acid, L-serine (Gengenbach and Green, 1975), and 0.5 g/l casein hydrolysate, adjusted to 530 mosm/kg H_2O with glucose. Both media contained 1.0 mg/l 2,4-D and 0.02 mg/l kinetin. No difference in the division and growth rate could be observed between these two media (Kranz *et al.*, 1990, 1991a); however, using a feeder-cell system with a maize cell suspension proved beneficial in cultivating single egg cells and fusion products.

The observation that unfertilized egg cells did not divide in culture but fertilized egg cells did divide under comparable experimental conditions may facilitate studies to identify factors involved in inducing division.

V. Concluding Remarks and Further Directions

We have presented in this review a model for IVF using single plant gametes and the subsequent development of their fusion products, the zygotes.

The bottleneck is the limited yield of viable isolated egg cells and nongametic cells of the embryo sac. However, the frequency of fusion and high

cell division rates allow this method to be used routinely, despite the relatively small amount of material that can be isolated. Further studies are directed to achieve sustained growth of multicellular structures and plant regeneration. With a natural competence for differentiation, the artificially created zygotes might be useful for regeneration, contrary to cells of suspension, which often lose their regenerative capacity during prolonged *in vitro* culture. Detailed research on cultured zygotes may reveal specific developmental patterns that are unique to *in vitro* zygotes compared with those generated by *in vivo* conditions; such research may lead to a better understanding of somatic embryogenesis and how it is induced.

Whereas comprehensive information is available on egg and zygote development in animals at the cellular and molecular level, there is a lack of such studies in higher plant egg cells and zygotes, which are more difficult to isolate, handle, and manipulate. Therefore, especially free-living zygotes of algae have been used, mainly to investigate polarization. The zygote of *Fucus* serves as a model system to study cellular polarity, which is formed within 16 hours after fertilization of the former symmetric fertilized egg (Kropf *et al.*, 1989). The possibility of culturing zygotes of higher plant cells individually will be an ideal system to study cell polarization. It is unknown whether the egg cell after *in vitro* fusion is asymmetric or symmetric, or whether it becomes symmetric. It can be assumed that the pear-shaped maize egg cell is asymmetric before isolation from the polar environment of the embryo sac; both its shape and symmetry may be altered by the isolation procedure. The cell polarity of *in vitro* produced zygotes may thus be triggered by external factors that force the undetermined zygote to enter an embryogenic pathway.

Now it has become possible to study the early events of zygote formation *in vitro*, for example, the cell cycle and the isolation and characterization of development-specific genes. Using the polymerase chain reaction method, the amplification of mRNAs from only a few egg cells, sperm cells, and zygotes might become feasible (Li *et al.*, 1988; Belyavsky *et al.*, 1989; Erlich, 1989). It has become possible to handle single gametes and to manipulate them using methods of fusion, microinjection, and electroporation. The possibility of making fusion combinations of sperm cells with central cells or synergids will allow studies of fertilization mechanisms, the effects of nongametic cells on the growth of the artificially produced zygotes, and *in vitro* production of endosperm.

The possible transfer of cytoplasmic organelles via the sperm cell as well as additional contributions by somatic cytoplasts through the *in vitro* fertilization process will facilitate studies on the fate of these organelles during *in vitro* zygote development with respect to cytoplasmic inheritance.

Sperm cells, which possess both high fusion potential and a haploid genome, are attractive partners in performing cell reconstruction–fusion experiments. The possibility of establishing haploid cell lines and of regenerating plants derived from sperm cells and cytoplasts would have scientific and practical importance for plant breeding. Performing IVF using isolated gametes of other genera will demonstrate whether this technique can be used to overcome breeding barriers in applied breeding programs.

Acknowledgments

The authors thank Dr. P. A. Lazzeri and Dr. L. Mogensen for reviewing the manuscript.

References

Allington, P. M. (1985). *In* "The Experimental Manipulation of Ovule Tissues" (G. P. Chapman, S. H. Mantell, and R. M. Daniels, eds.), pp. 39–51. Longman, New York.

Bednarska, E. (1991). *Sex. Plant Reprod.* **4**, 36–38.

Belyavsky, B., Vinogradova, T., and Rajewsky, K. (1989). *Nucleic Acids Res.* **17**, 2919–2932.

Bleil, J. D., and Wassarman, P. M. (1980a). *Dev. Biol.* **76**, 185–203.

Bleil, J. D., and Wassarman, P. M. (1980b). *Cell* **20**, 873–882.

Bleil, J. D., and Wassarman, P. M. (1983). *Dev. Biol.* **95**, 317–324.

Bleil, J. D., and Wassarman, P. M. (1986). *J. Cell Biol.* **102**, 1363–1371.

Bolwell, G. P., Callow, J. A., Callow, M. E., and Evans, L. V. (1979). *J. Cell Sci.* **36**, 19–30.

Callow, J. A. (1985). *J. Cell Sci.* (Suppl. 1: 6th John Innes Symposium), 219–232.

Cass, D. D., and Fabi, G. C. (1988). *Can. J. Bot.* **66**, 819–825.

Chaboud, A., Perez, R., Digonnet, C., and Dumas, C. (1991). *In* Society for Experimental Biology Seminar Series, "Perspectives in Plant Cell Recognition" (J. A. Callow and J. R. Green, eds.), Vol. 48, pp. 59–77. Cambridge University Press, Oxford.

De Angelis, P. L., and Glabe, C. G. (1987). *J. Biol. Chem.* **262**, 13946–13952.

Dupuis, I., and Dumas, C. (1989). *Sex. Plant Reprod.* **2**, 265–269.

Dupuis, I., Roeckel, P., Matthys-Rochon, E., and Dumas, C. (1987). *Plant Physiol.* **85**, 876–878.

Erlich, H. A. (1989). "PCR Technology." Stockton Press, New York.

Florman, H. M., and Wassarman, P. M. (1985). *Cell* **41**, 313–324.

Foltz, K. R., and Lennarz, W. J. (1990). *J. Cell Biol.* **111**, 2951–2959.

Frova, C., and Padoani, G. (1990). *In* "Isozymes: Structure, Function and Use in Biology and Medicine" (Z. I. Ogita and C. L. Markert, eds.), pp. 375–398. Wiley–Liss, New York.

Gengenbach, B. G., and Green, C. E. (1975). *Crop Sci.* **15**, 645–649.

Hairing, V., Gray, J. E., McClure, B. A., Anderson, M. A., and Clarke, A., (1990). *Science* **250**, 937–941.

Hammitt, D. G., Walker, D. L., and Williamson, R. A. (1990). *Int. J. Fertil.* **35**, 46–50.

Hoshi, K., Aita, T., Yanagida, K., Yoshimatsu, N., and Sato, A. (1990). *Hum. Reprod.* **5**, 71–74.

Hu, S. Y., Li, L. G., and Zhou, C. (1985). *Acta Bot. Sin.* **27**, 337–344.

Huang, B. Q., and Russell, S. D. (1989). *Plant Physiol.* **90**, 9–12.

Jensen, W. A. (1974). *In* "Dynamic Aspects of Plant Ultrastructure" (A. W. Robards, ed.), pp. 481–503. McGraw-Hill, New York.

Keijzer, C. J., Reinders, M. C., and Leferink-ten Klooster, H. B. (1988). *In* "Sexual Reproduction in Higher Plants" (M. Cresti, P. Gori, and E. Pacini, eds.), pp. 119–124. Springer, Berlin.

Koop, H. U., and Schweiger, H. G. (1985a). *J. Plant Physiol.* **121**, 245–257.

Koop, H. U., and Schweiger, H. G. (1985b). *Eur. J. Cell Biol.* **39**, 46–49.

Koop, H. U., Dirk, J., Wolff, D., and Schweiger, H. G. (1983). *Cell Biol. Int. Rep.* **7**, 1123–1128.

Kranz, E., and Lörz, H. (1990). *Sex. Plant Reprod.* **3**, 160–169.

Kranz, E., Bautor, J., and Lörz, H. (1990). *In* "Progress in Plant Cellular and Molecular Biology" (H. J. J. Nijkamp, L. H. W. Van der Plas, and J. Van Aartrijk, eds.), pp. 252–257. Kluwer Academic, Dordrecht.

Kranz, E., Bautor, J., and Lörz, H. (1991a). *Sex. Plant Reprod.* **4**, 12–16.

Kranz, E., Bautor, J., and Lörz, H. (1991b). *Sex. Plant Reprod.* **4**, 17–21.

Kropf, D. L., Berge, S. K., and Quatrano, R. S. (1989). *Plant Cell* **1**, 191–200.

Labarca, C., and Loewus, F. (1972). *Plant Physiol.* **50**, 7–14.

Labarca, C., and Loewus, F. (1973). *Plant Physiol.* **52**, 87–92.

Li, H., Gyllensten, U. B., Cui, X., Saiki, R. K., Erlich, H. A., and Arnheim, N. (1988). *Nature* **335**, 414–417.

Lopez, L. C., Bayna, E. M., Litoff, D., Shaper, N. L., Shaper, J. H., and Shur, B. D. (1985). *J. Cell Biol.* **101**, 1501–1510.

Macek, M. B., and Shur, B. D. (1988). *Gamete Res.* **20**, 93–109.

Mascarenhas, J. P. (1990). *Annu. Rev. Plant Physiol. Plant Mol. Biol.* **41**, 317–338.

Matthys-Rochon, E., Vergne, P., Detchepare, S., and Dumas, C. (1987). *Plant Physiol.* **83**, 464–466.

Matthys-Rochon, E., Detchepare, S., Wagner, V., Roeckel, P., and Dumas, C. (1988). *In* "Sexual Reproduction in Higher Plants" (M. Cresti, P. Gori, and E. Pacini, eds.), pp. 245–250. Springer, Berlin.

Menezo, Y. (1985). *Ann. Biol. Clin.* **43**, 27–31.

Menezo, Y., Testart, J., and Perrone, D. (1984). *Fertil. Steril.* **42**, 750–755.

Mogensen, H. L. (1982). *Carlsberg Res. Commun.* **47**, 313–354.

Mogensen, H. L. (1988). *Proc. Natl. Acad. Sci. U.S.A.* **85**, 2594–2597.

Mogensen, H. L., and Rushe, M. L. (1985). *Protoplasma* **128**, 1–23.

Mogensen, H. L., and Wagner, V. T. (1987). *Protoplasma* **138**, 161–172.

Mogensen, H. L., Wagner, V. T., and Dumas, C. (1990). *Protoplasma* **153**, 136–140.

Mól, R. (1986). *Plant Cell Rep.* **3**, 202–206.

Murashige, T., and Skoog, F. (1962). *Physiol. Plant* **15**, 473–497.

Oura, C., and Toshimori, K. (1990). *Int. Rev. Cytol.* **122**, 105–151.

Phelps, B. M., Koppel, D. E., Primakoff, P., and Myles, D. G. (1990). *J. Cell Biol.* **111**, 1839–1857.

Potrykus, I., Harms, C. T., and Lörz, H. (1979). *Theor. Appl. Genet.* **54**, 209–214.

Primakoff, P., Hyatt, H., and Tredick-Kline, J. (1987). *J. Cell Biol.* **104**, 141–149.

Roman, H. (1948). *Proc. Natl. Acad. Sci. U.S.A.* **34**, 36–42.

Rossignol, D. P., Earles, B. J., Decker, G. L., and Lennarz, W. J. (1984). *Dev. Biol.* **104**, 308–321.

Russell, S. D. (1985). *Proc. Natl. Acad. Sci. U.S.A.* **82**, 6129–6132.

Russell, S. D. (1986). *Plant Physiol.* **81**, 317–319.

Russell, S. D., Rougier, M., and Dumas, C. (1990). *Protoplasma* **155**, 153–165.

Sailing, P. (1986). *Dev. Biol.* **117**, 511–519.

Sailing, P., and Waibel, R. (1985). *Biol. Reprod.* **33**, 537–544.

Samson, M. R., Klis, F. M., Homan, W. L., Van Egmond, P., Musgrave, A., and Van Den Ende, H. (1987). *Planta* **170**, 314–321.

Saxena, N., Sharif, S., Saxena, N. K., Russell, L. D., and Peterson, R. N. (1986). *Fed. Am. Soc. Exp. Biol.* **45**, 305.

Schweiger, H. G., Dirk, J., Koop, H. U., Kranz, E., Neuhaus, G., Spangenberg, G., and Wolff, D. (1987). *Theor. Appl. Genet.* **73**, 769–783.

Sheridan, W. F. (1982). "Maize for Biological Research." Plant Molecular Biological Association University Press of North Dakota, Fargo, North Dakota.

Sidhu, K. S., and Guraya, S. S. (1989) *Int. Rev. Cytol.* **118**, 231–280.

Van Den Ende, H., Musgrave, A., and Klis, F. M. (1990). *In* "Ciliary and Flagellar Membranes" (R. A. Bloodgood, ed.), pp. 129–147. Plenum, New York.

Van Went, J. L., and Kwee, H. S. (1990). *Sex. Plant Reprod.* **3**, 252–262.

Wagner, V. T., Song, Y., Matthys-Rochon, E., and Dumas, C. (1988). *In* "Sexual Reproduction in Higher Plants" (M. Cresti, P. Gori, and E. Pacini, eds.), pp. 125–130. Springer, Berlin.

Wagner, V. T., Dumas, C., and Mogensen, H. L. (1989a). *J. Cell Sci.* **93**, 179–184.

Wagner, V. T., Song, Y. C., Matthys-Rochon, E., and Dumas, C. (1989b). *Plant Sci.* **59**, 127–132.

Wilms, H. J. (1981). *Acta Bot. Neerl.* **30**, 101–122.

Wilms, H. J., and Leferink-ten Klooster, H. D. (1983). *In* "Fertilization and Embryogenesis in Ovulated Plants" (O. Erdelská, ed.), pp. 239–240. VEDA, Bratislava, Czechoslovakia.

Yanagimachi, R. (1988). *In* "The Physiology of Reproduction" (E. Knobil and J. Neill, eds.), pp. 135–185. Raven Press, New York.

Yanagimachi, R., and Chang, M. C. (1963). *Nature* **200**, 281–282.

Yang, H. Y., and Zhou, C. (1989). *Chin. J. Bot.* **1**, 80–84.

Zhou, C. (1987). Plant Sci. **52**, 147–151.

Zhou, C., and Yang, H. Y. (1985). *Planta* **165**, 225–231.

Zhou, C., and Yang, H. Y. (1986). *In* "Haploids of Higher Plants *In Vitro*" (H. Hu and H. Y. Yang, eds.), pp. 192–203. Springer, Berlin.

Plant Transformation Using the Sexual Route

Patricia Roeckel,* Maurice M. Moloney,* and Joël R. Drevet†

* Department of Biological Sciences, University of Calgary, Calgary T2N 1N4, Canada
† Faculty of Medicine, Calgary T2N 4N1, Canada

I. Introduction

The ability to generate stably transformed eucaryotic cells is of great importance for applied as well as basic research. At the fundamental level genetic transformation has been used to study the regulation of gene expression. For example, in *Drosophila* it has been possible to demonstrate the role of specific DNA sequences, such as promoters and enhancers, in the regulation of the transcription (Pelham, 1982). In mammals shotgun experiments have allowed the identification of cellular oncogenes (Cooper, 1982). However, in plants the exploitation of transformation techniques has been slower. One explanation is the presence of pectocellulosic cell walls surrounding plant cells, which constitutes an efficient barrier to the penetration of external DNA. Nevertheless, many dicotyledonous species have been transformed, allowing basic plant processes, such as the action of transposable genetic elements (Finnegan *et al.*, 1989) and tissue-specific gene expression (Benfey and Chua, 1989), to be studied. For example, the effect of overproduction of *Agrobacterium* hormone biosynthetic gene products was studied in transgenic petunia plants (Klee *et al.*, 1987). In addition, the introduction of potentially useful agronomic traits has been accomplished in a number of dicotyledonous crop species. For example, tobacco plants resistant to the herbicide glyphosate have been obtained through genetically engineered overproduction of the target enzyme (Comai *et al.*, 1985). Also, tobacco plants resistant to *Manduca sexta* larvae were obtained through the introduction of the gene coding for the *Bacillus thuringiensis* toxin and its consequent expression (Vaeck *et al.*, 1987).

Unfortunately, most monocotyledonous species are recalcitrant to the *Agrobacterium*-mediated transformation system currently used to transform dicotyledonous species. Consequently, direct DNA delivery methods have been and are still being developed for monocotyledonous species. However, plant regeneration from cereal protoplasts is a delicate process. So far rice is the only cereal grain for which transformed protoplasts have been regenerated into fertile transgenic plants (Shimamoto *et al.*, 1989). Recently, Gordon-Kamm *et al.* (1990) regenerated fertile transgenic plants from embryogenic maize suspension cultures after microprojectile bombardment of these cultures. This transformation method, however, involves *in vitro* culture techniques that are time-consuming; therefore, alternative methods involving male and female sexual route will be presented and discussed. Specifically, the alternative methods include the following: (1) the use of microspore systems as a biotechnological tool for plant transformation; (2) the use of pollen grains as vectors of the transforming DNA, which would use the natural process of plant fertilization, therefore avoiding all *in vitro* regeneration steps; and (3) the introduction of exogenous DNA into ovules and ovaries.

II. Microspores

The manipulation of higher plant microspores under tissue culture conditions permits the use of parasexual methods of plant transformation. This route provides some unique benefits but has been limited to date by species- or even genotypic-specificity. The property of cultured anthers which may allow the formation of embryos capable of producing plants via a germination phase (Guha and Maheshwari, 1964, 1966) is critical to this method of gene transfer. Studies on anther culture and embryogenesis have shown that isolated microspores can also give rise to gametophytic embryos (Sunderland and Wicks, 1969, 1971). High frequency embryogenesis from isolated microspores has now been reported for several species, including *Hordeum vulgare* (Fouroughi-Wehr *et al.*, 1976), *Brassica* spp. (Lichter 1982; Fan *et al.*, 1988), *Secale cereale* (Wentzel *et al.*, 1975, 1977), *Saccharum spontaneum* (Hinchee and Fitch, 1984), and a wide variety of *Solanaceae* (Wernicke and Kohlenbach, 1977; Tyagi *et al.*, 1979).

The ontogeny of a gametophytic embryo from a single isolated microspore and the high frequencies of germination or regeneration to whole plants make isolated microspore cultures an interesting target system for transformation. This interest is enhanced by a number of unique properties of microspore-derived embryos. These include the following: (1) high

frequencies of regeneration, even among species that are recalcitrant using normal somatic cell culture (e.g., barley and rye); (2) the production of haploid embryos that may spontaneously double or be treated to generate homozygous doubled haploids (Kuhlmann and Fouroughi-Wehr, 1989); and (3) the minimization of somaclonal variation due to the time of maintenance of the cells in culture.

A. Approaches to Gene Transfer in Microspores and Microspore-Derived Embryos

In considering a potential target for transformation in this system, it might seem obvious to focus on the isolated microspores themselves. This way it should be possible to minimize the production of chimeric transformants with the attendant difficulties this implies.

This approach has not been uniformly followed. It is widely recognized that the reprogramming of microspores to undergo embryogenesis rather than gametogenesis is a delicate and poorly understood process. However, this reprogramming often requires specific treatments, such as elevated temperature (Keller and Armstrong, 1983) or nutritional starvation (Wei *et al*, 1986), to assure high-frequency embryogenesis from the microspores. These treatments place constraints upon the choice of transformation method and may, in some cases, render certain approaches impractical. However, it has been shown that the production of chimeric embryos by transformation of multicellular structures derived from microspores need not be a major problem. Neuhaus *et al.* (1987), working with microspore-derived embryos of *Brassica napus,* recovered large numbers of non-chimeric secondary embryos after microinjection of individual cells in globular embryoids. The exploitation of secondary embryogenesis on microspore-derived embryos therefore may be used to broaden the potential target for transformation to include multicellular units.

1. Agrobacterium-Mediated Gene Transfer

The potential of microspores or microspore-derived proembryos to act as targets for *Agrobacterium*-mediated gene transfer was investigated by Pechan (1989). In that study *Brassica napus* microspores and proembryos were co-cultivated with *Agrobacterium tumefaciens* strain GV3850 harboring the plasmid pCV730, which contains both neomycin phosphotransferases (*NPT*) II (kanamycin resistance) and IV (hygromycin resistance). Co-cultivation was begun after a heat treatment (32°C) for 8 hours (minimal) or for 72 hours (maximal). The duration of co-cultivation was 24 hours at 25°C. After this treatment the cultures were subjected to 32°C for

another 10–12 days after which they were returned to 25°C for 2 days. Selection using hygromycin (5–10 mg/l) or kanamycin (25–30 mg/l) was applied only 16 days after initial culture of the microspores. The numbers of embryos showing resistance to the antibiotics varied according to when the co-cultivation was attempted. The best results were with microspores pretreated at 32°C for 72 hours and then subjected to co-cultivation. Under these conditions almost one-third of the subsequent embryos showed hygromycin resistance, and approximately 15% were kanamycin-resistant. Enzyme assays for *NPT* II on plants regenerated from kanamycin-resistant embryos showed that only about 25% of them had detectable levels of *NPT* II activity. Overall about 7% of all embryos surviving the co-cultivation procedure exhibited *NPT* II activity. Although the efficiency of this procedure is lower than some published methods (Moloney *et al.*, 1989) using vegetative explants, it should be noted that in a culture containing several thousand embryos, this translates into a large number of independent transformation events. Furthermore, it should be possible to obtain homozygous transgenic plants from this procedure either by spontaneous doubling or colchicine treatment. Because the target cell for transformation is haploid, this approach may be of particular interest for homologous recombination experiments in which gene inactivation is sought (Paszkowski *et al.*, 1988).

There are, however, some serious limitations to the use of *Agrobacterium* in gene transfer to microspore-derived embryoids. The most significant of these is the host range of *Agrobacterium* and its apparent incompatibility with many important monocotyledonous plants. The report by Raineri *et al.* (1990) on *Agrobacterium*-mediated rice transformation using zygotic embryos as target tissue suggests that further investigation of *Agrobacterium* transformation of embryonic cells is warranted even among the monocots. More recently, Criessen *et al.* (1990) showed that microspore-derived embryos might be a suitable target for *Agrobacterium*-mediated gene transfer to certain monocots. They employed *Agrobacterium rhizogenes* strain LBA9402 harboring a Bin 19 plasmid containing a tandem dimer of wheat dwarf virus (WDV) as an Agroinfection system (Grimsley *et al.*, 1987). This was used to infect barley microspore-derived embryos. Despite a lack of disease symptoms in the regenerated plants, Criessen *et al.* (1990) showed by Southern blotting the presence of viral DNA in barley plants derived from infected microspore-derived embryos. The pattern of hybridization in these plants indicated the replicative form of WDV and was therefore taken to indicate that T-DNA–mediated DNA transfer had occurred.

2. Direct DNA Delivery to Microspore-Derived Embryos

a. Microinjection   The most complete study of microspore-derived embryos as a target for transformation was performed by Neuhaus *et al.*

(1987) who used microspore-derived embryos of *Brassica napus* and delivered DNA via microinjection. In these experiments globular embryoids of a diameter of approximately 200 μm and containing an average of 12 cells were used. Most of the individual cells of the embryoid can be distinguished at this stage; so it was possible to inject many or all of the members of the embryoid. The microinjection procedure itself took place after the reprogramming treatment of the microspore and thus did not interfere with the most critical steps of the development of the culture. These manipulations had little or no effect on embryo viability or development; regeneration of whole plants was possible from 80% of the injected embryos. Overall transformation frequencies as high as 60% were obtained.

Among the transgenic embryos were numerous chimeric events observed because integration occurred in some, but not all, of the cells of the embryoid. Neuhaus *et al.* (1987) showed that from these primary chimeric regenerants, it was possible to obtain secondary embryos (derived presumably from single cells) that were uniformly transformed. They demonstrated the predicted segregation of marker genes (*NPT* II) in the secondary embryos by performing DNA hybridization analyses on entire families of secondary embryos derived from a single primary regenerant. Integrations that occurred in these experiments all appeared to comprise at least one full-length, functional copy of *NPT* II, although in many cases evidence of additional scrambled copies was also found.

Despite the fairly rigorous technical requirements for the performance of these transformations, the high frequency of transformation and the high frequency of recovery of transgenic plants make this method appealing.

It is noteworthy that the method of Neuhaus *et al.* (1987) contains no species limitation except that the target species has the ability to undergo embryogenesis *in vitro*. As a method for monocot transformation, and in particular for barley, rye, and sugar cane, this procedure may have a number of advantages over the alternative methods currently under investigation.

b. Biolistic DNA Delivery   With the advent of the biolistic particle gun for DNA transfer (Sandford, 1988), there is a species-independent alternative for gene transfer into cells. Embryos, of zygotic or other origin, are obvious targets for such studies.

Numerous laboratories are involved in employing particle bombardment as a method for the production of stable transformants. Most of these studies, at the time of writing, have not yet reached a stage where unequivocal demonstration of stable transformation has been obtained.

The most notable success to date using this technique used embryogenic suspension cultures of maize as a target (Gordon-Kamm *et al.*, 1990); however, other target issues (including gametophytic embryos) should

have equivalent potential using this approach. The work of Gordon-Kamm *et al.* (1990) and others raises an interesting question in relation to selectable markers. Although aminoglycoside antibiotics have proven useful for the selection of transformants in many species, kanamycin has not proven to be a good selection agent in corn, which is the target of much of the work pertinent to this review. Gordon-Kamm *et al.* (1990) showed that the exploitation of a herbicide resistance marker, such as the bialaphos-resistance allele (*bar*, Thompson *et al.*, 1987), was a successful alternative to antibiotic resistance to elevate the efficiency of selection in transformed corn cultures.

The possibility of using pollen or microspores as targets for transformation with biolistic methods has been investigated by a few laboratories (McCabe *et al.*, 1988; Twell *et al.*, 1989; Criessen *et al.*, 1990). Using pollen transformation to disseminate genes through natural fertilization has thus far not produced any transformed plants, although transient expression of the introduced DNA is clearly evident (Twell *et al.*, 1989).

Criessen *et al.* (1990) bombarded microspore-derived embryos of barley with a procedure similar to that used for Agroinfection containing a dimer of the WDV genome. They were able to detect replicated WDV DNA in cells derived from bombarded microspore embryo cultures. However, this work appears limited because the use of an autonomously replicating marker was essential for detection of the introduced sequence and, as is generally the case for Agroinfection experiments, no evidence for stable integration was obtained.

To date, no unequivocal success has been reported using the biolistic method with microspore embryo cultures as targets. This is primarily a function of the novelty of the technique rather than any specific limitation or incompatibility between the culture and delivery system. It is important to note that biolistic approaches appear to be effective for obtaining transient expression using many cellular targets, but the average rate of stable integration appears to be 10^{-4} to 10^{-3} of the transient expression rate (Sandford, 1990). For this method to become a routine procedure in transformation of plant cells, it may be necessary to incorporate a more sophisticated subcellular targeting ability to the introduced DNA to increase the total DNA reaching the nucleus. Such a targeting capability may indeed be possible given the report by Herrera-Estrella *et al.* (1990) on the vir-D2 gene product of *Agrobacterium*. These workers demonstrated that vir-D2 protein has a specific nuclear-targeting capacity. It was capable of specifying nuclear transport of an lacZ protein expressed as a fusion polypeptide in tobacco plants. Because the vir-D2 protein is found attached to the 5' ends of T strands during *Agrobacterium*-mediated transformation, it appears that it functions to pilot foreign DNA to the nucleus. This report raises the possibility that vir-D2 protein could in part account

for the efficacy of *Agrobacterium* in producing stable transformants and thus the additional possibility that it could be exploited in techniques in which DNA is delivered nonbiologically to target cells.

B. Discussion

Microspores and microspore-derived embryos have a high potential as transformation targets for a number of important food crops for which *Agrobacterium*-mediated transformation is inappropriate. As with many other transformation approaches, the regenerability of the culture system is critical. Unfortunately, the production of vigorous embryo cultures from isolated microspores is still limited to a few species. This is probably not an insurmountable problem but will require continued and painstaking work to devise protocols for several monocot crop plants. Evidence from studies of *Brassica napus,* which has undergone the most detailed investigations to date, indicate that the ability to respond to a particular reprogramming protocol is a function of genotype (Keller *et al.,* 1987). It is, therefore, possible that certain genotypes of a given species may be incapable of responding or that new induction schemes for microspore embryogenesis will need to be identified for specific genotypes.

Of the procedures used for the transformation of microspore-derived embryo cultures, the most efficient to date appears to be microinjection into globular embryos (Neuhaus *et al.,* 1987), which is species-independent in theory. It is critical that other laboratories equipped to perform these procedures confirm the results of Neuhaus *et al.* (1987) and, if possible, extrapolate them to other systems. The major drawback of this approach is the delicate manipulation required and the expensive equipment involved. It is the view of the authors that the biolistic approach will probably supercede microinjection as the preferred method for embryo transformation because of its simpler experimental setup and the increasing availability of the required equipment. For many species the inefficiency of this procedure may be tolerable if large numbers of viable target embryos are easily obtained. Ultimately, however, attention will need to be directed to the efficiency of stable transformation so that more recalcitrant culture systems than *Brassica* and *Hordeum* may be routinely used as transformation targets.

III. Pollen Grains as Vectors of Transforming DNA

Microgametophytes can be seen as natural vectors of foreign gene material. Tupý *et al.* (1980) have shown that products of DNA hydrolysis are

used in the synthesis of tobacco pollen DNA. According to Huang *et al.* (1983), tubular virus particles 25 × 190 nm in length were detected through the cytoplasm of pollen grains in *Paeonia emodi*. In 1985 Huang and Kokko reported the infection of alfalfa pollen by *Verticillum albo-atrum* preferentially entering through the pollen aperture.

Transformation using pollen as vector for the transforming DNA was proposed in the early 1970s (Hess, 1975). Germinating pollen was incubated with exogenous DNA and then used for pollination and fertilization of the species from which the pollen has been collected. This postulated transformation system would use the normal fertilization cycle of plants. Consequently, no *in vitro* methodology and no regeneration steps are required; thus, somaclonal variation is avoided. A review article analyzing all the attempts to produce transgenic plants via pollen grains published before 1987 was presented in a previous volume of this book (Hess, 1987). Some phenotypically transformed progenies were obtained by several authors (Hess, 1975; Dewet *et al.*, 1985; Ohta, 1986). However, no molecular proof of gene transfer was achieved.

In this chapter we focus attention on the more recent reports, all of which concerned agriculturally important cereals. This orientation is justified by the apparent host limitations of *Agrobacterium*-mediated transformation and the relative difficulties encountered in obtaining transgenic plants from cereals using alternative procedures.

A. Strategies for Transformation

Several transformation procedures have been investigated to transform plants via pollen grains. Initially, we wish to discuss the general procedures of transformation and the putative pathways that the transforming DNA follows for each particular method. Then we will critically appraise the results obtained using these different approaches.

1. Direct DNA Transfer

The procedure developed by Dewet *et al.* (1985) consisted of collecting and germinating pollen grains from *Zea mays* on a thin layer of pollen germination medium. After 3 minutes, when 10% of the pollen grains germinated, DNA solutions were mixed with the pollen. The resulting mixture was then used to pollinate maize ears. Pollen in suspension was transferred to stigmas with a Pasteur pipette. One day before pollination silks were cut back to the tip of the cob and allowed to recover. The authors specified the amount of pollen and volume of liquid media they

used. They also described how they avoided contamination by foreign pollen grains before and after the pollination experiments.

In an alternative method, instead of adding the DNA to germinating pollen grains, *Zea mays* pollen was also germinated in the presence of DNA. The mixture was then placed onto the silks (Waldron, 1987).

In other studies the pollen was not germinated before being placed onto the silks. Ohta (1986) suspended the transforming DNA in 0.3 *M* sucrose. Maize pollen grains were immersed in the DNA solution and used immediately for pollination. A similar method was described by Roeckel *et al.* (1988b) for *Zea mays*. The pollen was immersed in different liquid media containing the transforming DNA. Classical pollen germination medium (Brewbaker and Kwack, 1963) was supplemented in some of the experiments with polyethylene glycol (PEG) and/or ethylenediaminetetraacetic acid (EDTA). The mixture pollen-DNA-liquid media was placed immediately onto the silks.

All the preceding approaches involved *in vivo* pollination in which the pollen mixed with the DNA solution was used to pollinate the ears of maize plants. Body *et al.* (1989) used such a mixture of pollen and DNA in an *in vitro* pollination strategy. The pollen was not pregerminated before pollination.

Besides the factor of pregermination, the aforementioned strategies are similar. They are direct methods of transformation. These strategies imply two possible ways by which the DNA may be taken up. The DNA could penetrate via the tip of the growing pollen tubes. This hypothesis is based on the observation that the pollen tube at germination lacks a cell wall near its tip (Picton and Steer, 1982). The DNA could also penetrate into pollen grains during the leaky phase. It has been shown that macromolecules could pass through the intine immediately after arrival on the stigma (Heslop-Harrison, 1979). No evidence of DNA uptake through the pore of the growing tubes or through the intine has been demonstrated. On the contrary, negative results have been obtained. Heslop-Harrison and Heslop-Harrison (1988) reported on the uptake of the high molecular weight substances FITC-dextran (average molecular weight, 19,400) and FITC-albumin (average molecular weight, 67,000). Neither FITC-dextran nor FITC-albumin entered the pollen wall or vegetative cells of *Helleborus foetidus* or *Galanthus nivalis*. The authors checked the penetration of these substances using both ungerminated and germinated pollen. In addition, by means of microincubation of single maize pollen grain, Kranz and Lörz (1990) studied the local uptake of defined amounts of dyes and fluorescent substances into the pollen grain, tube, and tube tip. Particularly, the authors used FITC-dextran with an average molecular weight of 71,600 and 14,8900, and a plasmid DNA of 6.7 kb labeled with ethidium bromide. When an aqueous solution of FITC-dextran was added locally to the exine and tube, fluorescence microscopy indicated that it did not

penetrate. This lack of penetration was observed when pollen grains were hydrated or partially dehydrated, germinated or not, and in the presence or absence of 20% dimethylsulfoxide or Triton X-100. Concerning these last experiments, we must point out the fact that starch grains are the most important storage compound in corn pollen grains and they represent 22.4% of the dry weight of pollen, according to Stanley and Linskens (1974). Thus, the observation of a fluorescent probe through these abundant starch grains is rather difficult, especially if a limited number of probe or DNA molecules succeed in penetrating.

2. Pollen Tube Pathway and Agrobacterium-Mediated Transformation

Other strategies of transformation do not involve the precedent putative routes for DNA penetration but are based on a different hypothesis.

Luo and Wu (1988) presented a method for the transformation of rice via the pollen tube pathway. The stigma of recently pollinated florets (2 hours) were excised just above the ovary, and a drop of plasmid solution was applied to the cut end of the style. According to the authors, the DNA presumably reached the ovule by flowing down the pollen tube. The DNA could also be guided along the space between the pollen tube and the transmitting tissue toward the ovule. The problem of callose plugs, which seal off the route through the pollen tube during its growth, must be considered because these plugs could be an obstacle to the passage of DNA molecules. We can suppose that the stigma incision is located just after the last plug formed. The alternative that DNA is guided between the tube wall and the transmitting tissue is, from our point of view, less probable. Indeed, the hypothesis of the DNA moving inside the pollen tube could be accepted because of the flux of the grain contents in the direction of the style, but the fundamental question that remains is how the DNA could move between the tube and the transmitting tissue.

Hess *et al.* (1990) have experimented with a different method for the transformation of wheat (*Triticum aestivum* L.). This is an indirect system of gene transfer using *Agrobacterium tumefaciens* containing a gene responsible for kanamycin resistance (*NPT* II) in higher plants. *Agrobacterium* was pipetted onto all spikelets when the spikelets in the middle of the ear were close to anthesis. The authors supposed that the transforming DNA is probably transferred from *Agrobacterium* to pollen grains. The hypothesis is based on the following observations. Flavonol glycosides diffuse from wheat pollen into a surrounding aqueous medium and induce the vir region of *Agrobacterium* suspended within it. Previously, Wiermann and Vieth (1983) showed that the outer pollen wall is an important accumulation site for flavonoids. It is questionable whether the inducibility

of the vir region by pollen wall-diffusible substances is sufficient to explain the gene transfer. Messens *et al.* (1990) demonstrated that monocotyledonous cells produce vir-inducing molecules in quantities giving significant levels of induction, but the *Triticum monococcum* suspension culture producing these molecules is resistant to *Agrobacterium* infection. However, pollen walls are different in structure and composition from walls of other parts of the plant. Can these differences render possible the transfer of DNA from *Agrobacterium* to pollen grains?

3. Microinjection and Biolistics

The microinjection of DNA into the pollen tube cytoplasm or the generative nucleus has been suggested as a potential means of transforming whole plants by Flavell and Mathias (1984). Hepher *et al.* (1985) tested 18 species with the goal of selecting a model species for studying pollen microinjection. Pollen grains were germinated on a culture medium, including the DNA-specific fluorochrome 4'6-diamidino-2-phenylindole (DAPI). As a consequence, nuclei within the pollen tubes were visualized. In many species tested, microinjection into the generative nuclei was difficult to accomplish because of the small nuclear size. Moreover, the pollen of important crop species (*Zea mays*) is tricellular at the time of germination, necessitating injection in both sperm cell nuclei.

Microinjections of plasmid DNA in solution through the pore of maize pollen grains have been carried out by Kranz and Lörz (1990). A mean of 26% of such injected grains successfully pollinated stigmas or cultured ear segments, resulting in plant development. The authors are now evaluating the usefulness of pollen as a vector for transformation in maize using this method.

Using the particle bombardment process, Twell *et al.* (1989) introduced the *GUS* gene under the control of a pollen-specific promoter from tomato or the *Ca*MV 35S promoter into tobacco (*Nicotiana tabacum*) pollen. They demonstrated that these genes were transiently expressed in the pollen grains. In addition, it appeared that *GUS*-expressing pollen was able to germinate. If fertilization with this pollen results in plant development, this biolistic method may offer an alternative route to the production of transformed plants. As for all of the other approaches presented, the key question is whether the DNA has to be delivered and integrated into the generative nucleus (bicellular pollen grains) or into the sperm cell nucleus that will fuse with the egg cell (tricellular pollen grains).

B. Results and Controls of Transformation Experiments

The results of the transformation experiments described herein are presented in Table 1.

TABLE I

Results of Transformation Experiments via Pollen Grains

Reference	Species	Percentage (%) of transformation	No. of F_1 plants screened
Dewet *et al.* (1985)	Maize	0.03	6303
Ohta (1986)	Maize	3.2	1086
Waldron (1987)	Maize	0	1500
Luo and Wu (1988)	Rice	20	54
Roeckel *et al.* (1988b)	Maize	0	1805(1) + 1723(2)
Body *et al.* (1989)	Maize	0	830
Hess *et al.* (1990)	Wheat	1.035	2415

1. Direct Method of Transfer

Dewet *et al.* (1985) and Ohta (1986) obtained some putative transformants supported only by phenotypic evidence. Particularly, Dewet *et al.* (1985) used total genomic DNA from a rust-resistant line of maize with red cobs. The inbred used as a recipient is susceptible to common rust and has white cobs. Rust resistance is dominant over susceptibility and is due to a single inherited trait, whereas red cobs are incompletely dominant over white cobs; 1.23% of the germinated kernels showed a transforming phenotype. Hess (1987), after analyzing these experiments, did not exclude contamination by foreign pollen to explain the results obtained by Dewet *et al.* (1985) because plants were selfed under field conditions. We must emphasize that no molecular controls of the transformation have been performed. The same criticism can be applied to Ohta's results (1986). This author used total genomic DNA from 1-month-old plants of maize possessing all homozygous dominant alleles of a few marker genes with endosperm characters. The recipient strain possessed all recessive alleles of the same marker genes. The highest frequency of phenotypically transformed endosperm was 9.29% per ear.

2. Pollen Tube Pathway

Luo and Wu (1988) gave both phenotypic and molecular controls of transformation. The transforming DNA they used was the plasmid p 35S *NPT* II with a gene coding for the *NPT* under the control of the promoter *Ca*MV 35S. The percentage of transformation reported was up to 20%. They checked the transformation by both Southern blot analysis and enzymatic assays. According to Potrykus (1990), the available data are equivocal and do not include a Southern blot carrying the required signals to prove integrative transformation, such as the linkage of the foreign gene to host

DNA borders. However, the Southern blot that was presented showed digestion of the genomic DNA by *Xba*I, which does not cut into the transforming DNA. Resulting from this digest, different hybridization patterns were found with different transgenic plants. These results suggested that the foreign gene has been integrated into the rice genome at different locations in different individual transgenic plants. Unfortunately, two hybridization signals situated between 21 and 12 kb are shown on undigested DNA patterns corresponding to putative transgenic plants. If the transforming DNA were integrated in the genome of these plants, we would expect a unique hybridization signal located in a high molecular weight region of the gel.

3. Agrobacterium-Mediated Transformation

Hess *et al.* (1990) presented transformation frequencies of 1–2.6% in wheat. The authors used a transforming DNA carrying a gene conferring kanamycin resistance to the plants, under the control of the promoter nopaline synthase (NOS). The Southern blot presented for F_1 plants shows one band of 6.5–8.5 kb, depending upon the plants tested. In some of the investigated plants, an additional signal at 2 kb is also shown. Actually, following the restriction digests performed, the expected size of the hybridization signal was 3 kb. It is unfortunate that no explanation was given concerning the appearance of the 2-kb hybridization signal. However, the authors explained the presence of a large 8-kb fragment as representing the loss of one *Hin*dIII site, either by rearrangement of the transferred T-DNA or by methylation of this site. To prove integrative transformation the authors showed Southern blot analyses of F_2 plants obtained. The 8-kb signal previously detected in F_1 following *Eco*RI and *Hin*dIII cleavage was still detected in the F_2 generation, but the 2-kb fragment was lost. It is unfortunate that the lanes corresponding to undigested DNA showed hybridization signal extending from a high molecular weight to 2.5 kb, probably due to DNA degradation. Otherwise this control would have totally eliminated the possibility of contamination or nonintegrative transformation events. Another point discussed by the authors concerned the absence of *NPT* II activity in part of the F_2. This might be the result of gene rearrangements or of methylation of the transforming DNA. The level of expression of *NPT* II gene could also be influenced by the ploidy level of the transformed plants (Ren and Lelley, 1989).

As a conclusion on this approach, some points remain to be clarified, such as the appearance of unexplained hybridization fragments (2 kb) and the absence of a tight correlation between Southern blot analyses and enzyme assay data. The reason for the disappearance of the *Hin*dIII cleavage on the left of the *NPT* II gene should be clearly demonstrated by

sequencing this region. This is an important control because all the plants analyzed by Southern blot showed the unexpectedly large fragment that was hypothetically generated by the lack of this particular *Hin*dIII cleavage. However, these results are promising concerning the possibility to transform plants via the male sexual route.

The other articles listed in Table I presented negative results. After screening a significant number of plants and after performing many types of controls (e.g., enzymatic assays, Southern blot analyses), no transformed plants were obtained (Waldron, 1987; Roeckel *et al.*, 1988b).

A common criticism should be formulated concerning the choice of reporter gene in most of the transformation experiments (Table I). The transforming DNA was usually the gene responsible for kanamycin resistance in higher plants. Actually, natural resistance to kanamycin has been seen in various graminaceous monocots (Hauptmann *et al.*, 1988), and false-positive plants may be obtained. For further attempts to transform monocot species via pollen grains the use of other selectable markers, such as *bar* gene, responsible for resistance to the herbicidal compound phosphinothricin (Gordon-Kamm *et al.*, 1990), should be evaluated. In addition, the promoter *CaM*V 35S appears be preferable to the NOS promoter used by Hess *et al.* (1990); as suggested by the authors themselves, the *CaM*V 35S promoter has been proven to work efficiently in some graminaceous monocots (Fromm *et al.*, 1986).

C. Pollen and Stigma Quality and Protocol of Pollination

1. Nuclease Activities

Among the studies discussed herein (i.e., direct transfer, pollen tube pathway), no results interpretable as gene transfer were obtained. The failure to obtain transformants was probably due to DNA degradation by nucleases. Some studies have been conducted concerning pollen nucleases in different species. According to Matousek and Tupý (1985), these diffusible nucleases may protect pollen and ovary genomes against foreign genetic information. They have shown that nucleases diffuse out of pollen during the first minutes of culture. Roeckel *et al.* (1988b) demonstrated that plasmid DNA and genomic maize DNA were degraded in less than 1 minute by pollen diffusates. This degradation occurred in classical pollen liquid germination medium (Brewbaker and Kwack, 1963) but was even increased when the medium was supplemented with 10–40 mM EDTA. Because EDTA is able to chelate divalent cations, the authors supposed that divalent cations could inactivate maize pollen nucleases. It appeared that magnesium could inhibit diffusible pollen nucleases in maize (Roeckel

et al., 1988b). In contrast, Negrutiu *et al.* (1986) found that pollen *Nicotiana* nuclease activities were inhibited by 10 mM EDTA. Body *et al.* (1989) tried to remove maize pollen nucleases by washing the pollen 5 or 10 minutes before mixing it with the transforming DNA, but DNA was still completely degraded after such treatment.

One should keep in mind that during pollination the solution of DNA is in contact with both pollen and stigma. As a consequence, both pollen and stigma nucleases can diffuse in the germination liquid media and degrade the transforming DNA. To our knowledge, in the context of pollen-mediated transformation, the only work taking into account both pollen and stigma nuclease activities is the one presented by Roeckel *et al.* (1988b). These authors found that a classic germination medium supplemented by 300 or 600 mM, or 20% PEG, is able to inhibit both pollen and stigma nuclease activities in maize. Moreover, they demonstrated that such media allowed fertilization. Unfortunately, the authors have not yet published the results of transformation assays performed under the optimal conditions they have determined.

The approach described by Hess *et al.* (1990) overcame the problem of DNA degradation by nucleases by using *Agrobacterium*-mediated transformation. This is also the case when the DNA penetrates instantaneously into pollen grains by using the biolistic method. The DNA was not completely degraded as shown by the transient expression of the foreign DNA into tomato pollen grains (Twell *et al.*, 1989).

In addition to the evaluation of nucleases, it is important to check pollen and stigma quality before transformation experiments because the efficiency of plant fertilization depends upon both male and female viability. Although numerous viability tests are available to estimate pollen quality (Stanley and Linskens, 1974; Knox, 1984; Knox *et al.*, 1986), in most of the attempts to transform plants via pollen grains pollen quality has not been checked. Fresh pollen was used, but no parameters, such as the germination *in vitro*, the ability of the membranes to retain the fluorescein (fluorochromatic reaction test, Heslop-Harrison and Heslop-Harrison, 1970), or the water content for maize pollen (Kerhoas *et al.*, 1987), were reported for the pollen samples used in transformation experiments. Under these conditions it is difficult to repeat the experiments and interpret the results obtained.

Checking stigma quality is more difficult, but it is feasible to specify, in the case of maize, for example, the day of pollination according to the number of days after silking and the length of the protruding silks.

2. Pollination Protocol

The pollination protocol is not always well detailed. For example, Ohta (1986) did not give the weight of pollen that they mixed into 200 μl of DNA

solution prior to pollination. The only indicator is that the resulting mix was pasty. Furthermore, the method of application of the mixture pollen–DNA onto the silks was not adequately described. Hess (1987) previously pointed out the importance of the pollination method for transformation experiments. Pollen tubes can be destroyed by the procedure used to concentrate the pollen (filtration and scraping from Millipore filter) or by the application (spraying using an air brush). Thus, if the DNA penetrates into pollen grains through growing tubes (Section III,A,1), the transformation frequency could be greatly diminished when excessively harsh methods of pollination or pollen preparation are used.

D. Conclusions

Until now, promising results have been obtained (Hess *et al.*, 1990, Twell *et al.*, 1989), but definite proof of plant transformation via pollen grains is still lacking. In the previous section some points were presented concerning the improvement of transformation experiments via pollen grains.

1. Pollen and stigma quality should be rigorously checked and defined to be able to repeat the experiments under similar conditions.

2. The transforming DNA should produce an easily detectable phenotype, and its presence should be checked at the molecular level. The *NPT* II gene responsible for the kanamycin resistance in higher plants does not seem to be universally appropriate because false-positives have occurred in maize and other monocotyledonous species (Hauptmann *et al.*, 1988).

3. For direct transformation experiments involving the application of pollen mixed with a solution of DNA onto the stigma, both pollen and stigma nuclease activities should be inhibited or eliminated.

4. The pollination procedure must be gentle enough to avoid changes in viability or pollen tube destruction.

5. If some plants show a transformed phenotype, integrative transformation must be proven. The requirements for such proof have been listed by Potrykus (1990).

For the moment many points remain to be elucidated. In the case of direct transformation experiments (Section III,A,1), we do not know how DNA penetrates into pollen grains, nor do we even know whether it does (pollen tube pathway experiments). The fate of the DNA within pollen grains is also unknown. The DNA could be degraded by cytoplasmic nucleases or it could penetrate into the sperm cells (tricellular pollen) or into the generative cell (bicellular pollen). If the DNA does not penetrate into the male gametes, it is unclear how the DNA can go through the synergids without being degraded during the fertilization process.

An alternative route for transforming plants may be the use of isolated sperm cells from pollen grains. By using isolated sperm cells instead of whole pollen grains, the target is directly accessible to the DNA because all of the customary barriers, such as the pollen wall with its diffusible nucleases, are eliminated. Sperm cells have been isolated from different species, such as *Plumbago zeylanica* (Russell, 1986) and *Zea mays* (Dupuis *et al.*, 1987; Cass and Fabi, 1988; Roeckel *et al.*, 1990). In certain species (e.g., maize) it has been shown that isolated sperm cells are surrounded only by a plasma membrane and no cell wall can be detected. Moreover, they have been shown to be viable according to fluorochromatic reaction test (Roeckel *et al.*, 1988a) or exclusion of Evans blue (Russell, 1986). Thus, isolated sperm cells can be seen as viable haploid protoplasts capable of being transformed similar to other diploid protoplasts (Negrutiu *et al.*, 1987). Ultimately, transformed sperm cells could be fused with isolated female gametes to form a transformed zygote. Electrofusion-mediated *in vitro* fertilization of maize using single sperm and egg cells was performed by Kranz *et al.* (1991). The development of a whole plant from such artificially produced zygotes has yet to be shown.

IV. Transformation *in Situ* of Female Sexual Cells

Natural fertilization in transformation experiments can also be approached by introducing exogenous DNA into the ovaries at the time of fertilization.

Such attempts have been reported. Hess (1972) was the first to demonstrate that radiolabeled exogenous DNA could be transferred into the egg cell using *Petunia,* presumably via the pollen tube, when injected 12 hours after the latter has reached the ovary. However, analysis of mutant progeny has shown that the partially restored character (anthocyanin synthesis) was lost after the F_1 generation.

Since then, other methods have been considered for the treatment of the ovaries. For example, microinjection experiments into ovaries were found to be unsuccessful in barley and *Salpiglossis sinuata* (Steinbiss *et al.*, 1985; Hepher *et al.*, 1985). However, positive results were reported by Zhou (1985) using cotton as recipient in which a variety of DNAs (from other cotton species, maize, or *Abutilon avicaenae*) were injected into the axile placentae of large flowers after the receptor has selfed. However, analysis of the progeny has revealed great instability and variability, and no molecular proof of gene transfer has been produced by the authors. Typical are the surprising results obtained in the restoration of wilt resistance via injection of wild-type DNA into ovaries of a *G. hirsutum* wilt-susceptible line, which has shown a high frequency of resistant individuals,

even after multiple generations. Unfortunately, restoration of wilt resistance (of unknown genetics) gave no molecular evidence for genetic transfer, as would have been possible if experiments had been conducted with a simple known gene model. Nevertheless, those experiments suggested that cotton plants are appropriate recipients for such ovarial microinjection experiments, plants with numerous, accessible ovaries more amenable to this procedure than cereal species, for example, which are uniovulate.

In 1987 De la Peña *et al.* reported successful injection experiments in immature floral meristems of rye 14 days before the occurrence of meiosis. After many failures in large-scale experiments using maize, wheat, barley, rice, and others species as well, it appears that this method is inefficient or inapplicable to a wider variety of monocotyledonous grains.

Thus, genetic manipulation of the ovule seems to be unpredictable. Results, if forthcoming, do not confirm that the *in situ* female sexual route is an efficient or convenient way to transform plants.

V. Conclusions

Sexual cells, either gametes or the gametophytes from which they are derived, have immense potential for genetic transfer. As natural agents for the delivery of DNA from one cell to another, they are obvious targets for the introduction of foreign DNA. The attractiveness of gametes and gametophytes in this respect is enhanced by the possibility of eliminating tissue culture manipulations altogether. A corollary of this is that such sexual cells might provide a general strategy for plant gene manipulation because most higher plants reproduce predominantly via a sexual route.

Approaches to transformation via direct DNA uptake into pollen grains or sperm cells followed by normal pollination have not yet provided a reliable method of stable transformation. Several technical problems remain to be investigated more thoroughly. These problems include the susceptibility of pollen grains to foreign DNA, the effect of diffusible nucleases, the relative merits of different DNA delivery methods, and the quality and the preparation of the pollen grains themselves.

The role of pollen aperture and pollen tubes in the DNA uptake–transfer process also remains obscure. There is conflicting information concerning the permeability of DNA across these barriers. Furthermore, it is not yet clear whether DNA must actually enter a sperm cell to effect a transformation in the subsequent pollination.

At this time no method of DNA delivery can be ruled out. Of the potential methods (DNA soaking, biolistics, *Agrobacterium*, microinjection, electroporation, etc.), none has been rigorously investigated in inde-

pendent laboratories using the same conditions. It must be concluded that much more work will be required to evaluate the potential of any of these systems before one becomes more attractive or advantageous than the others.

The use of sexual cells that have undergone developmental reprogramming, as described for microspore-derived embryos, provides an additional dimension to the potential of these cells to act as transformation targets.

Although direct manipulation of pollen grains or sperm cells has proven difficult, several successes have been reported involving transformation of microspore-derived embryoids. Among the species for which this culture procedure is routine, the approach may prove to be highly advantageous, and its value may be further enhanced by the ease with which homozygous transformants are obtained. The major problem is that there is no uniform procedure for reprogramming cultured microspores and most of the effects are genotype-specific. Until this genotypic limitation is better understood, the broader application of this technology is hampered. Using this approach several transformation methods have been shown to be effective. This raises the hope that if the developmental biology questions are resolved and the technique becomes more widely applicable, it will be possible to tailor one of several methods of transformation to the constraints of tissue culture.

Although the task of using sexual cells as transformation targets is a daunting one and one that will be plagued with many negative results, there are strong arguments for encouraging this kind of research. Success in one of these approaches for a grain monocotyledon will be a major step forward in our ability to manipulate food crops. However, equally important is the wealth of information emerging about male and female gametophytes of higher plants and the characterization of developmental switches that are revealed through the investigations.

References

Benfey, P. N., and Chua, N. H. (1989). *Science* **244**, 174–181.

Body, G., Krens, F. A., and Huizing, H. J. (1989). *J. Plant Physiol.* **135**, 319–324.

Brewbaker, J. L., and Kwack, B. H. (1963). *Am. J. Bot.* **50**, 859–865.

Cass, D. D., and Fabi, G. C. (1988). *Can. J. Bot.* **66**, 819–825.

Comai, L., Gacciotti, D., Hiatt, W. R., Thompson, G., Rose, R. E., and Stalker, D. (1985). *Nature* **317**, 741–744.

Cooper, G. (1982). *Science* **217**, 8010.

Criessen, G., Smith, C., Francis, R., Reynolds, H., and Mullineaux, P. (1990). *Plant Cell Rep.* **8**, 680–683.

De la Peña, A., Lörz, H., and Schell, J. (1987). *Nature* **325**, 274–276.

Dewet, J. M. J., Bergquist, R. R., Haplan, J. R., Brink, D. E., Cohen, C. E., Newell, C. A., and Dewet, A. E. (1985). *In* "Experimental Manipulation of Ovule Tissues" (G. P. Chapman, S. H. Mantell, and W. Daniels, eds.), pp. 197–209. Longman, New York.

Dupuis, I., Roeckel, P., Matthys-Rochon, E., and Dumas, C. (1987). *Plant Physiol.* **85,** 876–878.

Fan, Z., Armstrong, K. C., and Keller, W. A. (1988). *Protoplasma* **147,** 191–199.

Finnegan, E. J., Taylor, B. H., Craig, S., and Dennis, E. S. (1989). *Plant Cell* **1,** 757–764.

Flavell, R., and Mathias, R. (1984). *Nature* **307,** 108–109.

Fouroughi-Wehr, B., Mix, G., Gaul, M., and Wilson, H. M. (1976). *Z. Pflantzenzücht.* **77,** 198–204.

Fromm, M., Taylor, L. P., and Walbot, V. (1986). *Nature* **319,** 791–793.

Gordon-Kamm, W. J., Spencer, T. M., Mangano, M. L., Adams, T. R., Daines, R. J., Start, W. G., O'Brien, J. V., Chambers, S. A., Adams, W. R., Willets, N. G., Rice, T. B., Mackey, C. J., Krueger, R. W., Kausch, A. P., and Lemaux, P. G. (1990). *Plant Cell* **2,** 603–618.

Grimsley, N., Hohn, T., Davies, J. W., and Hohn, B. (1987). *Nature* **325,** 177–179.

Guha, S., and Maheshwari, S. C. (1964). *Nature* **204,** 497–499.

Guha, S., and Maheshwari, S. C. (1966). *Nature* **212,** 97–100.

Hauptmann, R. M., Vasil, V., Ozias-Akins, P., Tabaeizadeh, Z., Rogers, S. G., Fraley, R. T., Horsch, R. B., and Vasil, I. K. (1988). *Plant Physiol.* **86,** 602–606.

Hepher, A., Sherman, A., Gates, P., and Boulter, D. (1985). *In* "Experimental Manipulation of Ovule Tissues" (G. P. Chapman, S. H. Mantell, and W. Daniels, eds.), pp. 52–63. Longman, New York.

Herrera-Estrella, A., Van Montagu, M., and Want, K. (1990). *Proc. Natl. Acad. Sci. U.S.A.* **87,** 9534–9537.

Heslop-Harrison, J. (1979). *Ann. Bot.* **44**(Suppl.), 1–47.

Heslop-Harrison, J., and Heslop-Harrison, Y. (1970). *Stain Technol.* **45,** 115–120.

Heslop-Harrison, J., and Heslop-Harrison, Y. (1988). *Sex. Plant Reprod.* **1,** 65–73.

Hess, D. (1972). *Z. Pflanzenphysiol.* **66,** 155–166.

Hess, D. (1975). *In* "Genetic Manipulations with Plant Material" (L. Ledoux, ed.), pp. 519–537. Plenum, New York.

Hess, D. (1987). *Int. Rev. Cytol.* **107,** 367–395.

Hess, D., Dressler, K., and Nimmrichter, R. (1990). *Plant Sci.* **72,** 233–244.

Hinchee, M. A. W., and Fitch, M. M. M. (1984). *Z. Pflantzenphysiol.* **113,** 305–311.

Huang, B., Hills, J., and Sunderland, N. (1983). *J. Exp. Bot.* **147,** 1392–1398.

Huang, H. C., and Kokko, E. G. (1985). *Phytopathology* **75,** 859–865.

Keller, W. A., and Armstrong, K. C. (1983). *Euphytica* **32,** 151–159.

Keller, W. A., Armison, P. G., and Cardy, B. J. (1987). "Haploids from Gametophytic Cells in Plant Tissue and Cell Culture" (C. E. Green, D. A. Somers, W. P. Hackett, and D. D. Biesboer, eds.), pp. 223–241. Liss, New York.

Kerhoas, C., Gay, G., and Dumas, C. (1987). *Planta* **17,** 1–10.

Klee, H. J., Horsch, R. B., Hinchee, M. A., Hein, M. B., and Hoffmann, N. L. (1987). *Gene Dev.* **1,** 86–96.

Knox, R. B. (1984). *In* "Embryology of Angiosperms" (B. M. Johri, ed.), pp. 1–98. Springer-Verlag, Berlin.

Knox, R. B., Williams, E. G., and Dumas, C. (1986). *In* "Plant Breeding Review" (J. Janick, ed.), Vol. IV, pp. 9–79. Avi, Westport.

Kranz, E., and Lörz, H. (1990). *Sex. Plant Reprod.* **3,** 160–169.

Kranz, E., Bautor, J., and Lörz, H. (1991). *Sex. Plant Reprod.* **4,** 12–16.

Kuhlmann, U., and Fouroughi-Wehr, B. (1989). *Plant Cell Rep.* **8,** 78–81.

Lichter, R. (1982). *Z. Pflantzenzücht.* **105,** 427–434.

Luo, Z. X., and Wu, R. (1988). *Plant Mol. Biol. Rep.* **6**, 165–174.

Matousek, J., and Tupý, J. (1985). *J. Plant Physiol.* **119**, 169–178.

McCabe, D. E., Swain, W. F., and Martinell, B. J. (1988). "Pollen-Mediated Plant Transformation," Eur. Patent Appl. No. 87310612.4, 1988.

Messens, E., Dekeyser, R., and Stachel, S. E. (1990). *Proc. Natl. Acad. Sci. U.S.A.* **87**, 4368–4372.

Moloney, M. M., Walker, J. M., and Sharma, K. K. (1989). *Plant Cell Rep.* **8**, 238–242.

Negrutiu, I., Heberle-Bors, E., and Potrykus, I. (1986). In "Biotechnology and Ecology of Pollen" (D. L. Mulcahy, G. B. Mulcahy, and E. Ottaviano, eds.), pp. 65–70. Springer-Verlag, New York.

Negrutiu, I., Shillito, R., Potrykus, I., Biasini, G., and Sala, F. (1987). *Plant. Mol. Biol.* **8**, 363–373.

Neuhaus, G., Spangenberg, G., Mittelstein-Scheid, O., and Schweiger, H. G. (1987). *Theor. Appl. Genet.* **75**, 30–36.

Ohta, Y. (1986). *Proc. Natl. Acad. Sci. U.S.A.* **83**, 715–719.

Paszkowski, J., Baur, M., Bogucki, A., and Potrykus, I. (1988). *EMBO J.* **7**, 4021–4026.

Pechan, P. M. (1989). *Plant Cell Rep.* **8**, 387–390.

Pelham, H. (1982). *Cell* **30**, 517.

Picton, J. M., and Steer, M. W. (1982). *J. Theor. Biol.* **98**, 15–20.

Potrykus, I. (1990). *Physiol. Plant* **79**, 125–134.

Raineri, D. M., Bottino, P., Gordon, M. P., and Nester, E. W. (1990). *BioTechnology* **8**, 33–38.

Ren, Z. L., and Lelley, T. (1989). *Theor. Appl. Genet.* **77**, 742–748.

Roeckel, P., Dupuis, I., Detchepare, S., Matthys-Rochon, E., and Dumas, C. (1988a). In "Plant Sperm Cells as Tools for Biotechnology" (H. J. Wilms and C. J. Keijzer, eds.), pp. 105–110. Pudoc, Wageningen.

Roeckel, P., Heizmann, P., Dubois, M., and Dumas, C. (1988b). *Sex. Plant Reprod.* **1**, 156–163.

Roeckel, P., Chaboud, A., Matthys-Rochon, E., Russell, S., and Dumas, C. (1990). In "Microspores: Ontology and Evolution" (S. Blackmore and R. B. Knox, eds.), pp. 281–307. Academic Press, London.

Russell, S. D. (1986). *Plant Physiol.* **81**, 317–319.

Sandford, J. (1988). *Trends Biotechnol.* **6**, 299–302.

Sandford, J, (1990). *Physiol. Plant* **79**, 206–209.

Shimamoto, K., Terada, R., Izawa, T., and Fujimoto, H. (1989). *Nature* **338**, 274–276.

Stanley, R. G., and Linskens, H. F. (1974). "Pollen: Biology, Biochemistry and Management." Springer-Verlag, Berlin.

Steinbiss, H. H., Stabel, P., Toepfer, R., Hirtz, R. D., and Schell, J. (1985). In "Experimental Manipulation of Ovule Tissues" (G. P. Chapman, S. H. Mantell, and W. Daniels, eds.), pp. 64–75. Longman, New York.

Sunderland, N., and Wicks, F. M. (1969). *Nature* **224**, 1227–1229.

Sunderland, N., and Wicks, F. M. (1971). *J. Exp. Bot.* **22**, 213–221.

Thompson, C. J., Movva, N. R., Tizard, R., Crameri, R., Davies, J. E., Lauwereys, M., and Botterman, J. (1987). *EMBO J.* **6**, 2519–2523.

Tupý, J., Hrâbetová, E., and Cápková-Balatkova, V. (1980). *Physiol. Veg.* **18**, 677–687.

Twell, D., Klein, T. M., Fromm, M. E., and McCormick, S. (1989). *Plant Physiol.* **91**, 1270–1274.

Tyagi, A. K., Rashid, A., and Maheshwari, S. C. (1979). *Protoplasma* **99**, 11–19.

Vaeck, M., Reynaerts, A., Höfte, H., Jansens, S., De Beuckeleer, M., Dean, C., Zabeau, M., Van Montagu, M., and Leemans, J. (1987). *Nature* **328**, 33–37.

Waldron, J. C. (1987). *Maize Genet. Coop. News Lett.* **61**, 36–37.

Wei, Z. M., Kyo, M., and Harada, H. (1986). *Theor. Appl. Genet.* **72**, 252–255.
Wentzel, G., Hoffman, F., Potrykus, F., and Thomas, E. (1975). *Mol. Gen. Genet.* **138**, 293–299.
Wentzel, G., Hoffman, F., and Thomas, E. (1977). *Theor. Appl. Genet.* **51**, 81–88.
Wernicke, W., and Kohlenbach, H. W. (1977). *Z. Pflantzenphysiol.* **81**, 330–337.
Wiermann, R., and Vieth, K. (1983). *Protoplasma* **119**, 230–233.
Zhou, G. Y. (1985). *In* "Experimental Manipulation of Ovule Tissues" (G. P. Chapman, S. H. Mantell, and W. Daniels, eds.), pp. 240–250. Longman, New York.

Part V
Self-Incompatibility between Pollen and Stigma

Gametophytic Self-Incompatibility: Biochemical, Molecular Genetic, and Evolutionary Aspects

Anuradha Singh and Teh-Hui Kao

Department of Molecular and Cell Biology, Pennsylvania State University, University Park, Pennsylvania 16802

I. Introduction

In the latter half of the Nineteenth Century, Charles Darwin embarked on a study of the "extreme sensitiveness and delicate affinities of the reproductive system" of plants (Darwin, 1876). He noted, in a volume entitled "The Effects of Cross and Self Fertilization in the Vegetable Kingdom," that plants of some species are "completely sterile to their own pollen, but fertile with that of any other individual of the same species." Later, the term "self-incompatibility" (Stout, 1917) was used to describe this phenomenon, defined as a prezygotic barrier to self-fertilization in plants that otherwise produce fully functional gametes (Lundqvist, 1964).

About 100 years after the publication of Darwin's treatise, nearly all that was then known about the genetic, cytological, and physiological aspects of self-incompatibility in flowering plants was gathered into a definitive monograph by De Nettancourt (1977). In the closing chapter of his monograph, De Nettancourt predicted that the time would come when the biochemist, having cloned the genes controlling self-incompatibility, would have the means to characterize the function of the gene products, "and to reproduce *in vitro* . . . the sequence of events constituting the incompatibility reaction." Indeed, a significant breakthrough was made in the mid-1980s with the cloning of cDNAs encoding the pistil proteins involved in self-incompatibility in *Brassica oleracea* (Nasrallah *et al.*, 1985) and *Nicotiana alata* (Anderson *et al.*, 1986). Since then, a flurry of activity by an ever-increasing number of laboratories has resulted in a substantial increase in our understanding of the molecular and biochemical bases of self-incompatibility. Although De Nettancourt's prediction has

not yet been borne out entirely, the path has been laid for its eventual realization.

This chapter will attempt to provide an up-to-date review of recent advances in the study of gametophytic self-incompatibility. The aforementioned monograph by De Nettancourt (1977) is a comprehensive treatise on earlier work on self-incompatibility in flowering plants and provides background information for this review. A number of other reviews focusing on physiological, cytological, or molecular genetic aspects of gametophytic self-incompatibility may also be referred to (Heslop-Harrison, 1983; Linskens and Heslop-Harrison, 1984; Sussex *et al.*, 1985; Mulcahy *et al.*, 1986; Gaude and Dumas, 1987; Cornish *et al.*, 1988; Ebert *et al.*, 1989; Haring *et al.*, 1990).

II. Reproductive Barriers

A. Definitions

Two types of reproductive barriers are found to operate in plants: interspecific and intraspecific (Fig. 1). Interspecific reproductive barriers ensure the stability of each species, whereas intraspecific barriers permit a reasonable degree of variability within species. Interspecific reproductive barriers are a consequence of evolutionary divergence between species, and the failure to cross-fertilize is presumed to lie in the inability of pistil tissue

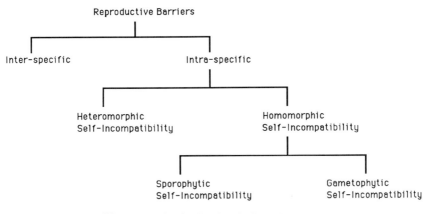

FIG. 1   Reproductive barriers in flowering plants.

to sustain the growth of pollen from other taxa, a phenomenon described as incongruity (Hogenboom, 1975).

Intraspecific barriers in plants are regulated by genetically determined self-recognition–self-rejection mechanisms, and this discrimination between self and nonself is exercised against pollen of the same taxon. Some authors have suggested that there may be mechanistic parallels between the two types of reproductive barriers (Lewis and Crowe, 1958), but in the absence of sufficient information about the genetic and biochemical bases of interspecific reproductive barriers, it would be prudent not to work from such an assumption.

Self-incompatibility is an intraspecific reproductive barrier. It is a system of cell-to-cell recognition that allows the pistil to recognize and reject pollen from genetically related individuals, thereby promoting outbreeding and heterozygosity in the population. Pollen that is not genetically related at the self-incompatibility locus (loci) is allowed to grow and effect fertilization. Plants have evolved various stratagems to circumvent the tendency toward self-fertilization that is created by the close proximity of male and female parts in a "perfect" flower. These include, for example, developmental mechanisms that ensure temporal separation of the sexes (protandry and protogyny) or spatial separation of the sexes (dichogamy and dioecy). However, the biochemically specified self-incompatibility system offers the most effective means of preventing inbreeding and homozygosity in flowering plants. This is evident from the prevalence of self-incompatibility: it has been reported in more than half of the angiosperm families (Brewbaker, 1959) and is estimated to occur in more than 3000 species of plants (East, 1940).

Self-incompatibility results in the failure of fertilization by viable pollen, that is, pollen that is fertile in crosses with individuals that are not related with respect to the self-incompatibilty locus (loci). This definition, as proposed by Hogenboom (1975), precludes postzygotic events that lead to an inviable zygote, which may constitute a delayed form of self-incompatibility, but are difficult to distinguish from embryo lethality. The term pseudo–self-compatibility has been used to describe individuals or lines of self-compatible plants identified in species that are generally self-incompatible (De Nettancourt, 1977; Ascher, 1984). However, it is not clear whether pseudo–self-compatibility is caused by genetic mechanisms different from those responsible for self-compatibility in species belonging to families, such as the *Solanaceae,* that tend to have large numbers of self-incompatible taxa. Until the genetic lesion responsible for the loss of self-incompatibility can be identified, it is perhaps more accurate to describe these cultivars simply as self-compatible. The reversible breakdown of self-incompatibility as a result of environmental conditions, such as

high temperature and humidity, or end-of-season effects, (Linskens, 1975) is known as physiological self-compatibility.

B. Types of Self-Incompatibility Systems

Two main types of self-incompatibility systems are recognized: heteromorphic and homomorphic (Fig. 1). In the heteromorphic type, plants of the same species produce more than one flower morphology and crosses are possible only between flowers of different morphological types. Species with homomorphic self-incompatibility produce only one morphological flower type.

Homomorphic self-incompatibility may be of the sporophytic or gametophytic type, depending on the genetic basis of the interaction between the pollen (gametophyte) and the pistil (sporophytic tissue). In gametophytic self-incompatibility (GSI) the genotypes of the haploid pollen (gametophyte) and the diploid pistil tissue determine whether a pollination will be compatible or incompatible. In sporophytic self-incompatibility (SSI) the outcome of pollen–pistil interactions is dictated by the genotypes of the female plant and the diploid pollen donor.

III. Genetics of Gametophytic Self-Incompatibility

A. Genetic Basis

GSI is a common outbreeding mechanism, occurring in more than 60 families of angiosperm. In the simplest case, GSI is determined by a single locus, called the S locus, which is highly polymorphic. For instance, Emerson identified 37 alleles in a population of 1000 plants of *Oenothera organensis* (Emerson, 1938, 1939). Thus, the S locus is likely the most polymorphic locus in plants. GSI in grasses is known to be controlled by two loci, designated S and Z; more complex systems involving three or four loci have been described in members of the *Ranunculaceae* and *Chenopodiaceae* (Lundqvist, 1975). To date, only the one-locus system has been extensively characterized at the molecular level.

In the one-locus system pollen bearing an S allele that is identical to one of the two S alleles of the pistil fails to effect fertilization. The outcome of pollen and pistil interactions in various S allelic combinations is illustrated in Figure 2. In the *Solanaceae,* the best studied of the families displaying GSI, incompatible pollen germinates normally on the surface of the stigma,

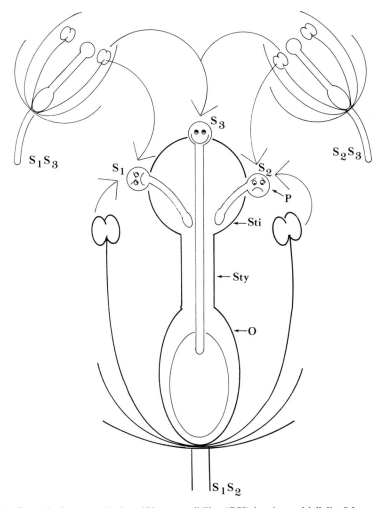

FIG. 2 Control of gametophytic self-incompatibility (GSI) by the multiallelic S locus. The figure shows the outcome of pollen–pistil interactions between a flower of S_1S_2 genotype (female parent) and pollen from the same flower or flowers of S_1S_3 or S_2S_3 genotype. Pollen that bears the S_1 or S_2 allele is incompatible on the S_1S_2 pistil. Pollen bearing the S_3 allele is compatible because it does not share a common S allele with the recipient pistil. Compatible pollen is able to grow into the ovary and effect fertilization, whereas the growth of incompatible pollen is arrested in the style. O, ovary; Sty, style; Sti, stigma; P, pollen.

but the growth of pollen tubes in the style is arrested before they traverse greater than half the length of the style.

B. Genetic Breakdown

1. *S* Locus Mutants: Pollen-Part and Style-Part Self-Compatible Mutants

GSI is more amenable to genetic analysis of self-compatible mutations than SSI because the self-incompatibility phenotype of pollen is determined by the haploid genome of each pollen grain itself, rather than by the genotype of the pollen mother plant as in SSI. Self-compatible mutants have been obtained by irradiation or treatment with chemical mutagens. Two types of self-compatible mutants resulting from mutations at the *S* locus have been identified: pollen-part and style-part (De Nettancourt, 1977). In pollen-part mutants pollen behavior in the incompatibility interaction is affected, but not stylar behavior. For example, if the genotype of a plant is S_1S_2, and $S_1^{PP}S_2$ denotes the genotype of a pollen-part mutant with S_1 being the affected allele, then pollen carrying the S_1^{PP} allele is not rejected by the wild-type S_1 allele in the pistil, whereas pollen carrying the wild-type S_1 allele is rejected by the S_1^{PP}-bearing pistil of the mutant. Similarly, a style-part mutation (i.e., S_1^{SP}) only affects the behavior of the pistil carrying the mutant *S* allele, but not the behavior of pollen carrying it.

On the basis of these two types of *S* locus mutants, Lewis proposed a tripartite structure for the *S* locus (Lewis, 1949, 1960). He suggested that the *S* locus is composed of a complex of three closely linked parts: the *S* allele specificity part, which determines the allelic specificity of pollen and pistil; the pollen activity part, which activates the *S* allele specificity of pollen; and the style activity part, which activates the *S* allele specificity of the style. Mutations in the pollen activity part affect pollen behavior without affecting stylar behavior in the self-incompatibility interaction, resulting in pollen-part mutants. Mutations in the style activity part affect stylar behavior only, resulting in style part mutants. The argument for pollen and style having a common *S* allele-specificity part is that when a new allele specificity arises, it must become operative in both pollen and style or there would be a loss of self-incompatibility.

The three parts of the *S* locus are postulated to be closely linked because of the failure to obtain recombinants between *S* allele specificity and pollen- and style-part mutations (De Nettancourt, 1977). Additional support was recently obtained from a study of pollen-part and style-part mutants identified in *Nicotiana alata* plants regenerated from anther and microspore cultures (Kheyr-Pour and Pernes, 1986). The mutations were

found to affect the S_{F11} allele. A glycoprotein of 27 kDa was found to be associated with the S_{F11} allele as well as the two mutant alleles $S_{F11}{}^{PP}$ (pollen-part) and $S_{F11}{}^{SP}$ (style-part). Taking advantage of the association of the 27 kDa protein with both mutant alleles, Kheyr-Pour and Pernes (1986) carried out a series of crosses to test the tripartite structure hypothesis by examining whether mutations in the pollen or style activity part of the S_{F11} allele could be transferred to a test allele, S_3, through recombination. Segregation of the S_3 allele in more than 500 plants did not reveal any $S_3{}^{PP}$ or $S_3{}^{SP}$ mutants. Thus, there appears to be no evidence of recombination between $S_{F11}{}^{PP}$ or $S_{F11}{}^{SP}$ and S_3, indicating a close linkage of the style-part and pollen-part elements to the S allele specificity part.

Although the tripartite structure hypothesis remains the most useful and credible working hypothesis for the organization of the S locus, it reveals very little about the structure of the S locus at the molecular level. According to this model, there could be one S gene, the protein product of which consists of three functional domains, or there could be two separate S genes, one encoding the pollen activity and the other encoding the style activity part, with both sharing a common S allele specificity part, or there could be three separate S genes, one for each of the three parts of the S locus.

One conceivable mechanistic interpretation of the tripartite structure of the S locus is presented in Figure 3. This model assumes that each of the three parts of the S locus constitutes a different gene and attempts to explain how the S locus gene products might interact to bring about S allele-specific recognition, if an identical S specificity gene is expressed in both pollen and pistil. The central tenet of this model is that pollen-part and style-part gene products activate the S specificity gene products of pollen and pistil or otherwise mediate their interaction. The products of the pollen-part and style-part genes could function either as molecular chaperones that order the pistil and pollen S specificity product into a requisite unique configuration, or they could act as complementary presenting molecules that are required for the interaction of the pistil and pollen S specificity products (Fig. 3). In the event of an incompatible pollination, the interaction of identical S specificity products would result in cellular events leading to the inhibition of the pollen. These events would fail to occur in a compatible pollination because the pollen- and style-specific gene products would present dissimilar S specificity proteins. A mutation in any component of this system would result in the inability to recognize and inhibit self-pollen. Various permutations of this model as well as other models could be devised to explain pollen-part and style-part mutants. In later sections we will weigh the evidence for and against the notion of identical S allele-specific sequences being expressed in pollen and pistil.

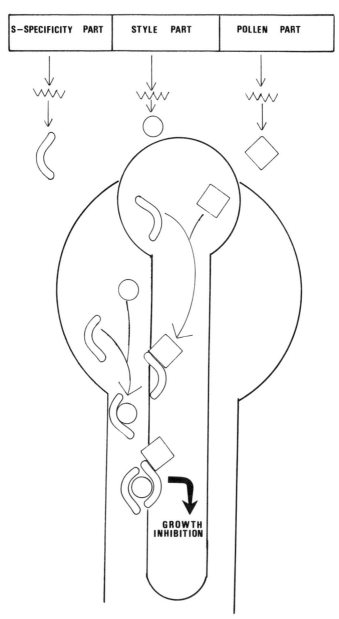

FIG. 3 Hypothesis of tripartite structure of S locus organization. This figure illustrates the tripartite model of the S locus envisaged by Lewis (1949, 1960) and extends the hypothesis to consider the interrelationship of S locus gene products. The model shown here assumes that each part of the S locus encodes a different protein, and self–nonself recognition results from the interaction of an S specificity part expressed in both pollen and pistil. The interaction

2. Self-Compatibility due to Mutations outside the *S* Locus

Although the recognition of self and nonself in GSI is known to be controlled, in the simplest case, by a single multiallelic locus, results from
numerous classical genetic studies suggest a complex, polygenic regulation
of self-incompatibility. The first line of evidence came from analysis of the
inheritance of self-incompatibility in hybrids between self-incompatible
species and closely related self-compatible species. Introgression of functional *S* alleles into a self-compatible genetic background was found
to be insufficient for conferring self-incompatibility to the compatible
species, suggesting the involvement of other genes in the control of self-
incompatibility (Mather, 1943; Martin, 1961, 1968). Taking advantage of
the identification of two *S* alleles and their associated proteins in a self-
compatible cultivar of *Petunia hybrida,* the inheritance of self-incompatibility in F_1 hybrids between *Petunia hybrida* and its self-incompatible
relative *Petunia inflata* was reexamined (Ai *et al.,* 1991). Again, the results
strongly suggest that functional *S* alleles are necessary but not sufficient
to elicit a self-incompatibility response.

The second line of evidence came from analysis of the self-incompatibility behavior of progeny obtained from crossing self-incompatible and self-
compatible lines of the same species. The results also imply the existence
of unlinked genes, which can affect the strength of self-incompatibility
(Townsend, 1969; Takahashi, 1973; Flaschenriem and Ascher, 1979a,b).
Although the nature of these modifier genes has not been determined, their
gene products are thought to act as "switches" for the expression of self-
incompatibility or as modulators (activators or suppressors) of the activity
of *S* allele products. It is therefore necessary to keep in mind the complex
genetic control of a fully functional self-incompatibility system when for-
mulating models to explain the molecular basis of GSI.

IV. Cellular and Physiological Aspects

Angiosperm pollen is a product of reduction division in the pollen mother
cell, and ensuing mitotic divisions result in a mature grain with two or

---

of the S specificity proteins is conditioned by the products of the pollen and style activity
parts. The pollen and style activity proteins may function as complementary presenting
proteins, perhaps as receptor and ligand. When identical S specificity products are involved
in the interaction, it leads to a chain of events culminating in the inhibition of pollen tube
growth.

three nuclei. In trinucleate pollen one nucleus constitutes a vegetative nucleus, whereas the other two function as sperm nuclei. Binucleate pollen bears, in addition to the vegetative nucleus, a single generative nucleus that divides in the pollen tube to give rise to two sperm nuclei during growth in the pistil.

In an extensive survey of diverse genera, Brewbaker (1959) observed a correlation of GSI with species that shed binucleate pollen and SSI with species that have trinucleate pollen. However, a number of departures from this norm have been found in species with bifactorial GSI. For instance, although monocot species, including the grasses, display GSI, most (though not all) shed pollen at the trinucleate stage.

Pollen tubes reach the ovary through the transmitting tissue, a specialized tract of cells that is continuous with the stigmatic tissue, and forms the central core of the style. The transmitting tissue consists of long, fusiform cells with plasmodesmatal connections across their end walls and a thick pectinaceous middle lamella (Herrero and Dickinson, 1979). Pollen tubes are believed to hydrolyze this pectinaceous intercellular matrix as they traverse the length of the style.

In many species the growth of pollen tubes in the pistil is biphasic. In *Petunia* compatible and incompatible tubes grow at the same rate during the first phase of growth (6–9 hours after pollination). Thereafter, compatible tubes show a marked acceleration in growth, while the rate of growth of incompatible tubes decreases over the next few hours until complete arrest (Herrero and Dickinson, 1981). The inhibition of self-pollen tubes occurs at the time of mitotic division in the pollen tube (Brewbaker and Majumder, 1961), when the generative nucleus gives rise to two sperm nuclei. The generative nucleus fails to divide in incompatible tubes (Pacini, 1981). This stage coincides with the point at which pollen tubes are thought to switch from autotrophic to heterotrophic growth (Rosen and Gawlik, 1966; Van der Donk, 1974). The concept of an initial autonomous phase of growth stems from the observation that pollen grains have large amounts of stored RNA and that pollen tubes are capable of growing to a considerable extent *in vitro* in the presence of transcription inhibitors (Mascarenhas et al., 1984). The idea is reinforced by experiments in which injection of inhibitors of RNA synthesis into the style caused compatible tubes to cease growth at the transition point where incompatible tubes are normally found to be inhibited (Ascher, 1972).

This description of progamic development led to the suggestion of the complementary model for the inhibition of incompatible pollen, predicated on the inability of such pollen to acquire precursors necessary for sustained growth (Bateman, 1952; Van der Donk, 1974). The oppositional model, in contrast, invokes the active inhibition of self-pollen by products of the *S* locus (De Nettancourt, 1977; Lewis, 1979). At the present time, the latter

model finds favor with most workers in this field (Lawrence *et al.*, 1985), although elements of the first model remain attractive as a means of explaining the well-documented polygenic influences in a functional self-incompatibility system (Mulcahy and Mulcahy, 1983, 1985).

The modulation of self-incompatibility by developmental and environmental factors has been noted for some time. High temperatures and low light have been shown to weaken or overcome self-incompatibility, and a loss of incompatibility at the end of the growing season has also been noted in several species (Linskens, 1975). These effects further underscore the facultative nature of the expression of the self-incompatibility phenotype.

Incompatible pollen tubes of *Petunia hybrida* show distinctive morphological features, the most prominent being dilated tips with thickened irregular walls and heavy subapical deposits of callose. They frequently burst open, releasing their contents into the intercellular matrix of the style (Herrero and Dickinson, 1980; De Nettancourt *et al.*, 1973). Several distinctive ultrastructural features of inhibited pollen tubes have been observed (Fig. 4). The most striking of these is the accumulation of concentrically arranged endoplasmic reticulum (ER) configurations in the apical zone of incompatible pollen tubes (De Nettancourt *et al.*, 1973, 1974); the authors noted that such configurations are correlated with inhibition of protein synthesis and are found in such tissues as resting potato tubers and overwintering buds. This observation becomes especially interesting in light of the recently discovered enzymatic activity associated with the *S* allele product in the pistil, discussed later in this review.

V. Biochemical Basis

A. Identification of *S* Allele-Associated Proteins

The strategy generally used for identifying *S* allele products involves finding those pistil- and pollen-specific proteins that co-segregate with their respective *S* alleles in genetic crosses. This method has been used successfully to identify a number of *S* allele–associated proteins (*S* proteins) from pistils of several solanaceous species. However, no *S* allele–associated proteins have yet been found in the pollen, using this or any other approach.

To identify *S* genotypes, diallele crosses are first carried out to determine mating interactions within a population of plants, so that each individual plant is assigned to a group in which all members are mutually incompatible, while members of different groups are cross-compatible. Because

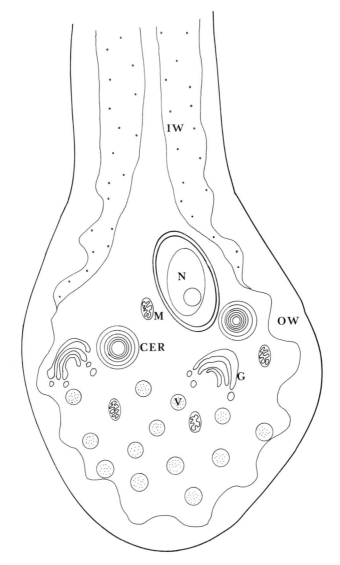

FIG. 4 Schematic representation of the ultrastructure of incompatible pollen tubes inhibited in the style. The features that distinguish them from compatible tubes are as follows: the thickened, convoluted outer wall (OW), the disappearance of the callosic inner wall (IW) at the tip, the accumulation of large numbers of vesicles (V) with granular or fibrillar contents, and the configuration of the endoplasmic reticulum into concentric tubules (CER). Other organelles shown: generative nucleus (N), mitochondria (M), and Golgi apparatus (G). This composite is based on the cytological and ultrastructural studies by De Nettancourt *et al.* (1973) and Herrero and Dickinson (1981).

plants occurring in natural outcrossing populations are generally heterozygous with respect to the S locus, plants in different groups could either have one allele in common or have two different alleles.

The use of homozygotes greatly facilitates identification of the S genotypes of each group of plants and makes it easier to track the co-segregation of pistil proteins with their respective alleles. Plants homozygous for the S alleles can usually be raised by selfing at the bud stage when self-incompatibility is not expressed to a sufficient degree. Homozygotes in the selfed progeny can be identified by backcrosses using the parent as pollen donor; plants that set seed in such backcrosses would be homozygotes. Crosses between homozygotes obtained from the same group or different groups would then allow one to determine the S genotype of each group of plants. Each group is assigned arbitrary S allele designations using numerals or letters (e.g., S_1, S_2, S_3 . . . etc., or S_a, S_b, S_c . . . etc.). Once the S genotypes are established, total stylar protein is extracted and electrophoresed on SDS-polyacrylamide gels. Proteins, the presence or absence of which correlates with S alleles, can then be identified. Figure 5 shows three S proteins of *Petunia inflata* that correlate with their respective S alleles.

The strategy outlined herein has worked well for most of the S alleles studied so far because their protein products are usually abundant and are sufficiently diverse to be separated by SDS-polyacrylamide gels or isoelectric-focusing gels. S proteins have now been identified in the following solanaceous species: *Lycopersicon peruvianum* (Mau *et al.*, 1986), *Nicotiana alata* (Bredemeijer and Blass, 1981; Kheyr-Pour *et al.*, 1990), *Petunia inflata* (Ai *et al.*, 1990), *Petunia hybrida* (Kamboj and Jackson, 1986; Broothaerts *et al.*, 1989), *Solanum chacoense* (Xu *et al.*, 1990a), and *Solanum tuberosum* (Kirch *et al.*, 1989). Their molecular weights range from 23 to 34 kDa. All that have been studied are glycosylated and are basic proteins, with pI higher than 9.0 for some S proteins.

Tan and Jackson (1988) have identified pistil proteins with molecular weight ranging from 43 to 97 kDa and pI ranging from 5 to 7, which differ among various incompatibility genotypes of *Phalaris coerulescens*, a species of grass with bifactorial self-incompatibility. In addition, pistil proteins that may be involved in self-incompatibility have been identified in *Lilium longiflorum* (Dickinson *et al.*, 1982) based on their biochemical properties, and in *Prunus avium* (Mau *et al.*, 1982), *Papaver rhoeas* (Franklin-Tong *et al.*, 1988), and *Malus domestica* (Speranza and Calzoni, 1990), based on their ability to inhibit the growth of self-pollen *in vitro*.

B. Temporal and Tissue-Specific Expression of *S* Alleles

S proteins have not been detected in tissues other than the pistil. Further, the S proteins appear to be localized mostly in the stigma and the upper

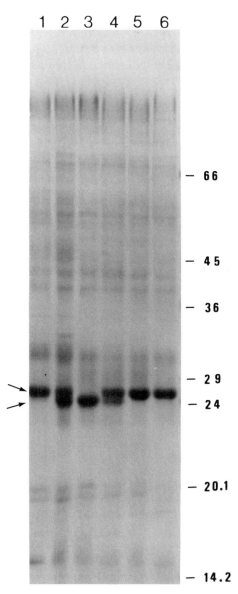

FIG. 5 Co-segregation of *Petunia inflata* S proteins with their respective *S* alleles. Pistil proteins from six different genotypes were analyzed by SDS-PAGE. S_2 and S_3 proteins migrate at 25 kDa (upper arrow) and S_1 protein at 24 kDa (lower arrow). The genotypes of the plants are as follows: lane 1, S_2S_2; lane 2, S_1S_2; lane 3, S_1S_1; lane 4, S_1S_3; lane 5, S_3S_3; lane 6, S_2S_3. Reproduced with permission from Ai *et al.* (1990).

third of the style (Fig. 6), which coincides with the site of pollen tube rejection. In the style S proteins are confined to the transmitting tissue, and lower amounts are also detected in the epidermal layer of the placenta (Cornish *et al.*, 1987; Anderson *et al.*, 1989).

As mentioned in the previous section, self-incompatibility can be overcome by selfing of immature buds because self-incompatibility is not expressed at this stage of floral development. S proteins are generally found at a lower level in immature buds compared with later stages. It is possible that a threshold amount of S proteins is required for the self-incompatibility interaction to occur or that the still uncharacterized modifier genes are not expressed to a sufficient level at the bud stage.

C. *In Vitro* Assays of S Protein Function

The co-segregation of S proteins with their respective *S* alleles suggests that these proteins are the products of the *S* alleles. In addition, the temporal appearance and spatial distribution of S proteins is correlated with the onset of self-incompatibility and the site of pollen tube rejection in the pistil. The claim that S proteins are causally involved in self-incompatibility is further reinforced by the success, albeit to varying degrees, of bioassays in which either purified S proteins or crude extracts containing them are found to inhibit the growth of self-pollen cultured *in vitro*. In a number of species, discrimination between self- and cross-pollen by crude pistil extracts has been demonstrated *in vitro*. However, as pointed out by Jackson and Linskens (1990), few of these assays are entirely satisfactory. In some studies, for example, cross-pollen is also inhibited by the addition of pistil extracts, although self-pollen is more drastically inhibited (e.g., Franklin-Tong *et al.*, 1988). In others an inexplicable effect on pollen germination is seen (Sharma and Shivanna, 1982), which is not congruent with the situation *in situ*.

Jahnen *et al.* (1989) showed that purified S proteins are effective in inhibiting the growth of self-pollen. In this study, some nonspecific inhibitory activity against cross-pollen was also observed, although the growth inhibition is more severe for self-pollen. The inhibitory effect was magnified and rendered totally nonspecific upon heat treatment of the S protein. Heat treatment of the S protein also inhibited its uptake by pollen tubes cultured *in vitro* (Grayr et al., 1991). Broothaerts *et al.* (1991) found that purified *Petunia hybrida* S proteins show highly specific inhibitory activity against self-pollen, reducing pollen tube growth by approximately 32% from that of cross-pollen or pollen cultured in the absence of S proteins. The inhibitory activity of S proteins was heat-sensitive in this case.

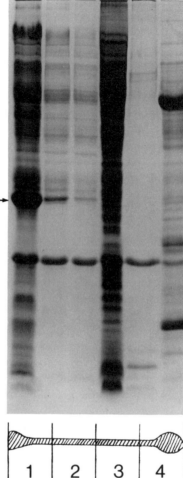

FIG. 6   Tissue-specific distribution of S proteins in *Petunia inflata* of S_2S_2 genotype. Total protein from petal and leaf tissue and from various segments of the pistil were analyzed by SDS-PAGE. The bottom diagram indicates the region of the pistil that the extracts in lanes 1–4 were derived from (stigmatic end is to the left in the diagram). S_2 protein (arrow) is found in the upper segment of the pistil (lane 1), with decreasing amounts in lower sections (lane 2 and 3). It is not detectable in petal (lane 5) and leaf (lane 6). Reproduced with permission from Ai *et al.* (1990).

The failure to reproduce *in vitro* the precise milieu encountered by pollen tubes growing *in situ* may be one element that limits many bioassays. The rates of pollen tube growth *in vitro* are typically 10% or less of the growth rates obtained *in situ,* and linear growth ceases after several hours of culture, generally before the division of the generative cell. For instance, Brewbaker and Majumder (1961) observed that *in vitro* growth of *Petunia inflata* pollen levels off 6–8 hours after germination, and the manifestation of self-incompatibility and the division of the generative cell in compatible pollen tubes *in situ* also occur approximately 6–8 hours after pollination. If these observations represent more than mere coincidence, it would seem that pollen tube growth *in vitro* starts to become limiting just when pollen tubes *in situ* would begin to show a marked response in terms of self-incompatibility.

The inhibitory effect of S proteins against incompatible pollen would be less pronounced if pollen tube growth *in vitro* were suboptimal. Broothaerts *et al.* (1991) succeeded in obtaining *in vitro* pollen tube growth rates that were as high as one-third of the *in situ* growth rates using an improved culture medium containing polyethylene glycol. In this study the observed 32% reduction in self-pollen tube growth could actually have been 96% or higher if *in vitro* growth was more comparable with those obtained *in situ.*

VI. Molecular Characterization of the S Gene

A. Cloning of DNA Encoding S Proteins

The first cDNA clone for an S protein of the GSI system was identified in *Nicotiana alata* by making use of the tissue-specific and developmental expression of the S gene (Anderson *et al.,* 1986). A cDNA library was constructed from mRNA isolated from mature styles of *Nicotiana alata* plants with S_2S_2 genotype and then screened for clones that hybridized to cDNA probes derived from mature stylar mRNA but did not hybridize to cDNA probes derived from immature stylar mRNA or ovary mRNA. cDNA clones encoding S_2 protein were identified from those style-specific cDNA clones by virtue of their ability to hybridize to synthetic oligonucleotides corresponding to the amino-terminal sequence of the S_2 protein.

Differential hybridization was also used in the isolation of cDNA clones encoding S_z protein of *Nicotiana alata* (Kheyr-Pour *et al.,* 1990). In this case a style cDNA library was separately screened with cDNA probes derived from mRNA isolated from flower buds 1 day before anthesis, and

from mRNA isolated from young leaves. Of the 40 randomly chosen style-specific cDNA clones analyzed, 22 were found to be derived from the same mRNA. These clones, which comprised the most abundant class of style-specific clones, were sequenced and their deduced amino acid sequences were found to match the sequence of the S_z protein determined by Mau et al. (1986). This latter strategy can be expected to work for S alleles, the protein products of which are the most abundant proteins in the pistil, as was the case for S_z protein. However, because there is sufficient sequence homology among S alleles, the cDNA for S_z allele has been used as a hybridization probe to isolate cDNA clones for a number of S alleles from Nicotiana alata (Kheyr-Pour et al., 1990), Petunia inflata (Ai et al., 1990), Petunia hybrida (Ai and Kao, unpublished results), Solanum chacoense (Xu et al., 1990b), and Lycopersicon peruvianum (Tsai and Kao, unpublished results).

So far, amino acid sequences of 18 S proteins from five solanaceous species have been reported, all of them derived from the corresponding cDNA sequences. These include seven S alleles of Nicotiana alata (Anderson et al., 1986, 1989; Kheyr-Pour et al., 1990), three of Petunia inflata (Ai et al., 1990), three of Petunia hybrida (Clark et al., 1990), two of Solanum chacoense (Xu et al., 1990b), and three of Solanum tuberosum (Kaufmann et al., 1991). The amino-terminal sequences of some of these S proteins have been determined and shown to match perfectly with the sequences deduced from corresponding cDNAs.

All the S alleles for which full-length cDNAs have been obtained predict a leader peptide of 22 amino acids not found in mature S proteins. This is consistent with the finding that S proteins are synthesized in the cells of the transmitting tissue and secreted into the transmitting tract of the style (Cornish et al., 1987; Anderson et al., 1989). The leader peptides exhibit properties typical of the signal sequence in eukaryotic proteins. Although there is considerable variation in the exact amino acid sequences, the leader peptides of all the S proteins contain a hydrophobic core, with a relatively polar sequence of about five amino acids at the carboxyl-terminus, and basic or neutral amino acids near the amino-terminus. A few key amino acids, including two serines and a leucine, and the proline believed to function as a helix breaker, are conserved in the leader peptides of all the S proteins for which information is available (Anderson et al., 1989; Ai et al., 1990).

B. Primary Structural Features of S Proteins

1. Extreme Sequence Diversity

Comparisons of the amino acid sequences of various subsets of the 18 S proteins have been made (Anderson et al., 1989; Ai et al., 1990; Clark et

al., 1990; Haring *et al.*, 1990; Kheyr-Pour *et al.*, 1990; Xu *et al.*, 1990b; Ioerger *et al.*, 1991; Kaufmann *et al.*, 1991). Figure 7 shows an alignment of the amino acid sequences of all 18 S proteins; all but the sequence of the S_a protein of *Nicotiana alata* are complete mature protein sequences. The amino acid sequence identity among these S proteins varies over a wide range: from as low as 38.74% to as high as 93.47% (Fig. 8). The tremendous sequence diversity between some alleles of the same species is in sharp contrast to other multiallelic genes in which differences between alleles usually lie in a few amino acid replacements.

2. Conserved Residues

Visual inspection of the aligned sequences reveals 34 perfectly conserved residues and 27 sites that accommodate only conservative replacements (Fig. 7). About half of the residues conserved among the S proteins, namely, 16 of the 34 residues, are also conserved in RNase T2 of *Aspergillus oryzae* and RNase Rh of *Rhizopus niveus*. In 10 additional sites only conservative replacements among the S proteins and two ribonucleases are observed. This sequence homology to the two fungal ribonucleases was first discovered in three *Nicotiana alata* S proteins (McClure *et al.*, 1989). The residues conserved between S proteins and the two fungal ribonucleases include His-38 and His-103, which have been shown to constitute the active site of RNase T2, and the members of two disulfide bonds in RNase T2, Cys-54, -106, -168, and -197. The S proteins have a total of eight conserved cysteines. The S_1 and S_2 protein of *Petunia inflata* have been found to contain four intramolecular disulfide bonds, suggesting that all eight conserved cysteines are involved in disulfide bridges (Ai and Kao, unpublished results).

The conserved residues among the S proteins, although dispersed throughout the protein, tend to fall into five distinct regions, C1–C5. A window-averaged composite hydrophobicity plot of 12 S proteins shows that C1, C4, and C5 are hydrophobic, suggesting that those regions may form the core structure of the S protein (Ioerger *et al.*, 1991). C2 and C3 regions are hydrophilic and are highly associated with residues shared among the S proteins and the two fungal ribonucleases. By analogy, these two regions may constitute the active site of the S protein ribonucleases.

3. Variable and Hypervariable Residues

The degree of variability for the nonconserved sites has been measured by a normed variability index (NVI). This index is defined as follows: log (variability index)/log (number of residues at the site). The variability index was defined by Bjorkman *et al.* (1987) as the number of different

```
              C1                              C2                    HVa
              20                              50
SF11    D FEYLQLVLTWP ASfCY-ANH---CER--IApN  nFTiHGLWP DN--VKTR1HNC  KPKP--TySYF--TGKML NDL
Sz      D FDYMQLVLTWP ASfCY-PKN--FCSR--IApK  nFTiHGLWP DK--VRGR1QFC  TSE---KyVNFAQDSPIL DDL
Sa      - ----------- -----------------I-pK  nFTiHGLWP DE--QHGM1NDC  GE----TfTKL--EPREK KEL
S1nic   N FEYMQLVLTWP TAfC--NVM--NCER--T-pT  nFTiHGLWP DN--VSTE1NYC  DRQK--KyKLF-EDDKKQ NDL
S2nic   A FEYMQLVLTWP ITfC--RIK--HCER--T-pT  nFTiHGLWP DN--HTTM1NYC  DRSK--PyNMF-TDGKKK NDL
S3nic   A FEYMQLVLQWP AAfC--HTTPSPCKR--I-pN  nFTiHGLWP DN--VSTM1NYC  SGED--EyEKL-DDDKKK KDL
S6nic   A FEYMQLVLQWP TAfC--HTT--PCKN--I-pS  nFTiHGLWP DN--VSTT1NFC  GKED--DyNII-MDGPEK NGL
S1pet   N FEYLQLVLTWP ASfCFRPKN--ICKR--P-aK  nFTiHGLWP EI--TGFR1EFC  TGSP--KyETF-KDNNIV DYL
S2pet   N FDYFQLVLTWP ASfCY-PKN--FCKR--K-sN  nFTiHGLWP KN--KHFR1EFC  TGD---KySRF-KEDNII NVL
S3pet   N FDYIQLVLTWP ASfCYRPKN--ICRR--I-pN  nFTiHGLWP EK--EHFR1EFC  DGD---KfVSFSLKDRIV NDL
PS3A    N FDYFQLVLTWP ASfCY-PKN--KCQR--R-sN  nFTiHGLWP EK--KRF-1EFC  TGD---KyKRFLEEDNII NVL
PS2A    N FDYFQLVLTWP ASfCY-PKN--KCQR--R-sN  nFTiHGLWP EK--KRF-1EFC  PGD---KfSRF-KEDNII NVL
PS1B    S FDHWQLVLTWP AGyC--KVK--GCPRPVI-pN  dFTiHGLWP DS--ISVImNNC  DPTK--TfATI-TEIKQI TEL
S2sol   T FDYMKLVLQWP PMyC--RNK--FCER--I-pR  nFTvHGLWP DN--KKYL1NNC  RSY---AyNAL-TNVREQ SKL
S3sol   T FEHLQLVLTWP TSfC--HKE--RCIR--S-sS  nFTiHGLWP DN--TSTR1NFC  KIV---KyNKI-EDEHKI SKL
S1stu   D FELLELVSTWP ATfCY-AYG---CSKRPI-pN  nFTiHGLWP DN--KSTV1NFC  NLAHEDEyIPI-TDHKIL TEL
Sr1stu  N FELLELVSTWP ATfCY-AYG---CSRRPI-pN  nFTiHGLWP DN--KSVI1NDC  KVVNKEGyVKI-TDPKQI TEL
S2stu   D FDYMQLVLTWP RSfCY-PYG--FCNR--IPpN  nFTiHGLWP DKKPMRGQ1QFC  TSD---DyIKF-TPGSVL DAL
          *        † †                     * *.* * * * * *  *   ††   ††  †         *


              HVb                            C3                    C4
              80                             120
SF11    DKHWMqlK FEQDYGRTEqP SWKYqyI kHGSCC QKRyNQNTYFGLaLR lkdkFDLL RTLQTHRiIPgSS-YTFQDI
Sz      DHHWMelK YHRDFGLENqF LWRGqyQ kHGTCC IPRyNQMQYFLLaMR lkdkFDLL ATLRTHGiTPgTK-HTFNET
Sa      TIRWPd1K RSRSDAQDVes FWEYeyN kHGTCC TELyDQAAYFDLaKN lkdkFDLL RNLKNEGiIPgST-YTVDEC
S1nic   DDRWPd1T LDRDDCKNGqG FWSYeyK kHGTCC LPSyNQEQYFDLaMA lkdkFDLL KSFRNHGiIPtKS-YTVQKY
S2nic   DERWPd1T KTKFDSLDKqA FWKDeyV kHGTCC SDKfDREQYFDLaMT lrdkFDLL SSLRNHGiSRgFS-YTVQNL
S3nic   DDRWPd1T IARADCIEHqV FWKHeyN kHGTCC SKSyNLTQYFDLaMA lkdkFDLL TSLRKHGiIPgNS-YTVQKI
S6nic   YVRWPd1I REKADCMKTqN FWRReyI kHGTCC SEIyNQVQYFDLaMA lkdkFDLL TSLKNHGiIRgYK-YTVQKI
S1pet   ERHWVqmK FDENYAKYHqP LWSYeyR kHGMCC SKIyNQAYFLLaTR  lkekFDLL TTLRTHGiTPgTK-HTFGDI
S2pet   ERHWIqmR FDEKYASTKqP LWEHeyN kHGTCC KNLyDQEAYFLLaiR lkdkLDLL TTLRTHGiTPgTK-HTFGEI
S3pet   ERHWVqmK FDEKFAKIKqP LWTHeyN kHGICS SNLyDQRAYFLLaMR vkdkFDLL TTLRTHGiTPgTK-HTFGEI
PS3A    ERHWIqmR FDETYANTKqP LWEHeyN rHGICC KNLyDQKAYFLLaMR lkdkLDLL TTLRTHGiTPgTK-HTFGEI
PS2A    ERHWIqmR FDEDYANAKqP LWQHeyN rHGICC KNLyDQKAYFLLaiR lkdkLDLL TTLRTHGiTPgTK-HTFGEI
PS1B    EKRWPelT TTAQFALTSqS FWRYqyE kHGTCC FPVySQSAYFDFaiK lkdkTDLL SILRSQGvTPgST-YTGERI
S2sol   DDRWPd1K SNKSMTMKEqK FWEYeyN kHGTCC EKLyNQAQYFNLtMN lkdkFDLL RILRNHGiVPgSL-ALLKQF
S3sol   EYGWPn1T TTEAVSKEDqV FWGKqyT kHGTCC TDLyDKDAYFDLaMN lkdrFDLL KILAMHGiTPgTSHHTSSNI
S1stu   DKRWPq1R YDYLYGIRKqY LWKNefI kHGSCS INRyKQPAYFDLaMK ikdkFDLL GTLRNHGiNPgST-YELDDI
Sr1stu  DKRWPq1R YEKLYGIDKqY LWKNefL kHGSCS INRyKQEAYFDLaMK ikdrFDLL GTLRNHGiNPgST-YELDDI
S2stu   DHHWIq1K FEREIGIRDqP LWKDqyK kHGTCC LPRyNQLQYFLLaMR lkekFDLL TTLRTHGiTPgTK-HTFKKI
          †  *      †      ††     *      .***  *    . †     .   .   .


              C5
              160                     200
SF11    FDAIKtvs-QEN-PDiKCAEVTK--GTPELY EigiCF TPNADSMFRCPQ-SDTCDKTAK--VLFRR
Sz      RDAIKtvt-NQVDPD1NCVGDPQ--GVRELY EigiCF TPTADSFFQCPH-SNTCDETGITKILFRR
Sa      EKQSEavt-QAY-PN1NCVGDPQ--KILELS EigiCF DRGATKVITCRR-RTTCNPINKKEISFPL
S1nic   NNTVKait-KGF-PN1TCNKQ-M-----ELQ EigiCF DQKVKNVIDCPR-PKTCK-ATRNGITFP-
S2nic   NNTIKait-GGF-PN1TCSRL-R-----ELK EigiCF DETVKNVIDCPN-PKTCK-PTNKGVMFP-
S3nic   NSTIKait-QGY-PN1SCTKR-Q-----MELL EigiCF DSKVKNVIDCPR-PKTCKPMGNRGIKFP-
S6nic   NNTIKvt-KGY-PN1SCTKG-Q-----ELW  EvgiCF DSTAKNVIDCPN-PKTCKTASNQGIMFP-
S1pet   QKAIKtvt-NQVDPD1KCVEHIK--GVQELN EigiCF NPAADNFYCPHH-SYTCDETDSKMILFR-
S2pet   QKAIKtvt-NNKDPD1KCVENIK--GVKELN EigiCF NPAADSFHDCRH-SKTCDETDSTQTLFRR
S3pet   QKAIKtvt-NNKDPD1KCVEHIK--GVKELK EvgiCF TPAADSFHDCRH-SNTCDETDSTKILFR-
PS3A    QKAIKtvt-SNNDPD1KCVENIK--GVMELN EigiCY TPAADFRDRCRH-SNTCDETSSTKILFRG
PS2A    QKAIKtvt-NNKDPD1KCVENIK--GVKELN EigiCF NPAADSFHDCRH-SYTCDETDSTQTLFRR
PS1B    NSSIAsvt-RVK-PN1KCLYY-R--GKLELT EigiCF DRTTVAMMSCPRISTSCKFGTNARITFRQ
S2sol   GEAIEtvtNKVF-PS1KCIDN-N--GIMELL EvgiCF DPAATKVIPCHR-PWICHADENTRIELVK
S3sol   QNAVKsvt-QGV-PHvCFNN-RFKGTSELL  EialCF DPQAQNVIHCPR-PKTCNSKGTKGITFP-
S1stu   ERAIMtvs-IEV-PS1KCIQKPL--GNVELN EigiCL DPEAKYMVPCPR-TGSCHNMGHK-IKFR-
Sr1stu  ERAIMtvs-IEV-PS1KCIQKPL--GNVELN EigiCL DPEAKYMVPCPR-TGSCHNMGHK-IKFR-
S2stu   QDAIKtvt-QEV-PD1KCVENIQ--GVLELY EigiCF TPEADSLFPCRQ-SKSCHPTENPLILFRL
           . . *       † *  †  †
```

FIG. 7   An alignment of the amino acid sequences of 18 mature S proteins from four solanaceous species. Dashes indicate gaps introduced to maximize homology. The conserved regions (C1–C5) and the hypervariable regions (HVa and HVb) are boxed. Perfectly conserved sites and sites with conservative replacements are shown in boldface, with the former

amino acids at a site divided by the frequency of the most common one(s). The index ranges from -1 for sites with identical residues to +1 for sites at which all residues are different, regardless of the number of sequences compared. The 17 sites with the highest NVIs are marked in Figure 7. A window-averaged plot of NVI over the length of the S protein shows two regions with the highest variability, which have been designated HVa and HVb (Ioerger *et al.*, 1991). These two regions together contain eight of the 17 most variable residues, and they are also the most hydrophilic regions in each S protein, despite the high degree of sequence diversity in these two regions.

The S protein is presumed to interact with the *S* allele product in the pollen in an allele-specific manner. Thus, the hypervariable sites may constitute the domain involved in recognition of pollen. It is interesting to note that there are only 13 amino acid differences between the most homologous pair of S proteins, S_2 allele of *Petunia inflata* and S_2 allele of *Petunia hybrida*. (The designation of both alleles as S_2 is purely coincidental.) Six of these 13 amino acids are among the 17 sites with the highest NVIs. The minimum number of amino acid changes required to alter *S* allelic specificity remains to be determined. It seems likely that a single amino acid replacement will not be sufficient to alter allelic specificity because standard mutagenesis fails to produce functionally distinct alleles (De Nettancourt, 1977). It should be interesting then to see whether two genetically identical *S* alleles might harbor small differences in their nucleotide and amino acid sequences.

C. Molecular Organization of the *S* Locus

Although there have been considerable advances in recent years in our understanding of the *S* locus component in the pistil, the question of the identity of the *S* locus component expressed in the pollen still remains unresolved. Messenger RNA homologous to the pistil *S* gene is not detectable in the pollen or in pollen tubes germinated *in vitro* despite the use of

---

indicated in upper case and the latter in lower case. Other symbols: the 21 most hypervariable sites (†); residues conserved in RNase T2 and RNase Rh (∗); sites with conservative replacements among the two fungal RNases and S proteins (·). The sequences used are as follows: SF11, Sz, Sa, and S1nic from *Nicotiana alata* (Kheyr-Pour *et al.*, 1990); S2nic, S3nic, and S6nic from *Nicotiana alata* (Anderson *et al.*, 1986, 1989); S1pet, S2pet, and S3pet from *Petunia inflata* (Ai *et al.*, 1990), S2sol and S3sol from *Solanum chacoense* (Xu *et al.*, 1990b), PS3A, PS2A, and PS1B from *Petunia hybrida* (Clark *et al.*, 1990), and S1stu, S2stu, and Sr1stu from *Solanum tuberosum* (Kaufmann *et al.*, 1991).

Matrix of pairwise amino acid identity among S proteins.

	Sz	Sa	S1nic	S2nic	S3nic	S6nic	S5lyc	S1pet	S2pet	S3pet	PS3A	PS2A	PS1B	S2sol	S3sol	S1Stb	Sr1St	S2Stub
SF11	62.05	40.00	46.81	45.74	48.42	46.56	60.31	55.67	55.15	54.40	53.89	54.12	43.52	43.75	45.03	51.81	51.03	60.82
Sz		40.70	47.12	44.50	48.70	47.92	71.21	65.15	63.32	67.34	63.82	62.81	43.88	44.39	46.67	48.45	47.18	70.71
Sa			48.80	46.99	52.98	51.50	40.94	39.18	40.12	42.11	39.77	40.12	46.20	49.71	43.53	42.60	44.71	42.44
S1nic				67.71	69.27	62.50	45.03	43.23	41.88	43.46	41.58	40.12	46.35	47.64	52.36	43.39	43.68	44.74
S2nic					65.10	61.98	42.93	40.10	40.84	41.36	38.95	41.36	46.35	47.12	51.31	43.92	45.26	43.68
S3nic						71.50	47.67	45.36	44.04	43.23	41.36	41.97	44.85	49.22	51.31	48.45	47.40	43.98
S6nic							47.92	44.04	42.19	43.23	40.10	40.10	42.49	50.00	45.13	48.45	47.18	48.96
S5lyc								59.90	58.88	60.41	60.71	57.87	44.10	45.64	48.21	45.64	44.90	48.69
S1pet									74.75	74.87	73.10	74.75	39.29	40.51	46.15	44.85	45.13	60.71
S2pet										81.31	74.87	74.75	41.33	41.33	44.10	44.85	44.62	60.41
S3pet											87.37	93.47	41.54	41.54	45.88	45.60	45.36	60.71
PS3A												82.32	40.51	41.33	45.88	44.62	46.94	61.22
PS2A													41.33	41.54	43.59	43.81	43.59	42.35
PS1B														42.86	43.59	44.62	46.94	58.88
S2sol															42.56	44.79	45.60	42.35
S3sol																43.75	46.11	46.67
S1Stub																	88.44	52.06
Sr1Stub																		50.26

FIG. 8 Matrix of pairwise amino acid identity among S proteins. Percent identity was calculated on the basis of a pairwise comparison of the 18 solanaceous S proteins shown in the alignment in Figure 7.

a variety of sensitive techniques, including Northern analysis, *in situ* hybridization, and ribonuclease protection assays (Clark *et al.*, 1990; Singh and Kao, unpublished results). It is still possible that mRNA homologous to the pistil *S* gene is present in the pollen, but at a low level, as might be expected if this pollen component comprises all or part of a membrane receptor and/or that the pollen component undergoes rapid turnover. Despite the failure to identify homologous sequences in the pollen, there is general reluctance to hew to the alternative model, viz., that the *S* locus component in pollen bears no sequence homology to the pistil *S* gene. The difficulty in this concept lies in explaining how mutations giving rise to new alleles would impart the same altered *S* allele identity to the pistil and pollen component if the two are determined by different genes.

Because pollen-part and style-part mutations were mapped to the *S* locus, the hope had been that an examination of genomic sequences encoding the pistil S proteins and sequences flanking it would yield some information about the organization of the *S* locus. The genomic clones for two *S* alleles each of *Solanum tuberosum* (Kaufmann *et al.*, 1991) and *Petunia inflata* (Coleman and Kao, 1992) have been isolated and characterized. Both *Petunia inflata* alleles were found to be embedded in a region rich in repetitive sequences, and no coding information was found in the flanking regions (7 kb upstream and 13 kb downstream). The 5′ noncoding regions of these alleles are extremely divergent, except for the putative TATA box sequence and a 8-bp motif that is found in all four alleles but at different locations (Kaufman *et al.*, 1991; Coleman and Kao, 1992). Each of the *S* alleles contains a single intron, ranging in size from 92 to 117 bp, located in the region of the sequence corresponding to the HVa hypervariable region of the S protein (Fig. 7).

None of the information on the *S* gene obtained so far has shed any light on the tripartite structure hypothesis of *S* locus organization. Because chromosome walking to linked sequences is vitiated by the presence of repetitive sequences in the flanking regions of the *S* gene (at least in *Petunia inflata*), other approaches may have to be used to delineate the structure of the *S* locus or sequences linked to the *S* locus that participate in self-incompatibility. For example, analysis of restriction fragment length polymorphisms of pollen- and style-part mutants and of mutants at unlinked modifier loci may yield information about the physical distance of these loci from the *S* gene and provide an alternative approach to locating and identifying other genes associated with self-incompatibility. Also, biochemical and cellular approaches to finding a receptor or gatekeeper that imparts *S* allele specificity to the response of pollen tubes to S proteins (discussed in the next section) may provide an alternate route to identifying other components of the self-incompatibility system, and perhaps to understanding *S* locus organization as well.

VII. Mechanism of S Protein Function

A. Ribonucleolytic Properties of S Proteins

The sequence similarity between S proteins and two fungal ribonucleases prompted immediate investigation into whether S proteins are indeed ribonucleases. A number of S proteins from *Nicotiana alata* and *Petunia inflata* have been studied and found to have ribonuclease activity (McClure *et al.*, 1989; Singh *et al.*, 1991; Broothaerts *et al.*, 1991). The enzymatic properties of three *Petunia inflata* S proteins have been extensively characterized (Singh *et al.*, 1991). The three proteins, despite having sequence diversity ranging from 20% to 25%, exhibit similar enzymatic properties, including pH and temperature optima and specific activity. When polyhomoribonucleotides are used as substrates, poly(C) is cleaved preferentially by all three S proteins. However, the extent of cleavage of poly(U) and poly(A) differs substantially: S_1 only cleaves poly(U); S_3 cleaves poly(A) and to a lesser degree poly(U); and S_2 does not cleave poly(A) or poly(U) to any degree. This difference among the three S proteins probably reflects the divergence in their primary sequences and may have no bearing on their biological activity in the self-incompatibility interaction because the *Petunia inflata* S proteins as well as the *Nicotiana alata* S proteins cleave all RNA substrates *in vitro* (McClure *et al.*, 1989; Singh *et al.*, 1991).

B. Biological Relevance of Ribonuclease Activity of
    S Proteins

The discovery that S proteins exhibit ribonuclease activity immediately raised the question of its relevance to the biological function of S proteins. The presence of catalytic activity in what is presumably a very specific recognition molecule is of interest. Although the similarity between S proteins and fungal ribonucleases is confined to a relatively small number of residues, the fact that all of these are conserved in each of the S proteins characterized thus far would argue against the catalytic activity of S proteins being entirely fortuitous. If the ribonuclease activity of S proteins is an integral part of the self-incompatibility reaction, then the most obvious biological role for this activity would be the degradation of pollen RNA. A simple mode of discrimination between self and nonself would be provided if S proteins could selectively degrade the RNA in self-pollen tubes, thereby arresting their growth, while having no deleterious effect on cross-pollen.

To determine whether S proteins specifically degrade RNA of self-pollen tubes, McClure *et al.* (1990) used ^{32}P to label nucleic acids of pollen *in situ*

and then followed the fate of the radiolabeled RNA in the pistils after self- or cross-pollination. They showed that 24 hours after pollination labeled ribosomal RNA recovered after cross-pollination was essentially intact, whereas ribosomal RNA obtained from self-pollinated pistils was degraded. However, because self-pollen tubes frequently burst in the style after their growth is arrested (De Nettancourt *et al.*, 1973), the cause-and-effect relationship between the ribonuclease activity of S proteins and degradation of pollen tube RNA is difficult to establish from this experiment.

In a follow-up study (Gray *et al.*, 1991), S proteins were shown to enter pollen tubes cultured *in vitro*. However, under these conditions no selectivity in uptake was discerned because S proteins were found in the cytoplasm of both self- and cross-pollen tubes, and the growth of both types of pollen was inhibited. Heat-treated S proteins were found to inhibit the growth of both self- and cross-pollen severely, although the protein failed to enter the cytoplasm and instead formed aggregates around the wall. It is difficult to reconcile these results with the situation obtained *in situ* and with the presumed role of S proteins in self-incompatibility because no selectivity of action was observed. It must be concluded that *in vitro* conditions deviate from the milieu encountered *in situ,* so that the interaction of pollen tubes and S proteins is aberrant. It is within the realm of possibility that proteins other than the S protein may be required for the effective manifestation of self-incompatibility.

Another model for the growth arrest of self-pollen posits that specific ribonuclease inhibitors residing in the pollen may disarm S proteins in an *S* allele-specific fashion. According to this model, each S protein would encounter a highly specific inhibitor in pollen tubes carrying a different *S* allele but not in pollen tubes carrying the same *S* allele, leading to the specific destruction of self-pollen tubes. Apart from the difficulty in envisioning a molecule that can interact with proteins over a wide range of sequence diversity (from 6.5% to 61.2% in the *Solanaceae*) while remaining ineffective against one species of molecule (the self S protein), initial experiments do not encourage acceptance of this model. McClure *et al.* (1990) failed to find any inhibitory effect of *Nicotiana alata* pollen tube extracts on ribonuclease activity of S proteins, irrespective of the *S* allelic combinations of pollen tube extracts and S proteins used. Results that corroborate this finding have also been obtained for *Petunia inflata* (Singh and Kao, unpublished data). However, it must be recognized that these experiments do not rule out the inhibitor model entirely. They do not preclude, for example, a pollen inhibitor acting in conjunction with a pistil protein, such that this pistil component is required for the specific inactivation of the S ribonuclease by the pollen inhibitor.

Despite the inconclusive results, the prevailing hypothesis at the time of this writing is that the specificity in S protein ribonuclease action lies

in the S protein being selectively taken up by self-pollen tubes, while either failing to enter cross-pollen tubes, or being compartmentalized in such a fashion that the cytoplasmic RNA pool is not accessible to them. In such a model, the nature of the receptor or gatekeeper molecule that confines or facilitates the movement of S proteins presents especially vexing theoretical problems. In order for it to discriminate among S proteins of different allelic specificities, it must itself be polymorphic with respect to the S locus. If such a gatekeeper, or a component of it, is identical to the pistil S protein, some version of the dimer hypothesis of Lewis (1949) may be invoked to explain its mode of action. In the model presented in Figure 3, for example, the interaction of identical pistil and pollen S gene products, mediated by a tissue-specific S locus product on each side, could be postulated to lead to the selective uptake of S proteins. However, experimental support for the expression in pollen of sequences with homology to the S proteins has not been forthcoming, as described in Section VI,C.

In contrast to other tissues (leaf and petal), pistils seem to possess a greater number of ribonucleases (Schrauwen and Linskens, 1972; Singh *et al.*, 1991). Schrauwen and Linskens (1972) found that species with solid styles have considerably greater amounts of these enzymes than species with hollow styles, with most of the activity localized in the transmitting tissue. Although the significance, if any, of this plethora of RNA degrading enzymes is not immediately clear, one may speculate that they could constitute part of a defense mechanism against colonization of the nutrient-rich stigmatic and transmitting tissues by fungal and bacterial pathogens. Alternatively, the ribonucleases could be components of a salvage mechanism acting to retrieve metabolites from senescent flowers. However, petals do not seem to be exceptionally rich in ribonucleases, as they might be expected to be if the second hypothesis were true, because they participate in the general movement of metabolites into the ovary in wilted, fertilized flowers (Linskens, 1975).

Ribonucleases have also been implicated in other key processes during plant growth and development, such as root growth and differentiation, light response, and water stress (Farkas, 1982). The RNA cleaving properties of certain plant lectins, in particular, offer an interesting parallel to the ribonuclease activity of S proteins. The ribosome-inactivating property of ricin A and phytolaccin, for example, is reported to be mediated by an associated ribonuclease activity (Obrig *et al.*, 1985). In view of the postulated role of plant lectins in defense against pathogens (Liener *et al.*, 1986), the question may now be asked whether some of them constitute a cytotoxic defense mechanism by virtue of their nucleolytic activity. With the emerging revelations of the role of ribonucleases in plant growth, differentiation, and defense, it becomes increasingly likely that an explication of the cytophysiological significance of S protein ribonuclease activity may have a bearing on other problems in plant development as well.

VIII. Evolutionary Aspects

Because self-incompatibility is a major outbreeding mechanism that is widespread in angiosperm but not found in gymnosperm, it is thought to have evolved very early in the evolution of angiosperm and to have been responsible for their rapid expansion (Whitehouse, 1950). The prevalence of self-incompatible taxa in centers of origin of species, and their self-compatible relatives in peripheral zones, is taken as a further indication that self-incompatibility is an ancient trait (Stebbins, 1957).

Phylogenetic analysis of S protein sequences from a number of solanaceous species has further revealed the antiquity of self-incompatibility. The genealogical tree of the S gene (Fig. 9) shows that S allele polymorphism predates the divergence of the genera *Nicotiana, Petunia,* and *Solanum* (Ioerger *et al.,* 1990). Because GSI has been found in more than 60 families, including both monocots and dicots, an extension of phylogenetic studies to include species in families other than *Solanaceae* would be especially useful in tracing the origin of self-incompatibility. This may soon become feasible because self-incompatibility is being investigated at the molecular level in a growing number of taxonomically diverse species.

New S alleles that appear in a population have a reproductive advantage over the extant S alleles because pollen bearing the new allele is less likely to land on stigma carrying the same allele and, thus, more likely to escape rejection (Wright, 1939). This results in a reduced rate of loss of new alleles, so that new alleles once generated can be expected to go to fixation and remain in a population for a long period of time. Consequently, S alleles would tend to accumulate greater sequence diversity than neutral alleles, as has been observed.

In view of the tremendous range of variation in amino acid similarity evident in pairwise comparisons of the 18 solanaceous S proteins (Fig. 8), it might be inferred that new S alleles are continuously being generated. One challenge that confronts both molecular biologists and population biologists is to determine how new S alleles are generated. Ebert *et al.* (1989) suggested that gene conversion might be responsible for generating allelic diversity at the S locus, although this inference was based on amino acid sequences of only three *Nicotiana alata* S proteins. However, recent examination of nucleotide sequences of the coding regions of 12 S proteins using sensitive statistical tests revealed an unusually low rate of intragenic recombination (Clark and Kao, 1991). Furthermore, as mentioned in Section VI,C, analysis of two alleles of the *Petunia inflata S* gene revealed extreme sequence heterogeneity in the flanking regions, suggesting a recombination suppression mechanism operating at the S locus, which prevents homogenization of S alleles (Coleman and Kao, 1992). The role of

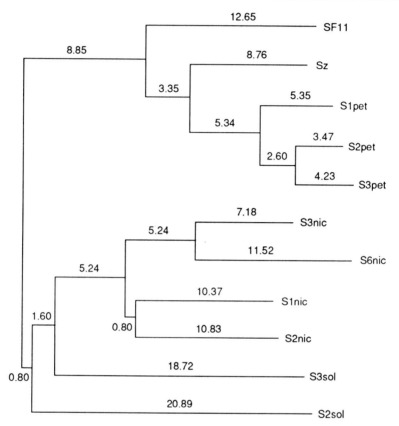

FIG. 9 Genealogical tree of *S* alleles. A distance matrix based on pairwise comparisons of 11 *S* proteins was used to construct the tree. An examination of the relationships among *S* alleles from *Nicotiana alata* (Sz, SF11, S1nic, S2nic, S3nic, and S6nic), *Petunia inflata* (S1pet, S2pet, and S3pet), and *Solanum chacoense* (S2sol and S3sol) shows that some alleles are more closely related across species than within species. For instance, the S_z and S_{F11} alleles of *Nicotiana alata* cluster with the three *S* alleles from *Petunia inflata* rather than with the other *S* alleles of *Nicotiana alata*. Reproduced with permission from Ioerger *et al.* (1990).

the *S* locus in self-incompatibility and its apparent lack of recombination are consistent with the highly polymorphic nature of the *S* locus and the extreme sequence diversity between alleles.

In the absence of definite information on the generation of *S* alleles, it may be speculated that a new *S* allele arises when the accumulation of point mutations in a critical subset of sites in an extant *S* gene reaches a level sufficient to alter its specificity. This model could be tested by using site-directed mutagenesis to change amino acids in the hypervariable sites

of an *S* gene and then examining the effect on its allelic specificity in transgenic plants carrying this mutagenized *S* gene.

The *Brassica S* gene has been shown to belong to a gene family comprised of genes with fairly high sequence homology (Nasrallah *et al.*, 1988). In species with GSI, however, the S-cDNA sequences do not seem to hybridize to genomic fragments other than the *S* gene itself, even when hybridizations are carried out at low stringency (Anderson *et al.*, 1989; Clark *et al.*, 1990; Coleman and Kao, 1992). This may be taken to signify that the *S* gene of GSI is not part of a gene family. However, it is entirely possible that the homology between other members of the *S* gene family and the *S* gene is too low to be detected, keeping in mind the degree of sequence diversity among S proteins themselves (Fig. 8).

Because pistils of *Petunia inflata* were found to contain a large number of ribonucleases (Singh *et al.*, 1991), the question might be asked: do any of these ribonuclease genes and the *S* gene constitute a ribonuclease gene family? The N terminal sequence of one such pistil ribonuclease, X_2 (Singh *et al.*, 1991), was sequenced and found to have substantial sequence homology to S proteins (Meager and Kao, unpublished results). It is unlikely that this homology is a result of convergent evolution or that it is merely a reflection of the underlying commonality in catalytic function because most other ribonucleases (RNase T1 or RNase A, for example) have little homology with S proteins.

It is tempting to speculate that an ancestral ribonuclease gene might have been recruited for a system of self and nonself recognition during reproduction, perhaps as an elaboration of polymorphisms developed in response to pathogen pressure. It has been suggested that the evolution of self-incompatibility could have been responsible for the great expansion of flowering plant species that occurred in the late *Cretaceous* (Whitehouse, 1950). Because GSI is thought to be an older form of self-incompatibility than SSI (De Nettancourt, 1977), the recruitment of the ancestral ribonuclease, together with all the other components that are presumed to be required for a functional self-incompatibility system, may well have occurred at this juncture in paleohistory.

IX. Potential Applications in Plant Breeding

In addition to providing a model system for basic research on cellular recognition during plant growth and development, self-incompatibility is also an important agronomic trait, presenting both opportunities and obstacles for plant breeding. Self-incompatibility could be used to advantage in the production of hybrid seed in some species, whereas in other species,

such as fruit trees (varieties of apples, cherries, and almonds, for example), it is usually seen as more of a nuisance because trees of a different, compatible variety are required to be planted to obtain fruit.

The reason for the limited use of self-incompatibility in hybrid seed production so far is that this trait was selected against early in the cultivation history of most crops for the more immediate benefit of inbreeding to achieve homozygosity of desirable traits (Frankel and Galun, 1977). Reintroduction of self-incompatibility from the genetic reservoir of wild relatives of crops is sometimes vitiated by the presence of reproductive barriers, including incongruity and unilateral incompatibility (De Nettancourt, 1977). At first sight the introduction of this trait into crops through gene transfer techniques may appear to be an attractive goal, but this optimism is tempered on contemplating the complexities of the self-incompatibility system and the growing realization that several genes may be required for the operation of a functional self-incompatibility system.

It is apparent that before the practical applications of self-incompatibility can be fully realized, one needs to determine the causes of the breakdown of self-incompatibility in commercial cultivars and identify all the genes involved in self-incompatibility. A study was recently initiated to address these questions in *Petunia*. F_1 hybrids were raised from crosses between a self-compatible cultivar of *Petunia hybrida* and self-incompatible *Petunia inflata* (with S_2S_2 genotype). Pistil protein patterns and the self-incompatibility behavior of 60 F_1 hybrids were analyzed. Results obtained so far reveal that a defective S allele and unlinked modifier gene(s) cause the breakdown of self-incompatibility in *Petunia hybrida* (Ai *et al.*, 1991). This study points to the importance of identifying modifier genes for the complete understanding of the operation of the self-incompatibility system and for its eventual application in plant breeding.

In many species with self-incompatible cultivars the expression of self-incompatibility is found to be too unreliable for extensive incorporation into plant breeding schemes. This has been reported to be a problem with using self-incompatibility in hybrid seed production of cole crops (broccoli, cabbage, cauliflower, and other *Brassica* species), which exhibit SSI. In some species, for instance, self-incompatibility is described as weak, although the underlying causes have not been investigated. With our increasing understanding of the molecular basis of self-incompatibility, and especially if the other elements participating in this phenomenon (modifier genes, for instance) come to be better understood, it may be possible to overcome the problems due to the erratic behavior of self-incompatibility in some species.

The study of the molecular genetic basis of GSI is now being extended to a number of families other than the *Solanaceae*. For example, in *Campanula* stylar proteins that cross-react with antiserum against *Petunia*

inflata S proteins have been detected (Stephenson and Kao, unpublished data). These proteins also have ribonuclease activity. *Campanula* is a member of the *Campanulaceae,* in the order *Campanulales,* while the *Solanaceae* belongs to the order *Polemoniales* (Benson, 1979). There is hope, therefore, that genes encoding S proteins in species outside the *Solanaceae* could be identified based on sequence homology with the solanaceous S proteins. If it emerges that the evolution of GSI is monophyletic and that the biochemical underpinnings are shared by all extant species, then it would be possible to use solanaceous *S* genes as probes to identify their counterparts in other species of interest. It would also be possible to use polymerase chain reaction primers homologous to the conserved regions of the solanaceous S proteins to directly amplify and sequence *S* genes from other species.

The value of understanding the mechanism of self-incompatibility in fruit trees and members of the *Gramineae* could be substantial, given the prominence of these taxa in world agriculture and the promise of potential practical applications over the longer term. The import for plant breeding strategies would be especially significant if the hypothesis that there is functional analogy between interspecific and intraspecific self-incompatibility is borne out because then the manipulation of interspecific barriers may also come within reach.

X. Conclusions

Our information on the molecular, biochemical, and cellular basis of self-incompatibility has grown by leaps and bounds since the mid-1980s. Although the most trying puzzles lie untouched, it must surely be deemed a measure of our progress that we can now begin to articulate some of the most complex and interesting questions about self-incompatibility. Some of these are questions raised by that insightful group of pioneers who laid the foundation for the wealth of data on the genetics of self-incompatibility upon which we now draw. Many of these can now be rephrased in molecular terms, as a first step in building hypotheses testable by the techniques of molecular genetics.

Amino acid sequences of 18 S proteins from five taxa of the *Solanaceae* have now been reported, and there is every reason to believe that this knowledge will lead us to the identification of more S proteins from more families. Comparison of the sequences available to date has already illuminated some significant properties of S proteins. Chief among these is the striking sequence similarity to the amino acids in the active site of two fungal ribonucleases. This has allowed us to ask meaningful questions

about the nature of the self-incompatibility reaction, and should, in the near future, allow us to determine conclusively the means by which S proteins thwart the growth of self-pollen tubes. The development of a convenient, reliable bioassay for S protein activity against self-pollen cultured *in vitro* would be beneficial, especially in assaying the activity of mutagenized S proteins produced in a suitable expression system.

Ultimately, the functional role of the S gene in self-incompatibility must be confirmed *in situ* by examining the ability of transgenic plants carrying a new S allele to reject pollen bearing the same allele. Efforts are underway in a number of laboratories to raise transgenic plants expressing an introduced S gene. Experiments are also in progress to identify the regulatory sequences that confer tissue-specific expression of the S gene.

Among the most daunting questions that confront us are those relating to the molecular organization of the S locus and the expression of the S allele in pollen. Despite the availability of genomic sequences for S alleles and several kilobases of their flanking sequences, we are no closer to verifying or refuting the tripartite structure hypothesis of S locus organization, nor can we boast an alternative model that explains the observed phenomena any better. Ingenious techniques are being brought to bear to identify S allele sequences in pollen, which best represents the quest for the "Holy Grail" in self-incompatibility research.

We do not understand why polyploidy leads to the breakdown of self-incompatibility by affecting its manifestation in pollen (pollen-part phenotype) or why this occurs in species with monofactorial, but not bifactorial, GSI (Lundqvist, 1975). It is not known why all attempts to generate new S allele specificities using standard means of mutagenesis fail, whereas a strategy of inbreeding over several generations appears to succeed (De Nettancourt, 1977). The role of unlinked loci that affect the manifestation of self-incompatibility has been addressed for the first time since the advent of the techniques of molecular genetics (Ai *et al.*, 1991), and the identification and cloning of such genes would advance the field enormously, with possible benefits for the use of self-incompatibility in plant breeding programs.

We do not know what mechanisms generate allelic diversity at the S locus, and what mechanisms operate to prevent recombination and consequent scrambling of S alleles on homologous chromosomes. The finding of mitochondrial sequences homologous to regions of the S gene (Bernatzky *et al.*, 1989) is another curious facet of the S gene system. The significance of this homology, beyond the well-known promiscuity of the mitochondrial genome (Palmer, 1985), cannot be fathomed at present.

It is clear that in understanding GSI a many-pronged approach will yield the best results. There is need, therefore, to consider the problem from the cellular, physiological, biochemical, and molecular genetic standpoint

and to integrate information from all areas. A reappraisal of the earlier work, especially the storehouse of information on genetics that was painstakingly built by earlier workers, should bring rich rewards. With an increasing number of researchers embarking on research in this field, there is every reason to believe that we are now poised at the threshold of a new and exciting era in research on self-incompatibility.

Acknowledgments

The authors thank Dr. Andrew Clark for the alignment of amino acid sequences and for generating the matrix of sequence identity. The work from T. H. Kao's laboratory described in this review was supported by grants from the National Science Foundation (DCB 86-16087 and DCB 89-04008), U.S. Department of Agriculture (90-37261-5560), and the Pennsylvania Agricultural Experiment Station (Project No. 2997).

References

Ai, Y., Singh, A., Coleman, C. E., Ioerger, T. R., Kheyr-Pour, A., and Kao, T. H. (1990). *Sex. Plant Reprod.* **3,** 130–138.

Ai, Y., Kron, E., and Kao, T. H. (1991). *Mol. Gen. Genet.* **230,** 353–358.

Anderson, M. A., Cornish, E. C., Mau, S. L., Williams, E. G., Hoggart, R., Atkinson, A., Bonig, I., Grego, B., Simpson, R., Roche, P. J., Haley, J. D., Penschow, J. D., Niall, H. D., Tregear, G. W., Coghian, J. P., Crawford, R. J., and Clarke, A. E. (1986). *Nature* **321,** 38–44.

Anderson, M. A., McFadden, G. I., Bernatzky, R., Atkinson, A., Orpin, T., Dedman, H., Tregear, G., Fernley, R., and Clarke, A. E. (1989). *Plant Cell* **1,** 483–491.

Ascher, P. D. (1972). *Am. J. Bot.* **59,** 647.

Ascher, P. D. (1984). *In* "Petunia" (K. C. Sink, ed.), pp. 92–110. Springer-Verlag, Berlin.

Bateman, A. J. (1952). *Heredity* **6,** 285–310.

Benson, L. (1979). "Plant Classification." D. C. Heath and Company, Lexington, Massachusetts.

Bernatzky, R., Mau, S. L., and Clarke, A. E. (1989). *Theor. Appl. Genet.* **77,** 320–324.

Bjorkman, P. J., Saper, M. A., Samraoui, B., Bennet, W. S., Strominger, J. L., and Wiley, D. C. (1987). *Nature* **329,** 512–518.

Bredemeijer, G. M., and Blass, J. (1981). *Theor. Appl. Genet.* **59,** 185–190.

Brewbaker, J. L. (1959). *Ind. J. Genet. Plant Breed.* **19,** 121–133.

Brewbaker, J. L., and Majumder, S. K. (1961). *Am. J. Bot.* **61,** 457–464.

Broothaerts, W. J., Laere, A. V., Witters, R., Preaux, G., Decock, B., Damme, J. V., and Vendrig, J. C. (1989). *Plant Mol. Biol.* **14,** 93–102.

Broothaerts, W. J., Decock, B., Damme, J. V., and Vendrig, J. C. (1991). *Sex. Plant Reprod.* **4,** 258–266.

Clark, A. G., and Kao, T. H. (1991). *Proc. Natl. Acad. Sci. U.S.A.* **88,** 9823–9827.

Clark, K. R., Okuley, J. J., Collins, P. D., and Sims, T. L. (1990). *Plant Cell* **2,** 815–826.

Coleman, C. E., and Kao, T. H. (1992). *Plant Mol. Biol.* **18,** 725–737.

Cornish, E. C., Pettitt, J. M., Bonig, I., and Clarke, A. E. (1987). *Nature* **326,** 99–102.

Cornish, E. C., Anderson, M. A., and Clarke, A. E. (1988). *Annu. Rev. Cell. Biol.* **4,** 209–228.

Darwin, C. (1876). "The Effect of Cross and Self Fertilization in the Plant Kingdom." John Murray, London.

De Nettancourt, D. (1977). "Incompatibility in Angiosperms." Springer-Verlag, Berlin.

De Nettancourt, D., Devereux, M., Laneri, U., Cresti, M., Pacini, E., and Sarfatti, G. (1973). J. Cell Sci. 12, 403–419.

De Nettancourt, D., Devreux, M., Laneri, U., Cresti, M., Pacini, E., and Sarfatti, G. (1974). Theor. Appl. Genet. 44, 278–288.

Dickinson, H. G., Moriarty, J., and Lawson, J. (1982). Proc. R. Soc. Lond. [Biol.] 215, 45–62.

East, E. M. (1940). Proc. Am. Philos. Soc. 82, 449–518.

Ebert, P. R., Anderson, M. A., Bernatzky, R., Altschuler, M., and Clarke, A. E. (1989). Cell 56, 255–262.

Emerson, S. (1938). Genetics 23, 190–202.

Emerson, S. (1939). Genetics 24, 524–537.

Farkas, G. L. (1982). In "Nucleic Acids and Proteins in Plants" (B. Parthier and D. Boulter, eds.), pp. 224–262. Springer-Verlag, Berlin.

Flaschenriem, D. R., and Ascher, P. D. (1979a). Theor. Appl. Genet. 54, 97–101.

Flaschenriem, D. R., and Ascher, P. D. (1979b). Theor. Appl. Genet. 55, 23–28.

Frankel, R., and Galun, E. (eds.) (1977). "Pollination Mechanisms, Reproduction and Plant Breeding." Springer-Verlag, Berlin.

Franklin-Tong, V. E., Lawrence, M. J., and Franklin, F. C. H. (1988). New Phytol. 110, 109–118.

Gaude, T., and Dumas, C. (1987). Int. Rev. Cytol. 107, 333–366.

Gray, L. E., McClure, B. A., Bonig, I., Anderson, M. A., and Clarke, A. E. (1991). Plant Cell 3, 271–283.

Haring, V., Gray, J. E., McClure, B. A., Anderson, M. A., and Clarke, A. E. (1990). Science 250, 937–941.

Herrero, M., and Dickinson, H. G. (1979). J. Cell Sci. 36, 1–18.

Herrero, M., and Dickinson, H. G. (1980). Planta 148, 217–221.

Herrero, M., and Dickinson, H. G. (1981). J. Cell Sci. 47, 365–383.

Heslop-Harrison, J. (1983). Proc. R. Soc. Lond. [Biol.] 218, 317–395.

Hogenboom, N. G. (1975). Proc. R. Soc. Lond. [Biol.] 188, 361–375.

Ioerger, T. R., Clark, A. G., and Kao, T. H. (1990). Proc. Natl. Acad. Sci. U.S.A. 87, 9732–9735.

Ioerger, T. R., Gohlke, J. R., Xu, B., and Kao, T. H. (1991). Sex. Plant Reprod. 4, 81–87.

Jackson, J. F., and Linskens, H. F. (1990). Sex. Plant Reprod. 3, 207–212.

Jahnen, W., Lush, W. M., and Clarke, A. E. (1989). Plant Cell 1, 501–510.

Kamboj, R. K., and Jackson, J. F. (1986). Theor. Appl. Genet. 71, 815–819.

Kaufmann, H., Salamini, F., and Thompson, R. D. (1991). Mol. Gen. Genet. 226, 457–466.

Kheyr-Pour, A., and Pernes, J. (1986). In "Biotechnology and Ecology of Pollen" (D. L. Mulcahy, G. B. Mulcahy, and E. Ottaviano, eds.), pp. 191–196. Springer-Verlag, Berlin.

Kheyr-Pour, A., Bintrim, S. B., Ioerger, T. R., Remy, R., Hammond, S. A., and Kao, T. H. (1990). Sex. Plant Reprod. 3, 88–97.

Kirch, H. H., Uhrig, H., Lottspeich, F., Salamini, F., and Thompson, R. D. (1989). Theor. Appl. Genet. 78, 581–588.

Lawrence, M. J., Marshall, D. F., Curtis, V. E., and Fearon, C. H. (1985). Heredity 54, 131–138.

Lewis, D. (1949). Heredity 3, 339–355.

Lewis, D. (1960). Proc. R. Soc. Lond. [Biol.] 151, 468–477.

Lewis, D. (1979). N. Z. J. Bot. 17, 637–644.

Lewis, D., and Crowe, L. K. (1958). Heredity 12, 233–246.

Liener, I. E., Sharon, N., and Goldstein, I. J. (eds.) (1986). "The Lectins: Properties, Functions and Applications in Biology and Medicine." Academic Press, San Diego.

Linskens, H. F. (1975). *Proc. R. Soc. Lond. [Biol.]* **188**, 299–311.

Linskens, H. F., and Heslop-Harrison, J. (eds.) (1984). "Encyclopedia of Plant Physiology, Cellular Interactions," New Series, Volume 17. Springer-Verlag, Berlin.

Lundqvist, A. (1964). *Hereditas* **52**, 221–234.

Lundqvist, A. (1975). *Proc. R. Soc. Lond. [Biol.]* **188**, 235–245.

Martin, F. W. (1961). *Genetics* **46**, 1443–1454.

Martin, F. W. (1968). *Genetics* **60**, 101–109.

Mascarenhas, N. T., Bashe, D., Eisenberg, A., Willing, R. P., Xiao, C. M., and Mascarenhas, J. P. (1984). *Theor. Appl. Genet.* **68**, 323–326.

Mather, K. (1943). *J. Genet.* **45**, 215–235.

Mau, S. L., Raff, J. W., and Clarke, A. E. (1982). *Planta* **156**, 505–516.

Mau, S. L., Williams, E. G., Atkinson, A., Anderson, M. A., Cornish, E. C., Grego, B., Simpson, R. J., Kheyr-Pour, A., and Clarke, A. E. (1986). *Planta* **169**, 184–191.

McClure, B. A., Haring, V., Ebert, P. R., Anderson, M. A., Simpson, R. J., Sakiyama, F., and Clarke, A. E. (1989). *Nature* **342**, 955–957.

McClure, B. A., Gray, J. E., Anderson, M. A., and Clarke, A. E. (1990). *Nature* **347**, 757–760.

Mulcahy, D. L., and Mulcahy, G. B. (1983). *Science* **220**, 1247–1251.

Mulcahy, D. L., and Mulcahy, G. B. (1985). *Heredity* **54**, 139–144

Mulcahy, D. L., Mulcahy, G. B., and Ottaviano, E. (eds.) (1986). "Biotechnology and Ecology of Pollen." Springer-Verlag, Berlin.

Nasrallah, J. B., Kao, T. H., Goldberg, M. L., and Nasrallah, M. E. (1985). *Nature* **318**, 263–267.

Nasrallah, J. B., Yu, S. M., and Nasrallah, M. E. (1988). *Proc. Natl. Acad. Sci. U.S.A.* **85**, 5551–5555.

Obrig, T. G., Moran, T. P., and Colinas, R. J. (1985). *Biochem. Biophys. Res. Commun.* **130**, 879–884.

Pacini, E. (1981). *Phytomorphology* **31**, 175–180.

Palmer, J. D. (1985). In "Molecular Evolutionary Genetics" (R. J. MacIntyre, ed.), pp. 131–240. Plenum Press, New York.

Rosen, W. G., and Gawlik, S. R. (1966). *Protoplasma* **61**, 181–191.

Schrauwen, J., and Linskens, H. F. (1972). *Planta* **102**, 277–285.

Sharma, N., and Shivanna, K. R. (1982). *Ind. J. Exp. Biol.* **20**, 255–256.

Singh, A., Ai, Y., and Kao, T. H. (1991). *Plant Physiol.* **96**, 61–68.

Speranza, A., and Calzoni, G. L. (1990). *Plant Physiol. Biochem.* **28**, 747–754.

Stebbins, G. L. (1957). *Am. Naturalist* **91**, 337–354.

Stout, A. B. (1917). *Am. J. Bot.* **4**, 375–395.

Sussex, I., Ellingboe, A., Crouch, M., and Malmberg, R. (eds.) (1985). "Plant Cell/Cell Interactions." Cold Spring Harbor Laboratory.

Takahashi, H. (1973). *Jpn. J. Genet.* **48**, 27–33.

Tan, L. W., and Jackson, J. F. (1988). *Sex. Plant Reprod.* **1**, 25–27.

Townsend, C. E. (1969). *Crop Sci.* **9**, 443–446.

Van der Donk, J. A. W. M. (1974). *Mol. Gen. Genet.* **134**, 93–98.

Whitehouse, H. L. K. (1950). *Ann. Bot. New Series* **14**, 198–216.

Wright, S. (1939). *Genetics* **24**, 538–552.

Xu, B., Grun, P., Kheyr-Pour, A., and Kao T. H. (1990a). *Sex. Plant Reprod.* **3**, 54–60.

Xu, B., Mu, J., Nevins, D. L., Grun, P., and Kao, T. H. (1990b). *Mol. Gen. Genet.* **224**, 341–346.

Sporophytic Self-Incompatibility Systems: *Brassica S* Gene Family

Martin Trick* and Philippe Heizmann†

* Cambridge Laboratory, Institute of Plant Science Research, John Innes Centre, Norwich NR4 7UJ, England

† Reconnaissance Cellulaire et Amélioration des Plantes, Université Claude Bernard-Lyon 1 Villeurbanne, France

I. Introduction

Incompatibility systems represent an important element of the control of sexual reproduction in higher plants (De Nettancourt, 1977). By preventing self-pollination they act to promote outbreeding and therefore may exert profound effects on the genetic structures of populations and perhaps the evolution of species. Many different systems in many different plant genera and species have been studied. Perhaps the simplest and most amenable to a genetic analysis are those termed the homomorphic, monofactorial self-incompatibility systems. In these the floral morphology exhibits no variation between types. Incompatibility phenotypes are genetically controlled by a single, major gene designated S. Examples include the gametophytically controlled system of the Solanaceae (Singh and Kao, this volume) and the sporophytic system that is characteristic of the Brassicaceae (Nasrallah *et al.*, 1991). Following a long tradition of genetic and physiological analysis, each of these is now being investigated at the level of the DNA sequences of the putative S alleles themselves.

In the species *Brassica oleracea,* a single highly polymorphic locus S, of which some 50 different alleles have been recorded (Thompson and Taylor, 1966; Ockendon, 1974), determines both the incompatibility phenotype of pollen via expression in the diploid tapetum of the anthers and also that of the stigma papillae cells. The system would thus appear to be founded on a specialized form of cell-to-cell interaction between the surface of the pollen grain and that of the style. Any identity in expressed S alleles between these two partners results in a rapid and specific inhibition of self-pollen germination at the stigmatic surface.

S locus-specific glycoproteins (SLSGs), which are secreted by the stigma papillae cells and deposited in their cell walls, have been proposed to be the female mediators of this recognition event. This proposal is based on several compelling lines of evidence. The SLSGs can be visualized as major bands, or groups of bands, on isoelectrofocusing gels of stigma protein extracts, and the mobilities of these molecules vary between different S genotypes (Nasrallah and Nasrallah, 1984). The inheritance of the glycoproteins co-segregates with S phenotype in F$_2$ populations (Nasrallah *et al.*, 1972), and the timing of their initial expression in the stigma, 1–2 days prior to anthesis, is coincident with that of the acquisition of the incompatibility system in *Brassica* buds (Nasrallah *et al.*, 1985a); before this time plants can be successfully pollinated with mature self-pollen. Finally, it had been noted that a suppressor mutation of self-incompatibility, itself unlinked to the S locus, reduced the expressed levels of the SLSG in a dosage-dependent manner (Nasrallah, 1974) and selfing of individuals with high, intermediate, and low levels of protein resulted in low, intermediate, and high seed sets, respectively.

The molecular analysis of the *Brassica* S system thus began with the cloning of a cDNA for one such SLSG (Nasrallah *et al.*, 1985b). It has since emerged that the S locus glycoprotein (*SLG*) gene, as it is now called, is but one member of a multigene family of related sequences comprising at least 10–15 copies, which is organized and expressed in a highly complex way. Intriguingly, as well as the *SLG* gene itself, at least two other copies are also abundantly expressed by the stigma papillae cells, but because these are unlinked to S, they can play no direct role in the determination of incompatibility specificity but could conceivably affect S phenotype in a pleiotropic manner. In this chapter we will give a short history of the isolation of *SLG* and such S locus-related sequences (*SLRs*) and then attempt to highlight some of the fascinating aspects of this gene family, including its unusual mode of evolution. More recent developments, in which *SLG*-homologous transcripts were detected in other tissues (Guilluy *et al.*, 1991) and homology demonstrated between the *Brassica SLGs* and the extracellular domain of a maize receptor protein kinase (Walker and Zhang, 1990), hinted at the possibility that the *SLG* gene may have evolved from a preexisting and ubiquitous cell-signaling system. Indeed, it has now been shown (Stein *et al.*, 1991) that the S locus comprises at least two related genes: *SLG* and an S receptor kinase (*SRK*). The latter is composed of an S domain, which appears to co-evolve with the *SLG*, juxtaposed to domains that encode a putative kinase function. At present, little progress has been made on the biology of the *SLGs* and *SRKs* at a functional level. However, it will become apparent that research into *Brassica* self-incompatibility is poised at an exciting stage because the tools are now available to test the hypothesis that the *SLG–SRK* genes do indeed encode a component of the incompatibility system.

II. Isolation of *Brassica S* Sequences

A. Nomenclature

The recent work on the molecular genetics of self-incompatibility in *Brassica* has involved several research groups isolating many examples of the putative *S* locus alleles themselves (*SLGs* and, more recently, *SLG–SRK* gene pairs) and also those of expressed *SLRs*. Inevitably, different systems for clone designations have been used in the original literature and these will occasionally be referred to. However, in this chapter we will mainly follow Chen and Nasrallah (1990) and retrospectively call those characterized *S*-linked sequences *SLG6*, *SRK6*, *SLG29*, and so on, thereby indicating the relevant *S* genotypes. We earnestly look forward to the proof that the *SLG–SRK* gene pairs do encode *S* specificity (at least in the female tissue) and thus the day we can legitimately refer to them as the S_6 and S_{29} alleles proper, or at least components thereof. Where required, we will refer to alleles of the first characterized *S*-related gene as *SLR1-6*, *SLR1-29*, etc., the latter part of these designations again referring to the particular *S* genotypes (for the lack of anything better) from which they were cloned. With no genetic linkage between *S* and the *SLR1* locus, this system will have its limitations; indeed, the same *SLR1* allele has clearly been isolated from more than one *S* homozygote (Lalonde *et al.*, 1989). The recent discovery of a second, expressed *S*-related gene (Boyes *et al.*, 1991), which is linked to *SLR1*, will now demand a similar system for *SLR2* alleles. At the time of writing, sequences are available for the *SLR2-5*, *SLR2-2*, and *SLR2-6* alleles.

B. Isolation of *SLG* and *SRK* Sequences

The high relative abundance and strict developmental and tissue-specific regulation of the *Brassica* SLSGs have enabled a simple strategy for the molecular cloning of the *SLG*-specific messages. Briefly, cDNA libraries were prepared from mRNA populations extracted from stigmas dissected from flower buds at 1–2 days before anthesis, the time at which SLSG synthesis is maximal and flowers become competent for the self-incompatible response (Nasrallah *et al.*, 1985a). The libraries were screened with radioactive stigma cDNA hybridization probes and then counterscreened with probes synthesized from mRNA from unrelated tissue types to subtract out nonstigma-specific sequences. Clones that were hybridized exclusively with the stigma probe would thus be expected to comprise the abundant stigma-specific messages, including those encoded by the *SLG* genes.

The first of the ever-growing family of *SLG* sequences to be isolated in this way, *SLG6* (Nasrallah *et al.*, 1985b), was derived from an S_6 homozygote of *Brassica oleracea* and was originally designated *pBOS5*. This clone was one of eight to be identified (apparently all homologous in sequence) from the differential screening of approximately 2000 plasmid cDNA clones. It was shown to hybridize to a stigma-specific mRNA species of about 2 kb, which exhibited a pattern of expression over time paralleling the synthesis of SLSG in the stigma. When used in Southern blot analysis of *Brassica* DNAs, however, the probe hybridized to multiple bands, hinting at a complex genomic organization and somewhat at odds with the hitherto simple genetic picture of the *S* locus. Nevertheless, with a subset of these hybridizing fragments, the probe did identify restriction fragment length polymorphisms (RFLPs) in an F_2 population that co-segregated with the *S* phenotype (as assayed by both crosses with tester *S* homozygotes and the analysis of stigma glycoprotein types). The cross-reaction of polyclonal antibodies, raised against purified S_6-specific glycoprotein, with *pBOS5-lacZ* fusion proteins, further supported the hypothesis that this cDNA clone encoded *SLG6*. However, the initial nucleotide sequence analysis of the cDNA insert predicted an incomplete protein with some inconsistencies with the known molecular weight of the SLSG. Subsequently, the partial protein sequencing of three different mature *S* glycoproteins extracted from the closely related species *Brassica campestris* (Takayama *et al.*, 1987) provided direct evidence for the expression of homologous molecules, which, in turn, strongly suggested that the *SLG6* DNA sequence had been misinterpreted.

The *SLG6* sequence was revised and extended through the isolation of a new full-length cDNA clone, together with near full-length clones corresponding to *SLG13* and *SLG14* transcripts. These new cDNA clones were isolated from λ *gt10* libraries prepared as before from homozygous S_{13} and S_{14} genotypes but screened with the *pBOS5* insert as a hybridization probe. All three predicted *SLG* protein sequences were then compared (Nasrallah *et al.*, 1987). A common feature to emerge was the presence of an N-terminal hydrophobic domain of about 30 amino acids, which preceded the N-terminal sequence homology determined from the *Brassica campestris* glycoproteins. This domain is thus assumed to form a signal peptide, which directs the SLSG to the secretory system (Section VI,A). Taken together with the *Brassica campestris* protein data, the *SLG6*, *SLG13*, and *SLG14* sequences displayed an overall structural homology but also remarkable levels of divergence for putative alleles of the *S* locus gene, an average of some 20% at the protein level. Such an interpretation of the data is complicated by the multigenic nature of the *S* family; therefore, the possibility that the expressed *SLG* sequences, although linked to the *S* locus, may not be strictly allelic. This factor may impinge on the

mechanism of evolution of *S* alleles (Section IV,E). Interestingly, it has emerged that in the gametophytic, *Nicotiana S* system, a single locus system in which the *S* sequence appears to be unique in the genome, is characterized by even higher DNA sequence divergences between alleles (Anderson *et al.*, 1989).

The *Brassica SLG* variation was found to comprise both nucleotide substitution and insertion–deletion differences, with a prominent feature being the occurrence of small blocks of sequence changes (Section IV). This variation was found to be partitioned within the molecules, such that four regions could be defined with high statistical significance (Nasrallah *et al.*, 1987); two variable domains, B and D, are interspersed with the relatively conserved regions A and C (Fig. 1). The C-terminal part of *SLG6* is characterized by the presence of 12 cysteine residues, 11 of which prove to be invariant in *SLG* sequence comparisons (Fig. 2). *SLG14* is the only *S*-linked sequence not to conserve all 12 cysteines, although the same amino acid substitution of glycine for cysteine in that same position is also a feature of the *SLR2* sequences (Section II,C). Another notable feature of the *SLGs* is variation in the position and number of potential N-glycosylation sites, at which carbohydrate moieties may be attached to the asparagine residue in either of the motifs asn-X-ser and asn-X-thr. Using the results of the *Brassica campestris S* glycoprotein sequencing as a guide (Takayama *et al.*, 1987; Isogai *et al.*, 1987), it is probable that all

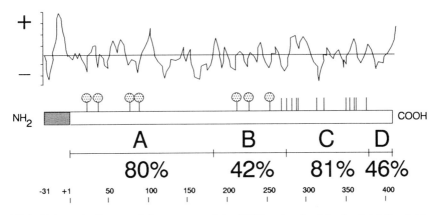

FIG. 1   Schematic diagram of the structure of the *SLG6* molecule. A hydrophobicity plot is drawn above the figure. The shaded box indicates the N-terminal hydrophobic signal peptide, the filled circles indicate the N-glycosylation sites most likely to be occupied *in vivo*, and the vertical bars denote the conserved cysteine residues. Regions A, B, C, and D are those of relative conservation and divergence identified by Nasrallah *et al.* (1987); the percentages are the average amino acid sequence identities in these regions between *SLG6*, *SLG13*, and *SLG14*.

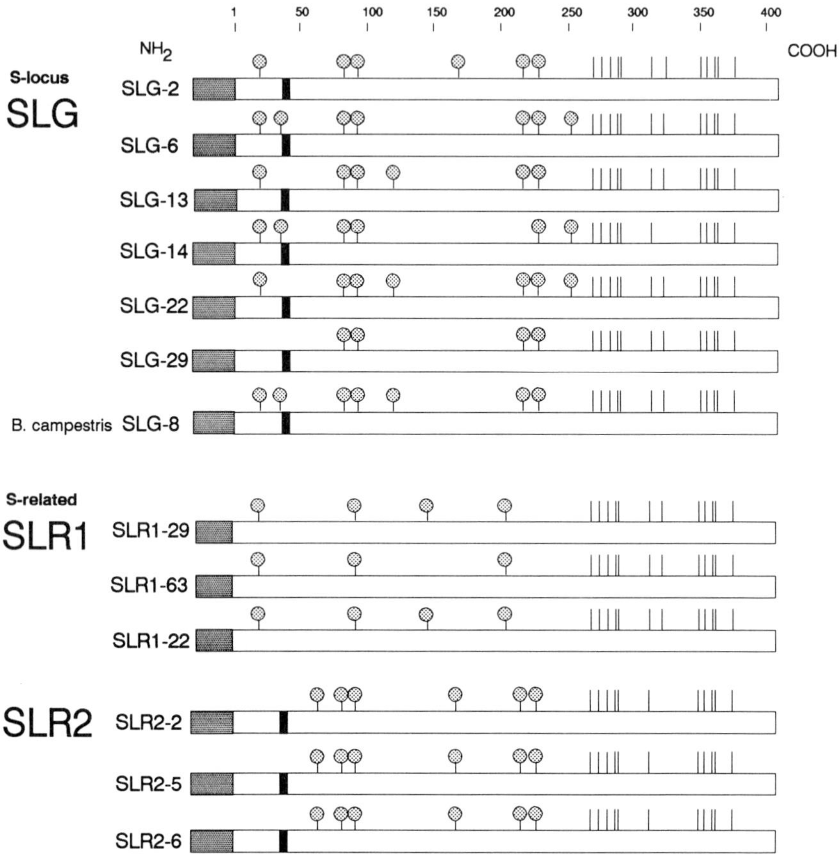

FIG. 2 Schematic diagram of all characterized, expressed *S* sequences drawn from the data of Nasrallah *et al.*, 1987; Takayama *et al.*, 1987; Lalonde *et al.*, 1989; Trick and Flavell, 1989; Trick, 1990; Scutt *et al.*, 1990; Chen and Nasrallah, 1990; Boyes *et al.*, 1991; and Dwyer *et al.*, 1991. The symbols for the signal peptides, potential N-glycosylation sites, and cysteine residues are as in Figure 1. The black bar denotes an insertion–deletion difference, which distinguishes *SLR1* molecules from the others.

such potential sites may be glycosylated *in vivo*, with the possible exception of those embedded within the cysteine-rich C-terminal region. However, the detection of multiple glycoforms of some *SLGs* (Section VI) indicates that posttranslational processing occurs involving the progressive occupation of such glycosylation sites. At present, each available *SLG* sequence is characterized by a unique glycosylation profile, raising the interesting possibility that this feature might form a component of allelic specificity, although it could be merely an inconsequential manifestation of the high underlying sequence variation.

It is therefore a major aim of the molecular description of the *Brassica* *S* system to relate its many biological features to elements of the *SLG* sequence data. The primary challenge, given these high levels of pervasive variation, is to ascribe, if possible, allelic specificities to certain sequence motifs. A related issue is the molecular rationale for the well-documented, complex dominance relationships that hold between *S* alleles in heterozygous combination (Thompson and Taylor, 1966). Sporophytic control of *S* in *Brassica* allows for such interactions to occur in both pollen and stigma. Although in detail these relationships are nonlinear and may sometimes even be reversed between the male and female tissues, groups of alleles may be broadly classified by dominance status. For instance, the S_2, S_5, and S_{15} alleles are recessive to all others in the pollen, although they do not occupy such a clear status in the stigma (Thompson and Taylor, 1966). That this property may have some correlates with the molecular features of the respective *SLGs* has been demonstrated by two observations. First, only weak hybridization signals were obtained in DNA blots on these genotypes using an *SLG13* probe (Lalonde *et al.*, 1989). Second, a monoclonal antibody raised against the polypeptide moiety of *SLG6* failed to react with these particular stigma proteins, although it does cross-react with the *SLG13*, *SLG14*, *SLG22*, and *SLG29* products (Kandasamy *et al.*, 1989). Some epitope polymorphism in the SLSGs may thus link these pollen-recessive *S* alleles, although such a connection would presuppose that the antigenic domain also contributes to pollen phenotype, something conspicuously not yet demonstrated. This area has now begun to be addressed at the DNA sequence level by the isolation and analysis of *SLGs* associated with the S_2 pollen-recessive allele (Chen and Nasrallah, 1990).

A stigma cDNA library from S_2 *Brassica oleracea* var. *alboglabra* (Chinese kale) plants was screened, at reduced stringency, with a full-length *SLG13* probe. Eight clones were isolated which contained the related *SLR1* sequence (Section II,C) and were not characterized further. In addition, one clone was found to be a partial cDNA for the putative *SLG2* sequence, which was then used to isolate a full-length clone from a library constructed from an S_2 *italica* (broccoli) line. Interestingly, these two cDNA clones apparently showed only 90% DNA sequence identity, although each represented a transcribed *SLG*-like message from plants of disparate genetic backgrounds, which nevertheless displayed the S_2 incompatibility phenotype (i.e., were cross-incompatible). Further investigations of this type may well help to define elements of *SLG* sequence variation, which are thus not important in the determination of allelic specificity. The *SLG2* cDNA sequences were found to be 30% diverged from *SLG13*, accounting for the previous hybridization results. An RFLP analysis of an F_2 population segregating for the S_2 and S_{13} alleles was then

employed to define three genomic *Eco*RI restriction fragments, hybridized by the cDNA probes, which were linked to the S_2 allele. These sequences were then isolated from a genomic library constructed from the S_2 *albogla-bra* line. It was found that one class of phage clones covered a contiguous region containing two of the hybridizing restriction fragments; the presence of the internal *Eco*RI site in the *alboglabra* transcription unit accounted for the recovery of truncated cDNA clones in the earlier experiments. The DNA sequence was identical to that of the cDNA and so this genomic copy was accordingly designated *SLG2A*. Interestingly, a second class of phage clones containing the third homologous *S*-linked *Eco*RI fragment was found to contain a gene copy, originally designated *SLG2B*, containing an open reading frame, which was 94% identical to that of *SLG2A*. It was shown recently that *SLG2B* corresponds to the extracellular receptor, or *S* domain, of the *SRK* gene. *SLG2A* and *SLG2B* were thus redesignated *SLG2* and *SRK2*, respectively (Stein *et al.*, 1991). An analysis of the predicted *SLG2* protein sequence revealed that the high divergence with respect to *SLG13* (and the other *SLGs*) was distributed more or less equally over the previously identified conserved and variable domains. Again, unique potential N-glycosylation profiles were encoded and the 12 cysteine residues conserved.

The most recent *SLG* DNA sequence to be reported is that of *SLG8*, which was isolated from a genomic library of a *Brassica campestris* S_8 homozygote using an *SLG13* 3' probe (Dwyer *et al.*, 1991). The deduced amino acid sequence differed by two residues from the previously directly determined S_8 glycoprotein sequence (Isogai *et al.*, 1987; Takayama *et al.*, 1987), the significance of which is unclear at present. Most interestingly, the degree of similarity of the *Brassica campestris SLG8* DNA and protein sequence to the *Brassica oleracea SLG13* was found to be significantly higher than that found for any characterized *Brassica oleracea SLGs* (Section IV,D), hinting that *SLG* gene polymorphisms predated speciation in the *Brassica* genus. A similar situation appears to hold in the gametophytic solanaceous system (Ioerger *et al.*, 1990).

A most exciting development in the molecular analysis of the *S* gene family has arisen with the recent isolation of genomic and cDNA clones for the *SRK* gene (Stein *et al.*, 1991). The *SRK6* gene was initially characterized by the analysis of a set of overlapping lambda clones from an S_6 genomic library (Nasrallah *et al.*, 1988), which defined a different, *S*-linked region with homology to *SLG*. Sequence analysis revealed that the region adjacent to this *SLG*-like domain contained motifs similar to those in the maize protein kinase *ZmPK1* (Walker and Zhang, 1990). *SRK6* cDNA clones were obtained by screening an S_6 stigma cDNA library with a kinase domain-specific probe (only two were recovered from some 10^5 plaques) or by polymerase chain

reaction (PCR) amplification from cDNA with appropriate primers. The *SRK* structure was then elucidated by comparison of genomic and cDNA sequences. The *SRK6* gene includes an *SLG*-like domain with 89% amino acid identity to *SLG6* (although with a less similar signal peptide). An 896-bp intron separates this domain from a kinase segment comprising another six exons, divided by small introns, which specify putative transmembrane and juxtamembrane domains, catalytic and C-terminal domains that are characteristic of receptor protein kinases. The *SRK6* sequence predicts a polypeptide of 857 amino acids (98 kDa). Certain sequence motifs within the catalytic domain would predict a serine–threonine substrate specificity; yet the presence of a glycosylated extracellular receptor has hitherto been confined to kinases with tyrosine specificity.

Further sequence analysis of the previously isolated *SLG2B* genomic clone (Chen and Nasrallah, 1990) revealed a similar *SRK* structure with an identical intron arrangement, only the first exon of which had been described earlier. A comparison of the *SRK6* and *SRK2* alleles yielded interesting results (Stein *et al.*, 1991). First, both genomic sequences have an in-frame stop codon two nucleotides downstream from the 5' splice site of intron 1, occupying in effect the same location as that utilized by the unspliced transcripts from the *SLG* gene. Second, these *SRK* alleles are only 68% conserved overall at the protein level, paralleling the divergence of their *SLG* counterparts (Chen and Nasrallah, 1990). That this level of variation exists between sequences that encode a putative catalytic function might be considered as somewhat surprising and would certainly have to be incorporated into any hypothesis of their role in incompatibility. However, when the *S* domain of each *SRK* allele was compared with its corresponding *SLG* allele, about 90% identity was revealed, indicating that the two are apparently undergoing concerted evolution (Section IV,D). The *SLG* and *SRK* thus constitute a gene pair or perhaps components of a broader *S*-specific haplotype.

RNA blotting experiments using discriminatory probes (Stein *et al.*, 1991) revealed that a fully spliced, 3-kb *SRK* transcript could be detected in stigmas at less than 0.75% of the level of the *SLG* message. Additional RNA species of different sizes could also be detected, some of which were interpreted as being derived from alternative splicing of the *SRK* gene and could encode truncated polypeptide products. For instance, because of the in-frame stop codon retained in intron 1, an unspliced *SRK* transcript could direct the synthesis of an *SLG*-like protein. Furthermore, fully spliced and unspliced *SRK* transcripts could be detected in mRNA from anthers at the uninucleate to binucleate stages. Finally, there is some evidence that *SRK*-related gene copies may be transcribed in leaves and cotyledons (Section VII).

C. Isolation of *SLR* Sequences

Lalonde *et al.* (1989) and Trick and Flavell (1989) reported further cDNA sequences, for *SLG22* and *SLG29*, but, perhaps more importantly, independently revealed the existence of a second *SLG*-like sequence that appeared to be co-expressed with the *SLG* gene. This so-called *SLR1* sequence was structurally homologous to the consensus of *SLGs* but showed substantial sequence divergence from it, significantly higher than that within the *SLGs* (an average of 30% at the nucleotide level and some 40% at the protein level), essentially conferring gene specificity to *SLR1* probes. It appeared to be highly conserved among the various *Brassica* S homozygotes and indeed other crucifers (Lalonde *et al.*, 1989). It is genetically unlinked to S and therefore not implicated in the determination of the specificity of self-incompatibility per se. *SLR1* gene transcription was also demonstrated in self-compatible lines of *Brassica campestris* (Lalonde *et al.*, 1989) and in the self-compatible amphidiploid species *Brassica napus* (Trick and Flavell, 1989), prompting the speculation that it may have a more general role in the physiology of the pollination event. Significantly, evidence had previously been reported for the existence of an identical S-like glycoprotein in two different S homozygotes of *Brassica campestris* (Isogai *et al.*, 1988); the partial sequence data presented here clearly placed it as an *SLR1* homolog, although certain sequence differences distinguished it as a *campestris* allele.

The *SLR1-22* and *SLR1-29* alleles differed by only three nucleotide substitutions in the coding sequences (each resulting in an amino acid replacement) and were identical in their nontranslated 3′ regions. The *SLR1-22* cDNA was then used to isolate genomic clones for *SLR1* genes from S_6 and S_{13} *Brassica oleracea* lines (Lalonde *et al.*, 1989). The S_{22} and S_{13} *SLR1* sequences were found to be identical, the implication being that strictly the same *SLR1* allele was present in both lines. The coding region of the *SLR1-6* sequence differed from *SLR1-22* by a single synonymous nucleotide substitution. The *SLR1-29* cDNA probe has also been used to isolate a genomic clone containing *SLR1-63* (Trick, 1990). This DNA sequence differed by 1% from *SLR1-29*, resulting in 10 amino acid substitutions between the two. Interestingly, *SLR1-29* is more similar to *SLR1-22* than it is to *SLR1-63* despite the fact that *SLR1-22* was in the *acephala* (kale) background and the others in the *alboglabra* (Chinese kale) background, indicating that *SLR1* alleles may be older than the racial differentiation of the *Brassica oleracea* species.

Scutt *et al.* (1990) reported an *SLG*-like cDNA sequence isolated from an S_5 stigma library by screening with an oligonucleotide probe specific for a region of *SLG6* highly conserved among other *SLG* sequences. This putative *SLG5* sequence was clearly S-homologous yet differed from *SLG6*

by 29% (36% at the protein level), with five mismatches over the conserved probe region. It thus occupied a somewhat anomalous position in the family of *SLG* alleles. At the same time, they reported the isolation of a presumptive *SLR1-5* cDNA by screening and counterscreening the library with stigma and leaf probes. The *SLR1-5* sequence was found to differ from *SLR1-22* by just one nucleotide, resulting in a conservative amino acid substitution, but exhibited significant divergence in its 3′ untranslated region.

The characterization of a second, expressed *SLR* gene (*SLR2*) from *Brassica oleracea* has been reported (Boyes *et al.*, 1991) and has put the S_5 results into a different perspective. Sequence analysis reveals that it is highly homologous (99% identical) to the sequence identified as *SLG5* (Scutt *et al.*, 1990), which should thus now be considered as an *SLR2* allele. This development concurs with the finding that oligonucleotide primers specific to hypervariable regions of that putative *SLG5* sequence hybridized to a common restriction fragment among *S* homozygotes. Indeed, PCR amplification of a fragment from an S_{29} stigma cDNA library (T. K. Bradshaw and C. G. Mousley, personal communication) indicates its transcription in other *S* genotypes. A new cDNA clone has since been isolated from the S_5 library, which appears to be the authentic *SLG5* sequence (C. P. Scutt *et al.*, personal communication). The *SLR2* work of Boyes *et al.* (1991) involved the isolation of cDNAs from S_2 and S_5 genotypes by screening methods designed to subtract out *SLR1*-specific clones. It was not reported whether clones for *SLG5* were also recovered. An apparently homogenous class of clones was thus recovered from each library, which cross-hybridized weakly with an *SLG6* 3′ probe. Representative clones from each class produced identical hybridization patterns on blots of *Brassica* DNAs, different from that obtained with an *SLG* probe, and were thus assumed to originate from a different locus, designated *SLR2*. The DNA sequence of the S_5 clone was not reported, so we do not know if it was identical to *SLR2-5* from Scutt *et al.* (1990); however the *SLR2-2* sequence did show 99% sequence identity with it. A 3′ probe from the *SLR2-2* cDNA clone was used to isolate a corresponding genomic sequence from the S_6 genotype. The sequence of this gene was again found to be highly homologous but to have a truncated 3′ terminus, lacking six amino acids with respect to *SLR2-2* and *SLR2-5*. A genetic analysis revealed that the *SLR2* gene was linked to the *SLR1* locus (Boyes *et al.*, 1991).

Figure 2 gives a schematic representation of the various *SLG* and *SLR* sequences so far characterized and summarizes the points discussed herein. The circumstances of the isolation of these various *S*-related messages may reveal fundamental data regarding the relative rates of transcription of the *SLG* versus *SLR* gene loci. For instance, originally no *SLR1*

sequences were found in 14 randomly picked stigma cDNA clones from an S_6 plant (Nasrallah *et al.*, 1985b). In later experiments *SLR1-29* and *SLG29* appeared to be equally abundant (Trick and Flavell, 1989) and *SLR1-22* seven times more prevalent than *SLG22* (Lalonde *et al.*, 1989). Differences in the level of *SLR2* transcription, inferred from Northern hybridization data, have also been noted between different *S* genotypes, being significantly higher in the pollen-recessive group of S_2, S_5, and S_{15} than the others (Boyes *et al.*, 1991).

III. Genomic Organization of the *S* Gene Family

SLR1 probes hybridize to single restriction fragments in *Brassica* genomic DNA (Lalonde *et al.*, 1989; Trick and Flavell, 1989), which copy-number reconstructions have estimated to represent single copies per haploid genome (M. Trick, unpublished data). Several such restriction fragments have been cloned from genomic libraries, and each was found to contain only one *SLR1* gene (Lalonde *et al.*, 1989; Trick, 1990; M. Trick, unpublished data). Thus, the *SLR1* gene appears to be a unique and substantially diverged member of the wider *S* family, which is now known to comprise the expressed *SLG* and a number of other sequences more related to the *SLG* gene, including the *SLR2* locus. The exact status of this latter gene within the *S* family is rather unclear. Although the evidence for limited allelic variation at the *SLR2* locus (Scutt *et al.*, 1990; Boyes *et al.*, 1991) is reminiscent of *SLR1*, a full-length *SLR2* cDNA probe hybridizes to many different genomic sequences (Scutt *et al.*, 1990; Boyes *et al.*, 1991), rather like *SLG6*. The *SLR2* probe appears to detect a substantially different set of sequences to those hybridized by *SLG6*, perhaps implying an even larger copy number for the *S* family. Furthermore, a 3' probe had to be used to isolate an authentic *SLR2* genomic clone (Boyes *et al.*, 1991). The *SLR2* sequence would thus appear to be more related to *SLG* sequences in general, and to *SLG2* in particular, than to *SLR1*.

Almost all the data on the genomic organization of the *S* gene family in *Brassica* come from experiments using *pBOS5* and related probes (Nasrallah *et al.*, 1985b,c). Analysis of *S* cDNA clones has revealed that *SLG* and *SLR1* messages make up the bulk of the class of abundant stigma-specific transcripts (Nasrallah *et al.*, 1985b, 1988; Trick and Flavell, 1989; Lalonde *et al.*, 1989); *SLR2* mRNAs may be somewhat rarer, varying between classes of *S* genotype (Boyes *et al.*, 1991). In addition, sequencing experiments have shown that there is no microheterogeneity among the cloned representatives of each class (Nasrallah *et al.*, 1988; J. Oldknow and M. Trick, unpublished data). However, as described, a full-length *pBOS5*

probe hybridizes to many restriction fragments in *Brassica* genomic DNA, indicating the presence of gene family members, which are either not transcribed at all or else at low levels, in the stigma, or perhaps are transcribed in other tissues. This property initially confounded the isolation of genomic clones, which contained the authentic, expressed *SLG* sequence. For instance, the full-length probe hybridized to 77 phage clones from a library of 10^6 recombinants made from homozygous S_6 plants and to 22 phage from a smaller library made from an S_{13} genotype (Nasrallah *et al.*, 1988). By restriction mapping the inserts in these hybridizing phage were assorted into nonoverlapping regions comprising a total of approximately 220 kb. It was thus estimated that the *Brassica* genome contains between 10 and 15 *S*-related gene copies (Nasrallah *et al.*, 1988; Dwyer *et al.*, 1989). Some of these copies have been sequenced and apparently are pseudogenes, having accumulated frameshift mutations (Nasrallah *et al.*, 1988).

A full-length *SLG29* probe hybridizes, at high stringency, to only one or two restriction fragments in *Brassica* genomic DNAs (Trick and Flavell, 1989). Thus, the *SLG29* cDNA has been used to isolate a putative *SLG* gene from the S_{63} genotype. Only two lambda clones from a library of some 5×10^5 recombinants were hybridized by the *SLG29* probe, and these were shown to have overlapping restriction maps that corresponded perfectly with the genomic data (T. Franklin, J. Oldknow, and M. Trick, unpublished data). However, transcription of this gene, at typical *SLG* mRNA levels, has yet to be demonstrated in S_{63} stigmas, and, in this respect, it is noteworthy that the *SLG29* probe failed to detect an RNA counterpart in this genotype (Trick and Flavell, 1989). The cloned gene contains a promoter region with low and intermittent sequence homology with the *SLR1-63* upstream region (Section V) and an open reading frame that predicts a slightly unusual carboxyl-terminus, and so its exact status within the family has yet to be determined. One hypothesis is that it might constitute an *SRK*-related gene transcribed in vegetative tissues. It is thus possible, reflecting the reported experience with the *SLG6* probe, that maximal sequence homology with a given *SLG* probe cannot be used to automatically isolate expressed *SLG* genes from any other *S* genotype.

Genetic analysis of segregating F_2 populations has shown that several of the *S*-related restriction fragments corresponding to the 10–12 cloned regions discussed herein co-segregate with the *SLG*-specific fragment (and therefore with *S*), indicating a genetic linkage (Nasrallah *et al.*, 1985b, 1988; Lalonde *et al.*, 1989). Other gene copies appear to be unlinked and dispersed elsewhere in the genome. It has been noted that, in general, those fragments that hybridize most strongly to *SLG* cDNA probes (being therefore the most related in sequence) are the ones that co-segregate with *S* (Dwyer *et al.*, 1989). The best studied example is that of the *SLG2A*

(*SLG2*) and *SLG2B* (*SRK2*) copies (Chen and Nasrallah, 1990), which show 94% sequence identity to one another, but other *S* phenotypes also co-segregate with at least two large hybridizing restriction fragments (Chen and Nasrallah, 1990). It had been suggested that there may be a clustering of *S* family members most similar to the expressed *SLG* sequence near the *S* locus, which could perhaps reflect a maintenance of relative homogeneity by unequal crossing-over (Dwyer *et al.*, 1989). It is now known that these observations relate to the physical contiguity of the co-evolved *SLG–SRK* gene pair at the *S* locus (Stein *et al.*, 1991).

No reconciliation is yet available between physical and genetic maps in *Brassica*. However, analysis of the relevant lambda clones had failed to reveal the presence of linked copies within the average 20 kb of DNA inserted in these vectors. More recently, pulsed field gel electrophoresis has shown the *SLG–SRK* gene pair to reside on the same 250-kb fragment (J. B. Nasrallah, personal communication). Some data are now becoming available on the long-range organization of the family in the *Brassica* genome. For instance, the *SLG6* probe identifies three RFLPs that map to chromosome 2 of *Brassica oleracea* (Kianian and Quiros, in press): two markers are coincident and thus presumably correspond to *SLG* and *SRK*, and the third is only 3 map units away. The *SLR1* and *SLR2* loci are 18.5 map units apart, this group segregating independently of the *S* locus (Boyes *et al.*, 1991).

IV. Molecular Evolution of the *S* Gene Family

A. Overview

Molecular evolutionary study is a discipline that proceeds essentially through the analysis of base or amino acid changes that have occurred during the divergence of related sequences. The comparison between the evolution actually observed for the sequences and a hypothetical random evolution enables one to demonstrate the occurrence and the nature of selective forces that direct the fixation of mutations in populations.

Among the 50 or so allelic forms of *Brassica SLG* genes identified by Mendelian genetics (Ockendon, 1985), several have now been cloned and sequenced: *SLG2* and *SLG6* (together with their *SRK* counterparts), *SLG8, 13, 14, 22,* and *29* as well as some S-related genes, *SLR1-5, 6, 22, 29,* and *63* and *SLR2-2, 5,* and *6* (Section II). Some of these allelic variants have only been published as polypeptides deduced from cDNAs instead of the nucleotide sequences themselves. Although a limited number of base substitutions can thus be compared, the analysis of the data presently

available still allows a significant description of the evolution of the S multigene family. The results we obtained will be compared here with those concerning two well-documented multigene families that we will use as references for two different modes of evolution. These are the hemoglobin genes, evolving neutrally, and the immunoglobulins, which are undergoing positive, diversifying selection. To allow this comparison, some of these sequences were briefly reanalyzed with the same procedures and software as those used for the S sequences.

B. Evolution of Globin Genes

Myoglobin and hemoglobin genes diverged about 650 million years (Myr) ago. Successive gene duplication events have since produced the α-globins (400 Myr), the adult β-globins and γ-globins (200 Myr), and more recently in higher primates, the δ-globins (40 Myr). Different evolutionary lineages have produced various genetic organizations of the globin gene family in different species (Nei, 1987).

These duplications allowed the diversification of the affinity of the globins for oxygen; myoglobin remained a monomeric transporter of oxygen in muscle tissues with a high substrate affinity. In contrast, hemoglobins have evolved into transporters of oxygen in the blood in the form of tetrameric assemblies displaying a refined affinity for oxygen. The associations of two α- and two non-α-chains are more efficient than the primitive chains alone, owing to cooperative allosteric changes of the conformation of these complexes with oxygen concentration. Moreover, the different non-α-globin genes, embryonic (ε), fetal ($^A\gamma$, $^G\gamma$), and adult (β, δ) are differentially expressed at various stages of development, presumably in response to different requirements in affinity for oxygen resulting from the changing environment during gestation.

In the terminology of molecular evolution, myoglobins and hemoglobins are paralogous sequences, whereas α-globins from mouse and rat, for instance, are orthologous. The comparison of paralogous sequences describes the evolution of the gene family since the divergence of its members, whereas that of orthologous sequences describes the evolution and the differentiation of the species to which these sequences belong.

As soon as the amino acid sequences of polypeptides became available for evolutionary comparison (during the early 1960s), it quickly appeared that proteins evolve at an approximately constant rate. Each protein species accumulates amino acid changes at a specific tempo, which is generally inversely related to its perceived biological importance, being thus subject to more functional constraints (Table I).

TABLE I

Amino Acid Substitution Rates for Various Proteins

Protein	Substitutions/site/1000 Myr
Histone H4	0.01[a]
Glutamate dehydrogenase	0.09
Myoglobin	0.89
Hemoglobin $\alpha-\beta$ chains	1.2
Ribonucleases, lysozymes, and lactalbumins	2–3
Fibrinopeptides	9
Immunoglobulin genes	
Gamma-lambda (constant region)	3
Lambda (variable region)	6.5
Gamma (variable region)	7.8
Pseudogenes	12–20
Brassica SLR1 and *SLR2*	2–4
Brassica SLG	30–35
SLG hypervariable region	60
Nicotiana S glycoproteins	Significantly higher values than for *Brassica SLGs*

[a] Data for non-S proteins from Nei (1987) and Kimura (1983).

The neutral theory was proposed by Kimura (1968) to account for this constant rate of change observed in proteins. This theory assumes that most of the base substitutions fixed during evolution are selectively neutral, that is, the various mutated forms of a given protein have the same functional efficiency as the parental form. Thus, nucleotide changes must accumulate randomly in DNA sequences at a constant rate (the so-called molecular clock), following a simple Poisson process analogous to that of radioactive decay. Deleterious mutations conferring a selective disadvantage, however, are eliminated through purifying selection.

According to the neutral theory, the strong conservation observed for histones (Table I) or for ribosomal genes, for example, is the consequence of their central function in the organization of the chromatin and of the protein translation machinery common to all organisms. In contrast, the rapid evolution of fibrinopeptide sequences is attributed to the minor role of these peptides per se after they have taken part in the maturation of fibrinogen during blood clotting. The rapid changes of albumins or lysozymes are correlated with the fact that they are largely dispensable. For the same reasons, the active sites of enzymes are more conserved than other domains having less functional importance: in both α- and β-globins, for instance, the heme pocket that directly interacts with oxygen evolves about 10 times more slowly than the surface of these proteins (Kimura, 1983).

The relaxation of such functional constraints would also explain the accelerated evolutionary rates of genes following duplication or inactivation events. Hemoglobins evolved more rapidly after their separation from myoglobins (Table I), and further accelerations were observed following successive duplications within the hemoglobin family. For instance, the β^A, β^C, and γ-globin genes in goat and sheep show greatly accelerated amino acid substitution rates after their separations, apparently due to relaxation of purifying selection and acquisition of new functionalities (Li and Gojobori, 1983). High substitution rates are observed in pseudogenes, or the "dead" gene copies produced by nonsense mutations, modifications of the control regions or retrotranspositional events. After their initial inactivation these sequences are completely free to evolve without any further functional constraint, and their evolutionary rates are the highest observed, as predicted by the neutral theory. A tenfold rate increase in pseudogenes versus active genes (Table I) has been estimated in the case of the globin gene family (Kimura, 1983; Li *et al.,* 1985).

The neutral theory has been further supported by the analysis of the relative fixation rates of nonsynonymous and synonymous base substitutions in nucleotide sequences, that is, of substitutions that do or do not bring about amino acid substitution. Due to the degeneracy of the genetic code, most first and second position changes in a codon are nonsynonymous changes, whereas many third position changes are synonymous. The observation of base changes shows that nonsynonymous changes are fixed at reduced rates as a result of purifying selection acting against protein modifications, whereas synonymous changes accumulate much more rapidly because they undergo much less counterselection. Thus, globin genes, similar to most genes analyzed so far, show a low proportion of nonsynonymous changes (32% on average). In contrast, globin pseudogenes that are not subject to purifying selection evolve with a ratio of 75% nonsynonymous base substitutions (Table II), close to the theoretical proportion of nonsynonymous changes predicted from all possible changes within the codon table.

In summary, globin genes, similar to many genes hitherto analyzed, would appear to have evolved through a series of gene duplications, with the daughter products differentiating from each other at an accelerated rate as one of them is free to acquire new functionalities. In addition, they display constant rates of nucleotide and amino acid changes, specific for each gene, with a predominance of synonymous over nonsynonymous changes. These features are consistent with the neutral theory and in opposition to the selectionist, synthetic or neo-Darwinian theory. The latter predicts that most mutations are fixed because they confer a selective advantage and that biologically important proteins or genes (e.g., histones, rDNA) should evolve more rapidly than less important ones (e.g., fibrino-

TABLE II

Quantitative Analysis of Base and Amino Acid Changes

Comparison	Total	Syn	Nonsyn	% Syn	% Nonsyn
α1-Hum/α1-Chim[a,b]	4	4	0	100	0
α1-Hum/α1-Ora	11	8	3	73	27
α1-Chim/α1-Ora	11	7	4	64	36
α2-Hum/α2-Chim	4	4	0	100	0
α2-Hum/α2-Ora	11	7	4	64	36
α2-Chim/α2-Ora	10	6	4	60	40
α2-Hum/α2-Mac	12	6	6	50	50
Summed changes (α-globins)	63	42	21	68	32
β-Hum/ψ-Hum	61	12	49	19	81
δ-Hum/$\psi\eta$-Hum	12	6	6	50	50
Summed changes (ψ-globins[c])	73	18	55	25	75
All possible base changes	549	134	415	24.4	75.6
Mouse V_H genes [d]	220	13	207	5.9	94

[a] Globin gene sequences were extracted from the EMBL and Genbank data bases, aligned with the PBestfit program using the algorithm of Smith and Waterman (1981) and analyzed with the COS command of the Analseq package (Gouy *et al.*, 1984).

[b] Abbreviations: Hum, α1- and α2-globin genes of human; Chim, chimpanzee; Ora, orangutan; Mac, macaque; $\psi\beta$-Hum, pseudogenes of human β-globin; $\psi\eta$-Hum, pseudogene of human η-globin.

[c] Pseudogenes often undergo copy correction by active genes through unequal crossing-over or gene conversion and do not appear to evolve without constraints in these circumstances. The evolutionary history of the pseudogenes of human β-globin and ψ-globin (Martin *et al.*, 1983; Vincent and Wilson, 1989) indicates that they have not undergone correction since their inactivation.

[d] Mouse immunoglobulin V_H gene sequences from Tanaka and Nei (1989).

peptides) due to the fundamental selective advantages modifications to these should provide. In addition, the rates of changes should vary with time, depending on environmental conditions; as a corollary, nonsynonymous base changes should be favored relative to synonymous changes (Nei, 1987).

C. Evolution of Immunoglobin Genes

Immunoglobulins are among the most rapidly evolving proteins (Table I); one may infer that this fast diversification of molecular variants generates the important polymorphism that might be selectively advantageous for the specificity of the immune system. This rapid evolution is correlated with exceptionally high ratios of nonsynonymous base substitutions (Table II); 94% of the mutations fixed in the complementary determining regions

(CDRs) of the immunoglobulin V_H genes in mouse are nonsynonymous mutations (Tanaka and Nei, 1989). In addition, the antigen recognition site regions of the various allelic forms of the Class I and Class II major histocompatibility complex (MHC) loci in mouse and humans show high proportions of nonsynonymous changes (Hughes and Nei, 1988, 1989). In contrast, the rate of accumulation of synonymous substitutions is not particularly fast and does not indicate an important genetic divergence between the molecular variants analyzed.

These rapid changes of immunoglobulins, together with the high ratio of nonsynonymous base changes, would seem to result from processes much more efficient and extensive than the simple accumulation of point mutations. Becker and Knight (1990) identified the mechanisms operating to produce the somatic, rearranged immunoglobulin V_H1 sequence in rabbit from the V_H2 to V_H7 germ-line sequences. They detected substitution clusters of nonsynonymous changes in the framework regions and complementary determining regions that correspond to sequence domains of 40–50 similar or identical nucleotides in the germ-line V_H2 to V_H7 genes, which are located upstream of the V_H1 gene. These sequences would be involved in the diversifying rearrangements of the V_H1 genes and would behave as donor sequences in gene conversion events. Similar gene conversions operating on short "hot spots" would participate to generate the polymorphism of the Class I MHC genes that encode the H-2 polypeptides in mouse (Weiss et al., 1983) or the human leukocyte antigen (HLA) polypeptides in humans (Seeman et al., 1986).

These molecular rearrangements are correlated to a positive selection termed "overdominant" or "diversifying" by Hughes and Nei (1988, 1989), totally different from the neutral selection observed for most genes, and specific to immunoglobulins and microglobulins, proteins also involved in immune defenses (Hughes and Nei, 1988, 1989).

D. The *Brassica S* Gene Family

1. The *SLR1* Subfamily

The three main subfamilies of the *S* family (*SLR1, SLR2,* and *SLG–SRK*) display evident homology relationships. An analysis using the Clustal algorithm (PC-Gene package) generated a dendrogram suggesting an evolutionary scheme that comprises several successive duplications followed by the differentiation of each subgroup (Fig. 3). A low but significant homology, now substantiated by recent developments, was also shown between the *Brassica S* family and the 5' extracellular domain region of the protein kinase *ZmPK1* cDNA cloned from maize by Walker and Zhang

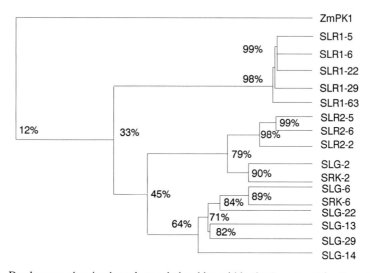

FIG. 3   Dendrogram showing homology relationships within the *Brassica S* family and with the extracellular domain of the maize *ZmPK1* sequence (Walker and Zhang, 1990). The dendrogram was generated from aligned sequences by the Clustal program of the PC-Gene package (Genofit, S.A.); the percentage figures are the calculated consensus amino acid identities between branches. It should be noted that each major cluster (*SLR1*, *SLR2*, and *SLG*) signifies an internally allelic group, but the *SLG–SRK* combinations represent nonallelic gene pairs or haplotypes.

(1990). Another representation of these homologies is given in Table III, as the numerical values of amino acid identities between pairs of subfamilies.

The *SLR1* genes form a group of sequences described as highly conserved among the *Brassica* species by Lalonde *et al.* (1989). This conservation is shown by a high degree of identity of all the *SLR1* sequences currently available in the literature, even if the *SLR1-63* variant reported

TABLE III

Matrix of Percentage Amino Acid Identity between
S Subfamilies[a]

	SLR1	*SLR2*	*SLG2–SRK2*	*SLG*
SLR1	98	—	—	—
SLR2	50	98	—	—
SLG2–SRK2	50	79	92	—
SLG	43	54	48	65

[a] Consensus values generated by the CLUSTAL program of the PC-Gene package.

TABLE IV

Quantitative Analysis of Nucleotide Changes among *SLR1* cDNAs

Comparison	Total	Syn	Nonsyn	% Syn	% Nonsyn
SLR1-63/SLR1-5	11	3	8	27	73
SLR1-63/SLR1-6	9	2	7	22	78
SLR1-63/SLR1-22	10	3	7	30	70
SLR1-63/SLR1-29	13	3	10	23	77
SLR1-5/SLR1-6	2	1	1	50	50
SLR1-5/SLR1-22	1	0	1	0	100
SLR1-5/SLR1-29	4	0	4	0	100
SLR1-6/SLR1-22	1	1	0	100	0
SLR1-29/SLR1-6	4	1	3	25	75
SLR1-29/SLR1-22	3	0	3	0	100
Total	58	14	44	24	76

(Trick, 1990) is slightly more divergent from the average of the other sequences. The conservation of *SLR1* gene sequences is also consistent with the low restriction polymorphism shown with *SLR1* probes among diverse *Brassica* lines and with the simple Southern patterns obtained, indicating that *SLR1s* are single-copy genes. The base or amino acid changes occurring among *SLR1s* are randomly scattered throughout the whole sequence, without any apparent localization at mutational hot spots.

The analysis and comparison of codon changes between the available *SLR1* sequences (Table IV) show an unusually high proportion of nonsynonymous versus synonymous substitutions (76% of nonsynonymous/24% of synonymous substitutions). As discussed, most genes evolve conservatively through high ratios of synonymous changes and low ratios of nonsynonymous changes (for example, 32% for α-globins), as a result of the purifying selection against deleterious mutations combined with the random occurrence of point mutations. The particular mode of base substitutions shown for the *SLR1* genes is prone to generate an intrinsically rapid diversification of the family of *SLR1* protein products, in contrast with its proposed strong conservation (based on hybridization data) among *Brassica* lines or species (Lalonde *et al.*, 1989). This proportion of 76% nonsynonymous base changes is close to that observed during the free evolution of pseudogenes (Table I) and the theoretical nonsynonymous proportion of all possible codon changes. *SLR1* genes thus seem to evolve through fixation of all base changes, in the complete absence of purifying selection. Given this preponderance of nonsynonymous substitutions among the *SLR1s*, it is perhaps even more noteworthy that a synonymous

change conserves one of the cysteine residues in the *SLR1-29–SLR1-63* comparison (Trick, 1990).

The rate of fixation of amino acid changes in the *SLR1* gene family can be estimated if one assumes that the *Brassica oleracea* species became rapidly established and isolated within the Brassicaceae, and thus is about as ancient as the family itself, which appeared some 5 Myr ago, according to the paleobotanical data of Muller (1981). We also have to assume a single common ancestor for extant *SLR1* alleles placed at 5 Myr (and, therefore, Kimura's 2T term is 10 Myr). Because 10 amino acid changes have been reported between the 444 residues of *SLR1-29* and *SLR1-63*, the two most distant *SLR1* sequences so far reported (Trick, 1990), the rate of amino acid substitution between these two is thus about 2.2 substitutions/100 sites/10 Myr or 2.2 substitutions/site/1000 Myr. Compared with the rates of amino acid changes reported from the literature for various proteins (Table I), this estimated maximal rate of *SLR1* evolution must be considered as rather high, although the rate of synonymous changes (0.7 changes/site/1000 Myr) is somewhat below the average rate cited for plant or vertebrate genes by Nasrallah and Nasrallah (1989) from the work of Wolfe *et al.* (1987). In conclusion, it seems that the underlying rate of nucleotide substitution for the *SLR1* genes may be rather low but that the fixation of nearly all such mutations without apparent selection for synonymous changes allows rapid evolution of the *SLR1* proteins.

2. The *SLR2* Subfamily

The *SLR2* sequences are also well conserved among the three *Brassica* lines (S_2, S_5, and S_6) analyzed so far (Scutt *et al.*, 1990; Boyes *et al.*, 1991) because they show about 98% homology between their amino acid sequences. There is 99% homology among the nucleotide sequences of *SLR2-2*, *SLR2-5*, and *SLR2-6*. In contrast to the *SLR1* genes, the *SLR2* sequences occur as multiple copies in the *Brassica* genome, as demonstrated by the complex Southern patterns obtained by Boyes *et al.* (1991). Moreover, these patterns show a high restriction polymorphism among the various *Brassica S* genotypes, yet clearly different from the RFLPs described for the *SLG* genes, and thus reveal a distinct and specific subset of the multigene family.

The analysis of variation between the two nucleotide sequences of *SLR2-2* and *SLR2-5* (Table V) show again, as for *SLR1* sequences, a high ratio of nonsynonymous base substitutions (70%), indicating the same rapid mode of diversification for the *SLR1* and *SLR2* proteins.

Having the same degree of amino acid conservation as the *SLR1* sequences, the *SLR2* proteins have evolved with the same relatively rapid rates of about two substitutions/site/1000 Myr. However, the *SLR1* and

TABLE V

Quantitative Analysis of Nucleotide Changes among *SLR2* and *SLG* cDNAs

Comparison	Total	Syn	Nonsyn	% Syn	% Nonsyn
SLR2-2/SLR2-5	20	6	14	30	70
SLG6/SLG8[a]	143	44	99	30	70
SLG6/SLG13	132	38	94	28	72
SLG6/SLG29	136	32	104	23	77
SLG8/SLG13	105	40	65	38	62
SLG8/SLG29	146	41	145	28	72
SLG13/SLG29	137	36	101	26	74
Summed changes					
Brassica oleracea SLGs	405	106	299	26	74
All *SLGs*	799	231	568	29	71
SLG6/SLG29	Hypervariable region			16	84
SLG2/SLG29	343	100	243	29	71
SLG2/SLG6	334	103	231	31	69
SLG2/SRK2	70	26	44	37	63
SLG6/SRK6	92	28	64	30	70

[a] This is a *Brassica campestris SLG* DNA sequence (Dwyer *et al.*, 1991).

SLR2 proteins have themselves diverged profoundly; they show only 50% amino acid homology, according to the calculations given in Table III, and their DNA sequences show low cross-hybridization in blotting experiments (Boyes *et al.*, 1991). These homologies are nevertheless highly indicative of a common gene lineage. Their extant amino acid divergence being thus around 0.5 substitutions/site, and the divergence rates between the two sequence classes being about $2 + 2 = 4$ substitutions/site/1000 Myr, they probably diverged about 120 Myr ago. This suggests that genes displaying structural homologies with the *SLR1* and *SLR2* families should be found in most angiosperms if the flowering plants appeared some 100–120 Myr ago, as proposed by Muller (1981).

3. The *SLG* and *SRK* Subfamilies

The deduced polypeptide sequences from cDNAs associated with five dominant *S* alleles from *Brassica oleracea* and one from *Brassica campestris* have been published. In addition, the recessive *Brassica oleracea SLG2* DNA and protein sequence is also available. Examination of these sequences demonstrates the extraordinarily high polymorphism of the *SLG* proteins in *Brassica*. The five dominant *Brassica oleracea* protein sequences show on average only 65% amino acid identity (Table III). For instance, the *SLG6* and *SLG29* proteins are 30% divergent. This variability

is correlated, as for the other members of the *S* family, with a very high proportion of nonsynonymous base changes (Table V), an average of 71–74%, depending on whether the *Brassica campestris SLG8* is included. Each pairwise comparison among the sequences shows an alternation of fairly conserved regions and hypervariable regions (Figs. 1 and 4). For these hypervariable regions the proportion of nonsynonymous base changes can be as high as 84% (for instance, in the region spanning from amino acid positions 258 to 344 for *SLG6* and *SLG29*).

As discussed in Section II,B, the higher *SLG8–SLG13* sequence homology may imply that *SLG* polymorphisms are older than the extant *Brassica* species. Nevertheless, to estimate the divergence rate of *SLG* proteins in *Brassica oleracea,* we again assumed that they began to diverge some 5 Myr ago in the *Brassica* lineage. This yields the average divergence rates of 30–35 substitutions/site/1000 Myr and the overwhelming rates of 60 substitutions/site/1000 Myr for the hypervariable regions (Fig. 4). Even the rate for synonymous changes (about 10–12 changes/site/1000 Myr) is exceptionally high for the *SLG* sequences with this assumption of divergence time. The combination of rapid base substitution and of apparent absence of purifying selection produces a pattern of positive, diversifying evolution, in perfect contrast with neutral evolution. These changes are so rapid that they probably cannot simply result from stochastic accumulation and fixation of point mutations; they are perhaps better understood if one accepts that they might result from radical changes, such as those occurring during gene conversion events, as demonstrated for the variable regions of immunoglobulins. Such hypotheses have already been proposed

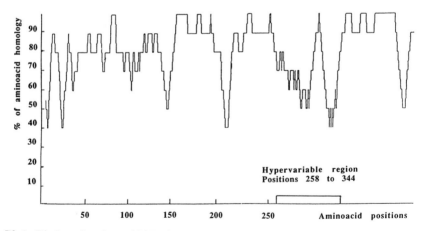

FIG. 4 Display of amino acid identity between the *SLG6* and *SLG29* sequences using a window length of 10 amino acids.

by Ebert *et al.* (1989) to explain the similarly high divergence rates of *S* alleles of the gametophytic self-incompatibility system in *Nicotiana*.

It is noteworthy that the *SLG* sequence cloned from a recessive *S* genotype, *SLG2*, shows a much higher homology with the *SLR2* genes than with the other *SLG* alleles, from dominant *S* genotypes. *SLG2* shares particularly with *SLR2* the same insertions and deletions relative to the *SLG* genes, suggesting that the functional properties of recessive SLSGs might be different from those of dominant SLSGs.

Finally, some observations may be made on the evolution of the *SLG–SRK* gene pair. Although the *SLGs* and the receptor *S* domains of their *SRK* counterparts are more closely related in sequence than are the various *SLG* alleles themselves (Table III), the percentage of nonsynonymous changes that have been fixed between the *SLG–SRK* pairs (63–70%) is as high as that observed between *SLGs* (Table V). The isolation of *SRK* genes from two *S* lines and the general observation of two strongly hybridizing RFLP markers co-segregating with *S* phenotype indicate that the duplication of the *S* domain between *SLG* and *SRK* was ancestral to the divergence of the *S* alleles. It thus appears that some mechanism of concerted evolution, through unequal crossing-over or perhaps an RNA intermediate, has maintained relative sequence identity between *SLG* and *SRK*. This mechanism must be imperfect, however, because it has not completely eradicated the apparent propensity of these genes to rapidly accumulate sequence changes.

E. Generation of New *S* Alleles

It has been well documented that mutagenesis experiments using ionizing radiation on gametophytic systems (which are intrinsically more amenable to such an analysis) have regularly failed to give rise to new *S*-allelic specificities (De Nettancourt, 1977). Such an approach has, however, generated mutations leading to loss of incompatibility function in pollen or style, a finding that has been interpreted as evidence of a tripartite structure for the *S* locus (Lewis, 1949, 1960). It is possible that this essentially genetic interpretation, extended to sporophytic systems, may well have some physical basis in the structure of the *SLG–SRK* gene pair at the *S* locus (Stein *et al.*, 1991), if these two genes do indeed cooperate in some way to effect incompatibility. However, the observation of natural, nonsynonymous nucleotide changes fixed between the *S* domains of the *SLG* and *SRK* genes leads us to expect that a proportion of induced point mutations should be without apparent consequence for *S* phenotype. Nevertheless, the phenomenon of high *SLG* polymorphisms, coupled with *SLG–SRK* co-evolution, may indicate why point mutations cannot con-

structively alter S specificity. However, it has been reported that new alleles have been apparently created at high frequencies through the intense, forced inbreeding of self-incompatible populations of plants.

Several reports of such events concerned the gametophytic systems of *Trifolium pratense* (Denward, 1963), *Nicotiana* (Pandey, 1970), and *Lycopersicon* (De Nettancourt *et al.*, 1971). The "new" alleles created here by inbreeding were generally found to be alleles already identified among S genotype collections: thus, inbreeding of an S_1S_2 genotype in *Lycopersicon* produced a stable allele identified as S_3 (De Nettancourt *et al.*, 1971). In a sporophytic system Zuberi and Lewis (1988) reported the creation of S_5, S_6, and S_x alleles from an S_1 allele in an S_1S_3 genotype of *Raphanus;* several cases of reversion to the original S_1 allele were subsequently observed. Although the execution of experiments designed to look for such events are clearly fraught with difficulty and these findings have no easy explanation, we should not be dismissive of them. We have seen how the level and nature of the variation between the SLG sequences are unusual and the existence of the wider family of S-related sequences could accommodate novel DNA rearrangements. The reason why inbreeding, which simply serves to decrease average heterozygosity, should trigger such genetic changes is unknown but could possibly be rationalized to the induction of homozygosity for a group of recessive genes controlling such a mechanism.

An important question about these allelic changes is whether a given S allele corresponds unequivocally to unique S sequences expressed in the stigma and in the anthers. At present, there are insufficient data to answer this comprehensively, although the finding of Chen and Nasrallah (1990) of rather different expressed SLG sequences, each purportedly linked to S_2 alleles, is suggestive of a greater complexity. However, if one-to-one correspondence were the case, the observations of allelic changes would support the hypothesis that extensive and probably frequent exchanges of S-related sequences occur, perhaps through unequal crossing-over or gene conversion (Zuberi and Lewis, 1988; De Nettancourt *et al.*, 1971). Because a limited set of S alleles is involved in the generation of these mutants instead of an infinite number of genetic combinations, the exchanges between the S-related sequences might consist of a real interchange between different molecular arrangements of S-related alleles. The large S multigene family in *Brassica* could provide the pool of S-related sequences for such recombinations. In the case of *Nicotiana,* a large multigene family that could sustain this explanation has not been described, but Ebert *et al.* (1989) pointed out that in *Nicotiana* the different functional S alleles are so divergent that they hardly cross-hybridize, perhaps preventing the detection of S-related sequences in the genome.

V. *S* Gene Structure

All of the *S*-homologous genomic sequences of expressed genes so far analyzed, whether of *SLG, SLR1,* or *SLR2* types, contain uninterrupted open reading frames. This rule extends to the *S* domains of *SRK* genes, even though their kinase components do contain introns. The conservation of internal organization across the gene classes (which appear to represent ancient divisions) underlines their perceived common phyletic lineage. At the same time, it would appear to deny one previously suggested possibility for the generation of new *S* alleles, namely, assembly from other sequences by exon shuffling. As we have seen, the observed variation between *S* sequences more resembles the inclusion of small substitution clusters, perhaps effected through gene conversion. However, it is noteworthy that the first and largest intron in the *SRK* gene, of different lengths in different *S* genotypes (Stein *et al.,* 1991), cleanly separates the *S* domain from the remainder, suggesting that the receptor has most recently evolved as a single unit.

SLG and SLR1 genes have the canonical eukaryotic signals for both the initiation and termination of transcription. In common with many plant genes, a comparison of the 3' untranslated regions of the various *SLG* and *SLR* cDNAs reveals the existence of several potential polyadenylation signals. *SLR1* genes are characterized by a single canonical 3' signal, AATAAA, 60 bp downstream from the termination codon, although the sequence AATAAT is present only 10 bp downstream. The *SLG* and *SLR2* alleles sequenced have twin AATAAA motifs, with the 3'-most of these two signals apparently utilized. At present, extensive 5' and 3' DNA sequences have only been published for the *SLR1-63* gene (Trick, 1990). These regions have also been sequenced for an *SLR1* allele from a self-compatible line of *Brassica oleracea* var. *alboglabra* and for the two homoeologous *SLR1* alleles present in *Brassica napus* (J. Oldknow and M. Trick, unpublished data), allowing some further comparisons to be made. The emerging picture is one of marked conservation, in keeping with the low variation between allelic coding regions, of the 5' regions as far as they were sequenced (− 150 from the ATG) but, surprisingly, almost total divergence in the 3' regions beyond the distal poly(A) addition signal. Such 3' divergence had also been noted between *SLR1-5* and *SLR1-22* (Scutt *et al.,* 1990).

The isolation of a genomic clone for *SLG13* (Nasrallah *et al.,* 1988) allowed the identification of polyadenylation signals and, interestingly, a putative TATA box element, TATAAAT, only 55 bp from the translation start. The upstream region of the *SLG8* genomic sequence (Dwyer *et al.,*

1991) contains the same motif. Again, the identical sequence was found similarly located in the *SLR1-63* genomic sequence (Trick, 1990), although, in this gene the sequence TATATAAT was also found in a somewhat more conventional location, 159 bp upstream of the ATG. Nevertheless, transcript mapping studies (J. Oldknow and M. Trick, unpublished data) indicate that the proximal TATA element is indeed utilized. A similarly short leader sequence has been reported for the potato S_2 self-incompatibility gene (Kaufmann *et al.*, 1991).

It is of particular interest to compare the promoter regions from *SLG* and *SLR1* genes because their expression would appear to be under identical developmental control; the high sequence conservation makes comparisons among *SLR1* alleles uninformative in this respect. Sequences from the *SLR1-63* and *SLG8* upstream regions can be legitimately compared. Little homology is found in the two 5' regions up to −90 from the ATG, apart from the putative TATA elements. An *SLG*-homologous genomic clone from the S_{63} genotype has recently been isolated and sequenced (T. Franklin, J. Oldknow and M. Trick, unpublished data). Again, in this three-way comparison, no homology is found up to −90. However, there is a region of perfect homology between the S_{63} sequence and the 5' end of the reported *SLG8* sequence, the motif CTGCTAGGTTC at −100. Interestingly, the first five nucleotides of this motif define the end of a 14-nucleotide domain conserved with only one mismatch between the *SLR1-63* and putative *SLG63* sequences. There is thus an indication of conserved domains between disparate *S* 5' regions. Promoter dissection studies will be required to gauge the significance of such sequence homologies.

VI. Expression of *S* Sequences

A. *SLG* Expression in *Brassica* Stigmas

A significant amount is known about the temporal and spatial expression of the *S* family genes in *Brassica*. Furthermore, the behavior of *Brassica* *S* transgenes in *Nicotiana* provides some evidence for a possible link in the mechanics of control of *S* gene expression between these two otherwise disparate, self-incompatibility systems. Expression of *S* transgenes has recently been obtained in *Brassica oleracea* but without straightforward results for incompatibility phenotype.

Earlier work had demonstrated the correlation of the levels of SLSG product and *SLG* transcript in the stigma (Roberts *et al.*, 1979; Nasrallah *et al.*, 1985b), with their reaching maximal levels upon the acquisition of competence for the self-incompatible response. Because this response is

characterized by a rapid interaction between the pollen grain and the stigmatic surface (Ockendon, 1972; Heslop-Harrison, 1975), with those rarely germinating pollen tubes being inhibited specifically at the cellulose–pectin layer of the papillar cell wall (Kanno and Hinata, 1969), this single cell layer had long been supposed to be the site of S gene action in *Brassica*. The previous work with extracts derived from whole stigmas could not give the necessary cellular resolution to confirm this. Localization of *SLG* transcription to the papillar cells was obtained by *in situ* hybridization of an *SLG6* probe to longitudinal stigma and style sections (Nasrallah *et al.*, 1988). Labeling was confined to the cytoplasm of the surface papillar cells, none being detected over the basal stigma cells or those of the transmitting tissue. The hybridization signal in the papillar cells was correlated with the developmental stage of the stigma, allowing a striking visual interpretation of the earlier RNA studies (Nasrallah *et al.*, 1985b).

Similar studies with an *SLR1* probe (Lalonde *et al.*, 1989) established that the same papillar cells also transcribe the *SLR1* gene, thus further defining the known co-expression of *SLG* and *SLR1* transcripts (Trick and Flavell, 1989). *SLR1*-directed transcription could be readily detected in the papillae of immature buds, 3 days before anthesis, a day earlier than with the previous *SLG* experiments (Nasrallah *et al.*, 1988). This effect could simply be related to allelic differences in promoter strengths because, in this case (an S_{22} genotype), maximal *SLR1* transcription was found to be about sevenfold higher than that of the *SLG* (Section II,C).

Further refinement to *SLG* localization, now at the subcellular level and directed toward the fate of the S glycoprotein products themselves, came with the use of immunogold labeling techniques using monoclonal antibodies raised against the polypeptide moieties of SLSGs (Kandasamy *et al.*, 1989). SLSGs were thus immunolocalized in the process of being exported to the cell walls of the stigma papillae. Their level of accumulation in the cell walls showed the expected correlation with bud development, *SLG* transcription, and the self-incompatibility response. In self-incompatible bud and mature, open flower stages, large amounts of gold particles were dispersed over the cell walls. In the cytoplasm gold particles also labeled the rough endoplasmic reticulum and Golgi apparatus, indicating that the S product is processed through an active secretory pathway, which is an obvious characteristic of these cells (Roberts *et al.*, 1984).

Immunolocalization work with polyclonal antibodies raised against a *trp*E-*SLR1* fusion protein (Umbach *et al.*, 1990) has now shown that the *SLG–SLR1* co-expression is complete in every detail, with the stigma papillar cells also processing and secreting *SLR1* glycoprotein to the cell wall. At present, knowledge of *SLR2* expression is limited to its transcription in stigma tissue (Boyes *et al.*, 1991), but there is no reason to doubt

that it, too, is processed and secreted in the same way. Thus, it is known that at least two, and probably three, related glycoproteins are present at the site of pollen–stigma interaction, making it all the more intriguing that only one is implicated in the specificity of the self-incompatibility response. The same studies (Umbach *et al.*, 1990) also clarified some aspects of the posttranslational processing of the SLSG and *SLR1* proteins, demonstrating that the multiple glycoforms resolved on SDS-PAGE gels represent different glycosylation products of the same polypeptide. Thus, much of the electrophoretic heterogeneity observed in stigma glycoprotein extracts (Nasrallah and Nasrallah, 1984) can probably be attributed to the posttranslational processing of a few proteins. In this respect it should be noted that glycosylation has been demonstrated to be necessary for the functioning of self-incompatibility in *Brassica* (Sarker *et al.*, 1988).

B. *SLG* Gene and Promoter Constructs in Transgenic Plants

Several pieces of work involving the insertion of *Brassica S* transgenes or constructs driven by *Brassica S* promoters into both *Nicotiana* and *Arabidopsis* have revealed an unexpected conservation in the regulation of *S* gene expression among these three species. In addition, complete *SLG* or *SRK* genes have been expressed in *Brassica*.

In the first study, an 11-kb *Eco*RI fragment, comprising the *SLG13* gene and approximately 4.2 kb of upstream and 5 kb of downstream sequence, was cloned into a Ti binary vector and then delivered by *Agrobacterium tumefaciens*-mediated transformation into *Nicotiana tabacum* (Moore and Nasrallah, 1990). *Brassica SLG* expression was subsequently detected at RNA and protein levels in both stigma and style extracts of various tobacco transformants. A broad correlation was observed between copy number of nonrearranged insertions and the levels of expression of RNA and protein among individual transformants. *SLG* transcripts were detectable in green buds but were found to be maximal in open flowers. Immunoblotting experiments on transgenic protein extracts revealed the presence of the three characteristic glycoforms, as in the homologous system (Umbach *et al.*, 1990), but with the lowest molecular weight glycoform being detectable in only the highest expressing transformants.

The tobacco heterologous host is thus capable of correctly processing the *Brassica* SLSG product, and this observation per se constitutes clear evidence of the multiple glycoforms arising from posttranslational processing of a single gene (Umbach *et al.*, 1990). Most interestingly, the *Brassica* SLSG was immunolocalized chiefly to the stylar transmitting tissue of the transgenic plants, with only low levels of staining over the stigmatic epidermal and papillar cells; anther tissue did not stain at all.

The *Nicotiana S* associated glycoprotein has been shown to be expressed in a similar pattern, the gametophytic self-incompatible response being characterized by a cessation of pollen tube growth in the transmitting tissue of the pistil (Cornish *et al.*, 1987).

These data thus indicate that the heterologous tobacco host recognizes some *cis*-acting elements in the *Brassica S* promoter but, in so doing, elicits a pattern of expression more closely resembling that of the endogenous, gametophytically controlled, *Nicotiana S* gene (Anderson *et al.*, 1989) than that of the homologous system. This points toward the conservation, between the Brassicaceae and Solanaceae, of some transcription factor, the expression of which itself is controlled in different cell type patterns in the two genera.

In another set of experiments, a fragment containing the *SLG22* gene was introduced into *Nicotiana* and the results examined by microscopic immunolocalization (Kandasamy *et al.*, 1990). Again, a high level of *Brassica* SLSG expression was detected in the transmitting tissue of the style, with lighter staining at the base of the stigma. The detailed pattern was slightly different from that produced in plants carrying the *SLG13* transgene, in that this time no staining at all could be detected in the stigmatic epidermis. At the ultrastructural level, immunostaining was particularly dense over the longitudinal walls in the intercellular matrix of the transmitting tissue. Labeling was also observed over the secretory matrix surrounding the placental epidermal cells adjoining the ovules, another point of detail that closely resembles the *Nicotiana* pattern of *S* expression (Anderson *et al.*, 1989). Again, staining could not be observed in anther tissue. In neither of these experiments did transgenic plants become self-incompatible.

Similar results have been observed with the expression of an *SLR1-63* transgene in tobacco (M. Trick, unpublished data), although in this case the stylar expression was assayed only crudely by Northern blotting experiments. Nevertheless, this further indicates the sharing of some regulatory promoter element, which a close comparison of *SLG* and *SLR1* upstream sequences may even identify.

Transformation experiments in which constructs containing an *SLG* promoter driving different reporter genes were delivered into *Nicotiana* and *Arabidopsis* have since yielded additional information. A 3.65-kb *Bam*HI promoter fragment from *SLG13* was fused to the diphtheria toxin A chain (DT-A) gene and also to the β-glucuronidase (GUS) reporter gene and these constructs transformed into *Nicotiana* (Thorsness *et al.*, 1991). Essentially, the same pattern of *Nicotiana S* expression was observed, with the *Brassica SLG* promoter directing GUS expression or DT-A mediated autonomous cell death (accompanied by developmental abnormalities) in the pistil. However, the increased sensitivity afforded by these

reporter gene techniques allowed *SLG* promoter activity to be detected in transgenic tobacco pollen. GUS expression or cell ablation was observed in a proportion of pollen grains from transgenic plants consistent with postmeiotic segregation for Ti insertion loci and subsequent gametophytic expression (Thorsness *et al.*, 1991). Similarly, weak transient expression of an *SLR1* promoter::GUS construct has been obtained in particle-bombarded tobacco pollen (M. Trick, unpublished data). However, when the *SLG*:GUS construct was transformed into *Arabidopsis* (Toriyama *et al.*, 1991a), GUS activity was observed in the stigma papillar cells only and not in the underlying style tissues. Significantly, the highest expressing transformants also showed weak staining in the tapetal cells of the anther at the uninucleate microspore stage. Mature pollen grains did not stain. A *Brassica*-like *S* pattern was thus obtained. The *SLG* promoter-directed anther–pollen activities in these experiments suggest that the earlier failure to detect *Brassica* SLSG in transgenic tobacco pollen grains (Moore and Nasrallah, 1990) was due to a lack of sensitivity.

A picture is thus emerging in which authentic *Brassica S* expression seems to be a function of the degree of taxonomic affinity of the recipient genome, with *S* transgenes being expressed gametophytically in *Nicotiana* and sporophytically in *Arabidopsis*. This interpretation may yet prove to be an oversimplification. The view has been expressed (Thorsness *et al.*, 1991) that this common regulation of *Brassica SLG*s, species-specific in detail, indicates an evolutionary link between gametophytic and sporophytic self-incompatibility, despite the nonhomology of the coding sequences of *Brassica SLG*s and *Nicotiana S* RNases. The hypothesis that sporophytic self-incompatibility is an evolutionary derivative of the gametophytic system had long been expounded before the availability of such contradictory evidence, being based on both phylogenetic (Brewbaker, 1967) and theoretical considerations (Whitehouse, 1950). However, it certainly seems likely that the two systems are related, by virtue of their being controlled by a regulation that is active in specific floral cell types. This regulatory mechanism could well have been recruited independently in a polyphyletic evolution of incompatibility systems.

Finally, two recent experiments have succeeded in transforming *S* sequences into *Brassica oleracea* plants, where expression was obtained. In the first of these studies, a 6.8-kb fragment containing the *SLG8* gene from *Brassica campestris* was introduced into two types of recipient plants by *Agrobacterium*-mediated transformation (Toriyama *et al.*, 1991b). Interpretation of the resultant transgenic phenotypes was complicated by the fact that these recipients were F_1 hybrids (with complex endogenous *S* genotypes) and that interspecific incompatibility would not allow testing for the new, introduced allelic specificity. Nevertheless, the self-incompatible phenotypes of the recipients were essentially modified in

stigma or in pollen. For one type of recipient, a change in stigma phenotype was accompanied by the observed synthesis of novel *SLG8* protein and a drastic reduction in the level of endogenous SLSG. This effect may have resulted from sense inhibition of endogenous transcripts by the introduced gene.

A genomic clone of the *SRK6* gene was introduced into *Brassica oleracea* plants homozygous for the S_2 allele (Stein *et al.*, 1991). Relying on the high sequence divergence between the two *SRK* alleles involved, an *SRK6* S domain probe was used to specifically detect expression of the transgene. Several transformed plants showed levels of both spliced and unspliced *SRK6* specific transcripts in stigma and anther mRNAs, but no phenotypic change was reported.

VII. Expression of *S*-Homologous Sequences in *Brassica* Anthers

A. Overview

Although the *S* transcripts and their protein products have been well characterized in *Brassica* stigmas, relatively little is known about their predicted counterparts in pollen or anther tissues. The protein dimer hypothesis of self-incompatibility function proposed by Lewis (1963) predicts that pollen is recognized as self-incompatible and rejected at the point of contact with female tissues when the two conjugating partners bear identical *S* products encoded by a common *S* allele. The autoassociation (or dimerization) of homologous male and female *S* glycoproteins would be the initial step in self-incompatible cell recognition. Moreover, the sporophytic control of self-incompatibility in *Brassica* species implies that the pollen *S* products must be synthesized in diploid tissue of the parent, either in pollen mother cells before meiosis (Pandey, 1960) or in the diploid tissue of the tapetum (Heslop-Harrison, 1968), which is known to contribute to the formation of the exinic outer layer. In plants with a gametophytic self-incompatibility system similar to *Nicotiana,* the production of pollen *S* products is predicted to occur in the male gametophyte itself after the meiotic divisions.

Clearly, the hypothesis that male and female tissues produce identical *S* products according to a homology model (rather than complementary *S* products following a lock and key model) is more easily experimentally falsifiable, given access to the putative female molecules, because the male *S* glycoproteins or *S* transcripts should then be structurally homologous. At the protein level immunochemical techniques have been used to detect

putative pollen *S* products with antigenic epitopes in common with the stigma *S* glycoproteins. Faint, but significant, responses were obtained with pollen protein extracts with a monoclonal antibody raised against a stigma *S* glycoprotein, supporting the hypothesis of Lewis (Gaude *et al.,* 1988; Dickinson *et al.,* in this volume). At the nucleic acid level molecular hybridization of anther RNAs with *SLG* cDNA probes has also given only very weak signals (Nasrallah and Nasrallah, 1986, 1989). Thus, both protein and nucleic acid approaches had indicated a possibly low level of expression of the *SLG* gene in male tissues, although these findings may have to be reevaluated in the light of the discovery of the *SRK* anther transcripts.

As we have seen, a completely different approach is now also in use, which has already demonstrated the potential of the *SLG* promoter to activate reporter genes conferring acute sensitivity, in the male tissues of transgenic *Arabidopsis* and *Nicotiana* plants (Thorsness *et al.,* 1991; Toriyama *et al.,* 1991a). However, as discussed, these experiments do not address the same question as those seeking male *S* products per se. More recently, the PCR has been used to specifically amplify sequences from anther cDNAs (Guilluy *et al.,* 1991), which are highly homologous to the *SLG* coding sequences, using primers synthesized to the *SLG6* cDNA sequence (Nasrallah *et al.,* 1987). These results will now be described in some detail.

B. PCR Detection of *SLG*-Homologous Transcripts in *Brassica* Anthers

1. Specificity of PCR Amplifications

Great precautions were taken to avoid artifactual amplifications resulting either from nucleic acid contamination or from nonspecific priming. First, the specificity of the PCR conditions was assessed by checking that the sizes of the PCR products were identical to those predicted from the sequences between the *SLG6* primers in the published *SLG6* (and *SLG29*) cDNA sequences. Second, two cDNA clones, which contained the sequences of the *SLG29* and *SLR1-29* genes from a *Brassica* S_{29} genotype (Trick and Flavell, 1989), were used as both template and probe controls. The PCR fragments obtained using the *SLG29* clone as a template hybridized strongly with an *SLG29* insert probe but not at all with the *SLR1-29* probe in Southern blot hybridizations and washes carried out at normal stringency. Furthermore, the *SLR1-29* template DNA did not permit any PCR synthesis with the *SLG6* primers, showing that the greater homology of these primers with the *SLG29* sequence (85–100%) than with the

SLR1-29 sequence (70–80%) was sufficient to discriminate between the two kinds of template. It was thus demonstrated that these experiments would exclusively amplify *SLG*-like sequences.

PCR products were amplified from cDNA templates synthesized by reverse transcription from mature stigma mRNAs. They had exactly the size expected for bona fide *SLG* sequences and hybridized with the cloned *SLG29* sequence but not with the *SLR1-29* clone. When used as a radioactive probe on Northern blots, these PCR fragments elicited the same hybridization patterns as those obtained with the *SLG29* probe, a single strong band corresponding to a stigma mRNA species of about 1.6 kb, as identified in the literature as the *SLG* stigma transcript (Nasrallah *et al.*, 1985b; Trick and Flavell, 1989). *In situ* hybridization of stigma sections with these PCR probes showed an exclusive labeling of papillar tissues, in agreement with the known transcription of the *SLG* in only these cells. No hybridization was obtained with RNAs from other tissues (roots, leaves, anthers). Thus, PCR amplifications performed with primers to constant domains of the *SLG* seem to produce authentic copies of the *SLG* sequence from stigma cDNA.

2. Amplification of *SLG*-Homologous Sequences from Anthers and Leaves

When anther cDNAs were used as templates, PCR products were also obtained, which again showed the strict specificity expected for *SLG* sequences, with the fragment sizes matching those predicted from the *SLG* cDNAs and exclusive hybridization with the *SLG29* probe. This supported the original hypothesis that *SLG* genes (or at least *SLG*-homologous sequences) were transcribed in the anthers of *Brassica* and can now be interpreted in terms of *SRK* or *SRK*-related products (Stein *et al.*, 1991). Higher levels of PCR product were observed with cDNA prepared from anthers at the microspore stage than from older, bicellular or tricellular pollen stage anthers.

Depending on the primers used, the PCR products synthesized from anther cDNA behaved differently when used as hybridization probes on Northern blots. PCR products made with primers specific to the 5′ region of the *SLG6* sequence (region A in Fig. 1) hybridized to RNAs from sexual tissues only; the typical *SLG* mRNA (1.6 kb long) was detected among stigma RNAs and, intriguingly, a 4-kb message was detected in low amounts in mRNA from anthers at the microspore stage of development. These results are thus consistent with the hypothesis of Lewis that homologous transcripts should be produced in both the anthers and stigmas of *Brassica* flowers. However, the two transcripts are not identical because their apparent molecular weights are different and probes synthesized

from PCR products from anther and stigma cDNAs do not behave reciprocally in Northern blots. Male PCR products hybridize both the 1.6-kb stigma mRNA and the 4-kb anther message, whereas the female PCR products hybridize only the 1.6-kb stigma message. No signal was obtained with mRNAs from older anther stages, in agreement with the hypothesis of Heslop-Harrison (1968) that the tapetum might be the site of synthesis of the male *S* products.

With primers specific to the more 3' constant region of the *SLG6* sequence (Fig. 1C), PCR amplifications from anther cDNAs evidently produced a much more heterogenous population of products. Such PCR probes hybridized to mRNAs from all sources, both sexual and vegetative. Yet the PCR products still displayed the main characteristics of *SLG* sequences (expected size and hybridization with *SLG29* but not *SLR1-29* clones). Similar results were observed for PCR amplifications from leaf cDNA templates. Questioning the accepted idea that *SLG* products are specific to the pollen–stigma interactions, PCR products were obtained from leaf RNA, often in low, but nevertheless significant, quantities. This observation of transcripts homologous to *SLG* sequences in vegetative tissues of *Brassica* was initially suspected to have resulted from mispriming in the PCRs. Its validity is, however, supported by the cloning from maize roots of the *ZmPK1* cDNA (Walker and Zhang, 1990), which encodes a putative protein kinase with an extracellular receptor domain showing a high protein homology (25% identity, 50% conservation) with *Brassica SLG* and the recent description of *SRK* transcripts in both stigma and anther tissues and of *SRK*-related transcripts in leaves and cotyledons (Stein *et al.*, 1991). These combined data strongly suggest that the expressed *S* family is larger than the *SLG*, *SLR1*, and *SLR2* subfamilies already shown to be transcribed in the stigma. It might be involved in several plant cell-signaling processes in sexual (i.e., the pollen–stigma interaction) as well as in vegetative tissues. It might be expected to be widely distributed among the flowering plants because some *S*-type motifs are now known to be shared by species as divergent as maize and *Brassica*.

An increasing number of *S* locus related sequences have now been shown to be expressed in at least the female sexual tissues of *Brassica* (Lalonde *et al.*, 1989; Trick and Flavell, 1989; Boyes *et al.*, 1991). Clearly, it will be necessary to determine whether the PCR products obtained from anther mRNAs correspond to transcripts of the characterized *SLG* or *SRK* genes in male tissues, rather than to transcripts of a hitherto unidentified *S* locus-related gene (the hybridization data make it unlikely that they could be *SLR1* or *SLR2* sequences). This point could be ascertained by the demonstration that allelic variants of these PCR products co-segregate strictly with the *S* genotypes in F_1 and F_2 progenies.

VIII. Conclusions

The molecular biology of the *Brassica* self-incompatibility system is still much at a descriptive stage. What is being attempted is to link a complex range of phenomena with the inheritance of certain DNA sequences, and it is probably fair to say that, so far, the pursuit at this level has not substantially illuminated the previously established links between SLSGs and incompatibility phenotypes. What it has done is to uncover some of the complexity of a fascinating gene family. We hope that some of the flavor of this complexity has been given in the course of this review. Although the recent research has amassed more detail on *SLG* genes and their expression in female tissues, perhaps the most exciting development has been the discovery of *S*-related expression in the male, in the form of the *SRK* transcripts and also of *SLG* and *SLR1* promoter-directed gene activities.

The transformations of *Arabidopsis* and *Nicotiana* plants with reporter genes under control of the *SLG* promoter and the transient expression of an *SLR1* construct in *Nicotiana* have clearly demonstrated that these promoters contain elements that are recognized in both female and male cell types. It is possible that a detailed study might allow the definition of such promoter subdomains, each specific to the tapetum at the microspore stage of development or to the stigma papillae in mature buds of crucifer flowers (and to pollen grains and stylar tissue, respectively, in solanaceous plants). A dissection of the *CaMV* 35S promoter revealed the presence of domains within the enhancer element that seem to stimulate such differential patterns of expression (Benfey *et al.*, 1989).

A potentially more significant finding for the mechanism of self-incompatibility itself is the expression of the *SRK* transcripts in each tissue. The discovery of the *S*-linked *SRK* gene, and its intimate physical and structural relationship with *SLG*, now allows for a model of a complex *S* locus, rather like a co-evolved haplotype. The classical genetic picture of *S* can accommodate such a complexity; indeed, some of its more unusual aspects may be more readily explicable in these terms. We now know that the site of pollen–stigma interaction is characterized by the presence of high levels of *SLG*, *SLR1*, and possibly *SLR2* products, together with much lower levels of products encoded by *SRK* transcripts. If all these products can interact and their relative stoichiometries are important, then there is much potential for cooperative and/or competitive effects that may, for instance, result in the observed dominance relationships among *S* alleles and in certain pleiotropic effects.

It may be significant that it is only in the stigma papillae of *Brassica* plants that high levels of *SLG* (and *SLR*) receptor-specific products are

observed, whereas low levels of *SRK*-related transcripts, which may encode authentic receptor kinases, are observed in a range of other tissues. In addition, a kinase molecule with homology to *SRK* is active in maize, a considerable taxonomic distance. Thus, self-incompatibility in *Brassica* may have evolved through an elaboration of a preexisting cell signaling system in which a receptor domain has been duplicated and brought under differential control of expression. In the stigma *SLG* and *SRK* might compete for the binding of some as yet unidentified ligand that may be *S*-encoded, or they might directly interact in a form of homophilic binding. It is difficult not to attribute great significance to the concerted evolution of *SLG* and the *SRK* receptor. However, if they do physically interact, absolute sequence identity is obviously not required for functionality and a critical analysis of observed *SLG–SRK* differences might be rewarding.

A working hypothesis can be formulated about the physiological basis of the incompatible pollen–stigma response in which the *SRK* is activated and the signal transduced through a cascade of metabolic events resulting in inhibition of pollen tube growth. The polymorphism of the *SRK2* and *SRK6* kinase domains might indicate that the *SRK*s may phosphorylate different substrates (perhaps themselves *S*-encoded), adding another complexity to the model. A crucial prerequisite for such speculations would be a clear demonstration that an *SLG–SRK* gene pair contains enough information to confer a change in allelic specificity. We now have at least some of the tools with which to test this important hypothesis, and the near future may yield very informative results.

Acknowledgments

The authors gratefully acknowledge the receipt of grants from INRA, MRT, and EC BRIDGE and thank Prof. Nigon and Prof. Gautier, Université Lyon-I, for the use of their PC-Gene and Analseq software packages.

References

Anderson, M. A., McFadden, G. I., Bernatzky, R., Atkinson, A., Orpin, T., Dedman, H., Tregear, G. W., Fernley, R., and Clarke, A. E. (1989). *Plant Cell* **1**, 483–491.
Becker, R. S., and Knight, K. L. (1990). *Cell* **63**, 987–997.
Benfey, P. N., Ren, L., and Chua, N. H. (1989). *EMBO J.* **8**, 2195–2202.
Boyes, D. C., Chen, C. H., Tantikanjana, T., Esch, J. J., and Nasrallah, J. B. (1991). *Genetics* **127**, 221–228.
Brewbaker, J. L. (1967). *Am. J. Bot.* **54**, 1069–1083.
Chen, C. H., and Nasrallah, J. B. (1990). *Mol. Gen. Genet.* **222**, 241–248.
Cornish, E. C., Pettitt, J. M., Bonig, I., and Clarke, A. E. (1987). *Nature* **326**, 99–102.

De Nettancourt, D. (1977). "Incompatibility in Angiosperms." Springer-Verlag, Berlin.

De Nettancourt, D., Ecochard, R., Perquin, M. D. G., Van der Drift, T., and Westerhof, M. (1971). *Theor. Appl. Genet.* **41**, 120–129.

Denward, T. (1963). *Hereditas* **49**, 189–334.

Dwyer, K. G., Chao, A., Cheng, B., Chen, C. H., and Nasrallah, J. B. (1989). *Genome* **31**, 969–972.

Dwyer, K. G., Balent, M. A., Nasrallah, J. B., and Nasrallah, M. E. (1991). *Plant Mol. Biol.* **16**, 481–486.

Ebert, P. R., Anderson, M. A., Bernatzky, R., Altschuler, M., and Clarke, A. E. (1989). *Cell* **56**, 255–262.

Gaude, T., Nasrallah, M. E., and Dumas, C. (1988). *Heredity* **61**, 317–318.

Gouy, M., Milleret, F., Mugnier, C., Jacobzone, M., and Gautier, C. (1984). *Nucleic Acids Res.* **12**, 121–127.

Guilluy, C. M., Trick, M., Heizmann, P., and Dumas, C. (1991). *Theor. Appl. Genet.* **82**, 466–472.

Heslop-Harrison, J. (1968). *Science* **161**, 230–237.

Heslop-Harrison, J. (1975). *Annu. Rev. Plant Physiol.* **26**, 403–425.

Hughes, A. L., and Nei, M. (1988). *Nature* **335**, 167–170.

Hughes, A. L., and Nei, M. (1989). *Proc. Natl. Acad. Sci. U.S.A.* **86**, 958–962.

Ioerger, T. R., Clark, A. G., and Kao, T. H. (1990). *Proc. Natl. Acad. Sci. U.S.A.* **87**, 9732–9735.

Isogai, A., Takayama, S., Tsukamoto, C., Ueda, Y., Shiozawa, H., Hinata, K., Okazaki, K., and Suzuki, A. (1987). *Plant Cell Physiol.* **28**, 1279–1291.

Isogai, A., Takayama, S., Shiozawa, H., Tsukamoto, C., Kanbara, T., Hinata, K., Okazaki, K., and Suzuki, A. (1988). *Plant Cell Physiol.* **29**, 1331–1336.

Kandasamy, M. K., Paolillo, D. J., Faraday, C. D., Nasrallah, J. B., and Nasrallah, M. E. (1989). *Dev. Biol.* **134**, 462–472.

Kandasamy, M. K., Dwyer, K. G., Paolillo, D. J., Doney, R. C., Nasrallah, J. B., and Nasrallah, M. E. (1990). *Plant Cell* **2**, 39–49.

Kanno, T., and Hinata, K. (1969). *Plant Cell Physiol.* **10**, 213–216.

Kaufmann, H., Salamini, F., and Thompson, R. D. (1991). *Mol. Gen. Genet.* **226**, 457–466.

Kianian, S. F., and Quiros, C. F. (in press). *Theor. Appl. Genet.*

Kimura, M. (1968). *Nature* **217**, 624–626.

Kimura, M. (1983). *In* "Evolution of Genes and Proteins" (M. Nei and R. K. Koehn, eds.), pp. 208–233. Sinauer Associates, Sunderland, Massachusetts.

Lalonde, B. A., Nasrallah, M. E., Dwyer, K. G., Chen, C. H., Barlow, B., and Nasrallah, J. B. (1989). *Plant Cell* **1**, 249–258.

Lewis, D. (1949). *Heredity* **3**, 339–355.

Lewis, D. (1960). *Proc. R. Soc. Lond.* [Biol.] **151**, 468–477.

Lewis, D. (1963). *In* "Genetics Today: Proceedings of the 11th International Congress on Genetics" (S. J. Geerts, ed.), pp. 657–663. The Hague.

Li, W. H., and Gojobori, T. (1983). *Mol. Biol. Evol.* **1**, 94–108.

Li, W. H., Luo, C. C., and Wu, C. I. (1985). *In* "Molecular Evolutionary Genetics" (R. J. McIntyre, ed.), pp. 1–93. Plenum Press, New York.

Martin, S. L., Vincent, K. A., and Wilson, A. C. (1983). *J. Mol. Biol.* **164**, 513–582.

Moore, H. M., and Nasrallah, J. B. (1990). *Plant Cell* **2**, 29–38.

Muller, J. (1981). *Bot. Rev.* **47**, 1–142.

Nasrallah, J. B., and Nasrallah, M. E. (1984). *Experientia* **40**, 279–281.

Nasrallah, J. B., and Nasrallah, M. E. (1989). *Annu. Rev. Genet.* **23**, 121–139.

Nasrallah, J. B., Doney, R. C., and Nasrallah, M. E. (1985a). *Planta* **165**, 100–107.

Nasrallah, J. B., Kao, T. H., Goldberg, M. L., and Nasrallah, M. E. (1985b). *Nature* **318**, 263–267.

Nasrallah, J. B., Kao, T. H., Goldberg, M. L., and Nasrallah, M. E. (1985c). *In* "Plant Cell–Cell Interactions" (I. Sussex, ed.), pp. 31–34. Cold Spring Harbor Laboratory, Cold Spring Harbor, New York.

Nasrallah, J. B., Kao, T. H., Chen, C. H., Goldberg, M. L., and Nasrallah, M. E. (1987). *Nature* **326,** 617–619.

Nasrallah, J. B., Yu, S. M., and Nasrallah, M. E. (1988). *Proc. Natl. Acad. Sci. U.S.A.* **85,** 5551–5555.

Nasrallah, J. B., Nishio, T., and Nasrallah, M. E. (1991). *Annu. Rev. Plant Physiol. Plant Mol. Biol.* **42,** 393–422.

Nasrallah, M. E. (1974). *Genetics* **76,** 45–50.

Nasrallah, M. E., and Nasrallah, J. B. (1986). *Trends Genet.* **2,** 239–244.

Nasrallah, M. E., Wallace, D. H., and Savo, R. M. (1972). *Genet. Res.* **20,** 151–160.

Nei, M. (1987). "Molecular Evolutionary Genetics." Columbia University Press, New York.

Ockendon, D. J. (1972). *New Phytol.* **71,** 519–522.

Ockendon, D. J. (1974). *Heredity* **33,** 159–171.

Ockendon, D. J. (1985). *In* "Plant Cell/Cell Interactions" (I. Sussex, ed.), pp. 1–6. Cold Spring Harbor Laboratory, Cold Spring Harbor, New York.

Pandey, K. K. (1960). *Evolution* **14,** 98–115.

Pandey, K. K. (1970). *Genetica* **41,** 477–516.

Roberts, I. N., Stead, A. D., Ockendon, D. J., and Dickinson, H. G. (1979). *Planta* **146,** 179–183.

Roberts, I. N., Harrod, G., and Dickinson, H. G. (1984). *J. Cell Sci.* **66,** 241–253.

Sarker, R. H., Elleman, C., and Dickinson, H. G. (1988). *Proc. Natl. Acad. Sci. U.S.A.* **85,** 4340–4344.

Scutt, C. P., Gates, P. J., Gatehouse, J. A., Boulter, D., and Croy, R. R. D. (1990). *Mol. Gen. Genet.* **220,** 409–413.

Seeman, G. H. A., Rein, S. A., Brown, C. S., and Ploegh, H. L. (1986). *EMBO J.* **5,** 547–552.

Smith, T. F., and Waterman, M. S. (1981). *Adv. Appl. Math.* **2,** 482–489.

Stein, J. C., Howlett, B., Boyes, D. C., Nasrallah, M. E., and Nasrallah, J. B. (1991). *Proc. Natl. Acad. Sci. U.S.A.* **88,** 8816–8820.

Takayama, S., Isogai, A., Tsukamoto, C., Ueda, Y., Hinata, K., Okazaki, K., and Suzuki, A. (1987). *Nature* **326,** 102–105.

Tanaka, T., and Nei, M. (1989). *Mol. Biol. Evol.* **6,** 447–459.

Thompson, K. F., and Taylor, J. P. (1966). *Heredity* **21,** 345–362.

Thorsness, M. K., Kandasamy, M. K., Nasrallah, M. E., and Nasrallah, J. B. (1991). *Dev. Biol.* **143,** 173–184.

Toriyama, K., Thorsness, M. K., Nasrallah, J. B., and Nasrallah, M. E. (1991a). *Dev. Biol.* **143,** 427–431.

Toriyama, K., Stein, J. C., Nasrallah, M. E., and Nasrallah, J. B. (1991b), *Theor. Appl. Genet.* **81,** 769–776.

Trick, M. (1990). *Plant Mol. Biol.* **15,** 203–205.

Trick, M., and Flavell, R. B. (1989). *Mol. Gen. Genet.* **218,** 112–117.

Umbach, A. L., Lalonde, B. A., Kandasamy, M. K., Nasrallah, J. B., and Nasrallah, M. E. (1990). *Plant Physiol.* **93,** 739–747.

Vincent, K. A., and Wilson, A. C. (1989). *J. Mol. Biol.* **207,** 465– 479.

Walker, J. C., and Zhang, R. (1990). *Nature* **345,** 743–746.

Weiss, E. H., Mellor, A., Golden, L., Fahrner, K., Simpson, E., Hurst, J., and Flavell, R. A. (1983). *Nature* **301,** 671–674.

Whitehouse, H. K. L. (1950). *Ann. Bot.* **14,** 198–216.

Wolfe, K. H., Li, W. H., and Sharpe, P. M. (1987). *Proc. Natl. Acad. Sci. U.S.A.* **84,** 9054–9058.

Zuberi, M. I., and Lewis, D. (1988). *Heredity* **61,** 367–377.

Sporophytic Self-Incompatibility Systems: S Gene Products

H. G. Dickinson,* M. J. C. Crabbe,† and T. Gaude‡

* Department of Plant Sciences, University of Oxford, Oxford, OX1 3RB, England
† Department of Microbiology, School of Animal and Microbial Sciences, University of Reading, Whiteknights, Reading, RG6 2AS, England
‡ Reconnaissance Cellulaire et Amélioration des Plantes, Université Claude Bernard-Lyon 1, Villeurbanne, France

I. Introduction

The ability of some plants to identify and reject their own pollen has been known for many years. Indeed, Darwin (1877) was able to describe differences in flower form by which self-pollination was discouraged in a number of species. The evolutionary significance of these systems of self-incompatibility (SI) is considerable, and the acquisition of SI early in the evolution of flowering plants has been held to be responsible for their rapid advance and current dominant position in the world's flora (Whitehouse, 1950). Subsequent study has revealed that in addition to possessing morphological mechanisms to prevent self-pollination (heteromorphic SI), a greater number of species appear to identify and reject self-pollen purely by physiological means (homomorphic SI) (De Nettancourt, 1977). East and Mangelsdorf (1925) examined the genetics of SI in *Nicotiana*, a plant possessing such a homomorphic system, and demonstrated it to be regulated by a single locus *(S)* possessing a large number of alleles. This model has been shown to hold true for most SI systems, except in the grasses where there are clearly two loci *(S* and *Z)* (Lundqvist, 1962) and *Ranunculus* and *Beta vulgaris*, where there appear to be three and four loci, respectively (Lundqvist *et al.*, 1973). However, another group of plants failed to fit this model in that the compatibility of the pollen with the pistil appeared to be regulated by two alleles. This group of species, which includes members of the Brassicaceae and the Asteraceae, was investigated by Bateman (1955) who proposed that the pollen of these plants was expressing the *S* alleles of the maternal plant, rather than the allele

possessed by the pollen grain itself. Interestingly, this sporophytic inheritance of SI is generally associated with a range of other physical and biochemical characteristics, including tricellular pollen, a complex exine coating, a dry stigma, and superficial rejection of self-pollen. In plants in which the pollen expresses its own allele and SI is said to be gametophytically regulated, pollen is frequently bicellular, has a simple coating, and germinates on a wet stigma. The stylar structure of this latter group of plants is generally complex and self-pollen tubes are rejected after a considerable period of growth. Thus, although sporophytic inheritance of SI may pose problems because of its genetic complexity, for example allelic dominance may occur either in pollen or stigma (Ockendon, 1975), the possession of maternal gene products by the pollen is interesting and may provide a useful clue in the search for the mechanism by which SI operates. Since the work of Bateman (1955) on the *Brassicaceae,* sporophytically controlled mechanisms have also been shown to be present in plants that already possess "mechanical" heteromorphic mechanisms to prevent outbreeding (De Nettancourt, 1977).

Not surprisingly, the molecular basis of the recognition and response step of SI has been the focus of considerable attention. Lewis (1954, 1965) proposed a dimer hypothesis in which the S locus is held to be tripartite in structure, containing male- and female-expressed domains together with an S specificity domain that is expressed in both sex organs. Lewis proposed that similar molecules are synthesized in both pollen and stigma, which on self-pollination form a dimer possessing biological activity. The nature of this biological activity was suggested to be analogous with some components of the immune response, and Lewis' early experiments on *Oenothera* (Lewis *et al.,* 1967) tended to support such an inference. However, more recent attempts to identify and characterize similar S allele products in the pollen and pistil have produced equivocal results. The S locus clearly encodes glycoproteins that are expressed in the styles of plants with gametophytic SI (Singh and Kao, this volume) and in the stigmatic papillae and rudimentary transmitting tissue of those with sporophytic SI, but our inability to identify a male S allele product in plants with either gametophytic or sporophytic SI has constituted a major barrier to progress. Some promising lines of inquiry are, however, currently under investigation (Dickinson, 1990). Models put forward to explain the operation of SI at a molecular level, including those of Lewis (1954, 1965) and numerous others (Ascher, 1975; Herrero and Dickinson, 1981; Heslop-Harrison, 1982), must therefore remain entirely speculative.

The genetics, physiology, and cell biology of sporophytic SI have principally been studied in the Brassicaceae, although there are some data for members of the Asteraceae (Howlett *et al.,* 1975). Within the Brassicaceae, *Brassica oleracea* and *Brassica campestris* have been subjected to

exhaustive molecular analysis, and results from these species, for which the genetic control of SI has also been well established (Section I; Thompson, 1965; Ockendon, 1974, 1980, 1982), form the basis of this review.

II. Proteins of the *Brassica* Stigma

A. History

The components of the stigma involved in the recognition and response steps of SI have been the subject of a number of investigations, the earliest of which (Nasrallah and Wallace, 1967; Nasrallah *et al.*, 1970, 1972) employed electrophoretic and immunological methods to identify a number of polypeptides that were specific to S alleles in *Brassica oleracea*. These findings were later confirmed by Sedgley (1974a,b) who noted that, of the different allelic products tested, only those of high or medium dominance were capable of inducing the production of antibodies. The antigenicity of S-linked polypeptides thus seemed to be correlated with their position in a dominance series. Further, the concentration of these polypeptides was generally doubled in stigmas homozygous for a particular S allele (Sedgley, 1974b). In a key series of experiments, Hinata and co-workers (Nishio and Hinata, 1977; Hinata and Nishio, 1978) demonstrated that the expression of individual S genes could be linked to the appearance of glycoproteins in stigmatic homogenates of *Brassica oleracea* and *Brassica campestris*. These authors subsequently determined the amino acid composition of four S-linked glycoproteins following purification by chromatography and preparative isoelectric-focusing (IEF) (Nishio and Hinata, 1979, 1982).

IEF has proved particularly suited to the study of stigmatic proteins, and using this method Roberts *et al.* (1979) detected the appearance of a major glycoprotein band in stigmatic tissue at the point in floral development when self-pollen could be first recognized and rejected. Because immature stigmas do not express SI, this glycoprotein thus constituted a potential S gene product. A glycoprotein specific to the S_2 allele was later isolated and partially characterized by Ferrari *et al.* (1981). Moreover, when pollen from a plant homozygous for the S_2 allele was pretreated with this glycoprotein, its germination on a compatible stigmatic surface was inhibited.

More recently a highly detailed study was carried out by Nasrallah and Nasrallah (1984) in which a number of SI lines of *Brassica oleracea* were analyzed using electrophoresis and IEF. Among the nine lines studied, each exhibited a glycoprotein profile, as detected by concanavalin

A–peroxidase, characteristic for the particular S allele carried by the plant. Interestingly, these S allele-specific glycoproteins are present only in the stigmatic tissue of the flower and may thus play a functional role in the SI response. Although differing in their isoelectric points, SDS-PAGE reveals all these glycoproteins to be of a molecular mass of between 57 and 65 kDa. Nasrallah *et al.* (1985a) also demonstrated the synthesis of these glycoproteins, now termed S locus-specific glycoproteins (SLSGs), to take place late in stigmatic development. The rate of their synthesis increases during development until the point at which self-compatible buds become self-incompatible (1 day before anthesis), after which it decreases. The peak of SLSG synthesis thus coincides with the acquisition of SI by the immature stigmas, indicating that a threshold level of the SLSG is required for the expression of SI in the stigma. Subsequent labeling and immunocytochemical studies (Kandasamy *et al.*, 1990) have identified the principal site of S allele product synthesis as being the stigmatic papillae themselves.

B. Primary Structure of S Locus-Specific Glycoproteins

Information on the primary structure of S glycoproteins in *Brassica* was first provided by Nasrallah *et al.* (1985b) who deduced the amino acid sequence of the S_6 allele glycoprotein in *Brassica oleracea* from the cDNA sequence. More recently, this cDNA sequence from the S_6 allele has been shown not to correspond with the amino acid sequence as determined by Takayama *et al.* (1986, 1987), who sequenced three S-linked glycoproteins from the stigmas of *Brassica campestris*. These authors pointed to the omission of three pairs of bases in the nucleotide sequence published by Nasrallah *et al.* (1985b), which introduced the nonsense codon TGA in the reading frame in position 370–372, and thus prevented complete analysis of the amino acid sequence of this S_6 glycoprotein. These sequence data from the Japanese group were obtained from gas-phase microsequencing of high-pressure liquid chromatography (HPLC)-purified peptides derived from enzymatic digestion or chemical cleavage of (SLSGs). This methodology has permitted the complete peptide sequencing of the S_8 SLSG and a partial determination of the S_9 and S_{12} allele products (Isogai *et al.*, 1987). Comparison of the peptide sequence of S_8 and S_6 (derived from the corrected nucleotide sequence of S_6 DNA) glycoproteins shows more than 81% conservation at the amino acid level. In a parallel approach, two cDNAs corresponding to the S_{13} and S_{14} alleles have been isolated from *Brassica oleracea* by Nasrallah *et al.* (1987). Comparing the sequences of these cDNAs with the corrected S_6, the peptide sequences deduced show between 79% and 85% homology (Table I). These differences observed within the primary structure of S allele products in *Brassica oleracea* are

TABLE I

Homology between the Amino Acid Sequences of SLSGs in
Brassica (in percentage of Identical Positions between Two
Aligned Sequences)

SLSG sequences[a]	Homology (%)
S_6-S_{13}	84.5
S_6-S_{14}	79.5
S_6-S_{22}	83.5
$S_{13}-S_{14}$	79.0
$S_{13}-S_{22}$	81.6
$S_{14}-S_{22}$	82.1
S_8-S_6	82.7
S_8-S_{13}	87.2
S_8-S_{14}	74.3
S_8-S_{22}	77.9

[a] S_6, S_{13}, S_{14}, and S_{22} SLSGs are from *Brassica oleracea* (Nasrallah *et al.*, 1987; Lalonde *et al.*, 1989); S_8 SLSG is from *Brassica campestris* (Takayama *et al.*, 1987).

of the same order as those observed between the SLSGs of *Brassica oleracea* and *Brassica campestris,* indicating a common origin for the two species. The differences between the polypeptides take the form of substitutions throughout the whole length of the sequence, and of deletions or additions of one or two residues at isolated points in the chain. It is difficult to predict which regions of the polypeptide chain are involved in the expression of allelic specificity, but a comparison by pairs of all the sequences published to date (three for *Brassica campestris* and six for *Brassica oleracea*) permits the identification of a number of sequences of between 3 and 7 residues in length, which seem specific to the different *S* alleles studied (Fig. 1).

For example, a domain in the middle part of the protein (amino acid positions 182–188) is clearly a region of high variability among the SLSG sequences studied and may be one of the elements responsible for S specificity (Takayama *et al.*, 1987). Nasrallah *et al.* (1987) considered the polypeptide sequences of the SLSGs of *Brassica oleracea* to be composed of alternating constant and variable regions (Fig. 2); these authors pointed to a highly variable domain situated between residues 182 and 274, which exhibits only 40% homology, and suggested that it may participate in the establishment of *S* allele specificity.

Among the different types of variability that might contribute to *S* allele specificity, Nasrallah *et al.* (1987) also identified the number and position of potential N-glycosylation sites as being characteristic for each allele.

```
          ***** **   *  **      ***     ***** *     *    **
SLSG-2A  1  IYVNTLSSSESLTISSNRTLVSPGGVFELGFFKPLGRSQWYLGIWYKKVS  50
SLSG-6   1  I--NTLSSTESLRISSNRTLVSPGNNFELGFFRTNSSSRWYLGIWYKKLL  48
SLSG-13  1  I--NTLSSTESLTISSNRTLVSPGNVFELGFFKTTSSSRWYLGIWYKKFP  48
SLSG-14  1  I--NTLSSIESLTISSNRTLVSPGNVFELGFFRTNSSSRWYLGIWYKKVS  48
SLSG-22  1  I--NTLSATESLTISSNRTLVSPGNVFELGFFRTTSSSRWYLGIWYKKLS  48
SLSG-29  1  I--NTLSSIESLKISNSRTLVSPGNVLELGFFRTPSSSRWYLGMWYKKLS  48
SLSG-8   1  V--NTLSSTESLTISNXRTLVSPGDVFELGFFRTXSSSPWYLGIWYKKLS  48
SLSG-9   1  L--STLSSTESLTISSXRTLVSPGNIFELGFXXXXXXXXXXXXXXXXXXX  48
SLSG-12  1  I--NILSSTESLTISSXRTLVSPGNVFELGFXXXXXXXXXXXXXXXXXLP  48

           **   *          ****    *** *** *** ***    ****** *
        51  QKTYAWVANRDNPLTNSIGTLKISGNNLVLLGQSNNTVWSTNLTRENVRS  100
        49  DRTYVWVANRDNPLSNAIGTLKISGNNLVLLGHTNKSVWSTNLTRGNERL  98
        49  YRTYVWVANRDNPLSNDIGTLKISGNNLVLLDHSNKSVWSTNVTRGNERS  98
        49  DRTYVWVANRDNPLSSSIGTLKISGNNPCHLDHSNKSVWSTNLTRGNERS  98
        49  NRTYVWVANRDNPLSNSTGTLKITSNNLVILGHSNKSIWSTNRTKGNERS  98
        49  ERTYVWVANRDNPLSCSIGTLKISNMNLVLLDHSNKSLWSTNHTRGNERS  98
        49  ERTYVWVANRDNPLSNSIGXL-ILGNNLVLLGHSXKSVWSTXVSRGYERS  97
        49  XXXXXXXXXXXXXXXXXXXXXXXXXXXXXXXXXXXXXSVWSTXLTXENVXS  98
        49  DRTYVWVANRDNPLSNSIGTL-ISNMNLVLLXQSXKSVWSTXITRGNERS  97

            *  ** *     *  *  *    **** **      **       *   *    **
       101  PVIAELLPNGNFVMRYS-SNKDISGFLWQSFDFPTDTLLPDMKLGYDLKT  149
        99  PVVAELLSNGNFVMRDS-SNNDASEYLWQSFDYPTDTLLPEMKLGYDLKT  147
        99  PVVAELLDNGNFVMRDSNSNN-ASQFLWQSFDYPTDTLLPEMKLGYDLKT  147
        99  PVVADVLANGNFVMRDSN-NNDASGFLWQSFDFPTDTLLPEMKLSYDLKT  147
        99  PVVAELLANGNFVMRDSN-NNRSSRFLWQSFDYPTDTLLPEMKLGYDLKT  147
        99  PVVAELLANGNFVLRDSN-KNDRSGFLWQSFDYPTDTLLPEMKLGYDLRT  147
        98  PVVAELLANGNFVMRDS-SNNXASQFLWQSFNYPTDTLLPEMKLGYDLKT  146
        99  PVVAELLANGNFVXXXXXXXXXXXXXXXXXXXXXXXXXXXXXLGYDLKD  148
        98  PVLAELLANGNLVIRDSN-NNDASXXXXXXXXXXXXXXXXXXXLGYDLK-  145

            *   *  * *  **  *   *  ***   ***   **  *********** *
       150  GRNRILTSWRSSDDPSSGNTTYKIDTQRGLPEFILNQGRYEMQRSGPWNG  199
       148  GLNRFLTSWRSSDDPSSGDFSYKLET-RSLPEFYLWHGIFPMHRSGPWNG  196
       148  GLNRFLTSWRSSDDPSSGDYSYKLEL-RRLPEFYLSSGSFRLHRSGPWNG  196
       148  GLNRFLTSRRSSDDPSSGDFSYKLEP-RRLPEFYLSSGVFLLYRSGPWNG  196
       148  GLNRFLTSWRSSDDPSSGDFSYKLRA-RRLPELYLSSGIFRVHRIGPWNG  196
       148  GLNRFLTSWRSSDTPSSGDFSYKLQT-RRLPEFYLFKDDFLVHRSGPWNG  196
       147  GLNRFLTSWRSYDDPSSGDFLYKLET-RRLPEFYLMQGDVREHRSGPXNG  195
       149  GLNRFLVSXXXSXXXXXXXXXXLDIQRGLPEFYTFKXXXXXXXXXXXXX  198
       146  GINRFLXSXXNSXDPSRGEFXXXLDTQRGMPEFYLLKDGLQGHRSGPXNG  195

            **** * **** * **  ********* * *   ***   *********
       200  MEFSGIPEVQGLNYMVYNYTENSEEISYTFHMTNQSIYSRLTVSDYTLN-  248
       197  VRFSGIPEDQKLSYMVYNFTENSEEVAYTFRMTNNSIYSRLTLSSEGYFQ  246
       197  FRISGIPEDQKLSYMVYNFTENSEEAAYTFLMTNNSFYSRLFVSFSTGYFE  246
       197  IRFSGLPDDQKLSYLVY--ISQDMRVAYKFRMTNNSFYSRLFVSFSGYIE  244
       197  IRFSGIPDDRKLSYLVYNFTENNEEVAYTFRMTNNTIYSRLTVSFSGYIE  246
       197  VGFSGMPEDQKLSYMVYNFTQNSEEVAYTFLMTNNSIYSRLTISSSGYFE  246
       196  IQFIGIPEDQKLSYMMYXFTENSEEVAYTFLMTXNSFYSXLTINSEGYLE  245
       199  XXXXXXXXXXXXXXXXXXXXXXXXXXXXXXXXXXXXXXXXXXXXXXXXXXX  248
       196  VQFSGIPEDQ-LNYMVYXFTENSEEVAYTFRMTXXXXXXXXXXXXXXXXX  244
```

```
       ** * **** **  ******* *****  * *  *** ** ** *
249   RLTWIPPSRAWSMFWTLP-TDVCDPLYLCGSYSYCDLITSPNCNCIRGFV   297
247   RLTWNPSIGIWNRFWSSPVDPQCDTYIMCGPYAYCGVNTSPVCNCIQGFN   296
247   RLTWAPSSVVWNVFWSSP-NHQCDMYRMCGPYSYCDVNTSPVCNCIQGFR   295
245   QQTWNPSSQMWNSFWAFPLDSQCYTYRACGPYSYCVVNTSAICNCIQGFN   294
247   RQTWNPSLGMWNVFWSFPLDSQCDAYRACGPYSYCDVNTSPICNCIQGFN   296
247   RLTWTPSSGMWNVFWSSDEDFQCDVYKICGAYSYCDVNTSPVCNCIQRFD   296
246   RLXXAPSSVV-NVFXSSPI-XQCDMY-TCGPYSYCDVNTSPVCNCIQGFN   292
249   XXXXXXXXXXXXXXXXXXXXXXXXXXXXXXXXXXXXXXXXXXXXXXXXXXX   298
245   XXXXXXXXXXXXXXXXXXXXXXXXXACGSYSYCDLNTSPVCNCIQGFK     294


      * *** ** ****  * * ****   * ***** *  * ** * * *
298   PKNPQQWDLRDGTQGCVRTTQMSCSGDGFLRLNNMNLPDTKTATVDRIID   347
297   PRNIQQWDQRVWAGGCIRRTRLSCSGDGFTRMKNMKLPETTMAIVDRSIG   346
296   PKNRQQWDLRIPTSGCIRRTRLSCSGDGFTRMKNMKLPETTMAIVHRSIG   345
295   PSNVQQWDQRVWAGGCIRRTRLSGSGDGFTRMKNMKLPETTMAIVDRSIG   344
297   PSNVEQWDQRVWANGCIRRTRLSCSGDRFTMMKNMKLPETTMAIVDRSIG   346
297   PSNVQEWGLRAWSGGCRRRTRLSCSGDGFTRMKKMKLPETTMAIVDRSIG   346
293   PKNRQQWDLRIPTSGCIRRTRLGCSGDGFTRMKNMKLPETTMAIVDRSIG   342
299   XXXXXXXXXXXXXXXXXXXXXXXXXXXXXXXXXXXXXXXXXXXXXXXXXXX   348
295   PLNVQQWDLRDGSSGCIRKXXXXXXXXXXXXXXXXXXXLPETMKAIVDRSID 344


      * *  *    *      *  *  *  ***   **  * *** ***** ****
348   VKKCEERCLSDCNCTSFAIADVRNGELGCVFWTGELVEIR-K--FAVGGQ   394
347   VKECEKRCLSDCNCTAFANADIRNGGTGCVIWTGRLDDMRNY--VA-HGQ   393
346   LKECEKRCLSDCNCTAFANADIRNRGTGCVIWTGELEDIRTY--FA-DGQ   392
345   VKECEKRCLNDCNCTAFANADIRNGGTGCVINTGELEDMRSYATGATDSQ   394
347   VKECEKRCLSDCNCTAFANADIRNGGAGCVIWTGRLDDMRNY--AADHGQ   394
347   LKECEKRCLSDCNCTAFANADIRNGGTGCVIWTGQLEDIRTY--FA-NGL   393
343   LKECEKRCLSDCNCTAFANADIRNRGTGCVIWTGELEDIRTY--FA-DGQ   389
349   XXXXXXXCLSDCNCTAFANADIXXXXXXXXXXXXXXXXXXXXXXXXXXXXXX 398
345   VKECENRCLSDCNCTAFANADIXXXXXXXXXXXXXXXXXXXXXXXXXXXXXX 394


      ** ***
395   DLYVRLNAADLG   406
394   DLYVRLAVADLV   405
393   DLYVRLAAADLV   404
395   DLYVRLAAADIV   406
395   DLYVRLAAANLV   406
394   DLYVRLAPADLV   405
390   DLYVRLAAADLV   401
399   XXXXXXXXXXXX   410
395   XXXXXXXXXXXX   406
```

FIG. 1 Comparison of the amino acid sequences of SLSGs in the *Brassica* species. From Gaude (1990). SLSG-2A, -6, -13, -14, -22, and -29 sequences are from *Brassica oleracea* (SLSG-2A from Chen and Nasrallah, 1990; SLSG-6, -13, and -14 from Nasrallah *et al.*, 1987; SLSG-22 from Lalonde *et al.*, 1989; SLSG-29 from Trick and Flavell, 1989). SLSG-8, -9, and -12 are from *Brassica campestris* (Takayama *et al.*, 1987). Gaps (-) have been introduced to maximize the homology. Amino acid residues identical for at least three SLSGs are represented in black, and the different residues are underlined and variation stressed above the alignment (*). X, undetermined amino acid.

This observation seems confirmed by the analysis of nine sequences of SLSGs presently available (Fig. 3, Table II). Examining these sequences together, it is possible to predict 13 potential N-glycosylation sites by using the general rule of Asn-X-Thr–Ser, where X is any amino acid other than proline (Struck and Lennarz, 1980). Among these sites four are conserved in the nine sequences listed (those at residues 85, 92, 233, and 360). It is possible that all potential sites for N-glycosylation may not be occupied, for in *Brassica campestris* microsequencing of the S_8 SLSG suggests that only seven of nine potential sites are glycosylated (Takayama *et al.*, 1987).

The structure of the oligosaccharide chains attached to the asparagine residues has been determined by Takayama *et al.* (1986, 1987) and two oligosaccharides, A and B, which are found distributed throughout plant glycoproteins, constitute the principal components of the glycosyl side chains of SLSGs (Fig. 4). Oligosaccharide B (molecular mass of 1391 daltons) is predominant and may correspond to the mature form of the glycosyl chain, whereas A is most probably its precursor. The molecular mass of the S_8 SLSG is estimated at 53 kDa; it is composed of 405 amino acids with a molecular mass of 46 kDa and seven oligosaccharide chains. The absence of galactose and arabinose from the acid hydrolysate of the S_8 SLSG indicates that there is unlikely to be any O-glycosylation in the region of the serine and threonine residues of the polypeptide chain (Takayama *et al.*, 1987).

Another peculiarity of the SLSGs is the presence of 11–12 cysteine residues in the carboxy-terminal regions. These residues are completely conserved in four SLSGs studied and may, through the formation of disulfide bridges, maintain the three-dimensional structure necessary for the function of the S allele products (Nasrallah *et al.*, 1987). The role of this domain of the SLSG is explored in Section II,D. A schematic representation of SLSG structure, based on the results of Takayama *et al.* (1987), is given in Figure 5.

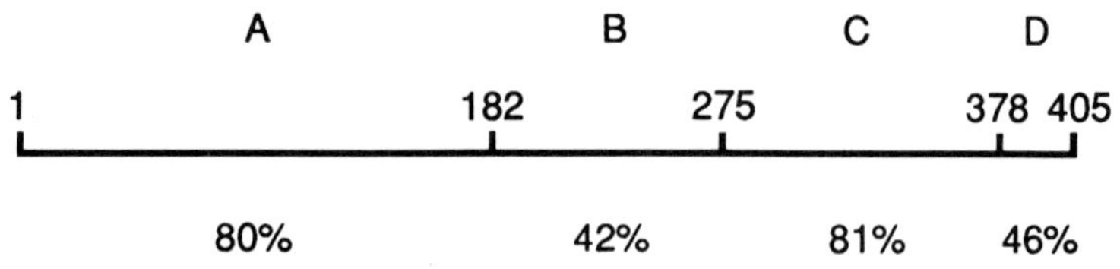

FIG. 2   Primary structure of *Brassica oleracea* SLSGs. By comparing the amino acid sequences of SLSG-6, -13, and -14, Nasrallah *et al.* (1987) described an alternation of conserved (A and C) and variable (B and D) domains throughout the length of the polypeptide chain. The highly variable B region is relatively hydrophilic, whereas the C region contains 10 conserved cysteine residues.

```
                             *                      *
SLSG-2A    1   IYVNTLSSSESLTISSNRTLVSPGGVFELGFFKPLGRSQWYLGIWYKKVS   50
SLSG-6     1   I--NTLSSTESLRISSNRTLVSPGNNFELGFFRTNSSSRWYLGIWYKKLL   48
SLSG-13    1   I--NTLSSTESLTISSNRTLVSPGNVFELGFFKTTSSSRWYLGIWYKKFP   48
SLSG-14    1   I--NTLSSIESLTISSNRTLVSPGNVFELGFFRTNSSSRWYLGIWYKKVS   48
SLSG-22    1   I--NTLSATESLTISSNRTLVSPGNVFELGFFRTTSSSRWYLGIWYKKLS   48
SLSG-29    1   I--NTLSSIESLKISNSRTLVSPGNVLELGFFRTPSSSRWYLGMWYKKLS   48
SLSG-8     1   V--NTLSSTESLTISNXRTLVSPGDVFELGFFRTXSSSPWYLGIWYKKLS   48
SLSG-9     1   L--STLSSTESLTISSXRTLVSPGNIFELGFXXXXXXXXXXXXXXXXXXX   48
SLSG-12    1   I--NILSSTESLTISSXRTLVSPGNVFELGFXXXXXXXXXXXXXXXXXLP   48


              *                *              *      *
          51  QKTYAWVANRDNPLTNSIGTLKISGNNLVLLGQSNNTVWSTNLTRENVRS  100
          49  DRTYVWVANRDNPLSNAIGTLKISGNNLVLLGHTNKSVWSTNLTRGNERL   98
          49  YRTYVWVANRDNPLSNDIGTLKISGNNLVLLDHSNKSVWSTNVTRGNERS   98
          49  DRTYVWVANRDNPLSSSIGTLKISGNNPCHLDHSNKSVWSTNLTRGNERS   98
          49  NRTYVWVANRDNPLSNSTGTLKITSNNLVILGHSNKSIWSTNRTKGNERS   98
          49  ERTYVWVANRDNPLSCSIGTLKISNMNLVLLDHSNKSLWSTNHTRGNERS   98
          49  ERTYVWVANRDNPLSNSIGXL-ILGNNLVLLGHSXKSVWSTXVSRGYERS   97
          49  XXXXXXXXXXXXXXXXXXXXXXXXXXXXXXXXXXXSVWSTXLTXENVXS   98
          49  DRTYVWVANRDNPLSNSIGTL-ISNMNLVLLXQSXKSVWSTXITRGNERS   97
                              o                  o
                        *
         101  PVIAELLPNGNFVMRYS-SNKDISGFLWQSFDFPTDTLLPDMKLGYDLKT  149
          99  PVVAELLSNGNFVMRDS-SNNDASEYLWQSFDYPTDTLLPEMKLGYDLKT  147
          99  PVVAELLDNGNFVMRDSNSNN-ASQFLWQSFDYPTDTLLPEMKLGYDLKT  147
          99  PVVADVLANGNFVMRDSN-NNDASGFLWQSFDFPTDTLLPEMKLSYDLKT  147
          99  PVVAELLANGNFVMRDSN-NNRSSRFLWQSFDYPTDTLLPEMKLGYDLKT  147
          99  PVVAELLANGNFVLRDSN-KNDRSGFLWQSFDYPTDTLLPEMKLGYDLRT  147
          98  PVVAELLANGNFVMRDS-SNNXASQFLWQSFNYPTDTLLPEMKLGYDLKT  146
          99  PVVAELLANGNFVXXXXXXXXXXXXXXXXXXXXXXXXXXXXXXLGYDLKD  148
          98  PVLAELLANGNLVIRDSN-NNDASXXXXXXXXXXXXXXXXXXXXLGYDLK-  145


                        *
         150  GRNRILTSWRSSDDPSSGNTTYKIDTQRGLPEFILNQGRYEMQRSGPWNG  199
         148  GLNRFLTSWRSSDDPSSGDFSYKLET-RSLPEFYLWHGIFPMHRSGPWNG  196
         148  GLNRFLTSWRSSDDPSSGDYSYKLEL-RRLPEFYLSSGSFRLHRSGPWNG  196
         148  GLNRFLTSRRSSDDPSSGDFSYKLEP-RRLPEFYLSSGVFLLYRSGPWNG  196
         148  GLNRFLTSWRSSDDPSSGDFSYKLRA-RRLPELYLSSGIFRVHRIGPWNG  196
         148  GLNRFLTSWRSSDTPSSGDFSYKLQT-RRLPEFYLFKDDFLVHRSGPWNG  196
         147  GLNRFLTSWRSYDDPSSGDFLYKLET-RRLPEFYLMQGDVREHRSGPXNG  195
         149  GLNRFLVSXXXSXXXXXXXXXXXXLDIQRGLPEFYTFKXXXXXXXXXXXX  198
         146  GINRFLXSXXNSXDPSRGEFXXXLDTQRGMPEFYLLKDGLQGHRSGPXNG  195


                        *                    *
         200  MEFSGIPEVQGLNYMVYNYTENSEEISYTFHMTNQSIYSRLTVSDYTLN-  248
         197  VRFSGIPEDQKLSYMVYNFTENSEEVAYTFRMTNNSIYSRLTLSSEGYFQ  246
         197  FRISGIPEDQKLSYMVYNFTENSEEAAYTFLMTNNSFYSRLTISSTGYFE  246
         197  IRFSGLPDDQKLSYLVY--ISQDMRVAYKFRMTNNSFYSRLFVSFSGYIE  244
         197  IRFSGIPDDRKLSYLVYNFTENNEEVAYTFRMTNNTIYSRLTVSFSGYIE  246
         197  VGFSGMPEDQKLSYMVYNFTQNSEEVAYTFLMTNNSIYSRLTISSSGYFE  246
         196  IQFIGIPEDQKLSYMMYXFTENSEEVAYTFLMTXNSFYSXLTINSEGYLE  245
         199  XXXXXXXXXXXXXXXXXXXXXXXXXXXXXXXXXXXXXXXXXXXXXXXXXX  248
         196  VQFSGIPEDQ-LNYMVYXFTENSEEVAYTFRMTXXXXXXXXXXXXXXXXX  244
```

FIG. 3   Potential sites of N-glycosylation (*, above the alignment) and cysteine residues (°, below the alignment) of 9 SLSGs in *Brassica* sp. From Gaude (1990).

```
                                              *
249  RLTWIPPSRAWSMFWTLP-TDVCDPLYLCGSYSYCDLITSPNCNCIRGFV  297
247  RLTWNPSIGIWNRFWSSPVDPQCDTYIMCGPYAYCGVNTSPVCNCIQGFN  296
247  RLTWAPSSVVWNVFWSSP-NHQCDMYRMCGPYSYCDVNTSPVCNCIQGFR  295
245  QQTWNPSSQMWNSFWAFPLDSQCYTYRACGPYSYCVVNTSAICNCIQGFN  294
247  RQTWNPSLGMWNVFWSFPLDSQCDAYRACGPYSYCDVNTSPICNCIQGFN  296
247  RLTWTPSSGMWNVFWSSDEDFQCDVYKICGAYSYCDVNTSPVCNCIQRFD  296
246  RLXXAPSSVV-NVFXSSPI-XQCDMY-TCGPYSYCDVNTSPVCNCIQGFN  292
249  XXXXXXXXXXXXXXXXXXXXXXXXXXXXXXXXXXXXXXXXXXXXXXXXXX  298
245  XXXXXXXXXXXXXXXXXXXXXXXXXXACGSYSYCDLNTSPVCNCIQGFK  294
                        o      o       o      o o


298  PKNPQQWDLRDGTQGCVRTTQMSCSGDGFLRLNNMNLPDTKTATVDRIID  347
297  PRNIQQWDQRVWAGGCIRRTRLSCSGDGFTRMKNMKLPETTMAIVDRSIG  346
296  PKNRQQWDLRIPTSGCIRRTRLSCSGDGFTRMKNMKLPETTMAIVHRSIG  345
295  PSNVQQWDQRVWAGGCIRRTRLSGSGDGFTRMKNMKLPETTMAIVDRSIG  344
297  PSNVEQWDQRVWANGCIRRTRLSCSGDRFTMMKNMKLPETTMAIVDRSIG  346
297  PSNVQEWGLRAWSGGCRRRTRLSCSGDGFTRMKKMKLPETTMAIVDRSIG  346
293  PKNRQQWDLRIPTSGCIRRTRLGCSGDGFTRMKNMKLPETTMAIVDRSIG  342
299  XXXXXXXXXXXXXXXXXXXXXXXXXXXXXXXXXXXXXXXXXXXXXXXXXXX  348
295  PLNVQQWDLRDGSSGCIRKXXXXXXXXXXXXXXXXXXXLPETMKAIVDRSID  344
                       o        o


                   *
348  VKKCEERCLSDCNCTSFAIADVRNGELGCVFWTGELVEIR-K--FAVGGQ  394
347  VKECEKRCLSDCNCTAFANADIRNGGTGCVIWTGRLDDMRNY--VA-HGQ  393
346  LKECEKRCLSDCNCTAFANADIRNRGTGCVIWTGELEDIRTY--FA-DGQ  392
345  VKECEKRCLNDCNCTAFANADIRNGGTGCVINTGELEDMRSYATGATDSQ  394
347  VKECEKRCLSDCNCTAFANADIRNGGAGCVIWTGRLDDMRNY--AADHGQ  394
347  LKECEKRCLSDCNCTAFANADIRNGGTGCVIWTGQLEDIRTY--FA-NGL  393
343  LKECEKRCLSDCNCTAFANADIRNRGTGCVIWTGELEDIRTY--FA-DGQ  389
349  XXXXXXXCLSDCNCTAFANADIXXXXXXXXXXXXXXXXXXXXXXXXXXXX  398
345  VKECENRCLSDCNCTAFANADIXXXXXXXXXXXXXXXXXXXXXXXXXXXX  394
           o   o   o o                    o


395  DLYVRLNAADLG  406
394  DLYVRLAVADLV  405
393  DLYVRLAAADLV  404
395  DLYVRLAAADIV  406
395  DLYVRLAAANLV  406
394  DLYVRLAPADLV  405
390  DLYVRLAAADLV  401
399  XXXXXXXXXXXX  410
395  XXXXXXXXXXXX  406
```

FIG. 3 (*continued*)

Most recently the Cornell group (Nasrallah *et al.*, 1991) has been able to correlate the peptide sequences of SLSGs with their position in the allelic dominance series. Dividing *S* alleles into two groups, class I (generally strong; e.g., S_6) and class II (generally weak; e.g., S_5), Nasrallah demonstrated that within the class I *SLSG* alleles there was 90% DNA sequence homology and 80% at the peptide level. However, class I and II alleles share only 70% DNA and 65% amino acid homology (Fig. 6). In addition, all class I SLSGs feature a conserved sequence at the 3′ terminus,

TABLE II

Potential Sites of N-Glycosylation at the Asparaginyl Residues among Nine SLSG Sequences from *Brassica* Species[a]

	Residue													No. of potential N-glycosylation sites	No. of sites probably glycosylated
SLSG	**17**	**35**	51	66	**85**	**92**	120	**121**	168	**217**	**233**	285	360		
−2A	+				+	+			+	+	+		+7	5	
−6	+	+			+	+				+	+	+	+	8	6
−13	+				+	+	+			+	+	+	+	8	5
−14	+	+			+	+					+	+	+	7	5
−22	+		+	+	+	+	+			+	+	+	+	10	5
−29					+	+				+	+	+	+	6	4
−8	+	+			+	+		+		+	+	+	+	9	7
−9	+	UD	UD	UD	UD	+	UD	UD	UD	UD	UD	UD	+	3	2
−12	+	UD			+	+				+	UD	+	+	6	4

[a] SLSG-2A, -6, -13, -14, -22, and -29 from *Brassica oleracea;* SLSG-8, -9, and -12 from *Brassica campestris*. Each *S* allele is characterized by a particular pattern of potential sites of N-glycosylation. Residues in boldface type correspond to glycosylated asparagines in *Brassica campestris* (Takayama *et al.*, 1987). The equivalent sites in *Brassica oleracea* might be glycosylated as well. The numbering of amino acid residues refers to the SLSG-2A sequence (Fig. 1); UD, undetermined amino acid. Modified from Gaude (1990).

A

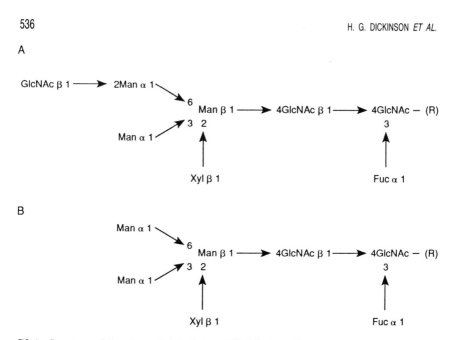

FIG. 4   Structure of the glycosylated chains of SLSGs from *Brassica campestris*. A and B oligosaccharides, according to Takayama *et al.* (1986). R, asparagine residue.

which is not found in class II alleles. This allelic difference can be distinguished at the protein level, for an antibody (MAbH8) will bind to class I SLSGs but not to class II. Interestingly, this type of allelic grouping is not restricted to *Brassica,* for in *Raphanus* two subsets of *S* alleles can be distinguished using the MAbH8 antibody. Further, class I and class II alleles clearly occur in *Brassica campestris*. Perhaps most significantly, there is greater nucleotide and peptide sequence homology between alleles of a particular class between species than there is between classes within a species, which indicates evolution and fixation of *S* alleles to have preceded recent speciation in this group of plants. These findings also call into question whether class I and II SLSGs can be considered as alleles, *sensu stricto*.

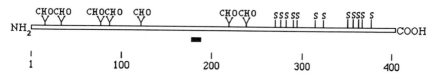

FIG. 5   Schematic representation of stigmatic SLSGs in *Brassica*. Adapted from Takayama *et al.* (1987). This example illustrates the SLSG-8 from *Brassica campestris*. The black box indicates the highly variable region of SLSGs (residues 182–188).

```
                                      *
SLSG-2A    1   IYVNTLSSSESLTISSNRTLVSPGGVFELGFFKPLGRSQWYLGIWYKKVS    50
               |  ||||| |||||||||||||| |||||||      | ||||||||||
SLSG-13    1   I--NTLSSTESLTISSNRTLVSPGNVFELGFFKTTSSSRWYLGIWYKKFP    48
                                                  *         *

          51   QKTYAWVANRDNPLTNSIGTLKISGNNLVLLGQSNNTVWSTNLTRENVRS   100
               || |||||||||| | ||||||||||||||| ||  ||||| || | ||
          49   YRTYVWVANRDNPLSNDIGTLKISGNNLVLLDHSNKSVWSTNVTRGNERS    98

         101   PVIAELLPNGNFVMRYS-SNKDISGFLWQSFDFPTDTLLPDMKLGYDLKT   149
               || |||| |||||||| | ||   | |||||| ||||||| |||||||||
          99   PVVAELLDNGNFVMRDSNSNN-ASQFLWQSFDYPTDTLLPEMKLGYDLKT   147
                                     o

         150   GRNRILTSWRSSDDPSSGNTTYKIDTQRGLPEFILNQGRYEMQRSGPWNG   199
               | || ||||||||||| ||        ||    | |  |    |||||||
         148   GLNRFLTSWRSSDDPSSGDYSYKLEL-RRLPEFYLSSGSFRLHRSGPWNG   196
                                *o                        *

         200   MEFSGIPEVQGLNYMVYNYTENSEEISYTFHMTNQSIYSRLTVSDYTLN-   248
               ||||| | | |||| |||||| ||| |||  ||| |  ||||||
         197   FRISGIPEDQKLSYMVYNFTENSEEAAYTFLMTNNSFYSRLTISSTGYFE   246

         249   RLTWIPPSRAWSMFWTLPTDVCDPLYLCGSYSYCDLITSPNCNCIRGFVP   298
               |||| | |  |  || |   ||  || |||||| |||  | |||| || |
         247   RLTWAPSSVVWNVFWSSPNHQCDMYRMCGPYSYCDVNTSPVCNCIQGFRP   296
                                                    o

         299   KNPQQWDLRDGTQGCVRTTQMSCSGDGFLRLNNMNLPDTKTATVDRIIDV   348
               || |||||| |  || | | |||||| | | || || |  ||  |
         297   KNRQQWDLRIPTSGCIRRTRLSCSGDGFTRMKNMKLPETTMAIVHRSIGL   346
                 *

         349   KKCEERCLSDCNCTSFAIADVRNGELGCVFWTGELVEIR-KFAVGGQDLY   397
               | ||  ||||||||| |  | || || ||  ||  ||  || || |||||
         347   KECEKRCLSDCNCTAFANADIRNRGTGCVIWTGELEDIRTYFA-DGQDLY   395

         398   VRLNAADLG    406
               ||| ||||
         396   VRLAAADLV    404
```

FIG. 6   Comparison between the primary structures of two SLSGs representative, respectively, of class I (SLSG-13) and class II (SLSG-2A) alleles, as defined by Nasrallah et al. (1991). The two peptide sequences share 69.2% of identical amino acid positions. Putative N-glycosylation sites common to the two proteins are indicated above the alignment (*), and unshared sites are noted below (°). Gaps (-) have been introduced to maximize the homology.

C. Other Families of Glycoproteins Homologous with the S Locus-Specific Glycoproteins

Analysis of *Brassica* genomic DNA has revealed the existence of a family of sequences possessing different degrees of homology with genes encoding the SLSGs (Nasrallah et al., 1985b, 1988). These sequences may thus be regarded as corresponding with pseudogenes. Lalonde et al. (1989) demonstrated that one of them is specifically expressed in the stigma. Based on this evidence, these authors proposed that the S gene family encodes at least two polypeptides, the SLSG, encoded by the *SLG* gene, and a second protein, defined as a product of the S locus-related *(SLR1)* gene.

Using a probe made from S_6 cDNA, two classes of clones were identified from a stigma-specific cDNA library, one corresponding with the *SLSG* DNA and the other, which hybridized less strongly to the probe, with *SLR1* DNA. The major difference between these two classes of cDNAs was that they exhibit different restriction patterns. For example, Southern analysis of different *S* genotypes of *Brassica oleracea* and *Brassica campestris* shows a restriction fragment length polymorphism (RFLP) between *S* allele homozygotes when probed with *SLSG* cDNA. The *SLR1* probe detects a single band of equivalent intensity in all the *S* genotypes studied, and shows limited polymorphism. Further, by means of segregation analysis using RFLPs, it has been possible to show that in F_2 populations, *SLR1*, in contrast to the *SLSG*, is not linked to the *S* locus. The *SLR1* gene cannot therefore be implicated in the determination of SI in *Brassica*.

However, the expression of the *SLR1* gene is in many ways analogous to that of the *SLSG*. It is expressed exclusively in the stigma, and transcripts accumulating during development achieve a maximum concentration at the time when the SI response is acquired by the flower. Lalonde *et al.* (1989) reported the *SLR1* gene to be highly expressed, with transcripts being seven times more abundant than those of the *SLSG*. Perhaps because of their high abundance, *SLR1* transcripts can be detected in the stigmatic papillae one day prior to the *SLSG* transcripts using *in situ* hybridization. The nucleic acid sequences of *SLR1* and *SLSGs* have been determined for a number of *S* alleles. For plants carrying the S_{22} allele, the sequences are approximately 68% conserved, and the deduced peptide sequences of the SLR1 and SLSG proteins exhibit about 58% homology. The predicted SLR1 protein possesses a number of characteristics in common with the SLSGs, including the presence of 12 cysteine residues in the region of the carboxy-terminus of the polypeptide chain (Fig. 7). Further, while the S_{22} SLSG has 12 potential sites of N-glycosylation, only a few of these sites are conserved in the SLR1. Sequence analysis of cDNA indicates the presence, in both genes, of a hydrophobic sequence composed of about 30 amino acids situated at the 5′ terminus (Fig. 8). Of the hydrophobic sequence, 42% is conserved between the *SLR1* and *SLSG* genes and corresponds to a peptide signal that is believed to be associated with the transport of their protein products toward the surface of the stigmatic papilla (Nasrallah *et al.*, 1987). Considering the evolutionary relationships between these different sequences, Lalonde *et al.* (1989) described this *S* gene family as being composed of a group of sequences that may play an important part in the cellular interactions occurring between pollen and stigma; it is suggested that the *SLR1* gene produces a product, which, on the basis of its continued presence and

```
                                    *
SLSG-22  1   INTLSATESLTISSNRTLVSPGNVFELGFFRTTSSS------RWYLGIWY   44
             ||||  | ||||||| ||||||| |||||||| ||        |||||||||
SLR1     1   TNTLSPNEALTISSNKTLVSPGTVFELGFFKTTTRNSPDGTDRWYLGIWY   50
              *                *                    *        *

        45   KKLSN-RTYVWVANRDNPLSNSTGTLKITSNNLVILGHSNKSIWSTNRTK   93
             |  |  |||||||||| || |||||   || ||    ||| |
        51   KTTSGHRTYVWVANRDNALHNSMGTLKISHASLVLLDHSNTPVWSTNFT-   99
                            *

        94   GNERSPVVAELLANGNFVMRDSNNNRSSRFLWQSFDYPTDTLLPEMKLGY   143
             |   || |||||||||| ||| | || ||||||| |||||||||||||
       100   GVAHLPVTAELLANGNFVLRDSKTNDLDRFMWQSFDYPVDTLLPEMKLGR   149
               *

       144   DLKTGLNRF-LTSWRSSDDPSSGDFSYKLRARR-LPELYLSSGIFRVHRI   191
               |    |||| |  |||||||  |    |  | ||    | | |
       150   NRNGSGNEKILTSWKSPTDPSSGDYSFILETEGFLHEFYLLNNEFKVYRT   199
                         *         *                  *

       192   GPWNGIRFSGIPDDRKLSYLVYNFTENNEEVAYTFRMTNNT-IYSRLTVS   240
             |||||  ||  |||          ||   |  || |||| |    ||  |  |
       200   GPWNGVRFNGIPKMQNWSYIDNSFITNNKEVAYSFQVNNNHNIHTRFRMS   249
                                                        *

       241   FSGYIERQTWNPSLGMWNVFWSFPLDSQCDAYRACGPYSYCDVNTSPICN   290
             ||   ||      | |||||  || |  ||| |||  || |||  ||
       250   STGYLQVITWTKTVPQRNMFWSFPEDT-CDLYKVCGPYAYCDMHTSPTCN   298

       291   CIQGFNPSNVEQWDQRVWANGCIRRTRLSCS-GDRFTMMKNMKLPETTMA   339
             || || ||     ||| ||   ||| || ||  |   ||||||| |
       299   CIKGFVPKNAGRWDLRDMSGGCVRSSKLSCGEGDGFLRMSQMKLPETSEA   348
                                     *

       340   IVDRSIGVKECEKRCLSDCNCTAFANADIRNGGAGCVIWTGRLDDMRNYA   389
             ||  || |||  |||||| || || ||| ||| ||| || |
       349   VVDKRIGLKECREKCVRDCNCTGYANMDIMNGGSGCVMWTGELDDMRKYN   398

       390   ADHGQDLYVRLAAANLV          406
             |  ||||||| ||| ||
       399   A-GGQDLYVKVAAASLVPS        416
```

FIG. 7   Comparison between the amino acid sequences of SLSG and SLR1 proteins of *Brassica oleracea*. The two peptide sequences have been deduced from cDNAs isolated from the S_{22} homozygous genotype (Lalonde *et al.*, 1989). Gaps (-) have been introduced to maximize the homology (>58%). Potential sites of N-glycosylation (*) are indicated above the alignment, and cysteine residues are underlined. From Gaude (1990).

```
                 -31                                      -1

          SLSG-6    MKGVRKPYDNSYTLSFLLVFFV-LILFCPAFS
          SLSG-13   EGVKKTYDISYTLSFLLVFFV-LILFRPAFS
          SLSG-14   VRKNYNNCYILSFLLVFFV-LILFPPAFT
          SLSG-22   MKGVRKTYDSSCTLSFLLVFFV-MIIFHPVFS
          SLR1      MRGVIPNYHHSYT---LL-FFVILVLFPHVFS
```

FIG. 8   Comparison of amino acid signal sequences of SLSG and SLR1 proteins in *Brassica oleracea*. These sequences were predicted from published cDNAs. The amino acids are numbered by negative figures starting with residue -1, which is followed by the first amino acid of the mature protein. From Gaude (1990).

lack of variability, is necessary for some fundamental aspect of pollination. The *SLG* gene is held to perform a similar function but also to possess a specificity component that modifies the behavior of its product depending upon the *S* alleles possessed by the pollen.

The work of Lalonde *et al.* (1989) suggests that the active transcription of the *SLR1* gene is accompanied by active synthesis of the SLR1 polypeptide. In an independent study in *Brassica campestris,* Isogai *et al.* (1988) have also demonstrated the existence of stigmatic glycoproteins that possess significant homologies with SLSG sequences. These proteins, termed NS glycoproteins (*S* allele nonspecific *S* glycoprotein-like substances), fulfill all of the criteria of *S* gene products, except that they are identical in two lines that differ in their *S* allele constitution. NS glycoproteins isolated from two homozygous genotypes, one for S_8 and the other for S_9, have been fully characterized with regard to amino acid composition, peptide mapping, and oligosaccharide chain determination. A comparison between the N-terminal sequences of the SLR1 protein, as described by Lalonde *et al.* (1989), and the N-terminal sequence of the *Brassica campestris* NS glycoprotein indicates the latter also to be the product of the *SLR1* gene (Fig. 9). Further, alignment of peptide sequences obtained after proteolytic digestion of the NS glycoprotein with the total SLR1 sequence shows a divergence of less than 20% between the two proteins (Fig. 10). This finding has been recently confirmed by the cloning and sequencing of a cDNA encoding the NS glycoprotein in *Brassica campestris* (Isogai *et al.*, 1991).

It is of particular interest to note that whereas the *SLR1* gene is highly conserved in *Brassica oleracea,* the equivalent gene product in *Brassica campestris* shares only 80% homology with the SLR1 protein of *Brassica oleracea,* a degree of homology similar to that observed between the *SLG* genes of the two species. This difference suggests that the *SLR1* gene may reflect a species specificity in the Brassicaceae not encountered at the *S* locus level and that *SLR1* may represent the ancestral gene from which the *S*-multigenic family has evolved.

In a second independent investigation, Trick and Flavell (1989) isolated two genes (termed BS_{29}-1 and BS_{29}-2) from a homozygous line of *Brassica oleracea* expressing the S_{29} allele. These two genes diverge by 30% in their

| SLR1 | T N T L S | P | N E A L T I S S | N | K T L U S P G D U F E L | G F |
| NS | T N T L S | S | N E A L T I S S | X | K T L U S P G D U F E L | |

FIG. 9 Comparison between the N-terminal sequences of SLR1 proteins and NS glycoproteins. X, undetermined amino acid corresponding probably to an asparaginyl residue carrying an oligosaccharide chain. From Gaude (1990).

```
SLR1    1   TNTLSPNEALTISSNKTLVSPGTVFELGFFKTTTRNSPDG-TDRWYLGIW          49
            |||||  ||||||||| |||||||| ||||     |||||| || ||| ||||
NS      1   TNTLSSNEALTISSXKTLVSPGDVFEL----TTTRNSQDGSTDRMYLGIX

            YKTTSGHRTYVWVANRDNALHNSMGTLKISHASLVLLDHSNTPVWSTNFT         99
            ||      ||| ||||||| | ||| |
            YKK    2 XTYVXVANRDNPLRNXMGXL

            GVAHLPVTAELLANGNFVLRDSKTNDLDRFMWQSFDYPVDTLLPEMKLGR        149
                                  |||||||| |||| ||
                             3 TNDLDRFMXQSFDFPV

            NRNGSGNEKILTSWKSPTDPSSGDYSFILETEGFLHEFYLLNNEFKVYRT        199
                     ||||||||    |                          ||||
                4 KILTSWKSMXXP                            5 VYRT

            GPWNGVRFNGIPKMQNWSYIDNSFITNNKEVAYSFQVNNNHNIHTRFRMS        249
            ||  ||||||||||    |
            GPDNGVRFNGIPNLQQ

            STGYLQVITWTKTVPQRNMFWSFPEDTCDLYKVCGPYAYCDMHTSPTCNC        299
                          |||||||  |||||| ||||   ||||| || ||
                        6 VVPQRNMFLSFPEDTXDLYKMXGPYAYXDMXTS

            IKGFVPKNAGRWDLRDMSGGCVRSSKLSCGEGDGFLRMSQMKLPETSEAV        349
                      ||||  ||||||||| |||||||   ||||||||||||||  |
                    7 NAGRXDLRDMSGGXVRSSKLSXX-GDGFLRMSQMKLPEQXXAX

            VDKRIGLKECREKCVRDCNCTGYANMDIMNGGSGCVMWTGELDDMRKYNA        399
            |                  |||||||||||||      |  |||||  |
            V                8 XXVTGYANMDIMNGG       9 LVDMRKYDA

            GGQDLYVKVAAASLVPS                                          416
            |||||||  |  |||||
            GGQDLYV--AEASLVP
```

FIG. 10   Comparison between the amino acid sequences of SLR1 protein (SLR1-22 from *Brassica oleracea;* Lalonde *et al.,* 1989) and NS glycoprotein (from *Brassica campestris;* Isogai *et al.,* 1988). The primary structure data of NS glycoprotein come from the amino acid sequences of peptides generated after proteolytic digestion of the NS glycoprotein (Isogai *et al.,* 1988). The total level of homology between all the NS peptide fragments and the SLR1 protein is about 79%. For each peptide fragment, the level of homology, expressed as the percentage of identical amino acid positions, is as follows: **1**, 77; **2**, 70; **3**, 87.5; **4**, 75; **5**, 80; **6**, 79; **7**, 79.5; **8**, 80; **9**, 80. X, undetermined amino acid residue. From Gaude (1990).

nucleic acid sequences and by 42% at the amino acid level. Information from Northern and Southern analyses combined with sequence homologies published by Nasrallah (Fig. 11) clearly indicates that the BS_{29}-1 and BS_{29}-2 genes correspond with the *SLR1* gene (Lalonde *et al.,* 1989) and the SLSG of this particular allele. Translation of *SLR1* transcripts in stigmas of *Brassica* has recently been confirmed with the aid of immunochemical and protein sequencing analyses (Umbach *et al.,* 1990; Gaude *et al.,* 1991).

a)

```
                                              *
BS29-1    1   TNTLSPNEALTISSNKTLVSPGTVFELGFFKTTTRNSPDGTDRWYLGIWY
                                                             *
         51   KTTSGHRTYVWVANRDNALHNSMGTLKISHASLVLLDHSNTPVWSTNFTG

        101   VAHLPVTAELLANGNFVLRDSKTTALDRFMWQSFDYPVDTLLPEMKLGRN
                                                  ::
                  *                               ND
        151   RNGSGNEKILTSWKSPTDPSSGDYSFILETEGFLHEFYLLNNEFKVYRTG
                              *
        201   PWNGVRFNGIPKMQNWSYIDNSFITNNKEVAYSFQVNNNHNIHTRFRMSS

        251   TGYLQVITWTKTVPQRNMFWSFPEDTCDLYKVCGPYAYCDMHTSPTCNCI

        301   KGFVPKNAGRWDLRDMSGGCVRSSKLSCGEGDGFLRMSQMKLPETSEAVV
                                          *
        351   DKRIGLKECREKCVRDCNCTGYANMDIMNGGSGCVMWTGELDDMRKYNAG

        401   GQDLYLKVAAASLVPS    416
                              :
                              V
```

FIG. 11 Comparison between the amino acid sequences of BS29-1 and BS29-2 proteins (Trick and Flavell, 1989), and SLR1 and SLSG (Lalonde *et al.*, 1989). The peptide sequences have been deduced from cDNAs isolated from plants homozygous for S_{29} (BS29-1 and -2 proteins) and S_{22} (SLR1 and SLSG proteins). Potential sites of N-glycosylation (*) are indicated above the alignment, and cysteine residues are underlined. From Gaude (1990). (a) Identity between BS29-1 and SLR1 is higher than 99%. The three different amino acids of SLR1 are indicated below the BS29-1 sequence. (b) Identity between BS29-2 and SLSG-22 (79%) is similar to that observed when SLSG proteins are compared by pair. BS29-2 contains only six potential sites of N-glycosylation, which are all conserved within SLSG-22. Among the 13 cysteine residues of BS29-2, 12 are conserved.

Evidence from Scutt *et al.* (1990) has pointed to the presence of further *SLR*-class transcripts in the stigmas of *Brassica* (Trick and Heizmann, this volume). Nasrallah has termed these *SLR2s* and has shown them to be present in most lines of *Brassica oleracea* and, like *SLR1s*, to be highly conserved (Umbach *et al.*, 1990; Boyes *et al.*, 1991; Fig. 12). This conservation of the *SLR* class of sequences extends outside the SI *Brassicas*, for they have been detected in self-compatible *Brassica*, in *Raphanus*, and in *Arabidopsis*. Interestingly, although SLR1 polypeptides have been detected immunologically in the stigmas of *Raphanus* (Umbach *et al.* 1990), there is no such evidence of *SLR* gene expression in *Arabidopsis* (Pruitt, 1990).

In a recent discovery by Walker and Zhang (1990), part of a putative receptor serine–threonine protein kinase from *Zea mays*, named ZmPK1, was found to possess about 27% amino acid sequence identity with the S_{13} SLSG from *Brassica oleracea*. Stein *et al.* (1991) have now shown an

b)

```
                        *                                              *
SLSG-22   1    INTLSATESLTISSNRTLVSPGNVFELGFFRTTSSSRWYLGIWYKKLSNR    50
               |||||   ||| ||  |||||||||| ||||||| |||||||| ||||||| |
BS29-2    1    INTLSSIESLKISNSRTLVSPGNVLELGFFRTPSSSRWYLGMWYKKLSER    50
                         *                              *         *

         51    TYVWVANRDNPLSNSTGTLKITSNNLVILGHSNKSIWSTNRTKGNERSPV   100
               |||||||||||||| | |||||    ||| | ||||| |||| | |||||||
         51    TYVWVANRDNPLSCSIGTLKISNMNLVLLDHSNKSLWSTNHTRGNERSPV   100
                           *

        101    VAELLANGNFVMRDSNNNRSSRFLWQSFDYPTDTLLPEMKLGYDLKTGLN   150
               |||||||||| |||| |  | |||||||||||||||||||||||  ||||
        101    VAELLANGNFVLRDSNKNDRSGFLWQSFDYPTDTLLPEMKLGYDLRTGLN   150

        151    RFLTSWRSSDDPSSGDFSYKLRARRLPELYLSSGIFRVHRIGPWNGIRFS   200
               |||||||||| |||||||||| || || |  | ||| |||| |||| ||
        151    RFLTSWRSSDTPSSGDFSYKLQTRRLPEFYLFKDDFLVHRSGPWNGVGFS   200
                         *                   *

        201    GIPDDRKLSYLVYNFTENNEEVAYTFRMTNNTIYSRLTVSFSGYIERQTW   250
               | | | |||| ||||| | |||||||| |||| |||||| | ||| || ||
        201    GMPEDQKLSYMVYNFTQNSEEVAYTFLMTNNSIYSRLTISSSGYFERLTW   250
                                                   *

        251    NPSLGMWNVFWSFPLDSQCDAYRACGPYSYCDVNTSPICNCIQGFNPSNV   300
               ||  |||||||||     | ||| |   || |||||||||| ||||| | ||||
        251    TPSSGMWNVFWSSDEDFQCDVYKICGAYSYCDVNTSPVCNCIQRFDPSNV   300

        301    EQWDQRVWANGCIRRTRLSCSGDRFTMMKNMKLPETTMAIVDRSIGVKEC   350
               |  | |  || |||||||||| || || |||||||||||||||||||| |||
        301    QEWGLRAWSGGCRRRTRLSCSGDGFTRMKKMKLPETTMAIVDRSIGLKEC   350
                         *

        351    EKRCLSDCNCTAFANADIRNGGAGCVIWTGRLDDMRNYAADHGQDLYVRL   400
               |||||||||||||||||||||||    || ||   | |  |||||||||
        351    EKRCLSDCNCTAFANADIRNGGTGCVIWTGQLEDIRTYFA-NGQDLYVRL   400

        401    AAANLV    406
               | | ||
        401    APADLV    406
```

FIG. 11 *(continued)*

equivalent gene, termed *SRK,* to be present in *Brassica oleracea,* and to be expressed in both stigmas and anthers. Further, the *SRK* gene is linked genetically to the *SLG* and, thereby, to the SI phenotype. Most interestingly, the *SLG* gene and the *SLG*-like component of the *SRK* gene of a particular allele share high levels of sequence similarity (94%), indicating that an effective mechanism must exist for maintaining this homology. The *ZmPK1* gene is not linked with SI in *Zea mays,* and appears to be expressed throughout the plant. The level of *SRK* expression in vegetative parts of the *Brassica* plant remains to be reported. Likewise, no data are available on the nature and cellular location of the translation product of this class of gene.

```
              *                                    *
SLR2     1    TYVNTMLSS-ESLTISSKRTLVSSGGVFELGFFKTSGRSRWYLGIWYKKV   49
              ||  ||| |||  ||| |||||| |  |||||| |  ||||||||||||
SLSG-6   1    I--NT-LSSTESLRISSNRTLVSPGNNFELGFFRTNSSSRWYLGIWYKKL   47
                *                          *         *

         50   PRRTYAWVANRDNPLPNSSGTLKISGNNLVLLGQSNNTVWSTNLTRCNLR   99
              |||  |||||||| |  |||||||||||  || |  ||||| ||  | |
         48   LDRTYVWVANRDNPLSNAIGTLKISGNNLVLLGHTNKSVWSTNLTRGNER   97
                                                         o

         100  SPVIAGSP-NGNFVMRYSNNKDSSGFLWQSFDSPTDTLLPDMKLGYDLKT   148
              || |     |||||||| ||||  |  |||||| |||||||  |||||||||
         98   LPVVAELLSNGNFVMRDSSNNDASEYLWQSFDYPTDTLLPEMKLGYDLKT   147
                                      *

         149  GRNRFLTSWRSYDDPSSGNTTYKLDIRRGLPEFILLINQRVEIQRSGPWN   198
              |  ||||||||| ||||||  ||||||   |   | |||| |        ||||||
         148  GLNRFLTSWRSSDDPSSGDFSYKLET-RSLPEFYLWHGIFPMH-RSGPWN   195
                                  *             *

         199  GIEFSVIPEVQGLNYMVYNYTENNKEIAYSFHMTNQSIHSRLTVSDYTLN   248
              |   || ||| |  |  |||||  ||| |  || |  |||  ||  |||| |
         196  GVRFSGIPEDQKLSYMVYNFTENSEEVAYTFRMTNNSIYSRLTLSSEGYF   245
                                               *

         249  -RFTWIPPSRG-WSLFWVLPTDV-CDSLYL-CGSYSYCDLTTSPSCNCIR   294
              | ||| || | |  ||  | |  ||  |  || |   |  |||  ||||
         246  QRLTWNP-SIGIWNRFWSSPVDPQCD-TYIMCGPYAYCGVNTSPVCNCIQ   293
                                       o        o         o   o o

         295  GFVPKNSQRWNLKDGSQGCVRRTRLSGSGDGFLRLNNMKLPDTKT-ATVD   343
              ||| | | ||  ||  ||||| |||||||||||||  |||||||  |||||
         294  GFNPRNIQQWDQRVWAGGCIRRTRLSCSGDGFTRMKNMKLPET-TMAIVD   342
                                      *o               o

         344  RTIDVRK-CEERCLSDCNCTSFAIADVRNGGLGCVFWTGELVEIRKYAVG   392
              | | |  || |||||||||||| || || |||| ||| ||| |  | | |
         343  RSIGV-KECEKRCLSDCNCTAFANADIRNGGTGCVIWTGRLDDMRNY-VA   390
                                o  o    o o                   o

         393  -GQDLYVRLNA-ADLGTG    408
              ||||||||| | |||
         391  HGQDLYVRL-AVADLV     405
```

FIG. 12   Amino acid sequence of the *SLR2* gene product deduced from the cDNA sequence isolated from the S_6 genotype (Boyes *et al.,* 1991) compared with SLSG-6; 66.4% of identical amino acid positions are conserved between the two sequences. Gaps (-) have been introduced to maximize the homology. Potential sites of N-glycosylation (*) and cysteine residues (°) are indicated above and below the alignment, respectively. Four N-glycosylation sites and 11 cysteine residues are common to the two sequences.

D. Homology Modeling of Female Glycoproteins

Because a number of domains are conserved between individual *SLGs,* and between SLSGs, SLRs, and the *SRK* gene product, it was considered that homology modeling might provide useful pointers to their function(s), if any, in pollination. A consensus sequence of the cysteine-rich region (residues 310–417) was therefore constructed using data from 8 SLSGs and 1 SLR (Fig. 13); no data on the *SRK* sequence were available at the time of study. When parts of this sequence were compared with sequences

RESIDUE 310

|

C DLYKVCGPY YCDVNT NCIQGFPKN VQQWDLRVWS GGCIRRTRLS CEGDGFTRM

KNMKLPET A *IVDRSIGLK ECEKRCLSDC NCTAFANADI RNGGTGC*

FIG. 13 Consensus sequence of the cysteine-rich region, beginning at residue 310, of the SLSG. The part of the sequence bracketed by asterisks (*) is that showing homology to other proteins and has been modeled to the putative structure shown in Figure 18.

in current data bases (Dickinson and Crabbe, 1990), homologies (Fig. 14) were noted with the deduced amino acid sequence of the *ZmPK1* gene from *Zea mays* (Walker and Zhang, 1990), two segments of the von Willebrand factor precursor (Verweij *et al.*, 1986), and one segment of Type VI collagen (Chu *et al.*, 1989). Interestingly, the domain of the *ZmPK1* gene homologous with the *SLSG–SLR* consensus sequence is reported to project from the plasma membrane into the extracellular domain, and perhaps to be involved in recognition (Walker and Zhang, 1990), whereas both collagen VI and the von Willebrand factor are located extracellularly. Further, the animal proteins are involved in protein–protein binding and possibly in forming extracellular matrices, both of which could form an important part of the pollen–stigma interaction. Sequence comparison also reveals that the SLSGs involved in sporophytic SI differ from those implicated in gametophytic SI, which display close homology to *Aspergillus* T_2 ribonuclease (McClure *et al.*, 1989).

Secondary structure predictions for the consensus SLSG–SLR sequence indicate helical sections bounded by turns. Similar analysis of the ZmPK1 polypeptide suggests that a similar helical sequence is bracketed on one side by a β-turn, but conclusive evidence is not found for a second turn at the carboxyl end of the sequence (Fig. 14). For the von Willebrand factor one of the helical sections is also preceded by such a turn. No secondary structure predictions are available for Type VI collagen. A systematic classification of β-hairpin turns in protein structures has been published (Sibanda *et al.*, 1989), and using this information, more than 100

For Figures 13–18 involving homology modeling, data base searches, sequence comparisons and secondary structure predictions were performed on the SEQNET VAX computer at Daresbury, Cheshire, UK, using UWGCG and PIR programs for searching and matching sequences in GENBANK, EMBL, NBRF, and SWISSPROT data bases maintained at Daresbury. Secondary structure predictions were made as described in Dickinson and Crabbe (1990). Desk Top Molecular Modeller (DTMM) version 1.2 (Oxford Electronic Publishing, Oxford, UK) and Alchemy II (Tripos Associates Inc., St. Louis) were used for molecular model building and quantitative calculations. Force field calculations were based on the COSMIC system described by Vinter *et al.* (1987). The validity of the modeling method for β-hairpins was checked as described previously (Crabbe and Bicknell, 1989).

```
CON.SEQ.(382):  IVDRSIGLK--ECEKRCLSDCNCTAFANADIRNGGTGC
(PRED)        :    tt      hhhhhhh      hhhhhhhh    TT
VW1 (700)     : LTCRSLSYPDEECNEACLEGCFC
(PRED)        :    tt      hhhhhh
VW2 (354)     :                CQERCVDGCSC
(PRED)        :                hhhhhh
COLL-VI (786):                 CREKKCPDYTCPITFANPADI
MAIZE ZmPK1:  QHLLSVSLR-TCRDICISDCTCQGF---- QYCEGTGC
BETA-TURNS:
FAB L.CH.(89) : SYDRSLRV
PROT. A (31) :                             ITTGGSRC
```

FIG. 14 Homologies between the region of the SLSG consensus sequence bracketed by asterisks (*) in Figure 13, starting at residue 382, and two sections of von Willebrand factor (VWI starting at residue 700; VW2 starting at residue 354), Type VI collagen (COLL-VI, starting at residue 786) and protein *ZmPK1* from *Zea mays* (starting at residue 366). The secondary structure predictions (h, helix; t or T, turn; where T is a given higher probability than t) are derived from Chou-Fasman and Garnier-Osguthorpe-Robson algorithms. The turn (β-hairpin) sequences are from immunoglobulin Fab NEW light chain (starting at residue 89) and protease A from *Streptomyces griseus* (starting at residue 31) (Sibanda *et al.*, 1989).

hairpin sequences have been compared with the consensus sequences of the *Brassica* SLSG–SLR, ZmPK1, von Willebrand factor, and collagen VI. Two hairpin sequences show a high level of homology with the SLSG–SLR sequence at positions on either side of the cysteine-rich region, where turns are predicted (Fig. 15). These are residues L89–L98 from immunoglobulin Fab NEW (light chain) and residues 31–42 from *Streptomyces griseus* protease A, both being loops of class 2 : 2. Likewise, hairpins are indicated at either side of the cysteine-rich region of ZmPK1 (residues 366–374 and 391–397) and at the amino-terminus end of the sequence of the von Willebrand factor. The actual and predicted structures of the β-hairpins are set out in Figure 15. The phi, psi angles have also been measured at the i_{-2}, i residue at each turn and the angles from the predicted β-hairpins are within ± 10 degrees of those measured on the x-ray structures of the Fab and protease A β-hairpins. Likewise, when the plots of predicted hydrogen-bonding distances are compared with x-ray–derived measurements from the equivalent Fab and protease A hairpin structures (Fig. 16), the values fall within 0.05 nm of each other. The only exception to this is the second, P_2 protease A type turn in the ZmPK1 protein, where the predicted second H-bonding position is different.

Although the cysteine-rich region of Type VI collagen gives no indication of β-hairpins, one section (FANPADI) shows interesting homology to a predicted helical section in the SLSG–SLR sequence (FANADI) immediately before the second β-hairpin. When subjected to energy minimization (Crabbe *et al.*, 1990; Crabbe, 1991) (Fig. 17), these sections give a distance rms deviation over residues F, A, N, D, and I of 0.082 nm.

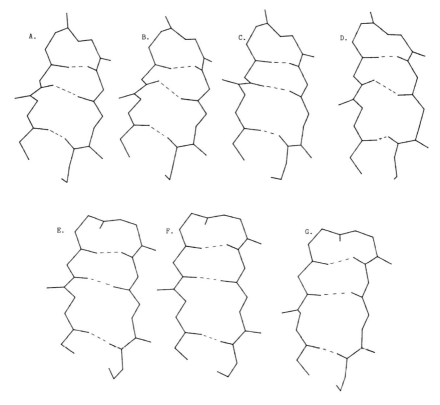

FIG. 15 Actual and predicted structures of β-hairpins. The dotted lines represent actual or predicted positions of hydrogen bonds. Apart from the hydrogens implicated in hydrogen bonding, hydrogens have been omitted for clarity. Distance root mean squares (rms) deviations are given in angstroms. Structures relating to homologies with SLSG residues 382–389: (A) X-ray structure of β-hairpin from FAB light chain. (B) Predicted structure of β-hairpin in SLSG (residues 382–389) (rms 0.127 Å). (C) Predicted structure of β-hairpin in von Willebrand factor (residues 700–707) (rms 0.43 Å). (D) Predicted structure of β-hairpin from *Zea mays* (residues 366–374) (rms 0.165 Å). Structures relating to homologies with SLSG residues 410–417: (E) X-ray structure of β-hairpin from *Streptomyces griseus* protease A. (F) Predicted structure of β-hairpin in SLSG (residues 410–417) (rms 0.179 Å). (G) Predicted structure of β-hairpin from *Zea mays* (residues 391–397) (rms 0.571 Å).

Interestingly, although some slight homology exists between the FANADI sequence of the SLSG–SLR sequence and the FQYCE of ZmPK1, there is clearly a greater similarity between the SLSG–SLR and collagen VI sequences. This disparity is however misleading, for in *Brassica*, the *SLG* domain of the *SRK* gene (equivalent to the *ZmPK1*) is almost completely homologous to the *SLG* gene with which it is linked (Stein *et al.*, 1991).

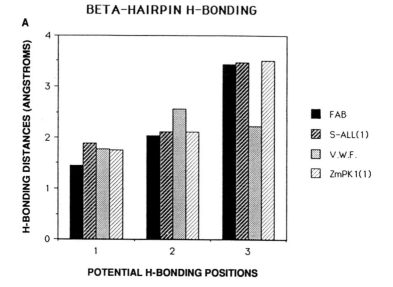

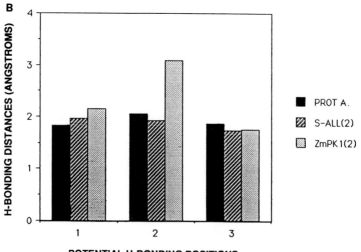

Combining the results from sequence matching and homology modeling enables a predictive model of the structure of SLSG–SLR proteins to be constructed from residues 382–417. This is shown in Figure 18 and comprises a region with helical character bounded by two β-hairpins. In this prediction we have not allowed for the possibility of S–S bridges within residues 382–417. This prediction of a helical region bound by turns permits a number of possibilities; for example, disulfide bridge formation may occur between one or more of the cysteines in this region and one or more of the six cysteines in the remainder of the cysteine-rich region (residues 310–382). This would stabilize the three-dimensional configuration of the protein, particularly in a reductive environment. As mentioned, Nasrallah *et al.* (1987) have proposed that such bridges do exist. Alternatively, the cysteines could interact, via S–S bridges, with a male determinant, other SLSG–SLRs, or even the *SRK* product.

This discovery of significant homology between the SLSG–SLR family from the stigma of *Brassica,* and extracellular matrix-forming proteins in animals (von Willebrand factor and Type VI collagen) is interesting and unexpected. The homology between the putative protein kinase from *Zea mays* (ZmPK1) and the SLSG–SLR class of proteins from *Brassica* has already been reported (Walker and Zhang, 1990), but its significance has been overtaken by the discovery of the SRK sequence by Stein *et al.* (1991). Comparison of homology levels between SLSGs and the ZmPK1 are thus not helpful because they should more usefully be made between an SLSG and the sequence of the SLSG-like component of its associated *SRK* gene product, where high levels of homology are found (Nasrallah *et al.,* 1991; Stein *et al.,* 1991). The polypeptide encoded by the gene from *Zea mays* and SRK contains three clearly defined domains, one extracellular that has homology with the SLSG sequence, transmembrane domain, and a third intracellular domain which is the kinase component of the molecule. Walker and Zhang (1990) suggested that these genes encode a new class of signaling molecules, probably peculiar to plants, which function to detect intercellular signals and to retransmit them intracellularly.

FIG. 16 Actual and predicted hydrogen-bonding distances in the β-hairpins. Position 1 is that closest to the turn, as shown in Figure 14. (A) Distances relating to homologies with SLSG residues 382–389. (B) Distances relating to homologies with SLSG residues 410–417. Abbreviations are as in Figure 14. H-bonding distances based on FAB were as follows: position 1, 1.72 ± 0.15; position 2, 2.2 ± 0.2; position 3, 3.16 ± 0.15. H-bonding distances based on protease A were as follows: position 1, 1.98 ± 0.13; position 2, 2.3 ± 0.50; position 3, 1.78 ± 0.06. (All distances are in angstroms.)

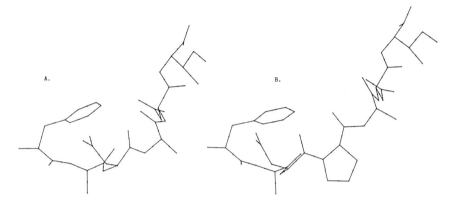

FIG. 17   Comparison between the predicted structure of (A) the FANADI sequence in SLSG and (b) the FANPADI sequence in Type VI collagen, shown in Figure 14. Hydrogens have been omitted for clarity.

The homology between the cysteine-rich regions of the *Brassica* SLR and SLSGs, the Zea ZmPK1 (and by implication the *SRK* sequence), and collagen VI is certainly worthy of further investigation. Sequence comparison and homology modeling indicate that this molecular domain could perform a similar function in these different molecules. Our present knowledge of collagen VI suggests that this domain is involved in intermo-

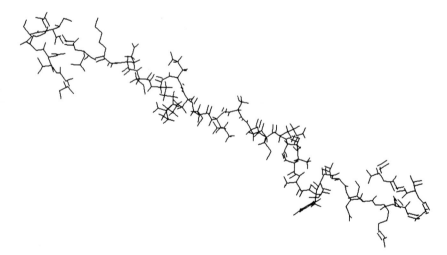

FIG. 18   View of the predicted structure of residues 382–417 of the cysteine-rich region of SLSG. Hydrogens have been omitted for clarity.

lecular binding, via disulfide bridges, to form multimers (Chu *et al.*, 1989). It must be emphasized that there is at present no evidence that the SLSGs or SLRs of *Brassica* form such multimers, and it also remains possible that the complementary sequences reside on the male determinant of SI. It may also be that the cysteine-rich region of the *SRK* gene product is involved in binding to a family of signaling molecules containing appropriate complementary sequences.

III. Proteins of the *Brassica* Pollen Grain

The dimer hypothesis of Lewis (1954, 1965) predicts that pollen should carry molecules nearly identical with those found in the stigma, and early investigations thus focused on pollen glycoproteins that resembled the SLSGs or SLRs of the stigma. Using IEF Nishio and Hinata (1978) separated 30 protein fractions of different pIs from homogenates of *Brassica* pollen. However, no unequivocal differences could be detected between pollen carrying different *S* alleles.

Surprisingly few pollen-specific proteins have been identified. Singh *et al.* (1983) have demonstrated that *Brassica* pollen deficient in a β-galactosidase gene adhered to the stigma surface less strongly than normal grains, suggesting that the enzyme could be involved in the pollen–stigma interaction. The authors demonstrated this particular enzyme to be gametophytically regulated, as are many intine-held enzymes (e.g., Howlett *et al.*, 1975). Thus, although research over the last decade seems to have established that identical *S* allele products are not present in pollen and stigma, there must nevertheless be some degree of commonality or complementarity between the male and female determinants of SI to enable recognition to take place. Indeed, determined attempts to search for male sequences with normal (Nasrallah *et al.*, 1991) or polymerase chain reaction (PCR)-generated (Guilluy *et al.*, 1990, 1991) probes have resulted in the detection of both *SLSG*- and *SRK*-like sequences, but at low levels. These results are further discussed in another chapter (Trick and Heizmann, this volume).

Adopting a sophisticated immunological approach, Gaude *et al.* (1988) used monoclonal antibodies raised to the S_6 SLSG to probe Western blots containing pollen proteins. Intriguingly, the antibody bound to the same family of protein bands, whatever the *S* allele possessed by the pollen (Gaude *et al.*, 1988; Gaude, 1990). The polypeptides involved were present in the pollen at extremely low levels, and possessed molecular masses of 36.5, 34, and 11–12 kDa (Fig. 19). Nasrallah and Nasrallah (1986) have predicted that the male component of SI would be present at relatively

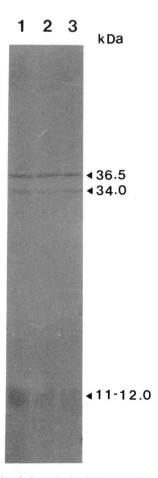

FIG. 19 Determination of antigenic homologies between stigma SLSGs and pollen proteins. From Gaude (1990). A monoclonal antibody (MAb 97-53) raised to S_6 SLSG of *Brassica oleracea* recognizes three pollen proteins with molecular masses of 36.5, 34, and 11–12 kDa. Pollen proteins (40 μg/lane) from S_6S_6, $S_{14}S_{14}$, and S_6S_{14} genotypes (lanes 1–3) were separated by SDS-PAGE, electrotransferred onto nitrocellulose, and probed with MAb 97–53 (1/100 dilution).

low levels, a view that is substantiated by the immunological data of Gaude *et al.* (1988). However, it remains possible that these determinants are present at relatively high concentrations but are focused at particular locations on or in the pollen grain. The sporophytic origin of these determinants supports such an inference, in that they may be derived from elements of the tapetal nurse cells, which are applied as a coating to the pollen grain in the later stages of maturation (Dickinson and Lewis, 1973a,b). The

superficial nature of the SI response in *Brassica* also indicates that both determinants are present at the interface between the grain and stigma surfaces. A number of models have now been proposed to explain the operation of SI based on molecular interactions taking place in the pollen grain coating (Dickinson, 1990), and the first analysis of this layer revealed it to contain a family of highly charged glycoproteins (Elleman *et al.*, 1989). More recent analyses using a highly effective method of coat removal have shown this family of polypeptides to consist of five principal members, two of which are highly glycosylated (Doughty *et al.*, in press). Interestingly, the molecular mass of four of these glycoproteins ranges between 32 and 36 kDa (Fig. 20a), corresponding with the masses of the antigens detected by antibodies raised to the S_6 SLSG in whole pollen extracts

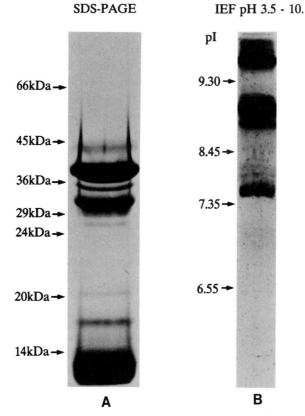

FIG. 20   Pollen coat proteins (PCPs) extracted and separated, as described by Doughty *et al.* (in press). (a) Separation according to molecular mass on SDS-PAGE (12% acrylamide). (b) Separation according to charge using isoelectric focusing, in the range pH 3.5–10.

(Gaude *et al.*, 1988; Gaude, 1990). Further, the pollen coating also contains a number of nonglycosylated polypeptides in the molecular mass range between 10 and 12 kDa, which is in accord with the second group of proteins recognized by the anti-S_6 SLSG antibody. The major pollen coat proteins (PCPs) fall into two groups with regard to their isoelectric points, one focusing at a pI of about 9, and the other between 10 and 11 (Fig. 20b). Preliminary sequence data indicate the major PCPs to be proline-rich. All the PCPs are clearly not involved in SI, for self-compatible *Brassicas,* such as *Brassica napus,* possess PCP profiles that are almost identical with *Brassica oleracea.* It is thus considered likely that these molecules are engaged in a more general pollen–stigma interaction that takes place after the contact of the pollen grain with the pistil. However, to explore the possibility that individual PCPs might interact with pistillar molecules, stigmatic extracts were challenged with purified PCPs. Strikingly, in all the experiments so far carried out, one or more of the PCPs have interacted with the glycoprotein band held to be the SLSG (Fig. 21) (Doughty *et al.,* in press). The identity of the individual PCP(s) involved in this binding has yet to be unequivocally identified. It is recognized that both SLSGs and PCPs are highly charged molecules, and the fact that some level of binding takes place is not surprising. However, it remains striking that of all the glycoproteins present in the stigma, only the SLSG interacts with the PCPs.

IV. Molecular Basis of Recognition and Response

The phenomenology of the SI response in *Brassica* is relatively straightforward. Following capture by the stigmatic papilla, the pollen grain commences hydration. In a cross-pollination the grain completes hydration within a matter of minutes and extends a tube into the papilla cell wall through the point of contact between grain and stigma (Elleman and Dickinson, 1990). Self-pollen, however, fails to hydrate, and in the presence of strong alleles, no further development takes place; there is even some preliminary evidence that the pollen may again become dehydrated and detach from the stigma surface. Weaker *S* alleles result in a range of morphologies, which may include growth of the pollen tube over the surface of the stigma. Interspecific pollinations between SI and self-compatible species (Lewis and Crowe, 1958; Hiscock and Dickinson, unpublished data) indicate that the stigmas of SI plants express an additional factor, which acts to prevent the entry of self-compatible pollen. Such an inference is supported by the observation that in SI plants immature stigmas fail to recognize self-pollen, only becoming competent to do

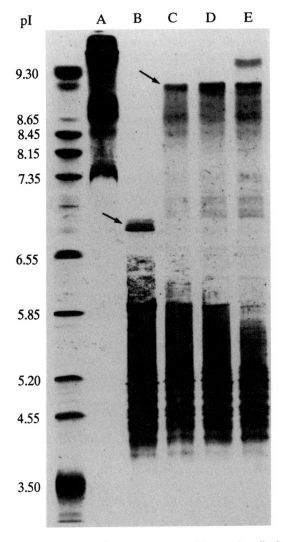

FIG. 21   Interaction between PCPs and stigmatic polypeptides, as described by Doughty *et al.* (in press). Lane pI, pI marker ladder (A) Extracted PCPs from S_{25} pollen (Doughty *et al.*, in press). (B) Stigmatic polypeptides; the SLSG is indicated by arrow. (C) Interaction between $S_{25}S_{25}$ PCPs and $S_{63}S_{63}$ stigmatic proteins. A new band is formed (arrow). (D) Interaction between $S_{25}S_{25}$ PCPs and $S_{25}S_{25}$ stigmatic proteins. (E) Interaction between $S_{25}S_{25}$ PCPs and $S_{29}S_{29}$ stigmatic proteins.

so in the later stages of maturity. This evidence points to SI being acquired once a threshold of a particular stigmatic molecule, probably the SLSG, is reached in the papillae. The direct involvement of SLSGs in SI is also indicated by their segregation with *S* alleles in breeding pedigrees (Nasrallah *et al.*, 1985b) and by the fact that SI in some self-compatible strains results from the action of genes that suppress or reduce the synthesis of SLSG in the stigmatic papilla (Nasrallah *et al.*, 1991).

The elegant immunolocalization work by Kandasamy *et al.* (1990) has demonstrated SLSG to be present in the cell walls of the papilla, and it is reasonable to assume that these molecules are drawn into the pollen grain coating during the first stages of hydration. The interactions that take place in the coat and in the grain itself are complicated and clearly do not only concern SI. For example, experiments with a range of protein synthesis inhibitors suggest that these early interactions control both pollen adhesion and the regulation of grain hydration (Stead *et al.*, 1980; Sarker *et al.*, 1988). Structural observations indicate the pollen coat to play a central part in these events (Elleman and Dickinson, 1986; Elleman *et al.*, 1988), for within minutes of pollination this layer is converted from a homogenous electron-lucent matrix to a complex, electron-opaque structure.

If the hypothesis is accepted that the SLSG and other female molecules interact with pollen coat constituents, perhaps the PCPs, the products of this interaction must either remain *in situ* or move to the pollen protoplast. Some diffusion back into the stigmatic papilla is also possible, but the SI response must be centered in the pollen grain, for it can easily be demonstrated that a compatible grain will develop successfully adjacent to, and contiguous with, an inhibited self-pollen grain on the same stigmatic papilla (Sarker *et al.*, 1988). The fact that part of the SI interaction may thus take place extracellularly and involve elements contained in a highly lipidic pollen coating suggests that SI in *Brassica* does not conform with the conventional view of intercellular signaling systems. Indeed, it seems probable that in the evolution of SI, *Brassica* has exploited pathogen-defense systems already active in the pistil.

The discovery by Walker and Zhang (1990) of considerable sequence homology between a domain of a transmembrane serine–threonine kinase and the *Brassica* SLSG indicates that SI in the crucifers may involve a previously undiscovered signaling system but one that is distributed throughout the plant. That this gene can act independently from sporophytic SI is underlined by the fact that it was characterized in *Zea,* which is self-compatible and a member of the grasses—where SI is regulated gametophytically. Further, no homology exists between gametophytic SLSGs and the *ZmPK1* gene. However, PCR experiments using *SLSG* and *SLR* sequences as primers have indicated the presence of *SLSG*-like transcripts throughout the plant body in *Brassica* (Trick and Heizmann,

this volume; Guilluy *et al.*, 1991), and the discovery of the *SRK* sequence (Stein *et al.*, 1991) has confirmed that an equivalent to the *ZmPK1* exists in *Brassica*. The strong homology between a domain of the *SRK* kinase and the SLSG indicates that, at some point in the operation of the SI system, SLSGs may interact with the transmembrane kinases to generate an intracellular signal. This is unlikely to be a simple process because a generalized response at the stigmatic papilla plasma membrane would result in the rejection of compatible pollen. Similarly, an *SRK*-encoded kinase in the pollen is unlikely to be the only male determinant because it would be gametophytically inherited. One of the few possibilities thus remaining is that a hybrid effector molecule is formed at the pollen on stigmatic papilla protoplast surface by the conjunction of an *SRK* kinase and an extracellular molecule that differs in form according to the compatibility of the pollination. This latter molecule may either be the female SLSG, modified or otherwise by sporophytic determinants in the pollen coat, or a tapetally synthesized *S* allele product, which reacts with the female SLSG in the manner of a conventional receptor. The activity of the kinase could then regulate the subsequent fate of the grain, perhaps through protein synthesis. The involvement of a protein synthesis step in the inhibition of the . pollen grain was first proposed by Ferrari and Wallace (1976) and has been confirmed subsequently by more detailed experiments using a range of protein synthesis inhibitors (Sarker *et al.*, 1988).

Little is known about the factors that might affect the specificity of binding of the *Brassica* SLSG to the male determinant, be it a PCP or the *SRK* kinase itself. Certainly, the sequences of a number of putative alleles are known, but our ignorance of the male molecules makes it impossible to relate the variation observed between putative alleles to potential binding sites. Trick proposed that SLSG S-specificity is effected through the position of glycosylation sites (M. Trick, personal communication), and to a certain extent, this inference is supported by work with the glycosylation inhibitor tunicamycin, which can completely suppress SI (Sarker *et al.*, 1988). Interestingly, in many of the *S* allele sequences, variation in glycosylation sites occurs adjacent to the cysteine-rich region discussed in a previous section.

Little of this recent information sheds any light on the evolutionary relationship between sporophytic and gametophytic SI. There is some acceptance that the former is generally the more effective system because self-pollen is inhibited on the stigma surface, but the gametophytically regulated system of the grasses probably approaches the efficiency of sporophytic mechanisms. Whether one system is derived from the other is a far more difficult question; certainly highly charged glycoproteins are common to the pistils of both types, and in both systems pollen development is arrested, at least in the first instance, by a biostatic system (Dickin-

son *et al.*, 1982; Sarker *et al.*, 1986). However, persuasive evidence is now accumulating that the SLSGs of many plants with gametophytic SI are functional RNAses (McClure *et al.*, 1989, 1990; Singh *et al.*, 1991), and this is certainly not the case for *Brassica*. Nevertheless, recent data on gene expression during pollen development in the anther (Scott *et al.*, 1991) confirm previous suspicions (Dickinson and Bell, 1976) that developing pollen cells and the tapetum share similar patterns of gene expression except with regard to meiosis. The possibility thus remains that *S*-specific sequences are expressed both gametophytically (in the pollen, pregermination and postgermination) and sporophytically (in the tapetum). The gene product in the tapetum would thus be deposited onto the pollen wall and well placed to interact with any female *S*-specific polypeptide emanating from the stigma surface. Because a more generalized expression of the *S* locus could have become focused in the sporogenous cells, it does not necessarily follow that the sporophytic system is derived from the gametophytic; however, it cannot be argued that the former system is characteristic of families held to be more advanced than those featuring the latter. In a recent series of reports, Toriyama *et al.* (1991) and Thorsness *et al.* (1991) demonstrated *Brassica SLSG* promoter sequences to direct the expression of reporter genes in the tapetum of *Arabidopsis* (Brassicaceae) but in the mature pollen of *Nicotiana*, a plant of the Solanaceae where SI is gametophytic. Further, this group has subsequently shown that the *SLSG* promoter sequences, when transferred to *Brassica* together with a β-glucuronidase reporter construct, will direct expression of the β-glucuronidase in both the tapetum and microspores, lending even further support to the hypothesis discussed herein.

Acknowledgments

The authors thank Valerie Norris and Ann Rogers for help in the preparation of the manuscript, and the Wellcome Trust, the Royal Society, the AFRC, the Humane Research Society, the Lord Dowding Fund, and the EEC BRIDGE program for support.

References

Ascher, P. D. (1975). *Bot. Gaz.* **136**, 317–321.

Bateman, A. J. (1955). *Heredity* **9**, 53–68.

Boyes, D. C., Chen, C. H., Tantikanjana, T., Esch, J. J., and Nasrallah, J. B. (1991). *Genetics* **127**, 221–228.

Chen, C. H., and Nasrallah, J. B. (1990). *Mol. Gen. Genet.* **222**, 241–248.

Chu, M. L., Pan, T. C., Conway, D. , Kuo, H. J., Glanville, R. W., Timpl, R., Mann, R., and Deutzmann, R. (1989). *EMBO J.* **8**, 1939–1946.

Crabbe, M. J. C. (1991). *Binary* **3**, 140–146.

Crabbe, M. J. C., and Bicknell, G. (1989). *Binary* **1**, 57.

Crabbe, M. J. C., Evans, D. J., and Almond, J. W. (1990). F.E.B.S. Lett. **271**, 194–198.

Darwin, C. (1877). "The Different Forms of Flowers on Plants of the Same Species", 2nd ed. John Murray, London.

De Nettancourt, D. (1977). "Incompatibility in Angiosperms." Springer-Verlag, Berlin.

Dickinson, H. G. (1990). *Bioessays* **12**, 155–161.

Dickinson, H. G., and Bell, P. R. (1976). *Ann. Bot.* **40**, 1101–1109.

Dickinson, H. G., and Crabbe, M. J. C. (1990). *In* "Mechanism of Fertilization" (B. Dale, ed.), NATO ASI Series, Vol. H45, pp. 219–237. Springer-Verlag, Berlin.

Dickinson, H. G., and Lewis, D. (1973a). *Proc. R. Soc. Lond. [Biol.]* **183**, 21–28.

Dickinson, H. G., and Lewis, D. (1973b). *Proc. R. Soc. Lond. [Biol.]* **184**, 149–165.

Dickinson, H. G., Moriarty, J. F., and Lawson, J. (1982). *Proc. R. Soc. Lond. [Biol.]* **215**, 45–65.

Doughty, J., McCubbin, A., Hedderson, F., Elleman, C., and Dickinson, H. G. (1992). *Proc. Natl. Acad. Sci. U.S.A.* in press.

East, E. M., and Mangelsdorf, A. J. (1925). *Proc. Natl. Acad. Sci. U.S.A.* **11**, 166–183.

Elleman, C. J., and Dickinson, H. G. (1986). *J. Cell Sci.* **80**, 141–154.

Elleman, C. J., and Dickinson, H. G. (1990). *New Phytol.* **114**, 511–518.

Elleman, C. J., Willson, C. E., Sarker, R. H., and Dickinson, H. G. (1988). *New Phytol.* **109**, 111–117.

Elleman, C. J., Sarker, R. H., Aivalakis, G., Slade, H., and Dickinson, H. G. (1989). *In* "Plant Reproduction: From Floral Induction to Pollination" (E. Lord and G. Bernier, eds.) *Am. Soc. Plant Physiol. Symp. Ser.* **1**, 136–145.

Ferrari, T. E., and Wallace, D. H. (1976). *Theor. Appl. Genet.* **48**, 243–249.

Ferrari, T. E., Bruns, D., and Wallace, D. H. (1981). *Plant Physiol.* **67**, 270–277.

Gaude, T. (1990). Thèse d'Etat, University of Lyon, Lyon.

Gaude, T., Nasrallah, M. E., and Dumas, C. (1988). *Heredity* **61**, 317–318.

Gaude, T., Denoroy, L., and Dumas, C. (1991). *Electrophoresis* **12**, 646–653.

Guilluy, C. M., Gaude, T., Digonnet-Kerhoas, C., Chaboud, A., Heizmann, P., and Dumas, C. (1990). *In* "Mechanism of Fertilization," NATO ASI Series, Vol. H45 (B. Dale, ed.), pp. 253–270. Springer-Verlag, Berlin.

Guilluy, C. M., Trick, M., Heizmann, P., and Dumas, C. (1991). *Theor. Appl. Genet.* **82**, 466–472.

Herrero, M., and Dickinson, H. G. (1981). *J. Cell Sci.* **47**, 365–383.

Heslop-Harrison, J. (1982). *Acta Bot. Neerl.* **31**, 429–439.

Hinata, K., and Nishio, T. (1978). *Heredity* **41**, 93–100.

Howlett, B. J., Knox, R. B., Paxton, J. D., and Heslop-Harrison, J. (1975). *Proc. R. Soc. Lond. [Biol.]* **188**, 167–182.

Isogai, A., Takayama, S., Tsukamoto, C., Ueda, Y., Shiozawa, H., Hinata, K., Okazaki, K., and Suzuki, A. (1987). *Plant Cell Physiol.* **28**, 1279–1291.

Isogai, A., Takayama, S., Shiozawa, H., Tsukamoto, C., Kanbara, T., Hinata, K., Okazaki, K., and Suzuki, A. (1988). *Plant Cell Physiol.* **29**, 1331–1336.

Isogai, A., Yamakawa, S., Shiozawa, H., Takayama, S., Tanaka, H., Kono, T., Watanabe, M., Hinata, K., and Suzuki, A. (1991). *Plant Mol. Biol.* **17**, 269–271.

Kandasamy, M. K., Dwyer, K. G., Paollilo, D. J., Doney, R. C., Nasrallah, J. B., and Nasrallah, M. E. (1990). *Plant Cell* **2**, 39–49.

Lalonde, B. A., Nasrallah, M. E., Dwyer, K. G., Chen, C. H., Barlow, B., and Nasrallah, J. B. (1989). *Plant Cell* **1**, 249–258.

Lewis, D. (1954). *Adv. Genet.* **6**, 235–285.

Lewis, D. (1965). *In* "Genetics Today." Proc. XI Int. Congress Genet., 1963. (S. J. Geerts, ed.), pp. 657–663. Pergamon, Oxford.

Lewis, D., and Crowe, L. K. (1958). *Heredity* **12,** 233–256.

Lewis, D., Burrage, S., and Walls, D. (1967). *J. Exp. Bot.* **18,** 371–378.

Lundqvist, A. (1962). *Hereditas* **48,** 153–168.

Lundqvist, A., Osterbye, U., Larsen, K., and Linde-Laursen, I. B. (1973). *Hereditas* **74,** 161–168.

McClure, B. A., Haring, V., Ebert, P. R., Anderson, M. A., Simpson, R. J., Sakiyama, F., and Clarke, A. E. (1989). *Nature* **342,** 955–957.

McClure, B. A., Gray, J. E., Anderson, M. A., and Clarke, A. E. (1990). *Nature* **347,** 757–760.

Nasrallah, J. B., and Nasrallah, M. E. (1984). *Experientia* **40,** 279–281.

Nasrallah, J. B., Doney, R. C., and Nasrallah, M. E. (1985a). *Planta* **165,** 100–107.

Nasrallah, J. B., Kao, T. H., Goldberg, M. L., and Nasrallah, M. E. (1985b). *Nature* **318,** 263–267.

Nasrallah, J. B., Kao, T. H., Chen, C. H., Goldberg, M. L., and Nasrallah, M. E. (1987). *Nature* **326,** 617–619.

Nasrallah, J. B., Yu, S. M., and Nasrallah, M. E. (1988). *Proc. Natl. Acad. Sci. U.S.A.* **85,** 5551–5555.

Nasrallah, J. B., Nishio, T., and Nasrallah, M. E. (1991). *Annu. Rev. Plant Physiol. Plant Mol. Biol.* **42,** 393–422.

Nasrallah, M. E., and Nasrallah, J. B. (1986). *In* "Biotechnology and Ecology of Pollen" (D. L. Mulcahy, G. B. Mulcahy and E. Ottaviano, eds.), pp. 197–201. Springer-Verlag, New York.

Nasrallah, M. E., and Wallace, D. H. (1967). *Heredity* **22,** 519–527.

Nasrallah, M. E., Barber, J. T., and Wallace, D. H. (1970). *Heredity* **25,** 23–27.

Nasrallah, M. E., Wallace, D. H., and Savo, R. M. (1972). *Genet. Res.* **20,** 151–160.

Nishio, T., and Hinata, K. (1977). *Heredity* **38,** 391–396.

Nishio, T., and Hinata, K. (1978). *Jpn. J. Genet.* **53,** 197–205.

Nishio, T., and Hinata, K. (1979). *Jpn. J. Genet.* **54,** 307–311.

Nishio, T., and Hinata, K. (1982). *Genetics* **100,** 641–647.

Ockendon, D. J. (1974). *Heredity* **33,** 159–171.

Ockendon, D. J. (1975). *Euphytica* **24,** 165–172.

Ockendon, D. J. (1980). *Theor. Appl. Genet.* **58,** 11–15.

Ockendon, D. J. (1982). *Euphytica* **31,** 325–331.

Pruitt, R. E. (1990). *In* "Abstracts of the 4th International Conference on *Arabidopsis* Research" (D. Schweitzer, K. Peuker, and J. Loidl, eds.), pg. 74, University of Vienna, Vienna.

Roberts, I. N., Stead, A. D., Ockendon, D. J., and Dickinson, H. G. (1979). Planta **146,** 179–183.

Sarker, R. H., Elleman, C. J., Harrod, G., and Dickinson, H. G. (1986). *In* "Biology of Reproduction and Cell Motility in Plants and Animals" (M. Cresti and R. Dallai, eds.), pp. 53–60. University of Siena, Siena, Italy.

Sarker, R. H., Elleman, C. J., and Dickinson, H. G. (1988). *Proc. Natl. Acad. Sci. U.S.A.* **85,** 4340–4344.

Scott, R., Daglass, E., Hodge, R., Paul, W., Souflers, I., and Draper, J. (1991). *Plant Mol. Biol.* **17,** 295–299.

Scutt, L. P., Gates, P. J., Gatehouse, J. A., Boulter, D., and Croy, R. R. D. (1990). *Mol. Gen. Genet.* **220,** 409–413.

Sedgley, M. (1974a). *Euphytica* **23,** 543–551.

Sedgley, M. (1974b). *Heredity* **33,** 412–416.

Sibanda, B. L., Blundell, T. L., and Thornton, J. M. (1989). *J. Mol. Biol.* **206,** 759–777.

Singh, A., Ai, Y., and Kao, T. H. (1991). *Plant Physiol.* **96,** 61–68.

Singh, M. B., Marginson, R., and Knox, R. B. (1983). *In* "Pollination '82" (E. G. Williams, R. B. Knox, J. Gilbert, and P. Bernhardt, eds.), pp. 135–144. University of Melbourne, Melbourne.

Stead, A. D., Roberts, I. N., and Dickinson, H. G. (1980). *J. Cell Sci.* **42**, 417–423.

Stein, J. C., Howlett, B., Boyes, D. C., Nasrallah, M. E., and Nasrallah, J. B. (1991). *Proc. Natl. Acad. Sci. U.S.A.* **88**, 8816–8820.

Struck, D. K., and Lennarz, W. J. (1980). *In* "The Biochemistry of Glycoproteins and Proteoglycans" (W. J. Lennarz, ed.), pp. 35–83. Plenum Press, New York.

Takayama, S., Isogai, A., Tsukamoto, C., Ueda, Y., Hinata, K., Okazaki, K., Koseki, K., and Suzuki, A. (1986). *Agr. Biol. Chem.* **50**, 1673–1676.

Takayama, S., Isogai, A., Tsukamoto, C., Ueda, Y., Hinata, K., Okazaki, K., and Suzuki, A. (1987). *Nature* **326**, 102–105.

Thompson, K. F. (1965). *Hort. Res.* **5**, 39–58.

Thorsness, M. K., Kandasamy, M. K., Nasrallah, M. E., and Nasrallah, J. B. (1991). *Dev. Biol.* **143**, 173–184.

Toriyama, K., Thorsness, M. K., Nasrallah, J. B., and Nasrallah, M. E. (1991). *Dev. Biol.* **143**, 427–431.

Trick, M., and Flavell, R. B. (1989). *Mol. Gen. Genet.* **218**, 112–117.

Umbach, A. L., Lalonde, B. A., Kandasamy, M. K., Nasrallah, J. B., and Nasrallah, M. E. (1990). *Plant Physiol.* **93**, 739–747.

Verweij, C. L., Diergaarde, P. J., Hart, M., and Pannekoek, H. (1986). *EMBO J.* **5**, 1839–1847.

Vinter, J. G., Davis, A., and Saunders, M. R. (1987). *Computer-Aided Molecular Design* **1**, 31–52.

Walker, J. C., and Zhang, R. (1990). *Nature* **345**, 743–746.

Whitehouse, H. L. K. (1950). *Ann. Bot.* **14**, 198–216.

Part VI
Postscript

Plant Reproductive Biology: Trends

Christian Dumas* and Scott D. Russell†
* Reconnaissance Cellulaire et Amélioration des Plantes,
Université Claude Bernard-Lyon 1
Villeurbanne, France
† Department of Botany and Microbiology, University of Oklahoma,
Norman, Oklahoma 73019

I. Introduction

Since a previous *International Review of Cytology* volume on "Pollen: Cytology and Development" appeared in 1987 (Giles and Prakash, 1987), fundamental progress has been made in studying the connected developmental programs of floral morphogenesis and embryogenesis. Several specific areas have progressed dramatically by the use of molecular analyses and mutants, for example:

1. The identification of some major genes (homeotic genes and the MADS box) isolated from both *Arabidopsis thaliana* and *Antirrhinum majus* (e.g., Coen and Meyerowitz, 1991).
2. The identification of a set of at least nine genes controlling pattern formation during embryogenesis in *Arabidopsis thaliana* (Mayer *et al.,* 1991).

Between these two connected developmental programs, significant progress has been made in increasing our knowledge of plant reproductive biology in the strict sense. Among them, we can point out a number of important contributions:

1. Pollen-specific gene expression (Mascarenhas, 1990a,b, this volume; Davies *et al.,* this volume)
2. Improvements in our understanding of pollen wall formation, biosynthesis, and biochemistry (Wiermann and Gubatz, this volume).
3. New data from compatible pollen–pistil interactions: the role of the extracellular matrix as an active component in the growth and guidance of pollen tubes (Sanders and Lord, 1989, this volume); the preprogramming of double fertilization and the characterization of sperm cells as having

unique functions within the pollen tube in a male germ unit (Roeckel *et al.*, 1990; Russell, 1991, this volume; Mogensen, this volume).

4. Identification and preliminary characterization of genes and gene products involved in self-incompatibility (SI) mechanisms (Haring *et al.*, 1990; Gaude and Dumas, 1990), the implication of these genes and products within the complex multigenic *S* family (Nasrallah *et al.*, 1991), and the putative involvement of enzymic activities, such as RNase in the gametophytic SI system (McClure *et al.*, 1989), and kinase in the sporophytic SI system (Stein *et al.*, 1991).

5. Attempts to transform plants using either pollen as a vector for introducing foreign DNA (Roeckel *et al.*, 1988) or introducing DNA through ovule manipulation (Chapman *et al.*, 1985; Roeckel *et al.*, this volume).

6. The isolation of gametophytic cells as an emerging technique for gamete characterization and experimentation (Russell, 1991; Chaboud and Perez, this volume; Huang and Russell, this volume), and the achievement of artificial zygotes *in vitro* (Kranz *et al.*, 1991, this volume).

7. New insights emerging on the evolution of double fertilization in angiosperm from detailed examinations of gymnosperms using modern techniques. These studies revealed that unappreciated parallels to double fertilization occur in gnetophytes and higher gymnosperms (Friedman, 1990, 1992, this volume).

8. The improved use of cellular markers to examine crucial questions related to the preprogramming of the process of double fertilization (Russell, 1985, this volume; Chaboud and Perez, this volume).

Perhaps the most significant of this progress is due to the application of new techniques in molecular genetic to long-standing problems in plant reproductive biology (e.g. Nasrallah *et al.*, 1985; Anderson *et al.*, 1986), the development of new cytological tools for cell investigation, such as *in situ* hybridization either using nucleic homologs (Cornish *et al.*, 1987) or heterologous probes (Sanders *et al.*, 1991), the use of immunological probes for cytolocalization (e.g., Kandasamy *et al.*, 1989), the possibility of constructing transformants to study gene regulation (Moore and Nasrallah, 1990), the three-dimensional reconstruction and quantification of organelles or cells with computer-aided techniques (e.g., Russell, 1985; Mogensen *et al.*, 1990), gamete isolation and manipulation *in vitro* of sperm cells (e.g., Russell, 1986; Dupuis *et al.*, 1987; Chaboud and Perez, this volume), and embryo sacs and egg cells: (e.g., Wagner *et al.*, 1989; Huang and Russell, this volume), and the use of techniques coming from such biophysical methods as nuclear magnetic resonance (NMR) (e.g., Kerhoas *et al.*, 1987) and electrofusion of gametes (Kranz *et al.*, 1991).

II. Advances in Compatibility and Male–Female Interactions

A. Pollen

1. Pollen Gene Expression and "Genetic Overlap"

Many of the genes expressed in the male gametophyte are also expressed in the sporophyte, comprising the so-called sporophytic–gametophytic genetic overlap (Mascarenhas, 1990a,b). However, the number of genes expressed specifically in the male gametophyte is surprisingly high in comparison with those expressed in the sporophyte. For instance, of approximately 20,000 different mRNAs found in mature pollen, most represent genes that are transcribed late in development; 60% of the isozymes examined in barley pollen were expressed in both gametophytic (pollen) and sporophytic (nonreproductive) tissues, 30% were specific to the sporophyte and 10% were specific to the pollen (Pedersen *et al.*, 1987). Similar data have also been obtained in poplar by Rajora and Zsuffa (1986).

It has been proposed that genetic selection could operate on pollen tubes during growth within the transmitting tissue (sporophytic tissue). Competitive selection of pollen (thus selecting the gametes) has been considered to be a crucial evolutionary factor that led to the domination of the angiosperms among the other plant groups (Mulcahy and Mulcahy, 1975, 1987).

Recently, attempts have been made to clone the genes expressed in the male gametophyte and to follow their pattern of transcription. Several cDNA libraries have been constructed to poly(A)RNA collected from pollen of *Tradescantia* and maize (Mascarenhas *et al.*, 1985; Stinson *et al.*, 1987), *Oenothera* (Brown and Crouch, 1990), and tomato (Ursin *et al.*, 1989), but only a few corresponding genomic clones have been isolated and sequenced. The predicted proteins of two cDNAs expressed in anthers and pollen of tomato show some amino acid sequence homology to pectate lyases from the plant pathogen *Erwinia* (Ursin *et al.*, 1989). Significant homology was also noticed between clones from *Oenothera* and maize pollen cDNA, and a polygalacturonase identified from tomato fruit (Brown and Crouch, 1990).

At least two sets of genes are activated at different times during pollen development, termed early and late genes. The gene for actin in *Tradescantia* is one of an early expressed set of genes (Mascarenhas *et al.*, 1985; Stinson *et al.*, 1987). The activity of β-galactosidase (in *Brassica*, Singh *et al.*, 1985) and alcohol dehydrogenase (in maize, Stinson and Mascarenhas, 1985) during pollen development appears comparable; therefore, these two genes also appear to have their mRNA synthesized in a similar manner.

The second or late set of genes are activated after microspore mitosis. Five cDNA clones from mature pollen of tomato appear to be among these late genes (Twell *et al.*, 1989; Ursin *et al.*, 1989) and three clones in *Oenothera* (Brown and Crouch, 1990). All the clones identified to date from *Tradescantia* pollen belong to the late set of genes (Stinson *et al.*, 1987). According to Mascarenhas (1990a), no evidence from the literature suggests the new translation of mRNAs not present at anthesis, and there is no evidence for the synthesis of new proteins during pollen germination and pollen tube growth. Taking these different data together, Mascarenhas (1990b) has suggested that the late genes may be involved in providing the enzymes necessary to facilitate pollen tube growth into the tissues of the style.

In addition, several reports describe the isolation of pollen-expressed genes and the identification of pollen promoters. This field is receiving increasing interest since the fascinating work done by Goldberg's group and the Plant Genetic Systems Company, who succeeded in creating transgenic male sterile plants by using an RNase gene construct under the control of a strong anther promoter (Mariani *et al.*, 1990). Other pollen-expressed genes have been reported in maize (*Zm13*, Hamilton *et al.*, 1989; *AdH-1*, Sachs *et al.*, 1986; starch glycosyltransferase, Klosgen *et al.*, 1986), in tomato (*LAT52*, Twell *et al.*, 1989; Ursin *et al.*, 1989), in *Oenothera* (Brown and Crouch, 1990), and in tobacco (Ursin *et al.*, 1989).

2. Pollen Wall as a Living Structure

The complex and resistant wall of the pollen grain has been the subject of considerable interest because of its unique external sculpturing and biochemistry. The characteristics of the pollen wall are often species-specific and have been used in various systematic treatments of specific groups, serving as a character in some taxonomic keys. The ornamentation of the wall is due to the pattern of polymerization of the resistant wall material called sporopollenin, which has itself defied traditional chemical description (Wiermann and Gubatz, this volume). How this material is synthesized and deposited on the microspore is of considerable interest to palynologists; the likelihood that such ornamentations are organized by some preexisting scaffolding has been the subject of considerable ultra-structural study and still needs to be resolved.

In terms of its essential role in reproduction, the pollen wall is a dynamic structure with notable biological activity. Numerous surface compounds are contained within the highly structured walls, including lipids, proteins, and numerous enzymes (Dumas *et al.*, 1984; Knox, 1984b; Wiermann and Gubatz, this volume). In the exine these arise from the sporophytic domain, derived from the tapetum. In the intine these are gametophytic

proteins (Heslop-Harrison, 1975), synthesized in the microspore and pollen grain. The highly diffusible components of the pollen wall have been separated by traditional methods, using analytical electrophoresis, or directly by depositing individual pollen grains on gels (Gay *et al.*, 1986). The latter method works well for the largest pollen grains (e.g., *Cucurbitaceae*) and allows genetic analysis of individual gametophytes from different lines and hybrids (Gay *et al.*, 1987).

Among the pollen wall compounds, some proteins appear to possess analogies with lectin-like molecules (e.g., in *Brassica* and poplar, Gaude *et al.*, 1983) and some display significant nuclease activity (Roeckel *et al.*, 1988). The former might be involved in recognition and adhesion, whereas nucleases in the pollen coat may serve to reduce the inadvertent acquisition of foreign DNA into the plant genome through viruses or other external vectors.

3. Cellular Organization and Quality

As discussed in detail elsewhere, generative cell division to form the two sperm cells may occur before or after anthesis, defining two different types of pollen in angiosperms: (1) bicellular pollen grains (shed with the undivided generative cell) are the most frequent (*Solanaceae, Rosaceae, Fabaceae, Liliaceae*, etc.); and (2) tricellular pollen grains (shed with sperm cells) are found in other families (*Asteraceae, Brassicaceae, Caryophyllaceae, Onagraceae, Poaceae*, etc.).

These two categories classify the pollen with regard to the isolation of reproductive cells (Russell, 1991; Chaboud and Perez, this volume), as well as their biochemistry, and usually the type of self-incompatibility mechanism, if present.

A common problem of pollen assessment is judging the fertility of the pollen. Three classical methods are currently widely used to estimate pollen quality:

1. Seed set. This is a widely used and obvious assay, but its reliability is affected by many factors, including pollen loading, environmental variability, incompatibility mechanisms in the pistil, and others.

2. Pollen tube culture. This assay is also common but is limited in usefulness when the optimal growth condition and media are unknown for the pollen tested. Also, tricellular species especially do not germinate easily *in vitro* using the classical Brewbaker and Kwack medium (Knox, 1984a,b; Kerhoas *et al.*, 1987).

3. Cytochemical staining tests are also frequently used, including the fluorochromatic reaction (FCR) introduced by Heslop-Harrison and Heslop-Harrison (1970). This test relies on two cellular parameters for a

positive result: the integrity of the plasma membrane and the presence of nonspecific esterases in the pollen cytoplasm.

A common problem of the methods described herein is that the pollen sample is destroyed by the testing process. NMR technology, however, has the advantage that the sample may be recovered after testing and successfully used for breeding experiments (Kerhoas *et al.*, 1987). This method can be used to assess the relative levels of both free and bound water (Dumas *et al.*, 1983). During aging the pollen grain loses part of its water content. At anthesis water content is variable in the different species examined (e.g., 60% for *Cucurbita,* 57–58% for corn, 8–10% for poplar; Dumas *et al.,* 1984). During dehydration the permeability of the plasma membrane is modified, but membranes still have a bilayered structure even at 3% water content in corn (Kerhoas *et al.,* 1987).

4. Sperm Cells

Because of their internal location within the pollen grain and growing tube, sperm cells are difficult to examine except by cytological techniques (Mogensen, this volume; Chaboud and Perez, this volume). Although progress has been made in isolating the male cells, with the eventual goal of understanding how the surface of these cells may regulate fertilization at the gamete level, both quantity and quality are a problem. The best suited species are wind-pollinated because of the relatively higher pollen yield per flower and should be tricellular because of the considerable problems culturing enough pollen tubes from bicellular species to obtain adequate sperm samples. Methods have now been established to isolate large quantities of maize sperm cells using a Percoll gradient (Dupuis *et al.,* 1987), yielding essentially a pure sperm fraction. The ultimate goal of plasma membrane fractionation and characterization, however, will require obtaining far more sperm cells than currently practiceable.

The quality of sperm cells has been examined using numerous different methods, including FCR, exclusion of polar dyes (e.g., Evans blue), adenosine triphosphate (ATP) content, phase-contrast microscopy, scanning electron microscopy (SEM), transmission electron microscopy (TEM) (Russell, 1991; Chaboud and Perez, this volume), and recently cell sorting (Zhang *et al.,* 1992). The use of membrane integrity tests is crucial to cell quality because the cells are easily damaged during their release from the pollen by osmotic shock, grinding, or enzymes. Phase-contrast microscopy is a rapid, albeit coarse method for examining a fraction quickly; intact sperm cells appear dense and have distinct, smooth edges. Their shape *in vitro* is similar to that of other protoplasts (i.e., essentially spherical), although Southworth and Knox (1989) have been able to isolate sperm

cells that appear to retain their *in vivo* appearance. SEM and TEM allow detailed examination of the cells but do not allow rapid assays for quality. Limited biochemical and immunological characterizations of the sperm cells have revealed that the sperm cells contain unique compounds (about 15%) that are not present in their surrounding environment (Pennell *et al.*, 1987; Geltz and Russell, 1988). This finding is a prelude to more detailed experimentation elsewhere (Chaboud and Perez, this volume).

One method to evaluate the surface of the sperm cell that has been omitted in the past is freeze-fracture, in which frozen material is cleaved and the resulting surface replicated for observation using TEM. This technique is one of the few methods available to obtain *en face* views of membranes and has been used to evaluate the presence of intramembrane proteins (IMPs), which are not easily demonstrated in TEM. Generative cells and sperm cells of a number of plants have been observed by Southworth (this volume) using freeze-fractured pollen (Southworth *et al.*, 1988, 1989; Southworth, 1990a). Isolated sperm cells have also been obtained from isolations and observed (Theunis *et al.*, 1991), revealing the presence of IMPs. Preliminary work has shown that the inner and outer surface of sperm cells may differ in IMP content. Although caution is required in interpreting such data, this seems to be an attractive approach in examining differences in dimorphic sperm cell surfaces.

B. Pistil Receptivity and Embryo Sac Isolation

Stigma, style, and ovule structure have been extensively studied and numerous reviews are available on this topic (e.g., Knox, 1984b; Knox *et al.*, 1986). Within the ovule the embryo sac represents the female gametophytic generation and is generally composed of seven cells and eight nuclei (*Polygonum* type). Among these two of the cells, the egg and central cells, play key roles in fertilization mechanisms as target cells for the two sperm cells that are delivered by the pollen tube during the fertilization process.

1. Pistil Receptivity and *in Vitro* Pollination

An important property of the pistil is the period when the stigma, style, and ovule are receptive and can be successfully fertilized. This period has often been determined by pollination experiments, and usually, receptivity of the stigma and ovule is synchronous. Nevertheless, in some cases a real gap exists between these two events. This is especially clear in some woody species, such as *Coryllus avellana,* and in orchids, in which the

period between stigma receptivity and ovule receptivity may extend several months (Knox et al., 1986).

Among the numerous methods currently used for assessing stigma receptivity are as follows:

1. Evaluation of seed set after pollination at different time intervals relative to anthesis (prior to and after anthesis).

2. Detection of stigmatic secretions on the papilla surface (e.g., Dumas et al., 1983; Knox et al., 1987).

3. Movement of stylar lobes in some species (e.g., *Malvaceae, Caryophyllaceae,* etc.) (Knox et al., 1986).

4. Cytochemical detection of nonspecific enzymes on the surface of the stigma (e.g., ATPase, peroxidase, etc.) (Gaude and Dumas, 1987). Although early studies suggested that esterase was a good marker enzyme for stigma receptivity (Mattsson et al., 1974), later research showed it to be detectable at all developmental stages of the flower without stage specificity.

Ovule receptivity, however, may be rapidly assessed using aniline blue fluorescence: nonfunctional and senescent ovules accumulate callose and consequently show a bright yellow fluorescence. By contrast, functional ovules exhibit low fluorescence or only autofluorescence (Dumas and Knox, 1983).

Because of the number of factors controlling the receptivity of the pistil and of the impracticability of field testing many of them, *in vitro* pollination appears to be a suitable technique allowing the nutritional and environmental conditions to be defined and assessing the behavior of the pistil after pollination (Zenkteler, 1980).

In vitro pollination analysis in the best studied system (maize) allows an expanded understanding of the fertilization process:

1. The precise kinetics of the *in vitro* fertilization process can be determined using radioactive labels added to pollen, such as ^{32}P, allowing autoradiographic analysis of pollen tube growth: 6–8 hours was required for crosses between A632 × W117 (Dupuis and Dumas, 1989).

2. Identification of biochemical markers of female receptivity: it was shown that the acquisition and loss of receptivity were related to a set of tissue and stage-specific gene products. It was observed that some new proteins appear to be correlated with the acquisition of silk (stigma) receptivity (Dupuis and Dumas, 1990a).

3. The effects of stress separately on pollen, pistil, or on the interaction of the pollen and pistil can be tested: using this method, the synthesis of heat-shock proteins was shown to be absent in mature pollen after heat stress but present in all other organs (Dupuis and Dumas, 1990b; Dupuis, this volume).

2. Embryo Sac and Gamete Analyses

Significant progress has also been made in embryo sac studies by using *in vitro* isolation procedures. Embryo sacs have been successfully isolated by means of either microdissection, enzymatic digestion, or both (Theunis *et al.*, 1991; Huang and Russell, this volume). The best procedure is generally a combination of gentle enzymatic digestion followed by microdissection under a dissecting microscope. This procedure is time-consuming, and even with good training the number of isolated embryo sacs is currently a limiting factor for further biochemical or molecular genetic studies (Chaboud *et al.*, in press). Three-dimensional reconstruction using computer-aided techniques allows the precise position of the embryo sac to be determined within the ovule of maize. This three-dimensional method combined with SEM of test ovules might be a good aid for further experimentation (Wagner *et al.*, 1989).

In a few cases, it has been possible to isolate the female gamete cells directly from embryo sacs *in vitro*, such as in *Nicotiana tabacum* (Hu *et al.*, 1985), *Torenia fourneri* (Mól, 1986), and more recently *Plumbago* (Huang and Russell, 1989) and *Zea mays* (Wagner *et al.*, 1989); observations of living embryo sac cells have been made using video-enhanced microscopy to examine cell behavior (Huang *et al.*, 1992).

C. Progamic Phase

The so-called progamic phase has been defined as events occurring from pollen arrival on the stigma to pollen tube discharge in the synergid. This phase is followed by the syngamic phase when fertilization leading to karyogamy strictly occurs.

1. Pollen Tube Wall Formation

Most pollen grains range in size from 20 to 200 μm; however, the pollen tube, which is a single cell, may extend hundreds of times its width (e.g., in lily, pollen grains less than 40 μm in diameter form tubes extending 10 cm through the length of the style). How the pollen grain obtains sufficient resources to sustain such growth has been the subject of considerable interest, as has the nature of the storage compounds involved.

Phytic acid is normally accumulated in the mature pollen grain and is hydrolyzed rapidly during germination (Lin *et al.*, 1987), releasing *myo*-inositol for incorporation into the pectic fraction of the pollen tube wall and contributing to the phosphatidylinositol in newly formed membranes (Helsper *et al.*, 1984). Because phytic acid is commonly found as a reserve

material in pollen, a partial purification and characterization of phytase from pollen have been conducted to understand its probable role in pollen nutrition (Lin *et al.*, 1990). These experiments on phytase and *myo*-inositol synthesis reinforce previous data obtained using *Lilium*, a hollow-style species (Loewus, 1973), and *Petunia*, a solid-style species (Kroh, 1973).

Based on data from *in vivo* pollen tube growth, Mulcahy and Mulcahy (1983) suggested that bicellular pollen exhibits two phases of growth when growing in a compatible pistil. In bicellular pollen the tube grows slowly without callose plugs until generative cell division. After division the tube grows rapidly and produces callose plugs at intervals until the completion of its growth. In contrast, tricellular pollen exhibits only a single phase of growth that is apparently similar to the later phase of growth in bicellular species; the pollen tube grows rapidly and forms callose plugs soon after pollen germination. Such data are consistent with heterotrophic nutrition of the pollen tube during the second growth phase.

In some regards and perhaps in an evolutionary sense as well, the pollen tube may be compared with a plant parasite. However, the female tissues do not reject this "parasite." Fetal acceptance during pregnancy in mammals displays a similar tolerance of nonself during reproduction, and both mammalian and seed habits are viviparous: the compatible pollen tube fertilizes an egg internally and develops an embryo *in situ* within the female tissues. Perhaps both systems are under a sophisticated genetic control, by the human leukocyte antigen (HLA) system in vertebrates, and by the complex *S* system in angiosperms (Section III).

2. Pollen Tube Growth

The pollen tube is a classical example of tip growth: the tip contains a highly polarized cytoplasm guiding exocytotic secretion from Golgi vesicles to the rapidly elongating tube. The membranes of these vesicles fuse with the plasma membrane of the vegetative cell, their vesicular contents are incorporated in the new wall, and their membranes are contributed to the expanding plasma membrane (Steer, 1990).

Significant progress has been made in understanding the role of the cytoskeletal components, specifically microtubules and microfilaments, during pollen tube growth. This work has included classical TEM (e.g., Theunis *et al.*, 1991), freeze-substituted pollen tubes (Lancelle *et al.*, 1987), and immunofluorescence observations (Traas, *et al.*, 1989; Pierson and Cresti, this volume) among other techniques. The observation of organelle movements in living pollen tubes reinforces the apparent role that the cytoskeleton possesses in orienting pollen tube growth and delivering male gametes (Heslop-Harrison and Heslop-Harrison, 1988, 1991). Apparently, microfilaments (i.e., F-actin) rather than microtubules pro-

vide the motive force for movement in the tube (Heslop-Harrison *et al.*, 1988), whereas the microtubules may provide a dynamic scaffolding. The role of other cytoskeleton-related proteins that are undoubtedly associated with pollen tube growth and gamete transmission have yet to be elucidated (Pierson and Cresti, this volume).

3. Pollen Tube Chemotropism: A New–Old Story

The guidance of the pollen tube down the style and into the ovary has been considered by many authors to be a typical case of chemotropic growth in angiosperms (Heslop-Harrison and Heslop-Harrison, 1986). Clearly, the tube must make several changes of direction within the style, and particularly to enter the ovule and the embryo sac, but a critical examination of the evidence has led several authors to disagree with this classical view (e.g., Mascarenhas, 1975). The mechanisms of pollen tube guidance and target finding have been completely reexamined by Heslop-Harrison and Heslop-Harrison (1986). These authors concluded their critical analysis on pollen tube chemotropism by questioning whether it is a "fact or delusion." They recognized only two points in pollen tube extension in which it might be necessary to postulate chemotropic guidance: (1) in the cortical tissues of the stigma during penetration into the transmitting tissue, and (2) in the guidance and entry of the pollen tube into the micropyle, where such a signal could be provided by a transfer cell, such as a synergid with its wall ingrowths, and in its penetration into the nucellus. Between these two points, an interesting set of data has recently been published by Sanders and Lord (1989) regarding the nature of the transmitting tract of the style. To evaluate whether the secretory matrix of the transmitting tissue is an active or passive component in different stylar types in angiosperm, they introduced latex spheres (of approximately the same diameter as the pollen tube) into the matrix and followed their behavior. In compact style species as well as on the mucilage of the stylar canal in hollow style species, the latex spheres were translocated to the ovary at rates statistically similar to those of pollen tubes (Sanders and Lord, 1989). In a previous review (Dumas *et al.*, 1983), the observation was made that the intercellular matrix of the transmitting tissue, mostly consisting of a fluid or mucilage, was the sole region in which the pollen tubes were usually observed; this lends support for the conclusion that this extracellular system is a "mechanical facilitation pathway" for the pollen tubes, often directly leading to the ovules themselves. The work performed by Sanders and Lord provides evidence that supports and extends this view. Further research on the transmitting tissue has provided evidence for substrate adhesion molecule (SAM)-like genes and gene products in flowering plants, specifically vitronectin or a vitronectin-like com-

pound. Sanders *et al.* (1991) proposed that vitronectin is involved in pollen tube guidance by an active and specific recognition mechanism relying on an integrin-like protein located in the plasma membrane (Sanders and Lord, this volume). This new area of research, based on concepts emerging from animal biology and termed topobiology, appears promising (Edelman, 1988). This concept may also apply to cell lineages in somatic tissues as well (Roberts, 1990).

The identification of concentrated calcium in the synergids of wheat embryo sacs is noteworthy in a more traditional sense (Chaubal and Reger, 1990) because many pollen tubes are known to reorient themselves in response to calcium gradients. This may provide evidence for chemotropic guidance at the micropyle, as postulated herein (Heslop-Harrison and Heslop-Harrison, 1986).

In summary, the problem of pollen tube chemotropism appears to be complex, involving positional determinants (the placement of the transmitting tissue), matrix determinants (e.g., the involvement of vitronectin as an active component similar to the SAM in animals), and calcium gradients. In addition, the involvement of electrotropism may occur as well and cannot be neglected.

D. Syngamic Phase

With regard to the syngamic phase, several particularly noteworthy observations have emerged. First, the male germ unit (MGU), composed of the two linked sperm cells and the physically associated vegetative nucleus, appears to be universal among flowering plants (Roeckel *et al.*, 1990), establishing a polarized unit with the implicit possibility of influencing later patterns of double fertilization (Mogensen, this volume). Evidence of preferential patterns of double fertilization in corn and *Plumbago* (Russell, this volume) also indicates preprogramming of the fertilization event. Additionally, there is a clearer understanding of how patterns of cytoplasmic inheritance are determined as a result of screening procedures for sperm and generative cell DNA (Corriveau and Coleman, 1988) and detailed ultrastructural observations of fertilization. Both uniparental and biparental cases of inheritance exhibit different patterns of cytoplasmic transmission, and mechanisms occur that may strip cytoplasm from the sperm cells at the time of fertilization (Mogensen, 1988).

1. MGU, Preferentiality, and the Synchronization of Double Fertilization

One of the principal findings of the last decade is that the two sperm cells are typically tightly associated, forming the MGU; the sperm cells usually

diverge from one another only once they have arrived within the embryo sac (Mogensen, this volume). This effectively synchronizes double fertilization. Although ultimately one gamete fuses slightly before the other, in the case of *Plumbago*, which is strongly preferential (95%) in its pattern of fertilization, the temporal advantage may be less than 1 minute. Both circumstances (i.e., the sperm fusing first with the egg or the sperm fusing first with the central cell) seem to occur equally often, suggesting that polarity of arrival is unimportant. However, synchrony of sperm arrival does serve one purpose well (i.e., that of reducing the likelihood of heterofertilization, *viz.*, fertilization by sperm cells arising from more than one pollen tube).

Increasingly, it appears that the sperm cell, itself, is dimorphic in the pollen in *Plumbago*, and in many other angiosperms as well (Russell, 1991; Mogensen, this volume). Cases have also been observed in which the sperm cells appear to be isomorphic. Interestingly, the association of the cells appears variable only in its temporal initiation. In the grasses, in particular, the sperms are usually found separately within the pollen, but in barley and presumably others, they clearly approach one another once they enter the pollen tube and are closely associated throughout the remainder of pollen tube growth.

2. Preprogrammed Fertilization

The basis of gametic recognition in angiosperms is not clearly known, as we currently lack an *in vitro* assay for identifying such compounds. Isolated gametes do not adhere to one another and must be carefully maneuvered to obtain electrofusion *in vitro* (Kranz *et al.*, this volume). Thus, there are conditions that occur within the embryo sac that cannot be easily repeated outside of the ovular environment. Underlying these events is the clear evidence that sperm cells do not fuse randomly with any but their target cells, and in this case appear to do so in a preferential manner in the plants in which this has been clearly examined (Russell, this volume). In *Plumbago*, the plastid-rich sperm fuses with the egg and the mitochondrion-rich sperm fuses with the central cell. In maize when B chromosomes are unequally apportioned between sperm cells, the one with supernumerary B chromosomes more frequently fuses with the egg (Russell, this volume).

3. Cytoplasmic DNA Inheritance

Cytoplasmic DNA inheritance is controlled at many stages during development, from generative cell initiation through fertilization (Russell *et al.*, 1990). At the time of gametic fusion, plants with biparental inheritance

undergo a form of fusion that results in essentially complete transmission of the sperm cell cytoplasm into the egg and central cells. However, in plants with uniparental maternal inheritance, the transmission of male cytoplasm may be only partial. In barley the male cytoplasmic organelles appear to be transmitted into the central cell, but those organelles belonging to the sperm cell that fuses with the egg remain within a cytoplasmic body at the exterior of the egg cell (Mogensen, 1988). Interestingly, this study revealed the presence of plastids in the sperm, which are not known to be transmitted in barley; apparently, this redundant mechanism ensures uniparental inheritance of plastids in the embryo. Similar patterns have been suggested in other plants, and it is possible in cotton that both of the sperm cytoplasms are excluded, based on the presence of two such cytoplasmic bodies next to both the fertilized egg and central cell (Jensen and Fisher, 1968). The recent description of uniparental paternal inheritance in alfalfa appears to be a restricted case but one in which the structure of the egg also influences inheritance by sequestering plastids to the suspensor during embryogenesis, thus eliminating maternal plastids from the heritable lineage (H. L. Mogensen, personal communication; Huang and Russell, this volume).

E. Experimental Fertilization

Controlled pollination and experimental fertilization are routine techniques for seed breeders but cannot ensure success in desired crosses. New experiments have been done to better control the sexual reproduction process and how to use it to transform plants by genetic engineering.

1. Pollen Selection

Pollen vigor and its variability within populations is crucial for evaluating whether male-to-male competition and/or female choice can lead to selection. Snow and Spira (1991) found that competition between pollen tubes is common in *Hibiscus* by measuring tube growth rates and relative seed set when different pollen mixtures were applied to stigmas. Significant differences between pairs of individuals in pollen tube growth rate were consistent in different maternal plants. Such data seemingly support a widely cited article in which Mulcahy (1979) proposed that gametophytic characters expressed in pollen tube competition played an active role in the rapid evolution of angiosperms.

Yet, differences in disease tolerance or resistance between the different male gametophytes can be reduced by specific protection mechanisms in the sporophytic tissue (Ottaviano *et al.*, 1990). For example, Hodgkin

(1988) successfully selected pollen for resistance to a phytotoxic compound of the fungal pathogen *Alternaria brassicola*.

In addition, it seems possible to apply a selected stress to a population of pollen grains and then separate germinable pollen from ungerminable pollen using density gradients (Mulcahy *et al.*, 1988). Based on the use of appropriate molecular markers in maize, it is possible to discriminate between differential gene transmission in the gametophyte and sporophyte (Wendel *et al.*, 1987).

Selection of stress-tolerant pollen has also been conducted by culturing microspores of tobacco under stress and then using the pollen grains that complete development to pollinate emasculated flowers and fertilize embryo sacs *in situ* (Moreno *et al.*, 1988). This technique could have a high potential for pollen conditioning to a variety of stresses in the natural environment.

2. Attempts to Obtain Artificial Zygotes

The first attempt to obtain artificial zygotes was reported in *Torenia fourneri*, which was chosen because of its unusual emergent embryo sac, which is exerted out of the micropyle (Keijzer *et al.*, 1988). Microinjection of sperm cells was performed into the side of the embryo sac. This trial, however, was unsuccessful because of some critical problems listed by the authors.

Procedures have now been developed to isolate viable sperm cells and female gametes from several species, including *Nicotiana, Lilium, Plumbago, Zea,* and *Cichorium* (Roeckel *et al.*, 1990; Russell, 1991). The first successful attempt to fuse male and female gametes was performed using corn (Kranz *et al.*, 1991). A micropipette connected with a computer-aided pump was used to pick up selected gametes and place them in an experimental microcell for electrofusion. The results demonstrated that sperm cells fuse efficiently with other sperm cells (mean frequency of one-to-one fusion is 85%). This clearly suggests that sperm cells exist as true protoplasts in the pollen and are comparable to somatic protoplasts in structure, as demonstrated previously by TEM (Dupuis *et al.*, 1987). This approach was successful; in 79% of the cases, electrofused gametes formed artificial zygotes (Kranz *et al.*, 1991).

3. Plant Transformation Using the Sexual Route

Because plant transformation is a challenge, especially in cereals, many attempts have been made and some spectacular reports have been published, but often the technique was not easily reproducible (e.g., De la

Peña et al., 1987), transformants were not stable, or no molecular data were shown (e.g., Ohta, 1986).

Since the original work of Hess (Hess, 1987), several reports have been published on direct plant transformation. The main cause of failure after culturing pollen with plasmids or mixing the pollen coating with exogenous DNA was likely due to the presence of DNases capable of digesting plasmids. Nucleic acid hydrolases can digest a plasmid within 1 minute in tobacco (Negrutiu et al., 1986) and in Zea mays (Roeckel et al., 1988); presumably, such hydrolases can digest naked DNA used for direct transformation with equal facility (Negrutiu et al., 1986).

Several patents have been obtained by using the microbiolistics gun to inject DNA into the pollen of some cereals (e.g., corn, rice; Agracetus San Francisco) or by microinjecting pollen tubes in rice. In the latter case around 10% transformation was claimed with controlled stability in the progeny until the F_3 generation (Institute of Crop Science, Beijing, China); however, few other details are available, and skepticism is warranted until molecular evidence is obtained. The possibility of microinjecting DNA directly through the pollen tube has also been the topic of controversial reports. The most interesting work using pollen tubes is based on the transformation of rice by the co-migration of exogenous DNA along with the pollen tubes during their growth within the style. Some molecular data have been published, but nothing is known about whether the progeny express stable transformation (Luo and Wu, 1988).

The possibility of successfully electroporating pollen grains of tobacco for plant transformation has also been reported (Abdul-Baki et al., 1990). Using this technique, DNA encoding the marker enzyme β-glucuronidase (GUS) was incorporated into electroporated pollen without significantly damaging the viability of the pollen. Another experiment has been conducted by directly fusing liposomes with the germinating pollen grains of watermelon. Such fusions were obtained using fusigenic agents, such as polyethylene glycol, $CaCl_2$ and polylysine; this procedure may constitute a new method for plant transformation (Gad et al., 1988). That macromolecules encapsulated in liposomes may penetrate through the exposed plasma membrane of elongating tube tip seems possible.

Finally, transformation of Petunia via pollen co-cultured with Agrobacterium tumefasciens has been reported to result in the expected production of tumors. A molecular analysis of DNA from the Petunia callus demonstrated, in a few cases, a distinct hybridization signal in plant material transformed with T-DNA grown on a hormone-free medium (Hess and Dressler, 1989).

Plant transformation via direct microinjection through the ovule or embryo sac has been reported in cotton (Zhou, 1985). In maize, because the precise stereolocalization of the embryo sac is known, such a technique

today seems feasible (Wagner *et al.*, 1989), but no significant data have been reported.

In conclusion, for all of these trials, few unequivocal successes in producing fertile progeny have been achieved that have withstood molecular analysis. Thus, the feasibility and reproducibility of these different techniques remain to be demonstrated.

III. Self-Incompatibility and Male–Female Interactions

A. Generalities

The fertilization process of flowering plants leading to seed set depends on recognition mechanisms that allow the pistil to discriminate between the different types of pollen grains that it may receive. During intraspecific matings self-pollen (i.e., pollen belonging to the same genotype as that of the pistil) is generally rejected, whereas allopollen (nonself-pollen) is accepted. This process enforces outbreeding and characterizes the self-incompatibility phenomenon, which is believed to have been responsible for the rapid advancement of flowering plants (Whitehouse, 1950). Self-incompatibility in plants, however, is an unusual system of recognition in biology because it is based on the rejection of self and acceptance of nonself, as opposed to the immune system of vertebrates (Gaude and Dumas, 1987). Most angiosperm possess a self-incompatibility system controlled by a single locus, the S locus, with many alleles (De Nettancourt, 1977; Gaude and Dumas, 1987, 1990; Haring *et al.*, 1990; Nasrallah *et al.*, 1991).

Two types of incompatibility systems occur in angiosperms. When the phenotype for incompatibility is determined by the haploid S genotype of the pollen, the system is defined as a gametophytic self-incompatibility system (GSI), and pollen rejection occurs in the style. Gametophytic systems are the most common, found in nearly half of the families of flowering plants and frequently found in plants that release pollen in the bicellular condition. The classical model of GSI used for molecular genetics is *Nicotiana*. In contrast, when the incompatibility phenotype is determined by the S genotype of the diploid pollen-producing plant (the sporophyte) and not the pollen itself, the system is defined as a sporophytic self-incompatibility system (SSI), and pollen rejection occurs at the stigma surface. The most widely used model for molecular analysis of SSI is *Brassica*. In both systems pollen–pistil recognition has generally been considered as involving a signal receptor-like interaction (Dumas *et al.*, 1983). The rejection of pollen grains by pistil tissues occurs when the same

S allele present in the pistil of the self-incompatible plant is also expressed by the pollen grain or pollen tube (De Nettancourt, 1977; Singh and Kao, this volume).

B. Molecular Analysis of Self-Incompatibility

Attempts to identify and characterize the molecules associated with S gene expression in the pistil as well as in the pollen have been the subject of considerable research (Gaude and Dumas, 1987; Cornish *et al.*, 1987; Ebert *et al.*, 1989). Indeed, identifying the S molecules and cloning the genes responsible for encoding them represent the most promising approach to improving our comprehension of pollen–pistil interactions. In this volume two model species, *Brassica oleracea* (SSI) (Dickinson *et al.*, this volume; Trick and Heizmann, this volume) and *Nicotiana alata* (GSI) (Singh and Kao, this volume) have been particularly intensely investigated for the study of the self-incompatibility phenomenon. By contrast, no molecular data are yet available for interspecific matings.

1. *S* Genes and *S* Products Identified from the Pistil

In GSI and SSI, the putative pistil S products have been identified, and their corresponding genes cloned and sequenced (Nasrallah *et al.*, 1985; Anderson *et al.*, 1986). These S product molecules appear at the same time that self-incompatibility develops in the pistil and are glycoproteins that bind the lectin concanavilin A. Further, the characterized S products segregate with the S alleles and have a distinct location at the site of action: the papillar cells of the stigma in *Brassica* (Nasrallah *et al.*, 1988; Kandasamy *et al.*, 1989) and the stylar transmitting tissue in *Nicotiana* (Cornish *et al.*, 1987; Anderson *et al.*, 1989).

These S molecules cannot be detected in overly immature stigmas of young buds, but their synthesis increases during development (i.e., apparently 3 days before anthesis synthesis of S products begins and increases to a maximum about 1 day before anthesis). This accumulation phase corresponds to the first acquisition of self-incompatibility, usually 1 day before anthesis, after which time synthesis decreases. At this time the threshold level of S molecules required for the expression of self-incompatibility is reached.

2. *S* Gene and *S* Products in Sporophytic Self-Incompatibility (SSI)

The most complete analysis of SSI that has been conducted since 1985 was by the group of Nasrallah *et al.* at Cornell University (Ithaca, New

York) using *Brassica oleracea* as a model species. A number of complementary DNA (cDNA) clones have now been isolated encoding different
S allele-specific glycoproteins (Nasrallah *et al.*, 1985, 1987; Trick and
Flavell, 1989; Scutt *et al.*, 1990). A comparison of the derived amino acid
sequences shows 79–85% identity in a comparison of different pairs of
sequences (Dickinson *et al.*, this volume). The primary structure of the S
glycoproteins has been considered to be composed of alternating constant
and variable regions; only 40% homology is exhibited in the high variability
domain situated between residues 182 and 189 (Nasrallah *et al.*, 1987). A
comparison of these sequences with known sequences in gene data bases
shows significant homology between the *S* locus gene (SLG) cysteine-
rich domain and extracellular matrix-forming proteins in animals (von
Willebrand factor and Type VI collagen), as well as the putative receptor
serine–threonine protein kinase from *Zea mays* (*ZmPK1*) (Dickinson *et
al.*, this volume).

Furthermore, Lalonde *et al.* (1989) have shown using molecular genetics
at the genomic level that the unique *S* gene encoding the S glycoproteins
(the *SLG* gene) in *Brassica oleracea* stigmas belongs to a multigenic family
of at least 12 related sequences. However, at least two other genes situated
at the same *S* locus have been described that seem to play a key role
in the SI response: the *SLG* gene and the *S* related kinase (SRK) gene.
The latter encodes a putative *S* locus receptor protein kinase (Stein
et al., 1991). In addition to this new set of data, two major classes of *S*
alleles (I and II) have been defined according to sequence data and self-
incompatibility phenotypes (Nasrallah *et al.*, 1991). Class I alleles exhibit
a strong self-incompatibility phenotype. In contrast, Class II alleles have
phenotypes with low activity and weak incompatibility, and are recessive
pollen genotypes.

Two other genes not genetically linked to the *S* locus are also expressed,
termed the *S* locus-related genes, *SLR1* and *SLR2*. These encoded proteins
possess 60% and 73% identity, respectively, with the *SLG* class I protein.
Evidence of the expression of *SLR1* proteins in the stigmatic tissue of
Brassica oleracea has recently been detected by the use of electrophoretic
methods and N-terminal amino acid sequence analysis (Gaude *et al.*, 1991).
Nevertheless, the role of these *SLR* molecules, apparently independent
of the self-incompatibility mechanism, is not yet known. Some workers
envisage the role of *SLRs* in more general events of the fertilization process, such as pollen adhesion, hydration, germination, or pollen tube
growth (Gaude *et al.*, in press; Nasrallah *et al.*, 1991). Additional proteins
required for the operation of the self-incompatibility system include those
required for pollen hydration and the glycosylation of stigma glycoproteins, which are required for the function of the self-incompatibility molecule (Sarker *et al.*, 1988).

The extracellular domain of a putative receptor protein kinase from maize presents surprisingly significant sequence homologies with the stigmatic *S* locus glycoproteins of *Brassica oleracea* (Walker and Zhang, 1990). That self-incompatibility mechanisms may share similarities with known cell-to-cell communication systems in animals based on kinase activity during signal transduction improves the outlook for understanding self-incompatibility systems in plants and cell-to-cell communication in general.

3. *S* Gene and *S* Products in Gametophytic Self-Incompatibility System (GSI)

The most complete analysis of the GSI system is being conducted by Adrienne Clarke's group at the Plant Cell Biology Centre (Excellence Centre of the Commonwealth) of the University of Melbourne using *Nicotiana alata* as a model species. This group was the first to clone the *S* gene of *Nicotiana* (Anderson *et al.*, 1986). The group has since reported on the developmental control of *S* gene expression using *in situ* hybridization (Cornish *et al.*, 1987), isolated and characterized stylar S glycoproteins (Jahnen *et al.*, 1989a), and localized the S molecules *in situ* (Anderson *et al.*, 1986, 1989).

The surprising lack of homology between the peptide sequences of *Brassica* and *Nicotiana alata* S glycoproteins, first mentioned by A. E. Clarke during the International Botanical Congress at Berlin in 1987, was confirmed by Takayama *et al.* (1987). This apparently indicates that the two systems evolved separately and are based on different mechanisms although sharing significant functional similarities. For instance, in both systems the primary structure of S glycoproteins shows an alternation of variable and conserved regions, and these regions of high variability could possibly be involved in the determination of allelic specificity of the S molecules (Ebert *et al.*, 1989). A certain level of homology, especially in the promoter regions, has been demonstrated between the two different self-incompatibility systems (SSI and GSI) using transformation experiments (Moore and Nasrallah, 1990), although the genes and gene products may have had different origins.

By analyzing the resulting amino acid sequences, McClure *et al.* (1989) recently demonstrated the existence of significant sequence homology between the S glycoproteins of three stylar S glycoproteins in *Nicotiana alata* and different extracellular ribonucleases (RNases) of fungi. These authors also provided evidence for a close structural relationship between the active site domain of fungal RNases and the homologous region of the S glycoproteins. In other self-compatible species within the *Solanaceae*, such as *Nicotiana tabacum*, stylar ribonuclease activity is low. These data

suggest that the incompatibility response in GSI only occurs when stylar S glycoproteins display sufficient RNase activity to penetrate self-pollen tubes and degrade their cytoplasmic RNA (McClure *et al.*, 1989). The inhibitory effect of S glycoproteins on *in vitro* pollen tube growth observed in *Nicotiana alata* seems to be consistent with this hypothesis (Harris *et al.*, 1989; Jahnen *et al.*, 1989b).

C. *S* Genes and *S* Products of the Pollen

The occurrence of complementary *S* genes and *S* products in the pollen, which has long been postulated (De Nettancourt, 1977), is not present in sufficient abundance to be detected using the different probes currently available (stigma or style S cDNAs as well as monoclonal antibodies from the different S molecules). A number of groups have reported unsuccessful attempts to identify male transcripts homologous to the pistil *S* products in *Brassica oleracea* (e.g., Scutt *et al.*, 1990) as well as in *Nicotiana alata* (Anderson *et al.*, 1986).

However, the application of the polymerase chain reaction (PCR) has provided sufficient sensitivity to detect an mRNA homologous to the self-incompatibility gene in anthers of *Brassica oleracea* (Guilluy *et al.*, 1991). These S transcripts were found only in anthers during early microsporo-genesis. They display an early expression pattern: a strong signal during the uninucleate microspore stage, becoming weaker at the bicellular pollen grain stage, and undetectable in mature tricellular pollen. Using PCR has also revealed that other *SLG*-like sequences shown to be transcribed in sexual tissues are transcribed as well as in vegetative tissues, including the roots and leaves. Considering these low levels of transcription in other tissues, Guilluy *et al.* (1991) suggested that the *SLG* gene family might be involved not only in pollen–stigma recognition but also in more general forms of plant cell-signaling processes. This hypothesis seems to be consistent with the emerging evidence that the *S* gene in maize displays homology with kinase, as reported by Walker and Zhang (1990). The use of a mono-clonal antibody raised against a purified S glycoprotein from *Brassica oleracea* stigmas has revealed the presence of cross-reacting material in pollen extracts (Gaude *et al.*, 1988). Nevertheless, the relationship be-tween these pollen proteins and the *SLG*-like transcripts described by Guilluy *et al.* (1991) remain to be investigated.

In *Nicotiana* pollen McClure *et al.* (1989) detected only faint amounts of RNase activity in pollen extracts or in pollen germinated *in vitro*, indicating that if the pollen *S* product is an RNase, it is not highly ex-pressed. These workers considered that the *S* gene products in the *Nicoti-*

ana alata pollen and style are not identical; however, the S molecules must share allelic specificity for recognition to occur (Haring *et al.*, 1990).

IV. Conclusions

During the course of several years (1985–1991) considerable progress has been made in understanding the reproductive biology of angiosperm in a number of separate areas.

A. Data from Pollen and the Progamic Phase

The pollen grain has clearly been shown to be a unique organism possessing a complex cellular organization both internally, as evidenced by the male germ unit (Mogensen, this volume), and externally, by the complex ornamentation and biochemistry of the pollen wall (Wiermann and Gubatz, this volume). Pollen is well protected against the introduction of foreign DNA with a set of efficient DNases (Roeckel *et al.*, 1988), ensuring the integrity of its DNA. The failure of pollen to respond to high heat stress is also notable (Dupuis and Dumas, 1990b).

With regard to genetic expression, it now appears that pollen-specific genes do occur and that others are found in common with the sporophyte (Mascarenhas, this volume; Mulcahy, 1979). It also seems evident that sperm cells have their own program, as expressed by specific polypeptides and their immunological determinants (Chaboud and Perez, this volume). After considerable research on the organization of microfilaments and microtubules forming the pollen cytoskeleton, it can be concluded that complex interactions occur in the pollen tube that guide and transmit the male gametes to their target cells (Pierson and Cresti, this volume). There also appear to be unique cell-biological phenomena occurring in the gametic lineage as well. Presently, it is controversial but possible that sperm cells lack F-actin. Also, the mode of sperm formation may involve an atypical form of mitosis that omits the formation of preprophase bands and employs an unconventional spindle composed of two opposed superbundles (Palevitz and Tiezzi, this volume).

B. Data from Gametes and the Syngamic Phase

Significant progress has been made in the isolation of gametes ([male]: Roeckel *et al.*, 1990; Russell, 1991; Chaboud and Perez, this volume;

[female]: Huang and Russell, this volume). The first attempt to fuse isolated male and female gametes has also been achieved (Kranz *et al.*, 1991, this volume), but it has not yet been demonstrated that the fused cells or putative artificial zygote is truly diploid or whether fertile plants can be regenerated from such cells. Similarly, attempts to transform plants through manipulating the sexual reproductive cells themselves (i.e., pollen, ovule, embryo sac or gamete) have, to date, been unsuccessful.

C. Data from Double Fertilization and Cytoplasmic DNA Inheritance

Double fertilization in plants by a polarized MGU implies that the assumption of random fertilization is not universally applicable and perhaps requires reevaluation (Mogensen, this volume). In *Plumbago* and corn, two angiosperm in which it has been possible to discriminate between the two sperm cells, preferential patterns of fertilization have been demonstrated that suggest that differences between gametes affect their ability to fuse with the egg (Russell, this volume).

The origin of double fertilization per se, which has been described in two species of *Ephedra*, may predate the origin of angiosperms (Friedman, 1990, 1992). Variants of this process may also be present in other gynmosperm as well (Friedman, this volume), suggesting that double fertilization may not have developed in isolation. Such a comparative basis for examining the behavior of reproductive cells may be illustrative of the developmental plasticity of seed plants in general.

There is also emerging evidence that complex mechanisms exist that influence cytoplasmic DNA inheritance throughout development extending from mechanisms of physical exclusion (Hagemann and Schröder, 1989), to cytoplasmic diminution in the pollen grain and tube (Mogensen and Rusche, 1985; Russell and Yu, 1991) and even exclusion at the site of gametic fusion (Russell, this volume). Such work suggests that the active mechanisms are not always obvious and, in many cases, may be redundant.

D. Data from Self-Incompatibility Mechanisms

The complexity of self–non-self recognition mechanisms in plants has provided a particularly exciting model. The molecular mechanisms of self-incompatibility have been highlighted by the discovery of the multigenic *S* family in *Brassica,* the rapidly changing molecular characterization of the self-incompatibility systems (Singh and Kao, this volume),

and the apparent divergence between sporophytic and gametophytic self-incompatibility systems (Trick and Heizmann, this volume).

A better understanding of self-incompatibility mechanisms will now depend on the isolation of the pollen S products and, for SSI systems, a determination of how the *Brassica* stigma S glycoproteins function (Dickinson *et al.*, this volume).

In addition, analogies between the self-incompatibility system and the immune system now appear to be increasingly informative. The structure of S genes, with their alternation of constant and variable regions, is similar to that of immunoglobulin genes, and interestingly, appear to display a similar rate of evolution (Trick and Heizmann, this volume).

If the *Nicotiana* model conforms to the recently proposed cytotoxic model in terms of cell-to-cell recognition (McClure *et al.*, 1989), new insights emerging from the *SRK* gene in the *Brassica* appear to suggest that the SSI system corresponds much more to a ligand receptor model (Stein *et al.*, 1991).

References

Abdul-Baki, A. A., Saunders, J. A., Matthews, B. F., and Pittarelli, G. W. (1990). *Plant Sci.* **70,** 181–190.

Anderson, M. A., Cornish, E. C., Mau, S. L., Williams, E. G., Hoggart, R., Atkinson, A., Bonig, I., Grego, B., Simpson, R., Roche, P. J., Haley, J. D., Penschow, J. D., Niall, H. D., Tregear, G. W., Coghlan, J. P., Crawford, R. J., and Clarke, A. E. (1986). *Nature* **321,** 38–44.

Anderson, M. A., McFadden, G. I., Bernatzky, R., Atkinson, A., Orpin, T., Dedman, H., Tregear, G., Fernley, R., and Clarke, A. E. (1989). *Plant Cell* **1,** 483–491.

Brown, S. M., and Crouch, M. L. (1990). *Plant Cell* **2,** 263–274.

Chaboud, A., Perez, R., Digonnet, C., and Dumas, C. (in press). *In* "Perspectives in Plant Cell Recognition" (J. Callow and J. R. Green, eds.). Cambridge University Press, Cambridge, Massachusetts.

Chapman, G. P., Mantell, S. H., and Daniels, R. W. (eds.) (1985). "The Experimental Manipulation of Ovule Tissues." Longman, New York.

Chaubal, R., and Reger, B. J. (1990). *Sex. Plant Reprod.* **3,** 98–102.

Coen, E. S., and Meyerowitz, E. M. (1991). *Nature* **353,** 31–36.

Cornish, E. C., Pettitt, J. M., Bonig, I., and Clarke, A. E. (1987). *Nature* **326,** 99–102.

Corriveau, J. L., and Coleman, A. W. (1988). *Am. J. Bot.* **75,** 1443–1458.

De la Peña, A., Lörz, H., and Schell, J. (1987). *Nature* **325,** 274–276.

De Nettancourt, D. (1977). "Incompatibility in Angiosperms." Springer-Verlag, Berlin.

Dumas, C., and Knox, R. B. (1983). *Theor. Appl. Genet.* **67,** 1–10.

Dumas, C., Duplan, J. C., Saïd, C., and Soulier, J. P. (1983). *In* "Pollen: Biology and Implications for Plant Breeding" (D. L. Mulcahy and E. Ottaviano, eds.), pp. 15–20. Elsevier Biomedical, New York.

Dumas, C., Knox, R. B., and Gaude, T. (1984). *Int. Rev. Cytol.* **90,** 239–272.

Dupuis, I., and Dumas, C. (1989). *Sex. Plant Reprod.* **2,** 265–269.

Dupuis, I., and Dumas, C. (1990a). *Plant Sci.* **70,** 11–19.

Dupuis, I., and Dumas, C. (1990b). *Plant Physiol.* **94**, 665–670.
Dupuis, I., Roeckel, P., Matthys-Rochon, E., and Dumas, C. (1987). *Plant Physiol.* **85**, 876–878.
Ebert, P. R., Anderson, M. A., Bernatzky, R., Altschuler, M., and Clarke, A. E. (1989). *Cell* **56**, 255–262.
Edelman, G. M. (1988). "Topobiology: An Introduction to Molecular Embryology." Basic Books, New York.
Friedman, W. E. (1990). *Science* **247**, 951–954.
Friedman, W. E. (1992). *Science* **255**, 336–339.
Gad, A. E., Zeevi, B. Z., and Altman, A. (1988). *Plant Sci.* **55**, 69–75.
Gaude, T., and Dumas, C. (1987). *Int. Rev. Cytol.* **107**, 333–366.
Gaude, T., and Dumas, C. (1990). *Bot. Acta* **103**, 323–326.
Gaude, T., Fumex, B., and Dumas, C. (1983). *In* "Pollen: Biology and Implications for Plant Breeding" (D. L. Mulcahy and E. Ottaviano, eds.), pp. 273–280. Elsevier Biomedical, New York.
Gaude, T., Nasrallah, M. E., and Dumas, C. (1988). *Heredity* **61**, 317–318.
Gaude, T., Boughaba, B., and Dumas, C. (in press). *In* "Proceedings of the XIth International Symposium on Embryology and Seed Reproduction." Komarov Botanical Institute USSR.
Gaude, T., Denoroy, L., and Dumas, C. (1991). *Electrophoresis* **12**, 646–653.
Gay, G., Kerhoas, C., and Dumas, C. (1986). *Electrophoresis* **7**, 148–149.
Gay, G., Kerhoas, C., and Dumas, C. (1987). *Theor. Appl. Genet.* **6**, 497–503.
Geltz, N. R., and Russell, S. D. (1988). *Plant Physiol.* **88**, 764–769.
Giles, K. L., and Prakash, J., eds. (1987). "Pollen: Cytology and Development." *Int. Rev. Cytol.,* **107**, Academic Press, Orlando, New York.
Guilluy, C. M., Trick, M., Heizmann, P., and Dumas, C. (1991). *Theor. Appl. Genet.* **82**, 466–472.
Hagemann, R., and Schröder, M. B. (1989). *Protoplasma* **153**, 57–64.
Hamilton, D. A., Bashe, D. M., Stinson, J. R., and Mascarenhas, J. P. (1989). *Sex. Plant Reprod.* **2**, 208–212.
Haring, V., Gray, J. E., McClure, B. A., Anderson, M. A., and Clarke, A. E. (1990). *Science* **250**, 937–941.
Harris, P. J., Weinhandl, J. A., and Clarke, A. E. (1989). *Plant Physiol.* **89**, 360–367.
Helsper, J. P. F. G., Linskens, H. F., and Jackson, J. F. (1984). *Phytochemistry* **21**, 1255–1258.
Heslop-Harrison, J. (1975). *Annu. Rev. Plant Physiol.* **26**, 403–425.
Heslop-Harrison, J., and Heslop-Harrison, Y. (1970). *Stain Technol.* **45**, 115–120.
Heslop-Harrison, J., and Heslop-Harrison, Y. (1986). *In* "Biology of Reproduction and Cell Motility in Plants and Animals" (M. Cresti and R. Dallai, eds.), pp. 169–174. Università di Sienna, Sienna, Italy.
Heslop-Harrison, J., and Heslop-Harrison, Y. (1988). *Sex. Plant Reprod.* **1**, 16–24
Heslop-Harrison, J., and Heslop-Harrison, Y. (1991). *Sex. Plant Reprod.* **4**, 6–11.
Heslop-Harrison, J., Heslop-Harrison, Y., Cresti, M., Tiezzi, A., and Moscatelli, A. (1988). *J. Cell Sci.* **91**, 49–60.
Hess, D. 1987. *Int. Rev. Cytol.* **107**, 169–190.
Hess, D., and Dressler, K. (1989). *Bot. Acta* **102**, 202–207.
Hodgkin, T. (1988). *In* "Sexual Plant Reproduction in Higher Plants" (M. Cresti, P. Gori, and E. Pacini, eds.), pp. 57–62, Springer-Verlag, Berlin.
Hu, S. Y., Li, L. G., and Zhou. C. (1985). *Acta Bot. Sin.* **27**, 337–344.
Huang, B. Q., and Russell, S. D. (1989). *Plant Physiol.* **90**, 9–12.
Huang, B. Q., Pierson, E., Russell, S. D., Tiezzi, A., and Cresti, M. (1992). *Sex. Plant Reprod.* **5**, 156–162.

Jahnen, W., Batterham, M. P., Clarke, A. E., Moritz, R. L., and Simpson, R. J. (1989a). *Plant Cell* **1**, 493–499.

Jahnen, W., Lush, W. M., and Clarke, A. E. (1989b). *Plant Cell* **1**, 501–510.

Jensen, W. A., and Fisher, D. B. (1968). *Planta* **78**, 158–183.

Kandasamy, M. K., Paolillo, D. J., Faraday, C. D., Nasrallah, J. B., and Nasrallah, M. E. (1989). *Dev. Biol.* **134**, 462–472.

Keijzer, C. J., Reinders, M. C., and Leferink-ten Klooster, H. B. (1988). *In* "Sexual Reproduction in Higher Plants" (M. Cresti, P. Gori, and E. Pacini, eds.), pp. 119–124. Springer Verlag, New York.

Kerhoas, C., Gay, G., and Dumas, C. (1987). *Planta* **171**, 1–10.

Klosgen, R. B., Gierl, A., Schwarz-Sommer, Z., and Saedler, H. (1986). *Mol. Gen. Genet.* **203**, 237–244.

Knox, R. B. (1984a). *In* "Cellular Interactions: Encyclopedia of Plant Physiology" (H. F. Linskens, ed.), vol. 17, pp. 508–608. Academic Press, San Diego.

Knox, R. B. (1984b). *In* "Embryology of Angiosperms" (B. M. Johri, ed.), pp. 197–272, Springer-Verlag, Heidelberg.

Knox, R. B., Williams, E. G., and Dumas, C. (1986). *In* "Plant Breeding Review" (J. Janick, ed.), pp. 9–80. Avi, Wesport.

Knox, R. B., Gaget, M., and Dumas, C. (1987). *Int. Rev. Cytol.* **107**, 315–332.

Kranz, E., Bautor, J., and Lörz, H. (1991). *Sex. Plant Reprod.* **3**, 160–169.

Kroh, M. (1973). *In* "Biogenesis of Plant Cell Wall Polysaccharides" (F. Loewus, ed.), pp. 195–206. Academic Press, San Diego.

Lalonde, B. A., Nasrallah, M. E., Dwyer, K. G., Chen, C. H., Barlow, B., and Nasrallah, J. B. (1989). *Plant Cell* **1**, 249–258.

Lancelle, S. A., Cresti, M., and Hepler, P. K. (1987). *Protoplasma* **140**, 141–150.

Lin, J. J., Dickinson, D. B., and Ho, T. H. D. (1987). *Plant Physiol.* **83**, 408–413.

Lin, J. J., Dickinson, D. B., and Ho, T. H. D. (1990). *Plant Cell Rep.* **9**, 211–215.

Loewus, F. (ed.) (1973). "Biogenesis of Plant Cell Wall Polysaccharides." Academic Press, San Diego.

Luo, Z. X., and Wu, R., (1988). *Plant Mol. Biol. Rep.* **6**, 165–174.

Mariani, C., De Beuckeler, M., Truettner, J., Leemans, J., and Goldberg, R. B. (1990). *Nature* **347**, 737–742.

Mascarenhas, J. P. (1975). *Bot. Rev.* **41**, 259–314.

Mascarenhas, J. P. (1990a). *Annu. Rev. Plant Physiol. Plant Mol. Biol.* **41**, 317–338.

Mascarenhas, J. P. (1990b). *In* "Microspores: Evolution and Ontogeny" (S. Blackmore and R. B. Knox, eds.), pp. 265–280. Academic Press, San Diego.

Mascarenhas, J. P., Eisenberg, A., Stinson, J. R., Willing, R. P., and Pe, M. E. (1985). *In* "Plant Cell/Cell Interactions" (I. Sussex, A. Ellingboe, M. Crouch, and R. Malmberg, eds.), pp. 19–23. Cold Spring Harbor Press, New York.

Mattsson, O., Knox, R. B., Heslop-Harrison, J., and Heslop-Harrison, Y. (1974). *Nature* **247**, 298–300.

Mayer, U., Torres Ruiz, R. A., Berleth, T., Miséra, S., and Jürgens, G. (1991). *Nature* **353**, 402–407.

McClure, B. A., Haring, V., Ebert, P. R., Anderson, M. A., Simpson, R. J., Sakiyama, F., and Clarke, A. E. (1989). *Nature* **342**, 955–957.

Mogensen, H. L. (1988). *Proc. Natl. Acad. Sci. U.S.A.* **85**, 2594–2597.

Mogensen, H. L., and Rusche, M. L. (1985). *Protoplasma* **128**, 1–13.

Mogensen, H. L., Wagner, V. T., and Dumas, C. (1990). *Protoplasma* **153**, 136–140.

Mól, R. (1986). *Plant Cell Rep.* **3**, 202–206.

Moore, H. M., and Nasrallah, J. B. (1990). *Plant Cell* **2**, 29–38.

Moreno, B. R. M., Macke, F., Hauser, M., Alwen, A., and Heberle-Bors, E. (1988). *In* "Sexual Plant Reproduction in Higher Plants" (M. Cresti, P. Gori, and E. Pacini, eds.), pp. 347–352. Springer-Verlag, Berlin.

Mulcahy, D. L. (1979). *Science* **206**, 20–23.

Mulcahy, D. L., and Mulcahy, G. B. (1975). *Theor. Appl. Genet.* **46**, 277–280.

Mulcahy, D. L., and Mulcahy, G. B. (1987). *Am. Sci.* **75**, 44–50.

Mulcahy, D. L., Mulcahy, G. B., Popp, R., Fong, N., Pallais, N., Kalinowski, A., and Marien, J. N. (1988). *In* "Sexual Plant Reproduction in Higher Plants" (M. Cresti, P. Gori, and E. Pacini, eds.), pp. 43–50, Springer-Verlag, Berlin.

Mulcahy, G. B., and Mulcahy, D. (1983). *In* "Pollen: Biology and Implications for Plant Breeding" (D. L. Mulcahy and E. Ottaviano, eds.), pp. 29–35. Elsevier Biomedical, New York.

Nakamura, N., Fukushima, A., Iwayama, H., and Suzuki, H. (1991). *Sex. Plant Reprod.* **4**, 138–143.

Nasrallah, J. B., Kao, T. H., Goldberg, M. L., and Nasrallah, M. E. (1985). *Nature* **316**, 263–267.

Nasrallah, J. B., Kao, T. H., Chen, C. H., Goldberg, M. L., and Nasrallah, M. E. (1987). *Nature* **326**, 617–619.

Nasrallah, J. B., Yu, S. M., and Nasrallah, M. E. (1988). *Proc. Natl. Acad. Sci. U.S.A.* **85**, 5551–5555.

Nasrallah, J. B., Nishio, T., and Nasrallah, M. E. (1991). *Annu. Rev. Plant Physiol. Plant Mol. Biol.* **42**, 393–422.

Negrutiu, I., Heberle-Bors, E., and Potrykus, I. (1986). *In* "Biotechnology and Ecology of Pollen" (D. L. Mulcahy, G. B. Mulcahy, and E. Ottaviano, eds.), pp. 65–70. Springer-Verlag, Berlin.

Ohta, Y. (1986). *Proc. Natl. Acad. Sci. U.S.A.* **83**, 715–719.

Ottaviano, E., Sari-Gorla, M., and Mulcahy, D. L. (1990). *In* "Isozymes: Structure, Function, and Use in Biology and Medicine" (Z. I. Ogita and C. L. Markert, eds.), pp. 575–588. Wiley–Liss, New York.

Pedersen, S., Simonsen, V., and Loeschcke, V. (1987). *Theor. Appl. Genet.* **75**, 200–206.

Pennell, R. I., Geltz, N. R., Koren, E., and Russell, S. D. (1987). *Bot. Gaz.* **148**, 401–406.

Rajora, O. P., and Zsuffa, L. (1986). *Can. J. Genet. Cytol.* **28**, 476–482.

Roberts, K. (1990). *Curr. Opin. Cell Biol.* **2**, 920–928.

Roeckel, P., Heizmann, P., Dubois, M., and Dumas, C. (1988). *Sex. Plant Reprod.* **1**, 156–163.

Roeckel, P., Chaboud, A., Matthys-Rochon, E., Russell, S., and Dumas, C. (1990). *In* "Microspores: Evolution and Ontogeny" (S. Blackmore and R. B. Knox, eds.), pp. 281–307. Academic Press, London.

Russell, S. D. (1985). *Proc. Natl. Acad. Sci. U.S.A.* **82**, 6129–6132.

Russell, S. D. (1986). *Plant Physiol.* **81**, 317–319.

Russell, S. D. (1991). *Annu. Rev. Plant Physiol. Plant Mol. Biol.* **42**, 189–204.

Russell, S. D., and Yu, H. S. (1991). *Am. J. Bot.* **78**(Suppl. 6), 34.

Russell, S. D., Rougier, M., and Dumas, C. (1990). *Protoplasma* **155**, 153–165.

Sachs, M. M., Dennis, E. S., Gerlach, W. L., and Peacock, W. J. (1986). *Genetics* **113**, 449–467.

Sanders, L. C., and Lord, E. M. (1989). *Science* **243**, 1606–1608.

Sanders, L. C., Wang, C. S., and Lord, E. M. (1991). *Plant Cell* **3**, 629–635.

Sarker, R. H., Elleman, C. J., and Dickinson, H. G. (1988). *Proc. Natl. Acad. Sci. U.S.A.* **85**, 4340–4344.

Scutt, C. P., Gates, P. J., Gatehouse, J. A., Boulter, D., and Croy, R. R. D. (1990). *Mol. Gen. Genet.* **220**, 409–413.

Singh, M. B., O'Neill, P., and Knox, R. B. (1985). *Plant Physiol.* **77**, 225–228.

Snow, A. A., and Spira, T. P. (1991). *Nature* **352**, 796–797.

Southworth, D. (1990a). *J. Ultrastruct. Mol. Str. Res.* **103**, 97–103.

Southworth, D. (1990b). *In* "Microspores: Evolution and Ontogeny" (S. Blackmore and R. B. Knox, eds.), pp. 193–212. Academic Press, San Diego.

Southworth, D., and Knox, R. B. (1989). *Plant Sci.* **60,** 273–277.

Southworth, D., Platt-Aloia, K. A., and Thomson, W. W. (1988). *J. Ultrastruct. Mol. Str. Res.* **101,** 165–172.

Southworth, D., Platt-Aloia, K. A., DeMason, D. A., and Thomson, W. W. (1989). *Sex. Plant Reprod.* **2,** 270–276.

Steer, M. (1990). *In* "Tip Growth in Plant and Fungal Cells" (M. Steer, ed.), pp. 119–145. Academic Press, San Diego.

Stein, J. C., Howlett, B., Boyes, D. C., Nasrallah, M. E., and Nasrallah, J. B. (1991). *Proc. Natl. Acad. Sci. U.S.A.* **88,** 8816–8820.

Stinson, J. R., and Mascarenhas, J. P. (1985). *Plant Physiol.* **77,** 222–224.

Stinson, J. R., Eisenberg, A. J., Willing, R. P., Pe, M. E., Hanson, D. D., and Mascarenhas, J. P. (1987). *Plant Physiol.* **83,** 442–447.

Takayama, S., Isogai, A., Tsukamoto, C., Ueda, Y., Hinata, K., Okasaki, K., and Suzuki, A. (1987). *Nature* **326,** 102–105.

Theunis, C. H., Pierson, E. S., and Cresti, M. (1991). *Sex. Plant Reprod.* **4,** 145–154.

Traas, J. A., Burgain, S., and Dumas De Vaulx, R. (1989). *J. Cell Sci.* **92,** 541–550.

Trick, M., and Flavell, R. B. (1989). *Mol. Gen. Genet.* **218,** 112–117.

Twell, D., Wing, R., Yamaguchi, J., and McCormick, S. (1989). *Mol. Gen. Genet.* **247,** 240–245.

Ursin, V. M., Yamaguschi, J., and McCormick, S. (1989). *Plant Cell* **1,** 727–736.

Wagner, V. T., Dumas, C., and Mogensen, H. L. (1989). *J. Cell Sci.* **93,** 179–184.

Walker, J. C., and Zhang, R. (1990). *Nature* **345,** 743–746.

Wendel, J. F., Edwards, M. D., and Stuber, C. W. (1987). *Heredity* **58,** 297–301.

Whitehouse, H. L. K. (1950). *Ann. Bot. New Series* **14,** 198–216.

Zenkteler, M. (1980). *Annu. Rev. Plant Physiol.* **15** (Suppl. 113), 137–155.

Zhang, G., Campenot, M. K., McGann, L. E., and Cass, D. D. (1992). *Plant Physiol.* in press.

Zhou, G. Y. (1985). *In* "Experimental Manipulation of Ovule Tissues" (G. P. Chapman, S. H. Mantell, and W. Daniels, eds.), pp. 240–250. Longman, New York.

Index